£13-40

WITHDRAWN
0251253-9
FROM STOCK

STEINBERG, DAVE S
VIBRATION ANALYSIS FOR ELECTRO
000251253

534.1 S81

VIBRATION ANALYSIS FOR ELECTRONIC EQUIPMENT

DAVE S. STEINBERG
CONSULTING ENGINEER

VIBRATION ANALYSIS FOR ELECTRONIC EQUIPMENT

A WILEY-INTERSCIENCE PUBLICATION

JOHN WILEY & SONS, NEW YORK · LONDON · SYDNEY · TORONTO

Copyright © 1973, by John Wiley & Sons, Inc.

All rights reserved. Published simultaneously in Canada.

No part of this book may be reproduced by any means, nor transmitted, nor translated into a machine language without the written permission of the publisher.

Library of Congress Cataloging in Publication Data:

Steinberg, Dave S 1923–
Vibration analysis for electronic equipment.

"A Wiley–Interscience publication."
Includes bibliographical references.
1. Electronic apparatus and appliances – Vibration.
I. Title.

TK7870.S8218——621.381——72-13763
ISBN 0-471-82100-4

Printed in the United States of America

10 9 8 7 6 5 4 3 2 1

To my wife Annette
and to my two daughters,
Cori and Stacie

PREFACE

Electronic equipment is being used more and more to provide comfort, safety, convenience, entertainment, and therapy. Many of these applications involve severe vibration environments that can seriously degrade the performance of the equipment and perhaps result in catastrophic failures. In a television set, an electronic failure is only an inconvenience. In mass transit systems such as airplanes, trains, and ships, an electronic failure can often mean the loss of many lives.

Fatigue failures are very common in electronic systems that are subjected to severe vibration environments. These failures are usually in the form of broken wires, broken component leads, cracked castings, cracked welds, and loose screws.

After I witnessed many of these failures and analyzed the reasons for them, it was obvious to me that many engineers require a better understanding of the basic fundamentals of vibration. In many cases they do not take the time and effort to analyze the electronic support structure to determine the critical dynamic load paths. Most of the engineering time and money goes into the electronic circuitry. This is logical since the purpose of an electronic system is to work electrically. However, unless a system is also designed to function in its dynamic environment, there can be many structural problems that may require an extensive amount of redesign and retesting.

There have been many cases where a carefully engineered electronic chassis will fail a vibration-qualification test when it should have passed easily. An evaluation of the vibration test fixture will often reveal a severe fixture resonance near a major chassis resonance. This condition can produce high acceleration G forces in the chassis, which can lead to structural fatigue failures.

Many vibration problems can be minimized by avoiding coincident resonances which can magnify acceleration G forces very rapidly. An examination of the books and literature available on vibration shows some very fine treatments on the subject, with classic frequency derivations for many different types of structures. However none of the published literature really gets down to the basic methods for analyzing the real hardware that is being used in today's sophisticated electronic systems. This book attempts to bridge that gap.

It is assumed that the reader has some basic understanding of elementary mechanics and strength of materials. An attempt was made to keep the book as simple as possible, while still showing the derivations of some equations that were considered important. The emphasis, however, is always on the actual hardware that is being manufactured in the electronics industry today.

The first two chapters are a quick review of the basic fundamentals and definitions relating to vibration. In the next two chapters the groundwork is laid for dealing with lumped springs, masses, and dampers. This method is very convenient for analyzing complex electronic systems with the use of a high-speed digital computer. Sample problems of this type are shown in Chapter 8.

Many approximate methods are shown in this book. These methods have all been verified by considering a wide range of parameters in tests and in analyses. Some of these methods may be considered unorthodox only because they do not, to my knowledge, appear anywhere else in the literature.

Although we try to avoid them, errors are bound to creep into a new book. If any errors are found, please bring them to our attention so they can be corrected as soon as possible.

Ft. Lee, N.J.
September 1972

DAVE S. STEINBERG

CONTENTS

CHAPTER 10. STRUCTURAL FATIGUE

LIST OF SYMBOLS

A	Area, in.^{2}
A	Amplification, dimensionless ratio
A	Deflection form-factor
a	length, in.
$\mathscr{A}_{ij}$	Influence coefficient
B	Length, in.
b	Width, in.
C	Dynamic constant
C	Length, in.
CG	Center of gravity
c	Distance from neutral axis to outer fiber
c	Length, in.
c	Damping coefficient, lb sec/in.
c_c	Critical damping, lb sec/in.
D	Dissipation energy, lb in./sec
D	Diameter, in.
D	Plate bending stiffness factor, lb in.
D_{XY}	Plate torsional stiffness factor, lb in.
d	Diameter, in.
d	Length, in.
E	Modulus of elasticity, lb/in.^{2}
e	Bolted efficiency factor, percent
F	Force, lb
f	Frequency, Hz
f_n	Natural frequency, Hz
f_r	Rotational natural frequency, Hz
G	Shear modulus, lb/in.^{2}
G	Acceleration in gravity units, dimensionless
G_{in}	Input acceleration G force, dimensionless
g	Acceleration of gravity, 386 in./sec^2
H	Horizontal force, lb
h	Height, in.
h	Thickness, in.
I	Moment of inertia (area), in.^{4}
I_m	Mass moment of inertia, lb in. sec^2

J	Torsional form factor, in.^{4}
K	Linear spring rate, lb/in.
K	Stiffness ratio, dimensionless
K	Buckling form factor, dimensionless
KIN	Kinetic energy
K_t	Theoretical stress-concentration factor, dimensionless
K_θ	Angular spring rate, in. lb/radian
L	Length, in.
M	Bending moment, lb in.
M_T	Total moment, lb in.
M_X	Bending moment per unit length along the X axis, in. lb/in.
M_X	Bending moment at point X, lb in.
Mo	Momentum, lb sec
MS	Margin of safety, dimensionless
m	mass, lb sec^2/in.
N	Number of fatigue cycles to fail
n	Actual number of fatigue cycles
NS	Number of sweeps through a resonance
P	Force, lb
P_d	Dynamic force, lb
p	Unit load, lb/in.
Q	Transmissibility, dimensionless ratio
q	Shear flow, lb/in.
q	Dynamic pressure, lb/in.^{2}
R	Radius, in.
R	Reaction, lb
R	Stress ratio, dimensionless
R_n	Fatigue-cycle ratio, dimensionless
R_c	Damping ratio, dimensionless
R_Ω	Frequency ratio, dimensionless
R	Sweep rate, octave/min
r	radius, in.
S_b	Bending stress, lb/in.^{2}
S_e	Endurance limit stress, lb/in.^{2}
S	Stress, lb/in.^{2}
T	Kinetic energy, lb in.
T	Torque, in. lb.
t	Time, min, sec
t	Thickness, in.
U	Strain energy, lb in.
U	Work, lb in.
V	Velocity, in./sec
V	Vertical force, lb
W	Weight, lb
W_d	Dynamic load, lb
w	Unit load, lb/in.
X	Displacement along X axis
X	Coordinate axis
Y	Coordinate axis
Z	Coordinate axis
$\dot{X}$	First derivative, velocity
$\ddot{X}$	Second derivative, acceleration

GREEK SYMBOLS

α	Angle, degrees, radians
δ	Displacement, in.
θ	Angular displacement, radians
μ	Poisson's ratio, dimensionless
ρ	Density, lb/in.^{3}
ρ	Mass per unit area, lb sec^2/in.^{3}
ϕ	Phase angle, degrees
Ω	Angular velocity, radians/sec
Ω_n	Natural frequency, radians/sec

SUBSCRIPTS

av	Average
b	Bending
c	Critical
d	Dynamic
e	Endurance
eq	Equivalent
max	Maximum
n	Natural
o	Maximum
st	Static
su	Shear ultimate
t	Tension
tu	Tensile ultimate
ty	Tensile yield
u	Ultimate
y	Yield

VIBRATION ANALYSIS FOR ELECTRONIC EQUIPMENT

CHAPTER 1

INTRODUCTION

1.1. VIBRATION SOURCES

Electronic equipment can be subjected to many different forms of vibration over wide frequency ranges and acceleration levels. It is probably safe to say that all electronic equipment will be subjected to some type of vibration at some time in its life. If the vibration is not due to an active association with some sort of a machine or a moving vehicle, then the vibration may be due to the act of transporting the equipment from the manufacturer to the customer. Vibrations encountered during transportation and handling can produce many different types of failures in electronic equipment unless the proper considerations are given to the mechanical design of the electronic structure and the shipping containers.

Vibration is usually considered to be an undesirable condition and in most cases it is. However there are many applications where vibration is deliberately imposed to improve a function. Some sophisticated applications of vibrations are in the use of ultrasonics which can be used to clean medical instruments, measure wall thickness, and find flaws in castings. Vibration is also used in the process of sorting different rock sizes by passing them over vibrating screens which have several different groups of graduated holes.

When the first jet airplanes were introduced, standard instruments from airplanes with piston engines and propellers were used. These instruments had a tendency to stick when they were used on the jet airplanes. The difference was due to the lower-frequency vibration levels developed in the piston-type engines with propellers. In order to make these same instruments work in the first jet airplanes, small vibrators had to be mounted on the instrument panels.

Mechanical vibrations can be developed from many different sources. In household products such as blenders and washing machines, the vibrations are due to the unbalance created by rotating and tumbling masses. In vehicles such as automobiles, trucks, and trains most of the vibration is due to the relatively rough surfaces over which these vehicles travel. In ships and submarines the vibration is due to the engines and to buffeting

FIGURE 1.1. An airborne electronic box with a cathode ray tube display, cooled by means of an exhaust fan at the rear of the box (courtesy Norden division of United Aircraft).

by the water. In airplanes, missiles, and rockets the vibration is due to jet and rocket engines and to aerodynamic buffeting. Most of the vibration in a missile, during subsonic flight, is due to the sound field developed by the rocket engines [1, 6].* This is due to the extreme turbulence of the jet exhaust downstream from the rocket engine.

1.2. DEFINITIONS

Vibration, in a broad sense, is taken to mean an oscillating motion where some structure or body moves back and forth. If the motion repeats itself, with all of the individual characteristics, after a certain period of time, it is called periodic motion. This motion can be quite complex, but as long as it repeats itself, it is still periodic. If continuous motion never seems to repeat itself, it is called random motion. Simple harmonic motion is the simplest form of periodic motion and it is usually represented by a continuous sine wave on a plot of displacement versus time, as shown in Fig. 1.2.

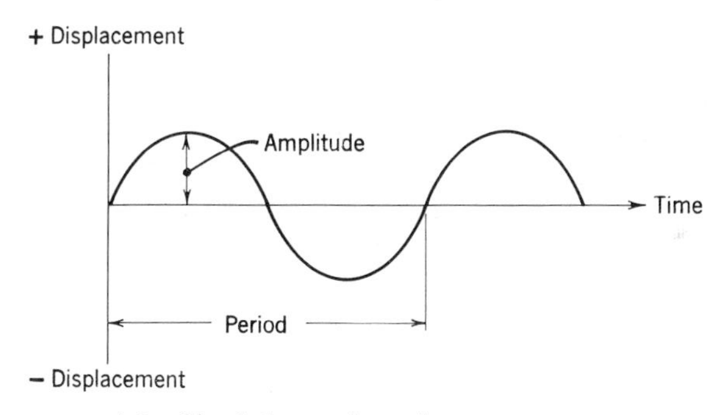

FIGURE 1.2. Simple harmonic motion.

The reciprocal of the period is known as the frequency of the vibration and is measured in cycles per second, or Hertz (Hz), in honor of the German who first experimented with radiowaves. The maximum displacement is called the amplitude of the vibration.

Crede [2, 3] defines shock as a transient condition where the equilibrium of a system is disrupted by a sudden applied force or increment of force, or by a sudden change in the direction or magnitude of a velocity vector.

**Note*. Numbers in brackets [ ] refer to references in the back of the book.

Shock in relatively light airframe structures, generally used in airplanes and missiles, is not transmitted easily. Impact forces often result in a transient type of vibration which is influenced by the natural frequencies of the airframe.

Only steady-state linear vibrations are considered in this book. Linear vibrations deal with linear springs where the displacements are directly proportional to the applied force. If the force is doubled, the displacement is doubled. However only stresses up to the elastic limit of any given material are considered. Higher stresses generally result in permanent deformations which extend into the plastic deformation region and are not considered here.

1.3. Vibration Representation

A rotating vector can be used to describe the simple harmonic vibration motion of a single mass suspended on a coil spring (Fig. 1.3).

The vector Y_0 rotates counterclockwise with a uniform angular velocity of Ω rad/sec. The projection of the vector on the vertical axis represents the instantaneous displacement Y of the mass as it vibrates up and down. This can be written as

$$Y = Y_0 \sin \Omega t \tag{1.1}$$

When the vector Y_0 rotates through one revolution, it rotates through an angle of 360°, which is 2π radians, for one complete cycle. The angular velocity is measured in radians per second and the frequency f is measured in cycles per second. This leads to the relation

$$\Omega = 2\pi f \tag{1.2}$$

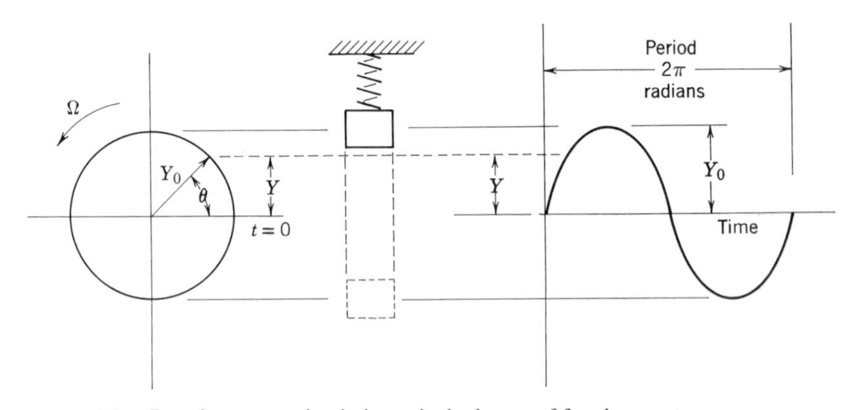

FIGURE 1.3. Rotating vector simulating a single-degree-of-freedom system.

1.4. Degrees of Freedom

A vibrating system requires some coordinates to describe the positions of the elements in the system. If there is only one element in the system that is restricted to move along only one axis, and only one dimension is required to locate the position of the element at any instant of time with respect to some initial starting point, then it is a single-degree-of-freedom system.

The same is true for a torsional system. If there is only one element that is restricted to rotate about only one axis so that only one dimension is required to locate the position of the element at any instant of time with respect to some initial starting point, it is a single-degree-of-freedom system. Some samples of systems with a single degree of freedom are shown in Fig. 1.4.

A two-degrees-of-freedom system will require two coordinates to describe the positions of the elements. Some samples of systems with two degrees of freedom are shown in Fig. 1.5.

Considering rigid body mechanics, a single mass can have six degrees of freedom. These are translation along each of its three mutually perpendicular X, Y, and Z axes, and rotation about each of its three axes, as shown in Fig. 1.6.

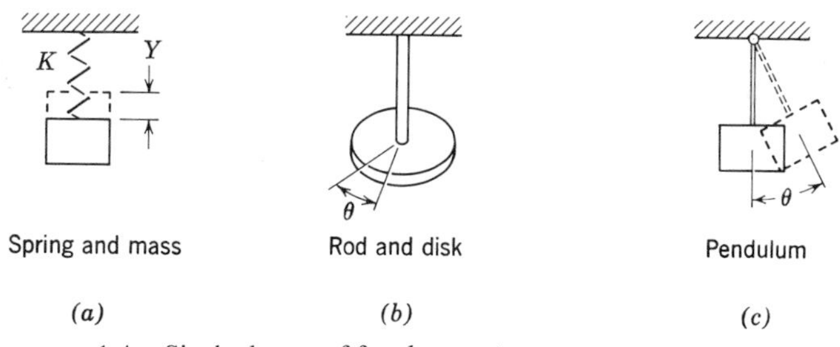

FIGURE 1.4. Single-degree-of-freedom systems.

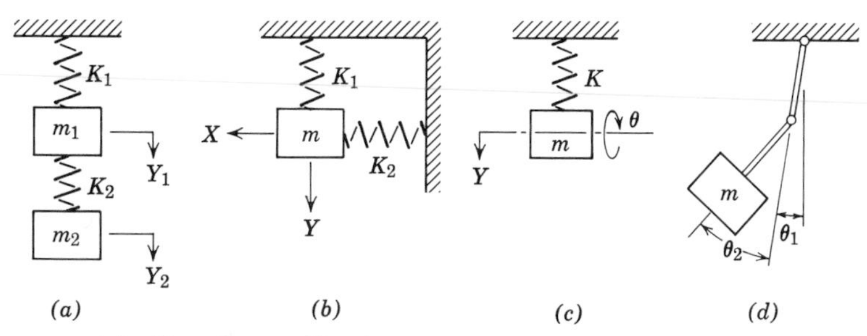

FIGURE 1.5. Two-degree-of-freedom systems.

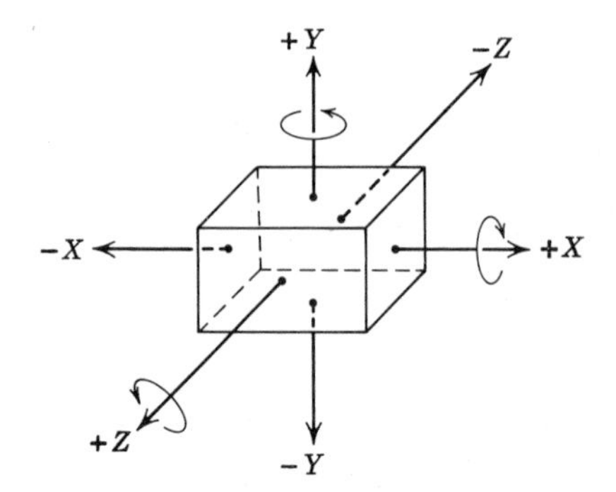

FIGURE 1.6. A single mass with six-degrees-of-freedom.

A typical beam can have an infinite number of degrees of freedom since it can bend in an infinite number of shapes, or modes, as shown in Fig. 1.7.

1.5. Vibration Modes

The manner in which a particular system is vibrating is known as the vibration mode. Each vibration mode is associated with a particular natural frequency and each vibration mode represents a degree of freedom. A single-degree-of-freedom system will have only one vibration mode and only one resonant frequency. The six-degrees-of-freedom system shown in Fig. 1.6 can have six vibration modes and six resonant frequencies. The simply supported beam shown in Fig. 1.7 can have an infinite number of vibration modes; this is the same as saying this beam can have an infinite number of different shapes for each of its resonances, which are also infinite.

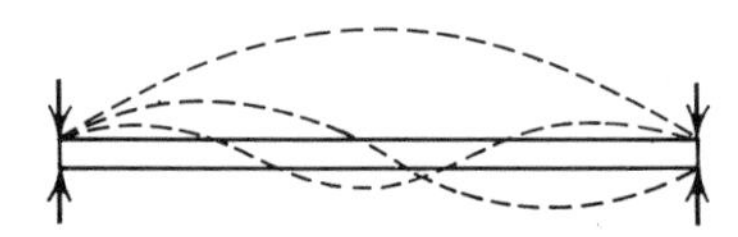

FIGURE 1.7. A simply supported beam showing several degrees-of-freedom.

The fundamental resonant mode of a vibrating system is usually called the natural frequency or the resonant frequency of the system. Sometimes it is called the first harmonic mode of the system. For example, a simply supported uniform beam vibrating at its fundamental resonant frequency would have the mode shape of a half sine wave as shown in Fig. 1.8*a*. When this beam is vibrating at its second natural frequency or second harmonic mode, it would have the mode shape of the full sine wave shown in Fig. 1.8*b*.

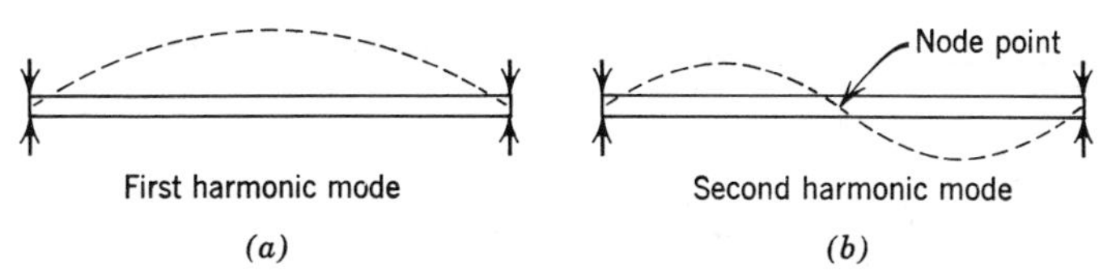

FIGURE 1.8. First and second harmonic modes for a simply supported beam.

The lowest natural frequency or first harmonic mode of a system is the fundamental resonant mode; this has the greatest displacement amplitudes and usually the greatest stresses. The second harmonic mode or second resonance has a smaller displacement than the first harmonic mode, so the stresses are usually smaller. The displacements continue to decrease for the higher resonant modes.

1.6. Vibration Nodes

Vibration nodes are unsupported points on a vibrating body that have zero displacements. Nodes are generally associated with bending or torsion modes. At the first harmonic bending resonant mode in a beam there are no node points. At the second harmonic mode there is one node point, the third harmonic mode has two node points, and so on. Figure 1.8*a* shows the bending mode of a vibrating beam with no node points and Fig. 1.8*b* shows a beam with one node point.

Vibrating plates can have straight-line bending nodes and circular nodes. The first four harmonic modes of a circular membrane are shown in Fig. 1.9.

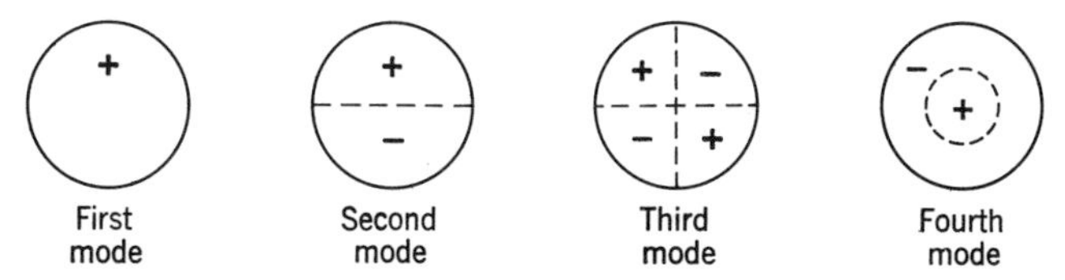

FIGURE 1.9. First four harmonic modes of a circular membrane.

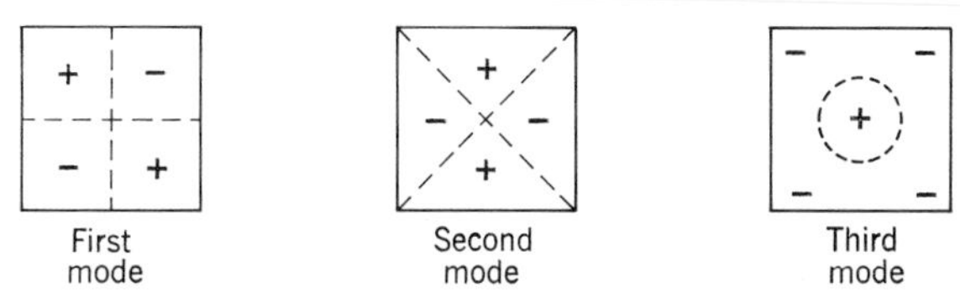

FIGURE 1.10. First three harmonic modes of a square plate.

The plus (+) sign shows positive displacements and the minus (−) sign shows negative displacements. The dashed lines represent nodal locations of zero displacement.

The first three harmonic bending modes of a square plate with free edges are shown in Fig. 1.10.

1.7. Coupled Modes

In a system with two or more degrees of freedom, the vibration mode of one degree often influences the vibration mode of the other degree. For example, in Fig. 1.5*a* if mass 2 is held rigidly while mass 1 is displaced in the vertical direction, mass 1 will oscillate up and down. Now if mass 2 is released, the motion of mass 1 will act upon mass 2 so mass 2 will begin to oscillate up and down. Because the motion of mass 1 has a direct effect on the motion of mass 2, these two vibration modes are defined as being coupled. In coupled modes, the vibration in one mode cannot occur independently of the vibration in the other mode.

Coupled modes can occur in translation, rotation, and combinations of translation and rotation for systems with more than one degree of freedom. For coupled modes in translation and rotation, it is often possible to determine whether the system is coupled or uncoupled by making a simple test. Apply a steady load to the body of the system at its center of gravity, in a specific direction. If the body moves in the direction of the applied load without rotation, then the translational mode is not coupled with the rotational mode for motion along the direction of the applied load.

Consider, for example, the mass with two springs as shown in Fig. 1.11. If a steady load is applied to the CG along the X axis, the mass will not only translate along the X axis, but it will also tend to rotate at the same time. This test shows that for vibration along the X axis, the translational mode will be coupled with the rotational mode.

If a steady load is applied to the CG along the Y axis, the mass will

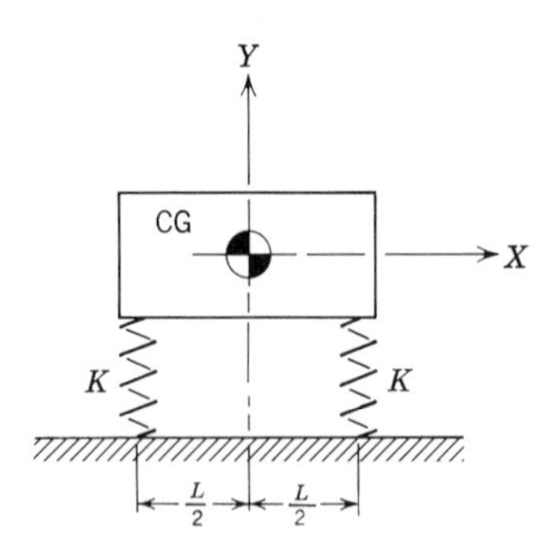

FIGURE 1.11. A single mass with two springs.

translate along the Y axis without rotation. Also, if a moment is applied to the mass about the CG, then rotation will take place without translation. These tests show that for vibration along the Y axis the translational mode is decoupled from the rotational mode.

1.8. Fasteners

Many different types of fasteners are used in electronic equipment. These include screws, nuts, rivets, clips, and so on. Fasteners are responsible for a very large percentage of field failures. In most shock tests it is the fasteners which are the largest single source of failure. Although fasteners have been used in very large quantities in electronic equipment for many years, most of the applications are based on static installation considerations. The fastening techniques, which are generally characterized by ease of installation and cost, are usually not satisfactory for severe shock and vibration environments. A contributing factor is that fasteners are such small items and their use so universal that their application tends to be on a semiautomatic basis without regard to the essential strength of the fasteners. This is particularly the case for machine screws. When larger screws are used, it is often because someone has had some association with the automotive or aircraft industry where very small screws are very seldom used.

Because fasteners play an important part in the overall reliability of the electronic equipment, extra consideration should be given in their selection as follows [5]:

1. Select the proper type of fastener, (screws, rivets, etc.), considering such trade-offs as environment, strength, maintenance, and cost.
2. Select the correct fastener size and location based on dynamic loads and geometry of the structure.
3. Select the correct locking device for screws and nuts.
4. Select the right installation technique.

Most electronics manufacturers will choose screws and rivets on a production basis. If he uses screws, he selects the size and locking device according to his tolerances as well as for ease in assembly. In installation, he depends upon the production personnel to use their own judgment in the proper installation of fasteners. The results of these conditions may be that the wrong fasteners are used, the wrong sizes are used, the wrong lock washers are used, and the wrong installation torques are used.

It has been found, generally speaking, that cold-driven shank-expanding rivets are very satisfactory and should be used more frequently in electronic assemblies. It has been found, too, that when screws are used, they should be of larger sizes than those customarily considered and they should be made of better materials. Many locking devices commonly used are unsatisfactory. Some are often the source of many severe problems. It has also been found that a mechanic's judgment in tightening a machine screw is usually faulty.

Many failures in electronic equipment have resulted because bolts have become loose. Consider what can happen to a sensitive electronic chassis when a large transformer becomes loose and rattles around during vibration.

Although many investigations have been made with high-speed films and strain gages to study the loosening action of screws and nuts, the mechanism by which this occurs is still not well established. It appears, however, that the bolt stretches slightly under the action of a dynamic load, so that the interface friction forces in the area of the bolt are suddenly and sharply reduced. Since the threads in the screw or nut tend to return to their original shape during the small time increments, the small geometric changes become a driving force which tends to loosen the screw and nut.

Some specific recommendations can be made to improve the quality of the fasteners in electronic equipment:

1. Steel screws should be used in all screw fastenings. The steel should have the minimum properties of SAE 1010.

2. The screws should be tightened by a torque device which can be preset to the required value.

3. The tightening torque should be 60–80% of the torque required to twist the head off the screw. These torque values for steel screws are given in Table 1.1 [11].

TABLE 1.1

Screw Size	Torque (in. lb)	Screw Size	Torque (in. lb)
2–64	3–3.5	12–24	45–56
4–40	5–6.5	12–28	50–64
6–32	10–12	$\frac{1}{4}$–20	65–80
8–32	20–24	$\frac{1}{4}$–28	85–100
10–24	22–27	$\frac{3}{8}$–16	250–320
10–32	34–42	$\frac{3}{8}$–24	330–415

4. The screw head should permit a positive nonslipping grip for the driving device and should also withstand the driving forces. A slotted-hex-head type of machine screw seems to be, for general purposes, the most satisfactory.

5. In all applications involving through-holes, locknuts should be used instead of lockwashers. Most of the standard steel locknuts are satisfactory.
6. Blind-tapped holes should be avoided, if possible. When they are necessary, locking devices such as lockwashers should be used under the screw head in order to prevent the screw from backing out during vibration. The tightening torque should be increased by an amount equal to the resisting frictional torque of the locking device.
7. The fasteners for a unit should be distributed so that the failure of one fastener will not free the unit or cause it to malfunction. Even for very small components there should be a minimum of two fasteners.

Lockwashers and screw-locking inserts should be used with care in electronic systems that will be used in the zero-g environment of outer space. These two devices create a binding friction in the screw by biting into the metal during installation. This action will very often shave metal particles from the screw. If these particles are not removed, they can float around in a zero-g environment and create electrical problems. An actual count was made of the metal particles after inserting 24 screws with external-star-type lockwashers and a total of about 1000 metal particles were counted.

The crossed-recess screwdriver slot (Phillips head) will also tend to be cut by the action of the screwdriver. When the screw is seated and the screwdriver is twisted, the screw driver will very often twist out of the slot and shave small bits of metal out of the screw head.

In order to avoid shaving small bits of metal from the screws with devices that bite into the metal, many electronic firms use liquids such as Loctite and Glyptal which bind the screws. Some firms use nylon inserts in the screws. Nylon inserts are usually good for about a dozen insertion and removal cycles before the nylon cold flows and reduces the binding torque.

NASA will not permit the use of products that are volatile in the area of displays and optical devices for their spacecraft. These products tend to outgas in a vacuum and then deposit themselves in a thin film; this can coat optical lenses and interfere with the operation of sensitive optical units such as cameras and telescopes.

1.9. Electronic Equipment for Airplanes and Missiles

Electronic boxes used in airplanes and missiles often have odd shapes that permit them to make maximum use of the volume available in tight spaces. Since volume and weight are generally quite critical, the electronic boxes usually have a high packaging density. This value normally ranges from about 0.03 to about 0.04 lb/in.^{3} volume, depending upon the severity of the environmental requirements. The average weight of a typical electronic box will range from about 10 lb for a small box to about 80 lb for a large box.

The vibration frequency spectrum for airplanes will vary from about 3 to about 1000 Hz with acceleration levels that can range from about 1 G to about 5 G peak. The highest-acceleration G levels appear to occur in the vertical direction in the frequency range of about 100–400 Hz. The lowest-acceleration G levels appear to occur in the longitudinal direction, with maximum levels of about 1 G in the same frequency range.

For helicopters, the frequency spectrum will vary from about 3 to about 500 Hz and acceleration levels will range from about 0.5 to about 4 G. The highest-acceleration G levels appear to occur in the vertical direction at frequencies near 500 Hz. The displacements at the low frequency are very large with values of about 0.20 in. double amplitude at about 10 Hz.

Missiles have the highest frequency range in this group with values that generally go up to 5000 Hz [7]. The lower-frequency limit is about 3 Hz and this appears to be due to bending modes in the airframe structure. Acceleration levels range from about 5 to about 30 G peak, with the maximum levels occuring during power plant ignition at frequencies above 1000 Hz.

The vibration environment in supersonic airplanes and missiles is actually more random in nature than it is periodic. However sinusoidal vibration tests are still being used to evaluate and to qualify electronic equipment that will be used in these vehicles.

Because the forcing frequencies in airplanes and missiles are so high, it is virtually impossible to design resonance-free electronic systems for these environments. Of course, it is always possible to completely encapsulate an entire electronic box with some expanding rigid type of foam, which could drive the resonant frequency well above 1000 Hz (possibly to 2000 Hz) for a small box. This is generally considered to be impractical, however, because it becomes too expensive to maintain, trouble-shoot, and repair such a system.

The obvious conclusion is that the forcing frequencies present in airplanes and missiles will excite many resonant modes in every electronic box. What becomes equally obvious is that extra care must be taken in the design and analysis of an electronic system or it can literally shake itself to pieces. The electronic support structure must actually be dynamically tuned with respect to the electronic components to prevent coincident resonances that can lead to rapid fatigue failures.

The first thought that comes to the mind of an experienced mechanical design engineer, when he is confronted with a severe vibration specification, is to mount the electronic equipment on vibration isolators. There is no doubt that a set of isolators, properly designed, can control shock and vibration. Figure 1.12 shows an airborne electronic box mounted on vibration isolators. There are four major factors that must be considered when isolation mounts are being discussed.

1. Sway space must be provided all around the electronic equipment to keep it from colliding with other objects. If volume is scarce, it might be more practical to pack more electronics into the same volume by using a larger electronic box with hard mounts.
2. Cold plates are being used more and more in electronic structures to remove the heat dissipated by the electronic equipment. If isolators are used, flexible couplings must be provided between the airframe structure and the electronic box to take care of the large displacements developed

FIGURE 1.12. An airborne electronic box mounted on vibration isolators (courtesy Norden division of United Aircraft).

by the isolators. Reliable flexible couplings must be used because cooling effectiveness may be sharply reduced if a coupling fails.

3. Electrical wire cables and harnesses must be used to connect the typical electronic box to the main electrical system in the airplane or missile. If isolators are used, these cables and harnesses will be forced through large amplitudes because of the sway space required by the isolators. Special precautions must be taken to prevent fatigue failures in the cables and harnesses.

4. A good vibration isolator is often a poor shock isolator and a good shock isolator is often a poor vibration isolator. The proper design must be incorporated into the isolator to satisfy both the vibration and the shock requirements.

Cold-plate designs for airborne electronic equipment have become more and more sophisticated with the use of air and liquids. Air heat exchangers make extensive use of multiple fins, wavy fins, split fins, and pin fins in order to improve the heat transfer characteristics. The multiple-fin type of heat exchangers are usually dip-brazed aluminum with as many as 22 fins per inch. These fins may be only 0.006 in. thick but the large number of fins in a typical heat exchanger makes it quite rigid for its weight. Airplanes make extensive use of electronic equipment where the cooling-air type of heat exchanger is built right into the electronic support structure. The heat exchanger is riveted, brazed or cemented to the major structural members in the electronic box so that the heat exchanger itself becomes a major load-carrying member of the system. The cooling air for the cold plate is usually taken from the first stage of the compressor on the jet engine that powers the airplane. This air must be conditioned before it can be used for cooling because the air temperature from the first stage is usually greater than 300°F.

Liquid-cooled cold plates are usually used to cool electronic equipment on spacecraft or very-high-flying research types of airplanes. The cooling liquid, usually a mixture of ethylene glycol and water, is very similar to the permanent type of antifreeze used in most automobile radiators in the wintertime.

The liquid-cooled cold plates are usually made part of the spacecraft airframe structure instead of the electronic box structure. When the electronic box must be removed from the spacecraft, it is not necessary to disconnect fluid lines, which can become quite messy. Since the cold plate stays with the airframe, the heat dissipated by the electronic box is usually transferred to the cold plate through a flat interface on the mounting surface of the electronic box that makes intimate contact with the cold plate.

The trend in commercial and military electronics is toward the line replaceable unit (LRU) with which it is possible to replace a defective electronic box right on the flight line in a matter of minutes. This is accomplished by providing all of the required interface connections, both mechanical and electrical, at the back end of the electronic box. The box becomes similar to a printed circuit board that can be plugged into its receptacle.

If there are several large electrical connectors on the back end of the electronic box, it may be quite difficult to insert the box and engage the electrical connectors properly. Some connectors may require a force of 0.50 lb/pin. for proper engagement. When there are 8 connectors, each with 100 pins, there is a total of 800 pins which will require a 400-lb force to engage them.

The plug-in electronic box is usually engaged and locked into position by some mechanism at the front of the box. Since the connectors are at the rear of the box, this means the force required to engage the connectors must pass through the box. When this type of electronic box is subjected to vibration, in many instances the vibration loads must be added to the installation loads to determine the total load acting on the structure.

There has been an attempt to standardize electronic equipment used in military and commercial airplanes by establishing certain sizes for modular electronic units. These modular units are then mounted in a standard 'air transport rack' (ATR) which provides rear-located dowel pins and connectors and a quick-release fastener at the front.

1.10. Electronic Equipment for Ships and Submarines

Ships and submarines will generally make use of a console type of cabinet to support their electronic equipment since there is usually more room available and weight is not very critical. The electronic components are usually mounted on panels and in sliding drawers. Panels are generally used to support dials, gages, manual controls, and test points. Only small masses are mounted on panels because they are fastened to a frame or rack in the cabinet and they cannot withstand large dynamic loads.

Drawers are often used to support the more massive electronic components such as those normally used in power supplies. The drawers are mounted on telescoping slides to provide access to the equipment. For safety, the drawers usually lock in the open and closed positions. For convenience, the drawers may also tilt to improve access and convenience in tight spaces.

The vibration frequency spectrum for ships and submarines varies from about 1 to about 50 Hz, but the most common range is from about 12 to about 33 Hz. The maximum acceleration level in this range is about 1 *G*, and appears to be due to vibrations set up by the engines and propellers.

In military ships, shock is an important factor due generally to various explosions which can do extensive damage to electronic equipment, unless proper consideration is given to the design and installation. For example, it is not desirable to have a very rigid structure supporting the electronics because a very rigid structure may not deform enough to absorb much strain energy. Theoretically any structure which does not deform when it is subjected to an impact load will receive an infinite acceleration. A large displacement is desirable since it can substantially reduce acceleration *G* levels. This displacement must either be confined to the structure or shock mounts must be used. In either case, provisions must be made in the design and installation to make sure parts will not collide and equipment will not break loose [36].

If shock isolators are used, they should be designed to deflect enough to absorb the shock energy without transmitting excessive loads to the electronic equipment. The shock mounts should have a minimum resonant frequency of about 25 Hz [2, 5]. Ideally the resonant frequency of the electronic components should be at least twice that of the shock mounts but never below 60 Hz. If the electronic components have resonances substantially below 60 Hz, it might bring them into the range of the most common vibration-forcing frequencies which, as previously mentioned, are as high as 33 Hz. If this should happen, the electronic components would be driven continuously near their resonance and could lead to fatigue failures.

When the forcing trequency of the ship's structure is near its higher frequency limit (around 25 Hz) the resonant frequency of the electronic equipment cabinet will be excited since the shock isolators also have their resonance at 25 Hz. This condition should not impose high stress levels on the electronic components mounted in the cabinet if the component resonance is twice that of the isolators. The cabinet support structure will, however, have to withstand the dynamic vibration loads. These loads will be determined by the amplification characteristics of the shock isolators during vibration. Shock isolators are available that provide a vibration amplification of about 3 for the conditions described above. Since the general vibration acceleration input levels in these frequency ranges is normally quite low, an amplification factor of 3 for the isolators does not result in high stresses in the equipment cabinet.

It is generally not desirable to increase the resonant frequency of the electronic equipment as high as possible. If the equipment is very stiff it may be good for the vibration condition but poor for the shock condition. A very high spring rate may result in very high shock stresses because of the high acceleration loads. Lee[5] recommends a maximum resonant frequency of about 100 Hz and a maximum acceleration of 200 *G* on the electronic components.

On tall narrow cabinets the load-carrying isolators should be at the base and stabilizing isolators should be at the top. A rigid structure must be used to support the stabilizing isolators at the top. If there is excessive deflection in the top support structure, it can change the characteristics of the entire system because of severe rocking modes.

If shock isolators are not used, the shock energy must be absorbed by deflections in the electronic equipment cabinet and in the structure of the ship supporting the cabinet. In this case the natural frequency of the assembly, which consists of the cabinet and the ship's structure, should be about 60 Hz. The natural frequency of the electronic components mounted on the equipment cabinet should be twice that of the assembly, or about 120 Hz. The ship's structure must provide a good part of the deflection required to attenuate the shock force, or the dynamic stresses in the equipment cabinet may be high enough to cause structural failures.

When the vibration-forcing frequency in the ship's structure is near its normal maximum limit of 33 Hz, the dynamic loads in the equipment cabinet will not be amplified to any great extent since the resonant frequency of the cabinet, at 60 Hz, is almost twice the forcing frequency. Furthermore, the electronic components mounted in the equipment cabinet have a still higher resonant frequency (120 Hz), so their dynamic vibration loads should be relatively small.

The materials that are best suited for shock are ductile materials with a high yield point, a high ultimate strength, and a high percentage elongation. In general, metals which are mechanically formed are more desirable than cast metals, which have a relatively low percentage elongation.

Since acceleration forces become progressively smaller as they propagate into the interior of the equipment cabinet, the electronic components that can withstand the highest *G* forces should be mounted near the exterior of the cabinet. Electronic components which have their own rigid structures should be used to lend additional strength to the outer structural elements in the cabinet.

Electronic components which cannot withstand high *G* forces should be mounted at the maximum elastic distance from the application points of the shock load. This will usually be at the center of the cabinet.

The load path for each mass element in the system should be examined closely to determine the path the load will take as it passes through the structure. For example, a large transformer should be mounted close to a major structural support in order to reduce the length of the load path. This will result in smaller deflections and stresses.

1.11. Electronic Equipment for Automobiles, Trucks, and Trains

Electronic equipment is very common in automobiles and trucks, and it is becoming more common in trains. The most common type of electronic equipment in these vehicles is the radio. Probably the next most common type of electronics is the two-way communication equipment being used by taxicabs, police cars, trucks, trains, ham operators, and emergency vehicles. Television is even being introduced for the back-seat area in automobiles. Some other applications of electronics for these vehicles have been in the use of transistorized ignition systems, fuel-metering devices, and nonskid braking devices.

Continuous vibration in automobiles, trucks, and trains is not very severe. On a specially constructed concrete washboard road, a 2.5-ton truck showed[5] 7 *G* at 2–4 Hz at the springs and 3 *G* at 8–12 Hz at the tires. In equipment-mounting areas the values were 2 *G* at 14–20 Hz and 0.5 *G* at 30–60 Hz. Dummer and Griffin[6] show 3–5 *G* at 2 to 4 Hz at the springs and 1 *G* at 8–15 Hz at the body of a typical wheeled vehicle.

A series of tests were run with a 2.5-ton 6×6 standard military cargo truck which consisted of driving the truck over a specially constructed steel-pipe washboard road at different speeds[8]. An electronic equipment cabinet 78 in. high × 45 in. wide × 24 in. deep, without drawers, was lashed to the side stakes of the truck near the center. Three triaxial accelerometers were placed on the truck bed and one on the top of the elec-

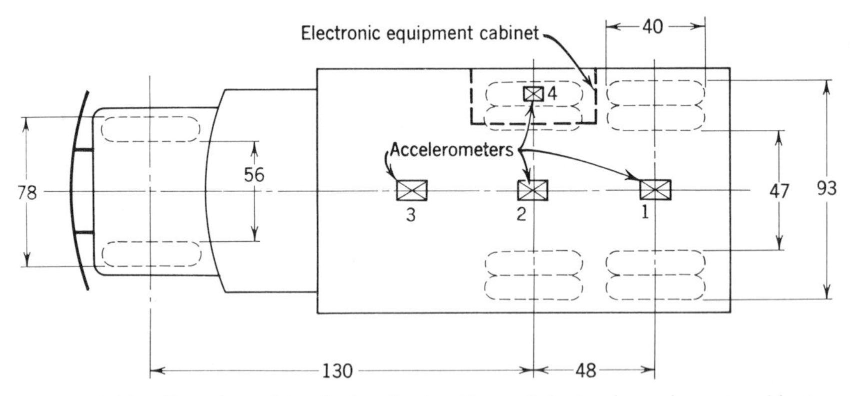

FIGURE 1.13. Top view of truck showing locations of electronic equipment, cabinets, and accelerometers.

tronic cabinet to monitor acceleration G levels and frequency ranges experienced in different sections of the truck, as shown in Fig. 1.13.

Triaxial accelerometers were mounted in four positions as follows:
Position 1: On truck bed over rear axle.
Position 2: On truck bed over inside rear axle.
Position 3: On truck bed midway between the cab and the inside rear axle.
Position 4: On the top of the electronic equipment cabinet.

A total of four different washboard roads were constructed of 2-in. diameter steel pipe spaced at different intervals over a 48-ft length. Periodic motion was observed only on the first washboard configuration, which had the pipes spaced 1 ft apart. The vibration frequencies recorded by the pickups are shown in Table 1.2. The characteristics of the vibration

TABLE 1.2. TRACK CONFIGURATION 1 ONLY

Speed (mph)	Frequency (Hz)	Speed (mph)	Frequency (Hz)
3	7.75	20	33.0
5	7.5	25	40.0
10	15.5	35	Random
15	22.0	40	Random

recorded by the pickups for the other three track configurations indicated they were all random in nature, not periodic.

The four different track configurations used in these tests are shown in Fig. 1.14; the peak accelerations recorded by the accelerometer pickups are shown in Table 1.3.

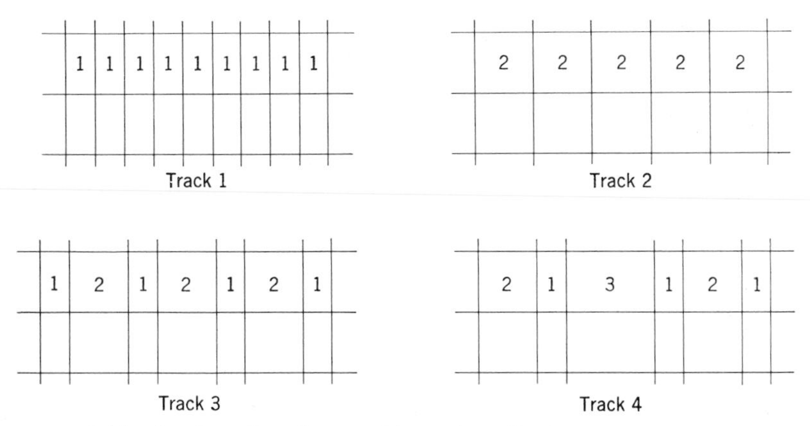

FIGURE 1.14. Track configurations used in rough road tests.

TABLE 1.3. PEAK ACCELERATIONS G's

Speed	Pickup 1			Pickup 2			Pickup 3			Pickup 4		
	Vert	Long	Lat	Vert	Long	Lat	Vert	Long	Lat	Vert	Long	Lat
Track Configuration 1												
3	6.2	3.4	2.9	2.4	1.5	1.2	2.7	1.1	1.5	1.2	1.8	1.0
5	6.0	2.9	2.9	4.3	1.6	1.9	3.1	1.9	2.0	2.0	1.3	1.4
10	2.6	2.7	2.2	2.0	1.7	1.2	1.7	1.6	0.8	1.6	1.8	1.0
15	9.0	4.2	5.4	4.0	3.9	2.0	4.0	2.9	1.5	2.9	1.8	1.1
20	3.6	2.5	2.8	3.1	2.4	2.4	3.6	1.9	1.6	1.1	1.2	0.6
25	3.3	3.0	2.7	3.0	2.9	2.8	3.3	2.5	1.7	1.0	1.0	0.6
35	4.5	3.2	3.3	5.2	2.7	2.9	3.4	2.7	1.8	1.0	1.3	0.6
40	5.4	3.3	3.0	6.8	3.4	3.0	4.6	2.9	2.1	1.1	1.0	0.8
Track Configuration 2												
3	11.4	6.3	7.1	8.8	5.6		5.5	4.1	5.0	1.9	2.5	1.5
5	12.6	7.7	8.3	10.0	5.0	6.4	7.0	3.7	4.5	2.5	1.9	1.8
10	18.5	10.6	13.8	16.7	8.4	10.3	13.5	6.7	7.2	9.0	8.1	3.9
15	6.8	5.2	3.9	6.0	3.6	3.7	5.8	2.7	2.3	2.7	2.2	1.9
Track Configuration 3												
3	7.8	5.1	3.0	3.2	2.0	1.5	2.9	1.6	1.4	3.2	2.7	1.6
5	10.0	5.4	4.4	3.7	2.7	3.1	4.6	2.6	2.0	4.9	5.8	1.6
10	12.3	5.9	6.6	8.8	3.8	3.4	8.3	4.3	2.6	5.4	4.1	1.6
15	16.8	7.2	12.2	7.1	4.0	5.8	7.7	3.8	4.0	11.2	4.6	4.3
Track Configuration 4												
3	9.3	4.9	6.6	4.8	2.1	2.7	5.5	2.2	2.3	3.0	2.5	2.1
5	10.4	4.9	6.6	4.8	3.2	4.2	5.2	3.2	2.7	5.8	4.2	2.5
10	13.5	7.2	7.6	7.1	5.0		9.7	5.1	5.1	3.3	4.8	2.2
15	16.1	9.5	13.0	6.7	3.3	4.5	7.1	3.6	3.1	7.5	4.7	4.3

Anyone who has ever removed a radio from an automobile knows that radios are hard-mounted to the dashboard of the auto. Two-way communication equipment used in police cars is generally hard-mounted to the floor in the trunk area of the auto. Two-way communication equipment used in taxicabs is now being hard-mounted under the dashboard of the automobile because the new solid-state equipment is quite compact. The older equipment, which required more volume, was generally hard-mounted in the trunk area of the auto. Ham radio operators also tend to hard-mount their electronic communication equipment under the dashboard of their autos.

An examination of the frequency ranges normally encountered in autos shows they are generally quite low. An examination of the commercial types of sheet-metal housings usually used for radio equipment shows they are stiff enough to have their resonant frequencies much higher than the forcing frequencies normally encountered in automobiles. This type of construction, however, is usually not rigid enough for airplanes or missiles, which have much higher forcing frequencies that can easily excite sheet-metal resonances.

Communication equipment on trains is often mounted on shock isolators because of high-impact shock loads that are developed when trains are shunted from one track to another. Impact velocities are as high as 18 ft/sec and deceleration levels often reach 20 G.

A series of tests were run on different vehicles at the Aberdeen Proving Ground in Maryland, and are generally referred to as the Munson Road Tests [35]. These tests are being used by many organizations as a basis for the qualification of electronic equipment that must be transported over rough roads.

CHAPTER 2

VIBRATIONS OF SIMPLE SYSTEMS

2.1. SINGLE-SPRING-MASS SYSTEM WITHOUT DAMPING

The natural frequency of many single-degree-of-freedom systems can be determined by evaluating the characteristics of the strain energy and the kinetic energy of each system. Considering a single-spring-mass system, for example, if there is no energy lost then the maximum kinetic energy of the mass must be equal to the maximum strain energy in the spring, if the spring mass is negligible.

All real systems have some damping. If a real mass is suspended on a coil spring and the spring is stretched and released, the mass will vibrate up and down. This free vibration may continue for a long time but eventually all free vibrations die out and the mass stops vibrating. This is because damping in the spring dissipates a little energy with each cycle and eventually the mass stops moving. If there is no damping, the mass will theoretically keep on vibrating up and down with the same amplitude and frequency forever.

In many systems, the damping is so small that it has very little effect on the natural frequency. Under these circumstances the natural frequency of the undamped system can be determined because it will be very close to the natural frequency of the damped system and will not require as much work.

The maximum kinetic energy of a vibrating-spring-mass system with no damping is at the point of maximum velocity. This occurs as the vibrating mass passes through the zero-displacement point, as shown in Fig. 2.1.

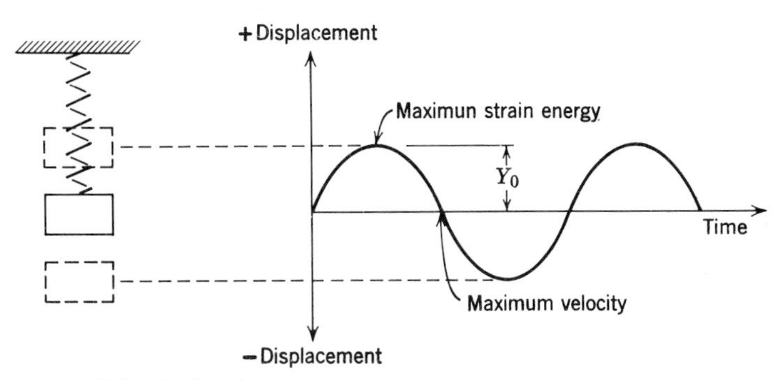

FIGURE 2.1. A vibrating spring mass system.

From elementary physics the maximum kinetic energy, T, of the vibrating mass is

$$T_0 = \tfrac{1}{2} m V^2 \tag{2.1}$$

The instantaneous tangential velocity, V, can be expressed in terms of rotational velocity as shown in Fig. 1.3.

$$V = Y_0 \Omega \tag{2.2}$$

Substituting Eq. 2.2 into 2.1, the kinetic energy of the system becomes

$$T_0 = \tfrac{1}{2} m Y_0^2 \Omega^2 \tag{2.3}$$

The maximum strain energy can be determined from the work done on the spring by the mass as it stretches and compresses the spring during vibration. Since the spring is linear, the deflection is directly proportional to the force as shown in Fig. 2.2.

The area under the curve represents the work done on the spring. Since the work is equal to the strain energy, the strain energy becomes

$$U_0 = \tfrac{1}{2} P_0 Y_0 \tag{2.4}$$

The spring rate, K, can be defined in terms of the maximum load P_0 and the maximum deflection Y_0 as $K = P_0/Y_0$ so

$$P_0 = K Y_0 \tag{2.5}$$

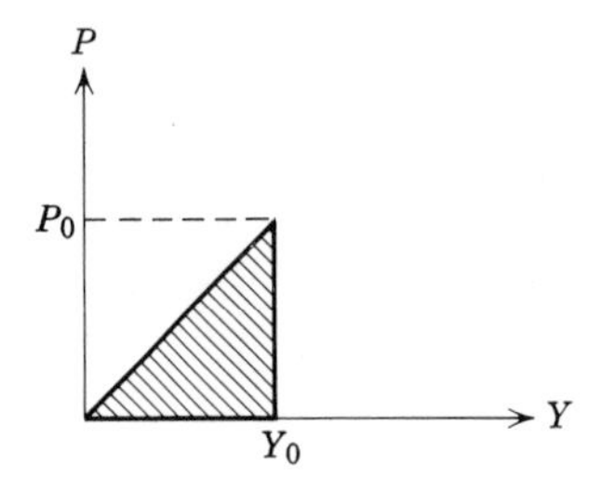

FIGURE 2.2. Typical load-deflection curve.

Substituting Eq. 2.5 into 2.4, the maximum strain energy becomes

$$U_0 = \tfrac{1}{2} K Y_0^2$$

If damping is assumed to be zero, then at resonance the kinetic energy will be equal to the strain energy.

$$\tfrac{1}{2} m Y_0^2 \Omega^2 = \tfrac{1}{2} K Y_0^2$$

$$\Omega_n = \left(\frac{K}{m}\right)^{1/2} \tag{2.6}$$

This is the natural frequency in radians per second. Using Eq. 1.2, the natural frequency in cycles per second (Hz) is

$$f_n = \frac{1}{2\pi}\left(\frac{K}{m}\right)^{1/2} \tag{2.7}$$

The natural frequency equation can be written in a slightly different form by considering that the static deflection δ_{st} of the spring is due to the action of the weight, W, acting on the spring. The spring rate can then be written as

$$K = \frac{W}{\delta_{st}} \tag{2.8}$$

Expressing the mass in terms of weight, W, and gravity, g

$$m = \frac{W}{g} \tag{2.9}$$

Substituting Eqs. 2.8 and 2.9 into 2.7 gives another relation for the natural frequency.

$$f_n = \frac{1}{2\pi}\left(\frac{g}{\delta_{st}}\right)^{1/2} \tag{2.10}$$

2.2. SAMPLE PROBLEM

The use of these natural-frequency equations can be demonstrated by considering a cantilever beam with an end mass as shown in Fig. 2.3.

If the weight of the aluminum beam is small compared to the weight of the end mass then the weight of the aluminum beam can be ignored without too much error. The cantilever beam can then be analyzed as a concentrated load on a massless beam which retains its elastic properties.

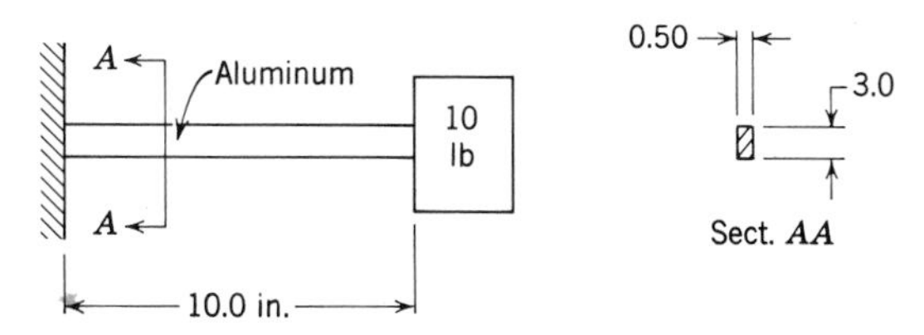

FIGURE 2.3. Cantilever beam with an end mass.

The static deflection at the end of this beam can be determined with the use of a structural handbook.

$$\delta_{st} = \frac{WL^3}{3EI} \tag{2.11}$$

where $W = 10.0$ lb end weight

$L = 10.0$ in. length

$E = 10.5 \times 10^6$ lb/in.^{2} modulus of elasticity, aluminum

$I = \frac{bh^3}{12} = \frac{(0.50)(3.0)^3}{12} = 1.125$ in.^{4} moment of inertia

$$\delta_{st} = \frac{(10.0)(10.0)^3}{3(10.5 \times 10^6)(1.125)} = 0.282 \times 10^{-3} \text{ in.} \tag{2.12}$$

Substitute Eq. 2.12 into Eq. 2.10 and note that the acceleration of gravity, g, is 386 in./sec^2. The natural frequency becomes

$$f_n = \frac{1}{2\pi}\left(\frac{386}{0.282 \times 10^{-3}}\right)^{1/2} = 187 \text{ Hz} \tag{2.13}$$

The natural frequency of the cantilever beam can also be determined from the spring rate, K, of the beam using Eq. 2.11.

$$K = \frac{W}{\delta_{st}} = \frac{3EI}{L^3} \tag{2.14}$$

Using the values previously calculated the spring rate becomes

$$K = \frac{3(10.5 \times 10^6)(1.125)}{(10.0)^3} = 3.54 \times 10^4 \text{ lb/in.}$$

Also

$$m = \frac{W}{g} = \frac{10}{386} = 0.0259 \text{ lb sec}^2\text{/in.}$$

Substituting into Eq. 2.7, the natural frequency becomes

$$f_n = \frac{1}{2\pi}\left(\frac{K}{m}\right)^{1/2} = \frac{1}{2\pi}\left(\frac{3.54 \times 10^4}{0.0259}\right)^{1/2}$$

$$f_n = 187 \text{ Hz} \tag{2.15}$$

The results are exactly the same as Eq. 2.13.

2.3. Force Balance Method

The natural-frequency equation for a single-spring-mass system without damping, can also be determined by considering a force balance using Newton's second law of motion ($F = m\ddot{Y}$). If the mass is displaced and then released, the forces acting on the mass are as shown in Fig. 2.4 if the spring mass is very small.

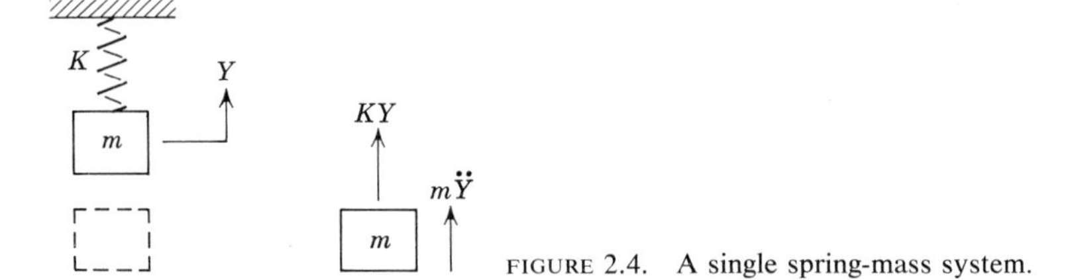

FIGURE 2.4. A single spring-mass system.

At the instant the mass is released, the force in the spring will tend to accelerate the mass in the vertical direction as shown. Considering these forces in the vertical direction,

$$m\ddot{Y} + KY = 0 \tag{2.16}$$

This is a differential equation which has a function Y and a second derivative of the same function $\ddot{Y}$.

The solution to this differential equation can be written in several different forms. One form, however, is relatively simple; it satisfies the equation and leads to a quick solution.

$$Y = a \sin \Omega t \tag{2.17}$$

$$\dot{Y} = \frac{dY}{dt} = \Omega a \cos \Omega t$$

$$\ddot{Y} = \frac{d^2Y}{dt^2} = -\Omega^2 a \sin \Omega t \tag{2.18}$$

Substituting Eqs. 2.17 and 2.18 into Eq. 2.16,

$$m(-\Omega^2 a \sin \Omega t) + K(a \sin \Omega t) = 0$$

Solving for the rotational natural frequency Ω_n,

$$\Omega_n = \left(\frac{K}{m}\right)^{1/2} \tag{2.19}$$

This is exactly the same as Eq. 2.6.

2.4. Single-Degree-of-Freedom—Torsional Systems

The energy method is convenient for determining the natural frequency of a torsional system with one degree of freedom, as shown in Fig. 2.5.

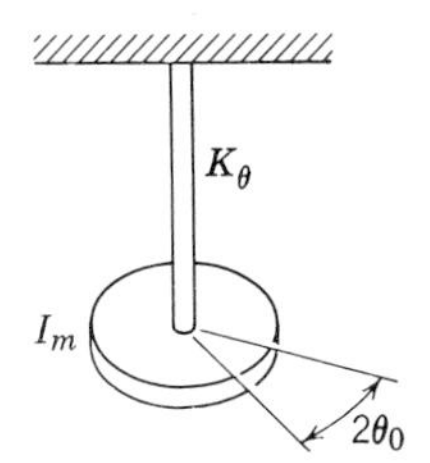

FIGURE 2.5. A single spring-mass torsional system.

The torsional system is similar to the spring-mass system shown in Fig. 2.4 where the spring action is due to the twisting of the rod and the inertia is due to the mass moment of inertia of the disk about an axis perpendicular to the plane of the disk. Assume the rod mass is small. The maximum kinetic energy of the oscillating disk is

$$\text{KIN} = \tfrac{1}{2} I_m \dot{\theta}^2 \tag{2.20}$$

where I_m = mass moment of inertia of disk
$\dot{\theta}$ = rotational angular velocity

The maximum angular velocity for the oscillating system moving through the angle θ_0 is

$$\dot{\theta}_0 = \theta_0 \Omega \tag{2.21}$$

Substituting Eq. 2.21 into Eq. 2.20, the maximum kinetic energy becomes (assuming a small rod mass):

$$\text{KIN} = \tfrac{1}{2} I_m \theta_0^2 \Omega^2 \tag{2.22}$$

The maximum strain energy can be determined from the work done on the twisting rod by the disk as it oscillates. Since the spring rate of the rod is linear, the angular deflection will be directly proportional to the torque, T, applied, as shown in Fig. 2.6. The area under the curve represents the

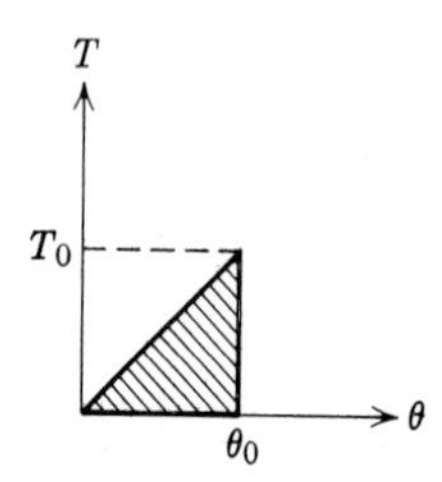

FIGURE 2.6. A torque-angular rotation curve.

work done on the rod. Since the work is equal to the strain energy, the strain energy becomes

$$U_0 = \tfrac{1}{2} T_0 \theta_0 \tag{2.23}$$

The torsional spring rate, K_θ, can be defined in terms of the torque, T_0, and the angular displacement, θ_0, as follows:

$$K_\theta = \frac{T_0}{\theta_0}$$

so

$$T_0 = K_\theta \theta_0 \tag{2.24}$$

Substituting Eq. 2.24 into Eq. 2.23, the maximum strain energy becomes

$$U_0 = \tfrac{1}{2} K_\theta \theta_0^2 \tag{2.25}$$

If damping is assumed to be zero, then at resonance the kinetic energy (Eq. 2.22) must equal the strain energy (Eq. 2.25).

$$\tfrac{1}{2} I_m \theta_0^2 \Omega^2 = \tfrac{1}{2} K_\theta \theta_0^2$$

$$\Omega_n = \left(\frac{K_\theta}{I_m}\right)^{1/2} \tag{2.26}$$

The natural frequency (in Hz) becomes

$$f_n = \frac{1}{2\pi}\left(\frac{K_\theta}{I_m}\right)^{1/2} \tag{2.27}$$

2.5. Sample Problem

The natural frequency of the torsional system shown in Fig. 2.7 can be determined from the geometry of the structure.

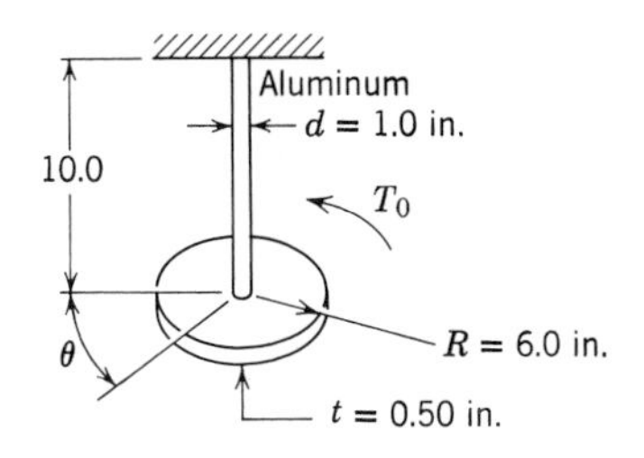

FIGURE 2.7. A single-degree-of-freedom torsional system.

The torsional spring rate K_θ of the rod can be determined using its angular displacement θ under the action of an external torque, T_0,

$$\theta = \frac{T_0 L}{GJ} \quad \text{and} \quad K_\theta = \frac{T_0}{\theta}$$

so

$$K_\theta = \frac{GJ}{L} \tag{2.28}$$

where $L = 10.0$-in. length of rod

$G = 4.0 \times 10^6$ lb/in.^{2} shear modulus of aluminum

$J = \dfrac{\pi d^4}{32} = \dfrac{\pi}{32}(1.0)^4 = 0.0981$ in.^{4} polar moment of inertia

$$K_\theta = \frac{(4.0 \times 10^6)(0.0981)}{10.0} = 3.92 \times 10^4 \text{ in. lb/rad}$$

The mass moment of inertia of the aluminum disk must be taken about the axis perpendicular to the plane of the disk.

$$I_m = \frac{WR^2}{2g}$$

where $W = \pi(6.0)^2(0.50)(0.10 \text{ lb/in.}^3) = 5.65$ lb disk weight

$R = 6.0$-in. disk radius

$$g = 386 \text{ in/sec}^2 \text{ gravity}$$

$$I_m = \frac{(5.65)(6.0)^2}{2(386)} = 0.264 \text{ lb in. sec}^2$$

The torsional resonant frequency can now be determined from Eq. 2.27.

$$f_n = \frac{1}{2\pi}\left(\frac{3.92 \times 10^4}{0.264}\right)^{1/2} = 61.5 \text{ Hz} \tag{2.29}$$

2.6. Oscillating Body as a Torsional System

There are cases where a cantilevered structure might be approximated as a torsional system so the natural frequency can be estimated. Consider a large transformer bolted to the center of a rectangular plate that is being vibrated in a direction parallel to the plane of the plate as shown in Fig. 2.8*a*.

The transformer can be approximated as a structure on a pivot and the stiffness can be approximated by springs as shown in Fig. 2.8*b*. In this view, the force, F, developed in each spring is proportional to the spring displacement, δ, in the vertical direction and the spring rate, K, as shown in Fig. 2.9.

For small angles, the displacement of the springs in the vertical direction is

$$\delta = \frac{a}{2}\theta$$

Substituting into the above equation,

$$F = \frac{Ka\theta}{2}$$

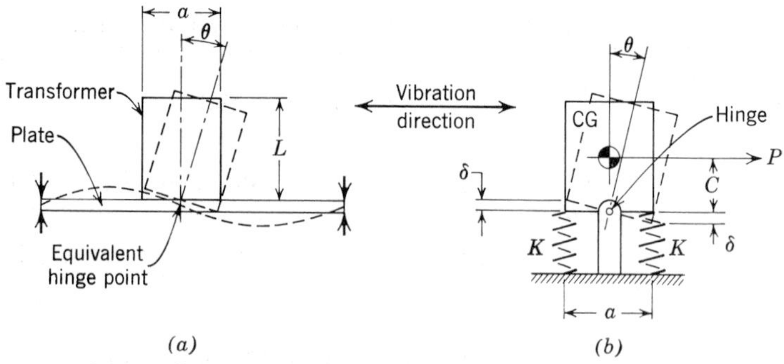

FIGURE 2.8. A flat plate with an overturning moment simulated by a pivoting system.

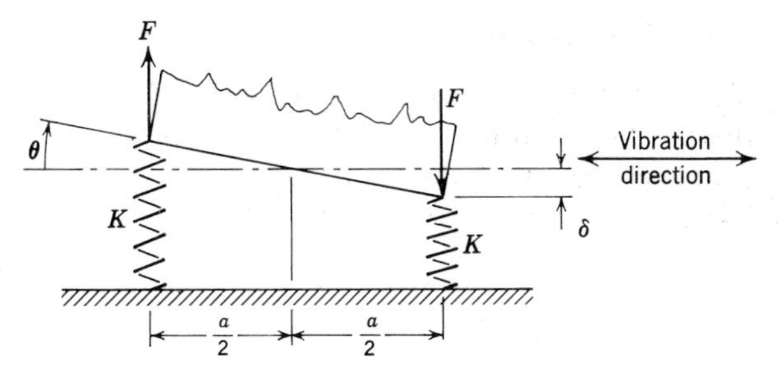

FIGURE 2.9. Displacements of a pivoting system.

The relation between the dynamic load, P, acting on the transformer and the force in the spring can be determined by taking moments about the pivot point.

$$PC = 2F\left(\frac{a}{2}\right) \qquad \text{or} \qquad F = \frac{PC}{a}$$

Substitute into the above equation and solve for the angle θ.

$$\theta = \frac{2\,PC}{Ka^2}$$

The rotational (or torsional) spring rate can now be defined in terms of the torque, T_0, and the angular rotation, θ.

$$K_\theta = \frac{T_0}{\theta} = \frac{PC}{\dfrac{2\,PC}{Ka^2}} = \frac{Ka^2}{2} \tag{2.30}$$

The mass moment of inertia in this case can be treated as a rectangular body pivoting about its base so that

$$I_m = \frac{W}{12\,g}(4\,L^2 + a^2) \tag{2.31}$$

The rotational natural frequency can then be determined by substituting Eqs. 2.30 and 2.31 into Eq. 2.27.

$$f_n = \frac{1}{2\pi}\left(\frac{K_\theta}{I_m}\right)^{1/2} = \frac{1}{2\pi}\left(\frac{6\,Ka^2 g}{W(4\,L^2 + a^2)}\right)^{1/2} \tag{2.32}$$

2.7. Sample Problem

A transformer is mounted at the center of a simply supported rectangular plate that is subjected to vibration in the plane of the plate. Determine the natural frequency of the transformer in its rotational mode. A sketch of the system is shown in Fig. 2.10.

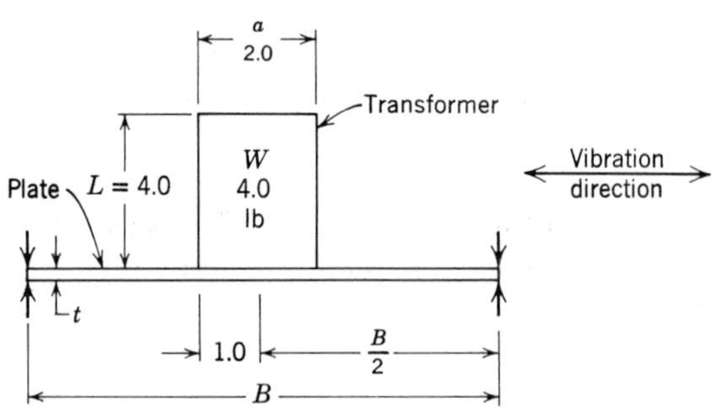

FIGURE 2.10. A transformer mounted at the center of a plate.

The spring rate, K, of the rectangular plate in the vertical direction can be approximated by considering the plate to be hinged at the center as shown in Fig. 2.8*a*. Using a unit load of 1 lb, the plate deflection can be determined at the edge of the transformer which is 1.0 in. from the plate center (Fig. 2.10). Assume these calculations result in a K value of 10,000 lb/in. Then substituting into Eq. 2.32,

$$f_n = \frac{1}{2\pi}\left\{\frac{6(1 \times 10^4)(2.0)^2(386)}{(4.0)\,[4(4.0)^2 + (2.0)^2]}\right\}^{1/2}$$

$$f_n = 93 \text{ Hz} \tag{2.33}$$

2.8. Single-Degree-of-Freedom—Simple Pendulum

The energy method can also be used to determine the natural frequency of a simple pendulum with a concentrated mass at the end of a rod with negligible mass, as shown in Fig. 2.11.

The pendulum is very similar to the torsional system since both systems have angular oscillations. In the pendulum, the restoring force is gravity instead of a torsion spring.

The maximum kinetic energy of the system will occur at the maximum angular velocity.

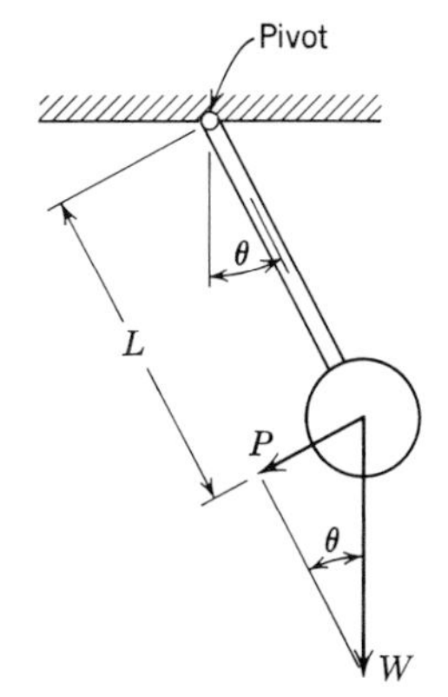

FIGURE 2.11. A simple pendulum as a single-degree-of-freedom.

$$\text{KIN} = \tfrac{1}{2} I_m \dot{\theta}_0^2$$

where $I_m = \dfrac{WL^2}{g}$ mass moment of inertia about pivot

$\dot{\theta}_0 = \theta_0 \Omega$ rotational angular velocity

Substituting into the above equation,

$$\text{KIN} = \frac{WL^2}{2g} \theta_0^2 \Omega^2 \tag{2.34}$$

The restoring torque, T_0, acting on the pendulum system can be determined from the angle θ at any position.

$$T_0 = PL \qquad \text{and} \qquad P = W \sin\theta$$

then

$$T_0 = WL \sin\theta$$

For small angles, less than about 9°, the sine of the angle is about equal to the angle in radians so the restoring torque becomes

$$T_0 = WL\theta$$

The maximum potential energy can be determined from the maximum work done on the pendulum mass, which is equal to the area under the curve shown in Fig. 2.6 when the maximum angle θ_0 is used. The work then becomes

$$U_0 = \tfrac{1}{2} T_0 \theta_0$$

Since the potential energy equals the work, the preceding equations can be substituted to give the potential energy as follows

$$U_0 = \tfrac{1}{2} WL\theta_0^2 \tag{2.35}$$

If no energy is lost through friction, the kinetic energy (Eq. 2.34), must equal the potential energy (Eq. 2.35).

$$\frac{WL^2}{2g}\theta_0{}^2\Omega^2 = \frac{WL}{2}\theta_0{}^2$$

Then

$$\Omega^2 = \frac{g}{L}$$

and

$$f_n = \frac{\Omega}{2\pi} = \frac{1}{2\pi}\left(\frac{g}{L}\right)^{1/2} \tag{2.36}$$

2.9. Springs in Series and Parallel

If a mass is suspended on two different springs in such a manner that the load path is directly from the mass through one spring, and then through the second spring before it reaches the support, the springs are said to be in series. The series spring, therefore, implies a series load path where the load must first pass through one of the springs before it can pass through the other spring. Cutting either spring would completely destroy the system. Some samples of springs in series are shown in Fig. 2.12.

Springs in series can be combined into one equivalent spring using the relation

$$\frac{1}{K_{eq}} = \frac{1}{K_1} + \frac{1}{K_2} + \frac{1}{K_3} + \cdots \tag{2.37}$$

If a mass is suspended on two different springs in such a manner that the load path from the mass to the support is split between the springs, then the springs are said to be in parallel. The parallel path then permits

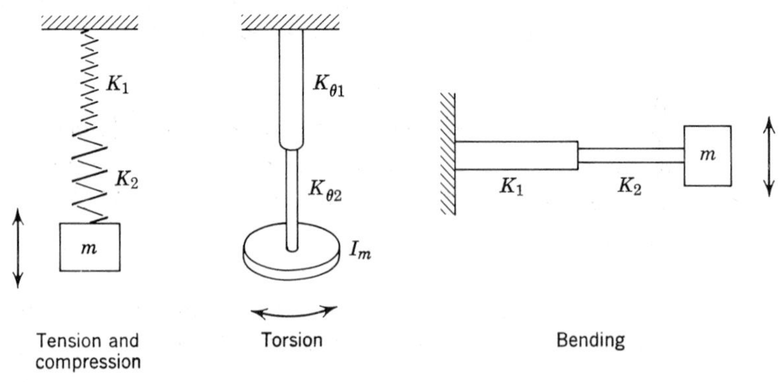

FIGURE 2.12. Springs in series.

the load to pass through one spring without passing through the other. Cutting one spring would still permit the other spring to carry part of the load. Some samples of springs in parallel are shown in Fig. 2.13.

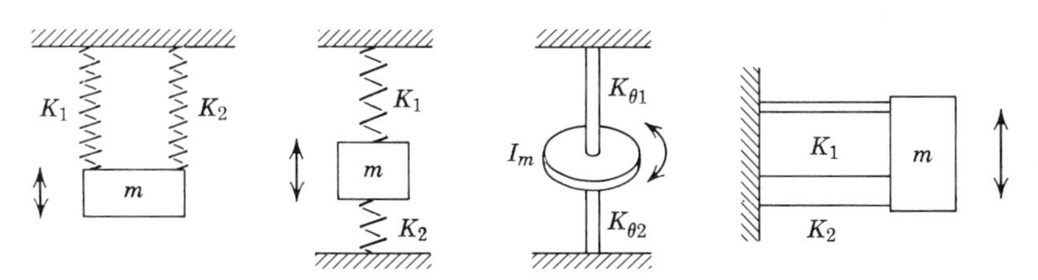

FIGURE 2.13. Springs in parallel.

Springs in parallel can be combined into one equivalent spring using the relation

$$K_{eq} = K_1 + K_2 + K_3 + \ldots \tag{2.38}$$

2.10. Sample Problem

The natural frequency of a mass supported by several springs in series and parallel combinations can be determined by obtaining the equivalent spring rate for the system.

Consider the system with spring constants as shown in the Fig. 2.14.

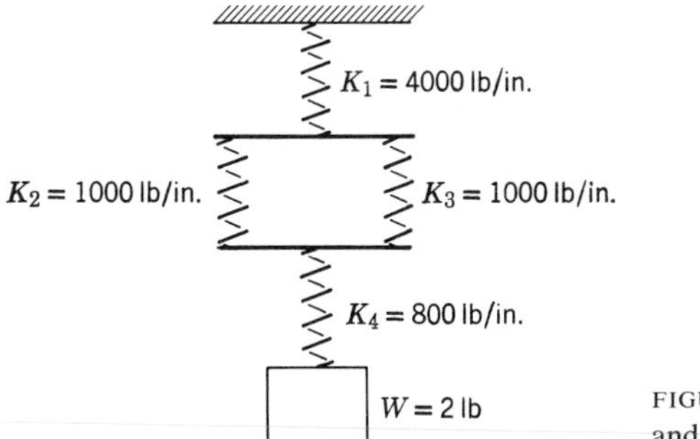

FIGURE 2.14. A combination of series and parallel springs.

Springs K_2 and K_3 are in parallel so they can be combined using Eq. 2.38.

$$K_5 = K_2 + K_3 = 1000 + 1000 = 2000 \quad \text{lb/in.}$$

Now there are three springs K_1, K_5, and K_4 in series, so they can be combined using Eq. 2.37.

$$\frac{1}{K_{eq}} = \frac{1}{K_1} + \frac{1}{K_5} + \frac{1}{K_4} = \frac{1}{4000} + \frac{1}{2000} + \frac{1}{800}$$

$$K_{eq} = \frac{1}{2.0 \times 10^{-3}} = 500 \text{ lb/in.}$$

The system now consists of one spring and one mass (Fig. 2.15).

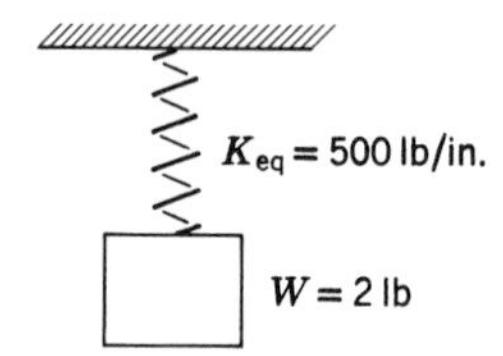

FIGURE 2.15. A single spring-mass system.

The natural frequency can be determined from Eqs. 2.7 and 2.9 as follows:

$$f_n = \frac{1}{2\pi}\left(\frac{Kg}{W}\right)^{1/2} = \frac{1}{2\pi}\left[\frac{500(386)}{2}\right]^{1/2}$$

$$f_n = 49.5 \text{ Hz} \tag{2.39}$$

2.11. Relation of Frequency and Acceleration to Displacement

Many important approximations can be related to the dynamic displacements developed during resonant conditions. For example, if the geometry of a specific structure can be defined, then dynamic bending moments and bending stresses can be calculated from the dynamic displacements. Since displacements are very seldom measured during vibration tests, because optical measurements are often difficult to make, accelerometers are generally used. This permits test data to be taken in terms of frequency and acceleration G forces. The relation of frequency and acceleration to the displacement must therefore be determined to use the test data.

Consider a rotating vector that is used to describe the simple harmonic motion of a single mass suspended by a spring. The vertical displacement, Y, of the mass can be determined by the projection of the vector Y_0 on the vertical axis as shown in Fig. 2.16.

The vertical displacement can be represented by the equation

$$Y = Y_0 \sin \Omega t \tag{2.40}$$

The velocity is the first derivative.

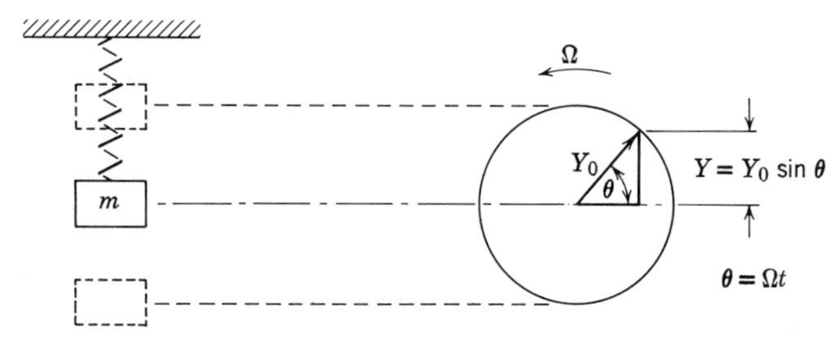

FIGURE 2.16. A rotating vector describing simple harmonic motion.

$$V = \dot{Y} = \frac{dY}{dt} = \Omega Y_0 \cos \Omega t$$

The acceleration is the second derivative

$$A = \ddot{Y} = \frac{d^2 Y}{dt^2} = -\Omega^2 Y_0 \sin \Omega t$$

The negative sign indicates that acceleration acts in the direction opposite to displacement.

The maximum acceleration will occur when $\sin \Omega t$ is 1.

$$A_{max} = \Omega^2 Y_0 \tag{2.41}$$

The acceleration, in gravity units G, can be determined by dividing the maximum acceleration by the acceleration of gravity, g.

$$g = 32.2 \text{ ft/sec}^2 = 386 \text{ in./sec}^2$$

Also, changing radians to cycles per second,

$$\Omega = 2\pi f$$

Substituting into Eq. 2.41,

$$G = \frac{A_{max}}{g} = \frac{4\pi^2 f^2 Y_0}{386}$$

$$G = \frac{f^2 Y_0}{9.8} \tag{2.42}$$

Note that the displacement, Y_0, is single-amplitude displacement.

2.12. Sample Problem

Vibration tests were run on a bracket that supported a large transformer. From the test data, determine the maximum dynamic stress in the bracket which is shown in Fig. 2.17.

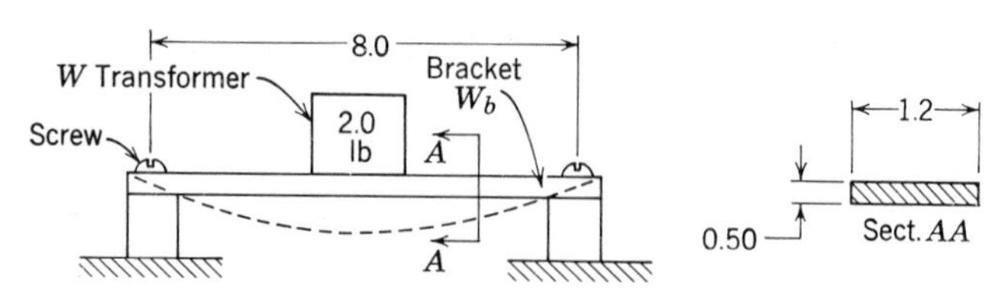

FIGURE 2.17. A transformer mounted on a bracket.

Test data: Natural frequency = 250 Hz
Transmissibility, $Q = 30$
Input G force = 5.0 G Peak

The weight of the aluminum bracket is:

$$W_b = 8(1.2)(0.50)(0.10\ \text{lb/in.}^3) = 0.48\ \text{lb}.$$

This is relatively small compared to the 2.0-lb transformer. As an approximation, ignore the bracket weight and consider the transformer as a concentrated load in the center of a massless beam.

Since relatively small screws are used to fasten the ends of the bracket, the end conditions are probably much closer to being simply supported than they are to being fixed. This can be checked by calculating the natural frequency of the bracket as a simply supported beam with a concentrated load and comparing the results with the test data.

The natural frequency of a simply supported beam with a concentrated load can be determined from Eq. 2.11 along with the static deflection from a standard handbook (Fig. 2.18).

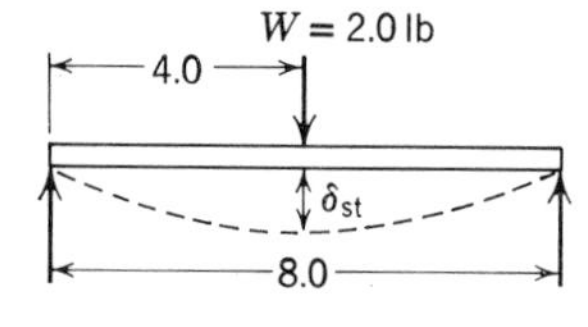

FIGURE 2.18. A simply supported beam with a concentrated load.

$$\delta_{st} = \frac{WL^3}{48\,EI} \tag{2.43}$$

where $W = 2.0$-lb transformer weight
$L = 8.0$-in. length
$E = 10.5 \times 10^6$ aluminum beam modulus of elasticity

$$I = \frac{bh^3}{12} = \frac{(1.2)(0.50)^3}{12} = 0.0125 \text{ in.}^4 \text{ moment of inertia}$$

$$g = 386 \text{ in./sec}^2$$

$$\delta_{st} = \frac{(2.0)(8.0)^3}{48(10.5 \times 10^6)(0.0125)} = 1.62 \times 10^{-4} \text{ in.}$$

Substitute into Eq. 2.10 for the natural frequency.

$$f_n = \frac{1}{2\pi}\left(\frac{g}{\delta_{st}}\right)^{1/2} = \frac{1}{2\pi}\left(\frac{3.86 \times 10^2}{1.62 \times 10^{-4}}\right)^{1/2}$$

$$f_n = 246 \text{ Hz (simply supported ends)} \tag{2.44}$$

If the bracket had fixed ends, the natural frequency would have been

$$f_n = 492 \text{ Hz} \tag{2.45}$$

Since the vibration test data showed a natural frequency of 250 Hz, it means the bolted ends on the transformer bracket are really much closer to a simply supported condition. The end conditions are required to determine the dynamic bending moments in the bracket.

Equation 2.42 can be used to determine the dynamic deflection at the center of the transformer bracket, using the 246-Hz resonance.

$$Y_0 = \frac{9.8\,G}{f^2} = \frac{9.8\,G_{in}Q}{f^2} \tag{2.46}$$

where $G_{in} = 5.0\,G$ input acceleration

$Q = 30$ transmissibility at resonance

$f_n = 246$-Hz simply supported beam

$$Y_0 = \frac{9.8(5.0)(30)}{(246)^2} = 0.0243 \text{ in.} \tag{2.47}$$

The dynamic load acting on the bracket can be approximated by letting Eq. 2.47 equal Eq. 2.43 and solving for the dynamic load W_d.

$$W_d = \frac{48\,EIY_0}{L^3}$$

$$W_d = \frac{48(10.5 \times 10^6)(0.0125)(0.0243)}{(8.0)^3}$$

$$W_d = 300 \text{ lb dynamic load} \tag{2.48}$$

If the acceleration loads were assumed to act directly on the transformer, the dynamic load would be

$$W_d = WG_{in}Q \tag{2.49}$$

where $W = 2.0$-lb. transformer weight
$G_{in} = 5.0\ G$ input (peak)
$Q = 30$ transmissibility at resonance

$$W_d = (2.0)\ (5.0)\ (30) = 300 \text{ lb} \tag{2.50}$$

This result is exactly the same as that shown in Eq. 2.48 which was obtained from the dynamic deflection of a simply supported beam with a concentrated load.

The dynamic bending stress in the simply supported transformer bracket can be determined from the geometry of the structure, using the dynamic loading as shown in Fig. 2.19.

The maximum dynamic bending moment occurs at the center of the bracket.

$$M = R\left(\frac{L}{2}\right) = 150\,(4) = 600 \text{ in. lb} \tag{2.51}$$

The maximum dynamic bending stress can be determined from the standard bending stress equation.

$$S_b = \frac{Mc}{I} \tag{2.52}$$

where $M = 600$ in. lb bending moment

$c = \dfrac{h}{2} = \dfrac{0.50}{2} = 0.25$ in. (Ref. sect. AA Fig. 2.17)

$I = \dfrac{bh^3}{12} = \dfrac{(1.2)\,(0.50)^3}{12} = 0.0125 \text{ in.}^4$

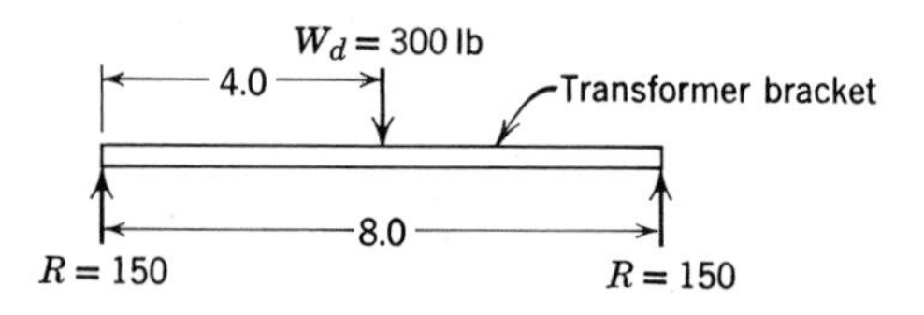

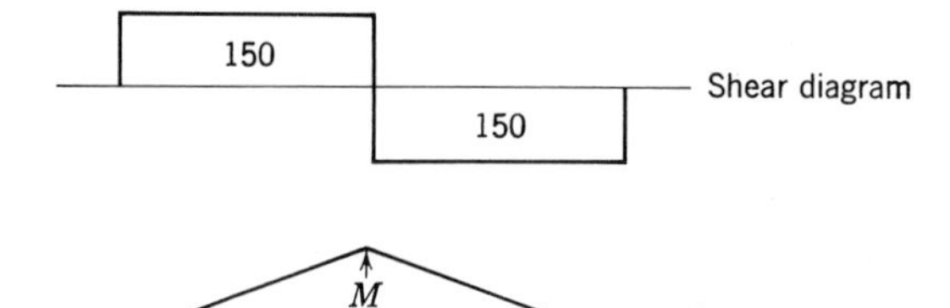

FIGURE 2.19. Shear and bending moment diagram for a beam.

Hence

$$S_b = \frac{(600)(0.25)}{0.0125} = 12{,}000 \text{ lb/in.}^2 \tag{2.53}$$

2.13. Free Vibrations with Viscous Damping

All real vibrating systems have some damping present, which eventually brings a freely vibrating system to rest. Damping is very complex in nature and there are many different types of damping. Viscous damping, which is damping that is proportional to velocity, is the most common type of damping used in vibration analysis. This does not mean to imply that viscous damping is the most common type of damping found in vibration, because it is not. It does mean, however, that viscous damping is very adaptable to mathematical analyses where other forms of damping are not.

Consider the single-degree-of-freedom system with the mass m, spring K, and damper c. The differential equation of motion can be determined by considering a force balance using Newton's second law of motion, $F = m\ddot{Y}$. When the displacement is downward and the motion is downward, the spring force and the damper force will act upward as shown in Fig. 2.20.

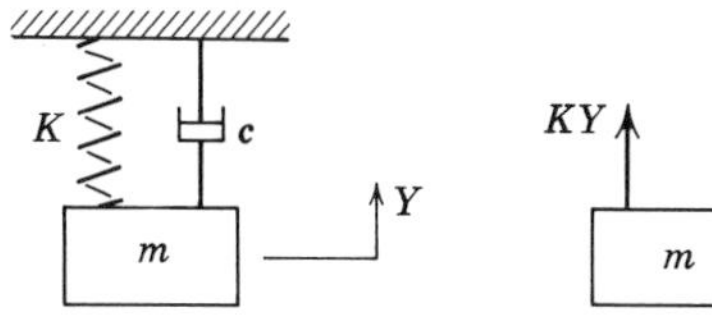

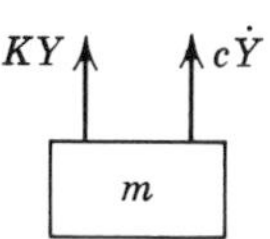

FIGURE 2.20. Free vibration forces acting on a simple system.

Considering the forces acting in the Y direction

$$m\ddot{Y} + c\dot{Y} + KY = 0 \tag{2.54}$$

Since all of the terms in the above equation must be in pounds, the units for the damper, c, must be (lb sec/in) because Y is the velocity (in/sec.)

The solution for the above equation can be taken in the form

$$Y = e^{Bt} \tag{2.55}$$

where t is time and B is an unknown constant.

Then

$$\dot{Y} = Be^{Bt} \tag{2.56}$$

$$\ddot{Y} = B^2 e^{Bt} \tag{2.57}$$

011717

Substituting back into Eq. 2.54,

$$mB^2e^{Bt} + cBe^{Bt} + Ke^{Bt} = 0$$

$$e^{Bt}(mB^2 + cB + K) = 0$$

Only the expression in the parenthesis can be equal to zero, and this expression is a quadratic equation in B. Solving this equation for B leads to two solutions.

$$B_{1,2} = -\frac{c}{2\,m} \pm \left[\left(\frac{c}{2\,m}\right)^2 - \frac{K}{m}\right]^{1/2} \tag{2.58}$$

Two values of B are obtained from the positive and negative signs shown in Eq. 2.58. If these values are designated B_1 and B_2 then the most general solution of Eq. 2.54 is

$$Y = D_1e^{B_1t} + D_2e^{B_2t} \tag{2.59}$$

where D_1 and D_2 are arbitrary constants.

The values of B_1 and B_2 in Eq. 2.58 may be real and distinct, real and equal, or complex conjugates.

2.14. The Damped Free-Vibration Equation

When $(c/2\text{ m})^2$ is greater than K/m all the terms are real. The value in the radical sign is then smaller than $c/2$ m so both values of B are negative. This represents an overdamped condition where there is no real oscillating motion, so this condition does not apply to the type of problems considered in this book.

When $(c/2\,m)^2$ is equal to K/m the condition is known as critical damping. Again there is no real oscillating motion. If the mass is displaced under these conditions, it will just creep back to its original starting point.

When $(c/2\,m)^2$ is less than K/m an oscillating motion will develop when the mass shown in Fig. 2.20 is displaced and released. The values of B in Eq. 2.58 then become complex conjugates. This condition leads to the damped free-vibration equation.

$$f_n = \frac{1}{2\pi}\left[\frac{K}{m} - \left(\frac{c}{2\,m}\right)^2\right]^{1/2} \tag{2.60}$$

Using the symbol of c_c for critical damping, which occurs when the expression for B changes from real to imaginary, then

$$\left(\frac{c_c}{2\,m}\right)^2 = \frac{K}{m}$$

$$c_c = 2(Km)^{1/2} \tag{2.61}$$

The damping ratio R_c is often used to define the amount of damping in a system. This is defined as the actual damping, c, in the system with respect to the critical damping c_c. This can be written several different ways, as shown below.

$$R_c = \frac{c}{c_c} = \frac{c}{2(Km)^{1/2}} = \frac{c\Omega_n}{2K} = \frac{c}{2m\Omega_n} \tag{2.62}$$

Here Ω_n is the natural frequency expressed in radians per second.

2.15. Forced Vibrations with Viscous Damping

When a harmonic shaking force acts on a spring-mass system with damping, the resulting forced vibration will also be harmonic. The final frequency of the mass will be the same as that of the shaking force because the initial transient vibrations will eventually be dissipated by the damping.

Consider the harmonic shaking force $P_0 \cos \Omega t$ acting on a damped spring-mass system. A free-body force diagram is shown in Fig. 2.21.

The differential equation of motion will be similar to Eq. 2.54 except now there is an external force exciting the system.

$$m\ddot{Y} + c\dot{Y} + KY = P_0 \cos \Omega t \tag{2.63}$$

The solution of this equation consists of a complementary function plus a particular function. The complementary solution is the free vibrations discussed in the preceding section and these will die out because of the damping. The particular solution can be taken in the form

$$Y = Y_0 \cos (\Omega t - \theta) \tag{2.64}$$

The maximum displacement, Y_0, can be expressed in terms of the maximum impressed force, P_0, as follows:

$$Y_0 = \frac{P_0}{[(K - m\Omega^2)^2 + c^2\Omega^2]^{1/2}} \tag{2.65}$$

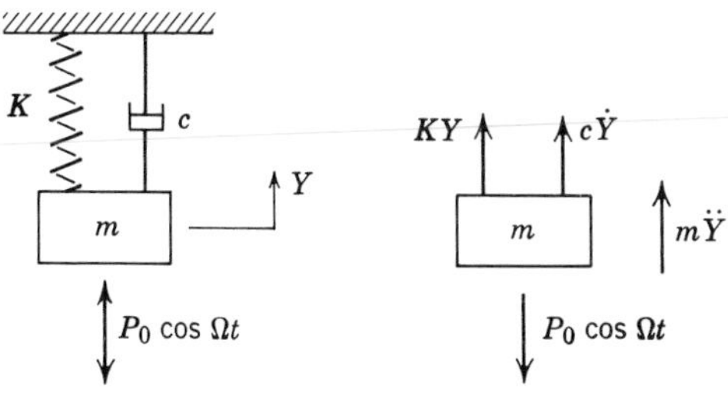

FIGURE 2.21. Forced vibration forces acting on a simple system.

Dividing the numerator and denominator of the above equation by K and substituting Eq. 2.6 leads to the following:

$$Y_0 = \frac{\dfrac{P_0}{K}}{\left\{\left[1-\left(\dfrac{\Omega}{\Omega_n}\right)^2\right]^2 + \left[2\dfrac{c}{c_c}\dfrac{\Omega}{\Omega_n}\right]^2\right\}^{1/2}} \tag{2.66}$$

Let $Y_{st} = P_0/K$, where Y_{st} is the deflection of the system due to the maximum dynamic input load acting as a static load.

For additional simplification, let

$$R_\Omega = \frac{\Omega}{\Omega_n} \quad \text{and} \quad R_c = \frac{c}{c_c} \tag{2.67}$$

This leads to the general amplification (not transmissibility) equation

$$A = \frac{Y_0}{Y_{st}} = \frac{1}{[(1-R_\Omega^2)^2 + (2R_c R_\Omega)^2]^{1/2}} \tag{2.68}$$

A plot of the dynamic amplification ratio, A, against the frequency ratio, R_Ω, is shown in Fig. 2.22. This is *not* a force transmissibility curve. A force transmissibility curve is shown in Fig. 2.23.

In Fig. 2.22 notice that the amplification factor increases very rapidly in the resonant area where $R_\Omega = 1$ when the damping ratio R_c is very small. As the damping ratio increases, the amplification factor, A, falls off rapidly.

For lightly damped systems R_c is small. When R_c becomes zero the amplification equation reduces to the following:

$$A = \frac{1}{1-R_\Omega^2} \tag{2.69}$$

The instantaneous magnitude of the force experienced by the support is the vector sum of the spring force and the damper force. These two forces have a 90° phase angle between them so the force becomes

$$F_0 = Y_0(K^2 + c^2\Omega^2)^{1/2}$$

Substituting Eq. 2.65 into the above expression,

$$F_0 = \frac{P_0(K^2 + c^2\Omega^2)^{1/2}}{[(K-m\Omega^2)^2 + c^2\Omega^2]^{1/2}}$$

The transmissibility, Q, which respresents the ratio of the maximum output force, F_0, to the maximum input force, P_0, can be determined by

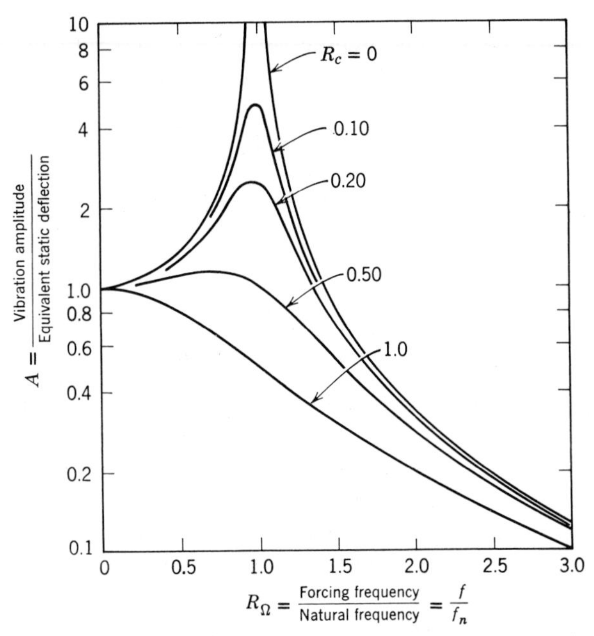

FIGURE 2.22. A dynamic amplification curve for a simple system.

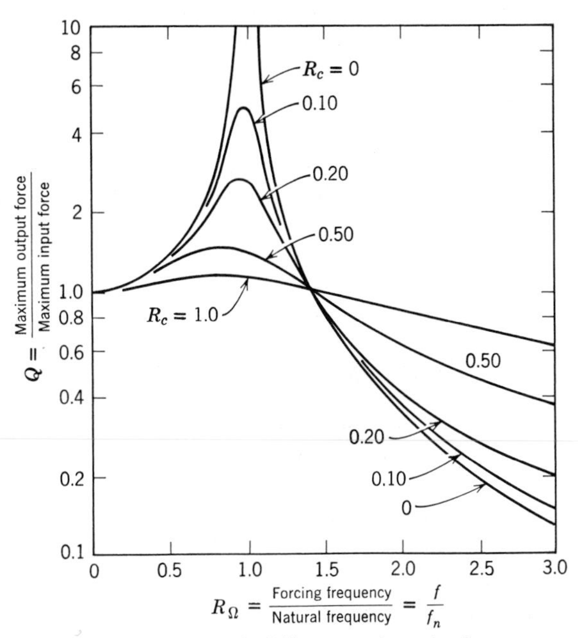

FIGURE 2.23. A transmissibility curve for a simple system.

dividing the numerator and denominator by K and substituting Eqs. 2.6 and 2.61 into the above equation.

$$Q = \frac{F_0}{P_0} = \left\{ \frac{1 + \left(2 \frac{\Omega}{\Omega_n} \frac{c}{c_c}\right)}{\left[1 - \left(\frac{\Omega}{\Omega_n}\right)^2\right]^2 + \left(2 \frac{\Omega}{\Omega_n} \frac{c}{c_c}\right)^2} \right\}^{1/2} \tag{2.70}$$

By substituting Eq. 2.67 into 2.70, the transmissibility, Q, becomes

$$Q = \left[\frac{1 + (2 R_\Omega R_c)^2}{(1 - R_\Omega{}^2)^2 + (2 R_\Omega R_c)^2}\right]^{1/2} \tag{2.71}$$

A plot of the force transmissibility, Q, against the frequency ratio, R_Ω, is shown in Fig. 2.23. This curve is very similar to the amplification curve shown in Fig. 2.22. In the resonant area, where $R_\Omega = 1.0$, the two curves are almost identical for systems with a low damping ratio.

For lightly damped systems, where $R_c{}^2$ is small enough to be assumed zero, the transmissibility expression reduces to the same value as Eq. 2.69.

$$Q = \frac{1}{1 - R_\Omega{}^2} \tag{2.72}$$

A very convenient relation can be obtained from Eq. 2.71 by considering the transmissibility at resonance, where $R_\Omega = 1.0$.

$$Q = \sqrt{\frac{1 + (2 R_c)^2}{(2 R_c)^2}}$$

For lightly damped systems, the ratio $R_c{}^2$ is very small compared to the value of 1 in the numerator, thus the transmissibility equation can be reduced to

$$Q = \frac{1}{2 R_c} \tag{2.73}$$

2.16. Sample Problem

The sample problem in Section 2.12 showed a bending stress of 12,000 psi (Eq. 2.53) at a resonance of 246 Hz in a transformer bracket that was subjected to a sinusoidal vibration test with a 5.0 G peak input. If this transformer and bracket are mounted in a rigid chassis that requires a vibration dwell test at a frequency of 224 Hz (for example, to dwell at the resonance of a cathode-ray tube mounted in the same electronic chassis), what will be the dynamic bending stress in the transformer bracket at 224 Hz?

Using Eq. 2.73, the damping ratio of the transformer bracket at resonance can be determined from the test data in the previous problem.

$$R_c = \frac{1}{2Q} = \frac{1}{2(30)} = 0.0166$$

At a forcing frequency of 224 Hz the frequency ratio becomes

$$R_\Omega = \frac{f}{f_n} = \frac{224}{246} = 0.911$$

Substitute these two values into Eq. 2.71 to determine the new transmissibility at 224 Hz.

$$Q = \left\{ \frac{1 + [2(0.911)(0.0166)]^2}{[1 - (0.911)^2]^2 + [2(0.911)(0.0166)]^2} \right\}^{1/2}$$

$$Q = 5.80 \tag{2.74}$$

If Eq. 2.72 is used to determine the approximate transmissibility at 224 Hz, the results will be almost the same

$$Q = \frac{1}{1 - R_\Omega^2} = \frac{1}{1 - (0.911)^2} = 5.88$$

Substitute Eq. 2.74 into Eq. 2.49 to determine the new dynamic load acting on the transformer bracket at 224 Hz.

$$W_d = (2)(5)(5.80) = 58.0 \text{ lb}$$

The bending stress shown in Eq. 2.53 can now be modified by a direct ratio of the dynamic loads using Eq. 2.50.

$$S_b = 12{,}000 \left(\frac{58.0}{300} \right) = 2320 \text{ lb/in.}^2 \tag{2.75}$$

This shows that if the forcing frequency is shifted about 10% from the resonant frequency, the bending stress in the transformer bracket is reduced about 80%.

2.17. Transmissibility as a Function of Frequency

An examination of Eqs. 2.73 and 2.62 indicates that the transmissibility and the natural frequency of a system may be closely related if the system has light damping. For example, substitute the second expression of Eq. 2.62 into Eq. 2.73.

$$Q = \frac{1}{2\left(\dfrac{c}{2\sqrt{Km}}\right)}$$

$$Q = \frac{(Km)^{1/2}}{c} \tag{2.76}$$

In the above expression, if the stiffness of the system, K, is increased while the mass, m, and the damping, c, are kept constant, the transmissibility, Q, must increase directly as the square root of the change in stiffness.

To expand upon this further, consider the case of a printed-circuit board that has a natural frequency of 100 Hz with a transmissibility of about 10. If lightweight ribs are added to the printed-circuit board so that its resonant frequency is doubled to 200 Hz, the transmissibility can also be expected to double to about 20 *if there is no change in the damping characteristics*. This is because the spring rate must be increased by a factor of four to double the natural frequency as shown by Eq. 2.7, and the transmissibility then increases by the square root of the spring rate, as shown by Eq. 2.76.

2.18. Sample Problem

The natural frequency of a printed-circuit board is 100 Hz with a transmissibility of 10 at the center of the circuit board. It is desirable to keep the maximum dynamic deflection at the center of the circuit board to 0.020 in. single-amplitude, in order to keep from overstressing the electrical lead wires on the resistors, capacitors, and diodes mounted on the circuit board. What should the natural frequency of the circuit board be for a 5-G-peak sinusoidal-vibration input?

The printed-circuit board can be approximated as a single-degree-of-freedom system when it is vibrating at its fundamental resonant mode. Therefore Eq. 2.46 can be used to estimate the single-amplitude displacement for the 100-Hz-resonant frequency.

$$Y_0 = \frac{9.8 G_{\text{in}} Q}{f^2}$$

where $G_{\text{in}} = 5.0G$ input (peak)

$Q = 10$ transmissibility at resonance

$f_n = 100$-Hz resonant frequency

$$Y_0 = \frac{9.8(5.0)(10)}{(100)^2} = 0.0490 \text{ in. single amplitude} \tag{2.77}$$

This displacement is much too high. If the resonant frequency is increased, the displacement will decrease very rapidly as shown by Eq. 2.46. However Eq. 2.76 shows that the transmissibility will also increase as the resonant frequency increases.

Extensive test data on printed-circuit boards, with various edge restraints, indicate that many epoxy-fiberglass circuit boards, with closely spaced electronic component parts, have a transmissibility that is often related to the natural frequency as

$$Q \propto (f_n)^{1/2} \tag{2.78}$$

Assume this approximation holds true for this particular circuit board, then substitute Eq. 2.78 into Eq. 2.46, and solve for f_n.

$$Y_0 = \frac{9.8 G_{\text{in}} (f_n)^{1/2}}{f_n^2}$$

$$f_n = \left(\frac{9.8 G_{\text{in}}}{Y_0} \right)^{2/3} \tag{2.79}$$

To obtain a single-amplitude displacement of 0.020 in. with a 5-G peak vibration input, the circuit-board resonant frequency must be increased to the following:

$$f_n = \left(\frac{9.8(5.0)}{0.020} \right)^{2/3} = 182 \text{ Hz} \tag{2.80}$$

2.19. Multiple-Spring–Mass Systems without Damping

The fundamental resonant frequency of a multiple-spring–mass system, without damping, can be determined by considering the strain energy and the kinetic energy of the system. If there is no energy lost, then the total kinetic energy of the masses must be equal to the total strain energy of the springs.

Consider the multiple-spring–mass system as shown in Fig. 2.24. The total kinetic energy of the system is the sum of the individual kinetic energies of each mass as shown by Eq. 2.3.

$$T_{\text{total}} = \tfrac{1}{2} m_1 Y_1^2 \Omega^2 + \tfrac{1}{2} m_2 Y_2^2 \Omega^2 + \tfrac{1}{2} m_3 Y_3^2 \Omega^2$$

This can be written as

$$T_{\text{total}} = \frac{\Omega^2}{2g} \sum_{i=1}^{i=n} W_i Y_i^2 \tag{2.81}$$

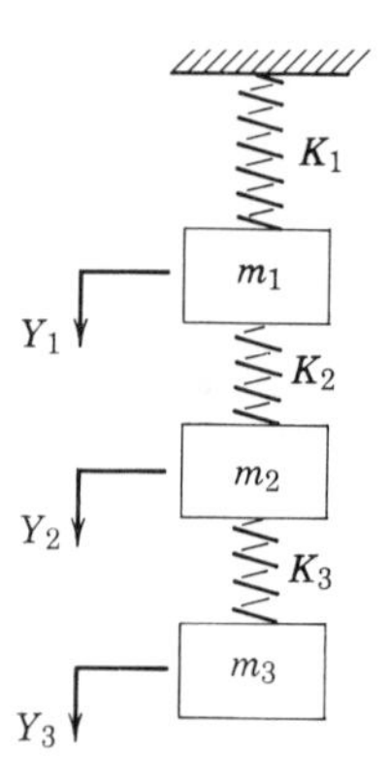

FIGURE 2.24. A multiple spring-mass system without damping.

The total strain energy of the system is the sum of the individual strain energies of each spring, as shown by Eq. 2.4.

$$U_{\text{total}} = \tfrac{1}{2}W_1Y_1 + \tfrac{1}{2}W_2Y_2 + \tfrac{1}{2}W_3Y_3$$

This can be written as

$$U_{\text{total}} = \frac{1}{2}\sum_{i=1}^{i=n} W_iY_i \tag{2.82}$$

Since the total kinetic energy must equal the total strain energy, set Eq. 2.81 equal to Eq. 2.82 and solve for the natural frequency.

$$f_n = \frac{\Omega}{2\pi} = \frac{1}{2\pi}\left(\frac{g\sum_{i=1}^{i=n} W_iY_i}{\sum_{i=1}^{i=n} W_iY_i^2}\right)^{1/2} \tag{2.83}$$

Notice that the deflections in the above equation are the total deflections. For example, in Fig. 2.24 the total deflection of mass 3 must include the deflections of masses 1 and 2, and the total deflection of mass 2 must include the deflection of mass 1. Also, the effective weight of mass 1 on spring 1 must include all three weights.

2.20. Sample Problem

Calculate the fundamental resonant frequency of the two-spring-mass system shown in Fig. 2.25.

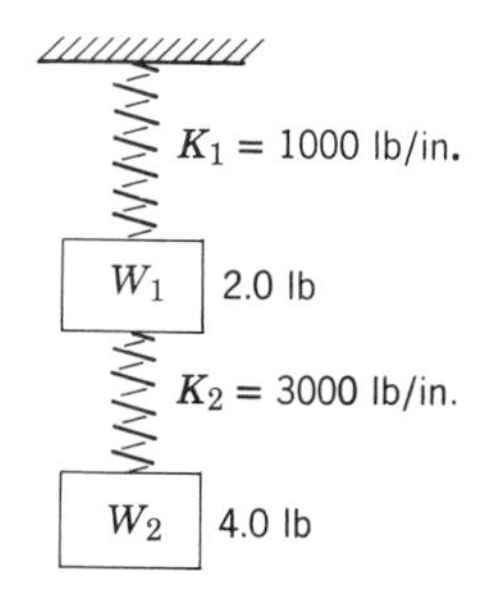

FIGURE 2.25. A two-degree-of-freedom system without damping.

The static deflection of weight 1 will be influenced by weight 2 since weight 2 is attached to weight 1, thus both weights will act on spring 1.

$$Y_1 = \frac{W_1 + W_2}{K_1} = \frac{2+4}{1000} = 0.006 \text{ in.} \tag{2.84}$$

The static deflection of weight 2 must include the static deflection of weight 1 since weight 2 is attached to weight 1.

$$Y_2 = Y_1 + \frac{W_2}{K_2} = 0.006 + \frac{4}{3000} = 0.00733 \text{ in.} \tag{2.85}$$

Substitute into Eq. 2.83

$$f_n = \frac{1}{2\pi}\left[\frac{(W_1Y_1 + W_2Y_2)g}{W_1Y_1^2 + W_2Y_2^2}\right]^{1/2}$$

$$f_n = \frac{1}{2\pi}\left\{\frac{[(2)(0.006) + (4)(0.00733)](386)}{(2)(0.006)^2 + (4)(0.00733)^2}\right\}^{1/2}$$

$$f_n = 37.3 \text{ Hz} \tag{2.86}$$

The static deflections of weights 1 and 2 in Fig. 2.25 could also have been determined by the simultaneous solution of the deflection equations for each mass. A free-body force diagram of the system is shown in Fig. 2.26.

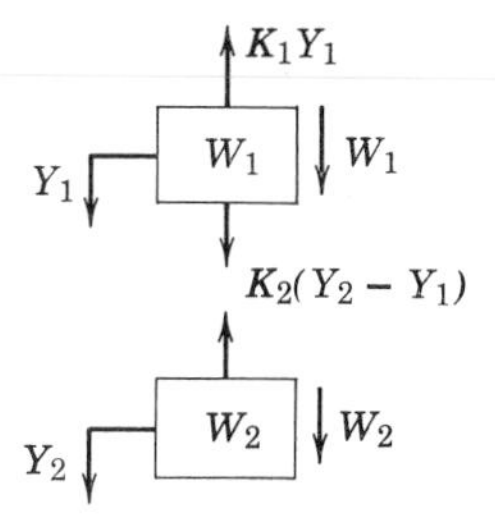

FIGURE 2.26. A free body force diagram for a two mass system.

Considering the forces in the vertical direction:

For weight 1.

$$-K_1Y_1+W_1+K_2(Y_2-Y_1)=0$$

$$-(K_1+K_2)Y_1+K_2Y_2+W_1=0 \tag{2.87}$$

For weight 2:

$$-K_2(Y_2-Y_1)+W_2=0$$

$$-K_2Y_2+K_2Y_1+W_2=0 \tag{2.88}$$

Substituting values for K_1, K_2, W_1, and W_2 into the above equations and solving

$$\begin{array}{r} -4000\,Y_1+3000\,Y_2+2=0 \\ 3000\,Y_1-3000\,Y_2+4=0 \\ \hline -1000\,Y_1 \qquad\qquad +6=0 \end{array}$$

$$Y_1=\frac{6}{1000}=0.006\text{ in.} \tag{2.89}$$

$$Y_2=\frac{3000(0.006)+4}{3000}=0.00733\text{ in.} \tag{2.90}$$

These results are exactly the same as those shown by Eqs. 2.84 and 2.85.

CHAPTER 3

LUMPED-MASS SYSTEMS

3.1. Superposition with Concentrated Loads

Many problems associated with the vibration analysis of beams can be simplified by using a concentrated load (or mass) in the place of a uniform load. If the entire weight of a uniformly loaded beam is assumed to be concentrated at the center of the beam, the deflection of the beam with the concentrated load will be greater than the beam with the uniform load. A larger static deflection means a lower resonant frequency. This means that if a concentrated load is used to approximate a uniform load of the same magnitude, the calculated resonant frequency will be lower than the true resonant frequency. This is desirable when the fundamental resonant mode of a structure is being determined, since experience shows that many structures tend to have resonances that are lower than those shown by the calculations. This is usually due to overestimating the stiffness of bolted and riveted structures and to the coupling effects of systems that may have several degrees of freedom.

The amount of error involved in a poor estimate of the boundary conditions will often be much greater than the error involved with the use of concentrated loads in place of uniform loads. For example, just changing the ends of a uniform beam from a fixed condition to a simply supported condition will reduce the resonant frequency about 55%. However if a uniformly distributed load is replaced by a concentrated load at the center of a simply supported beam, the resonant frequency of the beam will be reduced only about 30%.

Complex structures can very often be approximated by using many different concentrated loads or masses. In general, a greater number of individual loads will result in a more accurate model of the system. A greater number of loads will also require more work, since there will be more loads and more deflections to calculate for more positions.

The accuracy of a multiple-load system can be demonstrated by comparing the fundamental resonant frequencies that are obtained for a simply supported beam that is simulated with one concentrated load, two concentrated loads, then finally with three concentrated loads.

Starting with one concentrated load, a uniform beam can be approximated by concentrating the entire weight of the beam at the center. This is equivalent to a single-degree-of freedom system where the elastic properties of the beam become the spring. The approximate fundamental frequency can be determined from the static-deflection curve for the system shown in Fig. 3.1. All of the beam mass is concentrated at the center so that the beam itself now has no weight. The beam does, however, maintain its elastic properties characterized by its length, modulus of elasticity, and the moment of inertia of its cross-section.

The static deflection for a simply supported beam with a center load can be obtained from a structural handbook as follows:

$$\delta_{st} = \frac{WL^3}{48EI} \tag{3.1}$$

where $W = 7.72$ lb weight

$L = 10.0$ in. length

$E = 10.5 \times 10^6$ lb/in.^{2} (aluminum)

$$I = \frac{bh^3}{12} = \frac{(1)(2.29)^3}{12} = 1.0 \text{ in.}^4$$

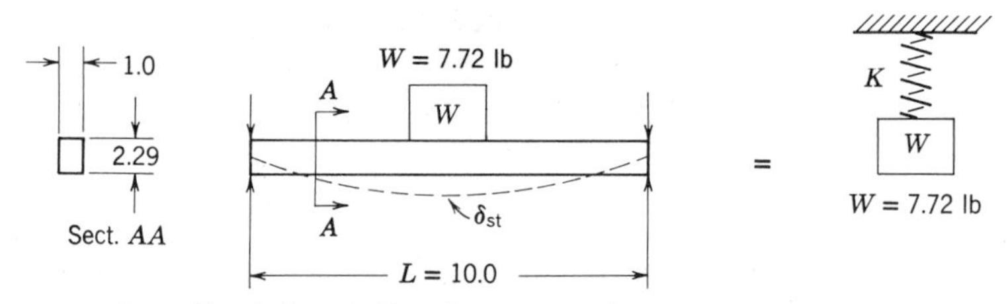

FIGURE 3.1. Simulating a uniform beam as a spring mass system.

Substituting into Eq. (3.1)

$$\delta_{st} = \frac{(7.72)(10.0)^3}{48(10.5 \times 10^6)(1.0)} = 0.153 \times 10^{-4} \text{ in.} \tag{3.2}$$

The natural frequency can be computed using the static deflection equation shown in Eq. 2.10.

$$f_n = \frac{1}{2\pi}\left(\frac{g}{\delta_{st}}\right)^{1/2} \tag{3.3}$$

where $g = 386$ in./sec^2 acceleration of gravity

$\delta_{st} = 0.153 \times 10^{-4}$ in. static deflection

Substituting into Eq. 3.3.

$$f_n = \frac{1}{2\pi}\left(\frac{386}{0.153 \times 10^{-4}}\right)^{1/2}$$

$$f_n = 798 \text{ Hz} \tag{3.4}$$

For a more accurate approximation of a uniform beam, two loads (or masses) can be used in place of the single load, keeping the same total weight and physical properties of the beam. The static deflection of the beam can again be used to approximate the fundamental natural frequency of the two-mass system. Using superposition, along with standard handbook formulas, the static deflection under each load can be computed.

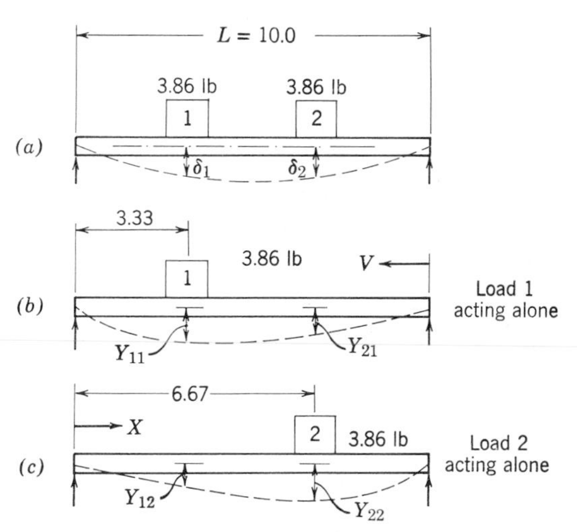

FIGURE 3.2 Two concentrated masses on a simply supported beam.

It is convenient to use a double subscript notation when more than one mass is involved in order to provide a convenient reference for the loads and deflections.

Considering load number 1 acting by itself, the static deflection at position number 1 and position number 2 can be determined. This is written as Y_{11} and Y_{21}. Then considering load number 2 acting by itself, the static deflection at position number 1 and position number 2 can again be determined. This is written as Y_{12} and Y_{22} as shown in Fig. 3.2. Since all of the beam mass is concentrated at two points, the beam is weightless. However the beam is assumed to retain its elastic properties.

The deflection at point 1 due to the load at point 1 can be determined from the standard handbook equation

$$Y_{11} = \frac{W_1 a^2 b^2}{3EIL} \tag{3.5}$$

where $W_1 = 3.86$ lb

$a = 3.33$ in.

$b = 6.67$ in.

$I = 1.0$ in.^{4} moment of inertia

$E = 10.5 \times 10^6$ (aluminum)

$L = 10.0$ in. length

Substituting into Eq. 3.5.

$$Y_{11} = \frac{(3.86)(3.33)^2(6.67)^2}{3(10.5 \times 10^6)(1)(10)} = 6.04 \times 10^{-6} \text{ in.} \tag{3.6}$$

The deflection at point 2 due to the load at point 1 can be determined from the standard handbook equation

$$Y_{21} = \frac{W_1 a V}{6EIL}(L^2 - V^2 - a^2) \tag{3.7}$$

where $W_1 = 3.86$ lb

$a = 3.33$ in.

$V = 3.33$ in.

Substituting into Eq. 3.7

$$Y_{21} = \frac{(3.86)(3.33)(3.33)}{6(10.5 \times 10^6)(1)(10)}[(10)^2 - (3.33)^2 - (3.33)^2] = 5.27 \times 10^{-6} \text{ in.} \tag{3.8}$$

The deflection at point 1 due to the load at point 2 can be determined from the standard handbook equation

$$Y_{12} = \frac{W_2 bX}{6EIL}(L^2 - X^2 - b^2) \tag{3.9}$$

where $W_2 = 3.86$ lb
$b = 3.33$ in.
$X = 3.33$ in.

Substituting into the equation

$$Y_{12} = \frac{(3.86)(3.33)(3.33)}{6(10.5 \times 10^6)(1)(10)}[(10)^2 - (3.33)^2 - (3.33)^2] = 5.27 \times 10^{-6} \text{ in.} \tag{3.10}$$

The deflection Y_{12} is exactly the same as deflection Y_{21} shown by Eq. 3.8. In fact, it was not necessary to calculate the deflection Y_{12}, since it could have been determined from the deflection Y_{21} with the use of Maxwell's theorem of reciprocity. This simply means that the deflection at point 1 due to a load at point 2 is the same as the deflection at point 2 due to the *same* load at point 1.

The total deflection at point 1 now becomes the sum of the two individual deflections.

$$\delta_1 = Y_{11} + Y_{12} \tag{3.11}$$

Substitute Eqs. 3.6 and 3.10 into Eq. 3.11

$$\delta_1 = 6.04 \times 10^{-6} + 5.27 \times 10^{-6} = 11.31 \times 10^{-6} \text{ in.} \tag{3.12}$$

The total deflection at point 2 is also the sum of the two individual deflections.

$$\delta_2 = Y_{21} + Y_{22} \tag{3.13}$$

Because of symmetry the static deflection will be the same at load 1 and load 2. Therefore, the static deflection at load 2 becomes

$$\delta_2 = (5.27 + 6.04) \times 10^6 = 11.31 \times 10^{-6} \text{ in.} \tag{3.14}$$

The approximate fundamental resonance of the two-mass system can again be determined from the static deflection equation, with a slight modification to include all of the weights and their associated deflections (see Chapter 2, Eq. 2.83).

$$f_n = \frac{1}{2\pi}\left[\frac{(W_1\delta_1 + W_2\delta_2)g}{W_1\delta_1^2 + W_2\delta_2^2}\right]^{1/2} \tag{3.15}$$

Substituting Eqs. 3.12 and 3.14 into Eq. 3.15

$$f_n = \frac{1}{2\pi}\left\{\frac{[(3.86)(11.31\times 10^{-6}) + (3.86)(11.31\times 10^{-6})](386)}{(3.86)(11.31\times 10^{-6})^2 + (3.86)(11.31\times 10^{-6})^2}\right\}^{1/2}$$

$$f_n = 930 \text{ Hz} \tag{3.16}$$

For an even more accurate approximation of a uniform beam, three masses can be used in place of the two masses, keeping the same weight and physical properties of the beam. Again the static deflections can be determined under each mass using superposition with standard handbook formulas.

Considering load number 1 acting by itself, the static deflection at positions 1, 2, and 3 can be determined. The same is done for load number 2 acting alone, then load number 3 acting alone, as shown in Fig. 3.3. The deflection at point 1 due to the load at point 1, using Eq. 3.5 can be determined from the following:

where $W_1 = 2.57$ lb

$a = 2.50$ in.

$b = 7.50$ in.

$I = 1.0$ in.^{4}

$E = 10.5 \times 10^6$ (lb/in.^{2} aluminum)

$L = 10.0$ in.

Substituting into Eq. 3.5:

$$Y_{11} = \frac{(2.57)(2.50)^2(7.50)^2}{3(10.5\times 10^6)(1)(10)} = 2.86 \times 10^{-6} \text{ in.} \tag{3.17}$$

The deflection at point 2 due to the load at point 1, using Eq. 3.7 where V is 2.50 in., is as follows:

$$Y_{21} = \frac{(2.57)(2.50)(5.0)}{6(10.5\times 10^6)(1)(10)}[(10.0)^2 - (5.0)^2 - (2.5)^2]$$

$$Y_{21} = 3.51 \times 10^{-6} \text{ in.} \tag{3.18}$$

The deflection at point 3 due to the load at point 1, using Eq. 3.7 where V is 2.50 in., is as follows:

$$Y_{31} = \frac{(2.57)(2.50)(2.50)}{6(10.5\times 10^6)(1)(10)}[(10.0)^2 - (2.5)^2 - (2.5)^2]$$

$$Y_{31} = 2.23 \times 10^{-6} \text{ in.} \tag{3.19}$$

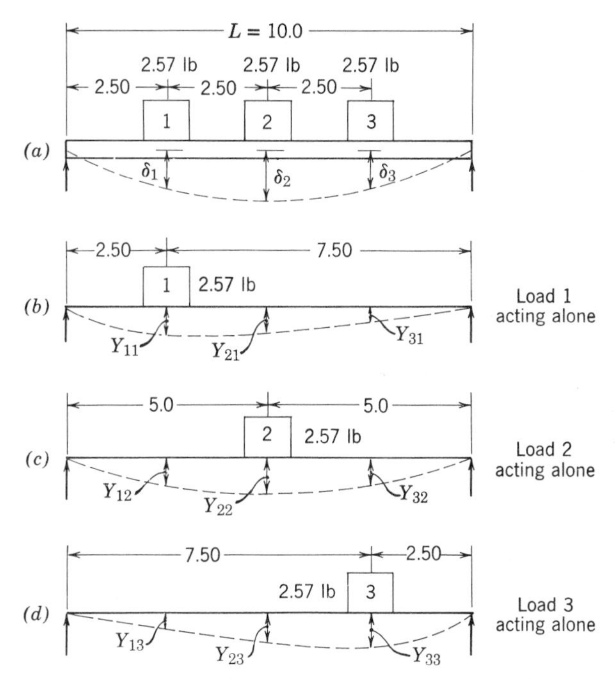

FIGURE 3.3. Three concentrated masses on a simply supported beam.

The deflection at point 1 due to the load at point 2 can be determined from the standard handbook equation

$$Y_{12} = \frac{W_2 X}{48\,EI}\,(3\,L^2 - 4\,X^2) \tag{3.20}$$

where $W_2 = 2.57$ lb weight
$X = 2.50$ in.
$L = 10.0$ in. length

Substituting into the above equation

$$Y_{12} = \frac{(2.57)(2.50)}{48(10.5 \times 10^6)(1)}\,[3(10.0)^2 - 4(2.50)^2]$$

$$Y_{12} = 3.51 \times 10^{-6} \text{ in.} \tag{3.21}$$

The deflection at point 2 due to the load at point 2 can be determined from the standard handbook equation

$$Y_{22} = \frac{W_2 L^3}{48\,EI} \tag{3.22}$$

where $W_2 = 2.57$ lb.
$L = 10.0$ in. length
$E = 10.5 \times 10^6$ (lb/in.^{2} aluminum)
$I = 1.0$ in.^{4} moment of inertia

Substituting into the above equation

$$Y_{22} = \frac{(2.57)(10.0)^3}{48(10.5 \times 10^6)(1.0)} = 5.10 \times 10^{-6} \text{ in.} \tag{3.23}$$

Since the loads at points 1 and 3 are the same, Maxwell's theorem of reciprocity can be used. The deflection of the beam at point 1 due to the load at point 3 is the same as the deflection of the beam at point 3 due to the load at point 1.

Then using Eq. 3.19

$$Y_{13} = Y_{31} = 2.23 \times 10^{-6} \text{ in.} \tag{3.24}$$

The remaining deflections can be determined from the symmetry of the beam.

From Eq. 3.21

$$Y_{12} = Y_{32} = 3.51 \times 10^{-6} \text{ in.} \tag{3.25}$$

From Eq. 3.17

$$Y_{33} = Y_{11} = 2.86 \times 10^{-6} \text{ in.} \tag{3.26}$$

From Eq. 3.18

$$Y_{23} = Y_{21} = 3.51 \times 10^{-6} \text{ in.} \tag{3.27}$$

The total deflection under the load at point 1 is

$$\delta_1 = Y_{11} + Y_{12} + Y_{13} \tag{3.28}$$

Substituting Eqs. 3.17, 3.21, and 3.24 into Eq. 3.28

$$\delta_1 = (2.86 + 3.51 + 2.23) \times 10^{-6} = 8.60 \times 10^{-6} \text{ in.} \tag{3.29}$$

The total deflection under the load at point 2 is

$$\delta_2 = Y_{21} + Y_{22} + Y_{23} \tag{3.30}$$

Substituting Eqs. 3.18, 3.23, and 3.27 into Eq. 3.30

$$\delta_2 = (3.51 + 5.10 + 3.51) \times 10^{-6} = 12.12 \times 10^{-6} \text{ in.} \tag{3.31}$$

The total deflection under the load at point 3 is

$$\delta_3 = Y_{31} + Y_{32} + Y_{33} \tag{3.32}$$

Substituting Eqs. 3.19, 3.25, and 3.26 into Eq. 3.32

$$\delta_3 = (2.23 + 3.51 + 2.86) \times 10^{-6} = 8.60 \times 10^{-6} \text{ in.} \tag{3.33}$$

The approximate fundamental frequency of the three-mass system can be determined from the static deflection Eq. 3.15.

$$f_n = \frac{1}{2\pi}\left[\frac{(W_1\delta_1 + W_2\delta_2 + W_3\delta_3)g}{W_1\delta_1^2 + W_2\delta_2^2 + W_3\delta_3^2}\right]^{1/2} \tag{3.34}$$

Substituting Eqs. 3.29, 3.31, and 3.33 into Eq. 3.34:

$$f_n = \frac{1}{2\pi}$$

$$\times \left\{\frac{[(2.57)(8.60\times10^{-6})+(2.57)(12.12\times10^{-6})+(2.57)(8.60\times10^{-6})](386)}{(2.57)(8.60\times10^{-6})^2+(2.57)(12.12\times10^{-6})^2+(2.57)(8.60\times10^{-6})^2}\right\}^{1/2}$$

$$f_n = \frac{1}{2\pi}\left[\frac{(75.4\times10^{-6})(386)}{758\times10^{-12}}\right]^{1/2}$$

$$f_n = 985 \text{ Hz} \tag{3.35}$$

The most accurate approximation for the natural frequency of a simply supported uniform beam is the exact equation for the beam

$$f_n = \frac{\pi}{2}\left(\frac{EIg}{WL^3}\right)^{1/2} \tag{3.36}$$

where $W = 7.72$ lb beam weight
$E = 10.5 \times 10^6$ (aluminum)
$I = 1.0$ in.^{4}
$g = 386$ in./sec^2 gravity
$L = 10.0$ in. length

Substituting into Eq. 3.36

$$f_n = \frac{\pi}{2}\left[\frac{(10.5\times10^6)(1.0)(386)}{7.72(10.0)^3}\right]^{1/2}$$

$$f_n = 1140 \text{ Hz} \tag{3.37}$$

A summary of the natural frequencies obtained by approximating a simply supported uniform beam with a one-mass system, a two-mass system, and a three-mass system is shown in Fig. 3.4. The summary for a uniform beam with fixed ends is shown in Fig. 3.5.

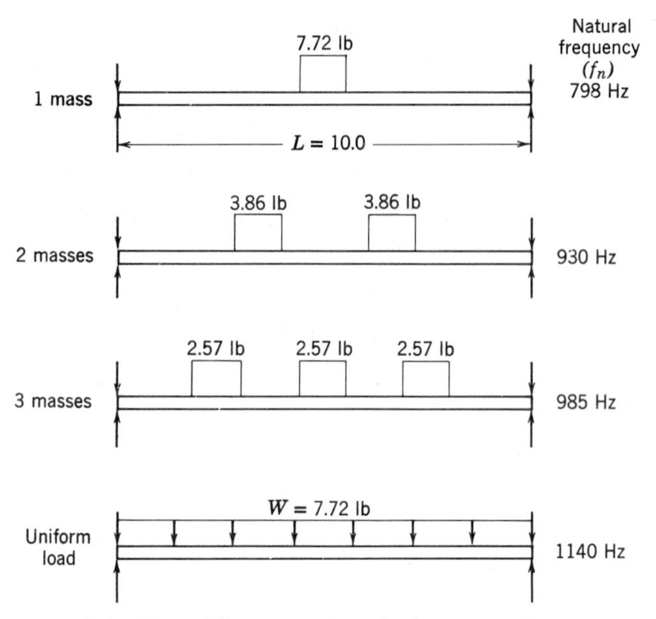

FIGURE 3.4. Natural frequencies for a simply supported beam.

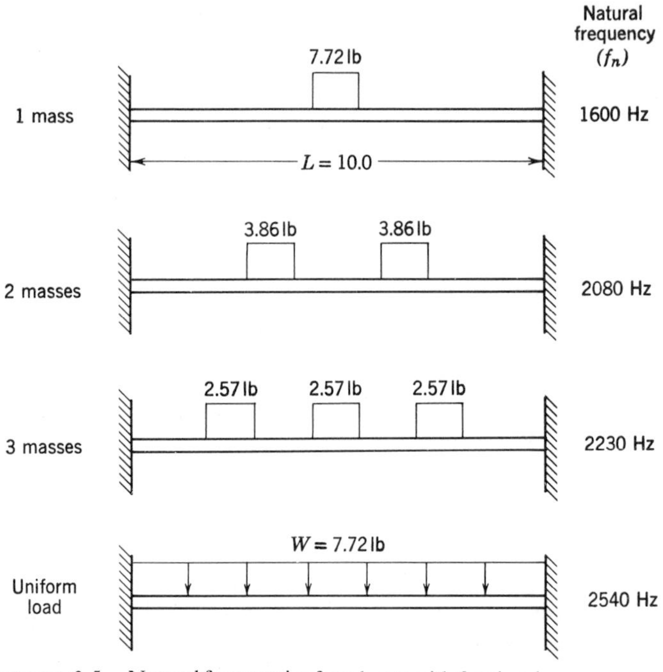

FIGURE 3.5. Natural frequencies for a beam with fixed ends.

3.2. Dunkerley's Method

Another approximate method for determining the natural frequency of a lumped-mass system is with the use of Dunkerley's equation. Using this method, the fundamental resonant frequency of a beam, with several loads, can be approximated by considering the natural frequency of the beam when each load acts separately. The individual resonances for each load are then combined to give the fundamental resonant mode of the system using the expression

$$\frac{1}{f_n^2} = \frac{1}{f_1^2} + \frac{1}{f_2^2} + \frac{1}{f_3^2} + \cdots + \frac{1}{f_i^2} \tag{3.38}$$

For example, consider the simply supported beam with two loads shown in Fig. 3.2. When only one load is considered, as shown in Fig. 3.6, the static deflection under the load is shown by Eq. 3.6 to be 6.04×10^{-6} in.

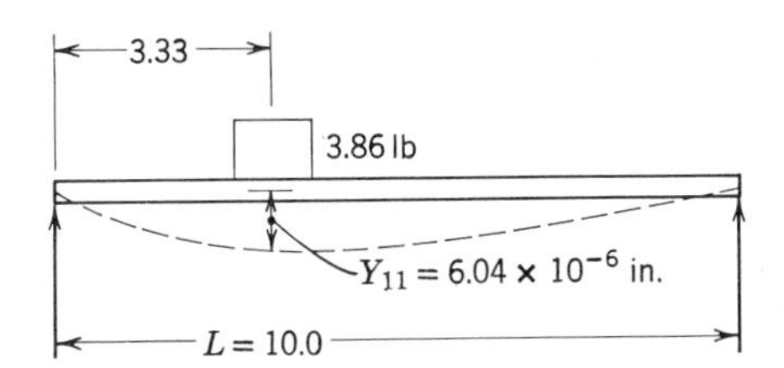

FIGURE 3.6. Static deflection of a beam with one mass.

The natural frequency for this system can be determined from Eq. 3.3.

$$f_1 = \frac{1}{2\pi}\left(\frac{g}{\delta_{st}}\right)^{1/2} = \frac{1}{2\pi}\left(\frac{386}{6.04 \times 10^{-6}}\right)^{1/2}$$

$$f_1 = 1275 \text{ Hz} \tag{3.39}$$

Because the system is symmetrical, the natural frequency for the system with the second load acting alone will be the same, thus $f_2 = 1275$ Hz.

The natural frequency of the beam with both loads can now be determined by substituting Eq. 3.39 into Eq. 3.38

$$\frac{1}{f_n^2} = \frac{1}{(1275)^2} + \frac{1}{(1275)^2} = \frac{2}{(1275)^2}$$

$$f_n = 901 \text{ Hz} \tag{3.40}$$

Consider the simply supported beam with three concentrated loads (Fig. 3.3). When the first load is considered to be acting by itself, the static deflection under the load was shown by Eq. 3.17 to be 2.86×10^{-6} in.

Using Eq. 3.3 to determine the resonant frequency for the first load acting alone

$$f_1 = \frac{1}{2\pi}\left(\frac{386}{2.86 \times 10^{-6}}\right)^{1/2} = 1850 \text{ Hz} \tag{3.41}$$

When the second load is considered acting by itself, the static deflection under the load was shown by Eq. 3.23 to be 5.10×10^{-6} in. Using Eq. 3.3 for the resonant frequency

$$f_2 = \frac{1}{2\pi}\left(\frac{386}{5.10 \times 10^{-6}}\right)^{1/2} = 1390 \text{ Hz} \tag{3.42}$$

Because the system is symmetrical, the natural frequency with the third load acting by itself will be the same as the first load. Thus:

$$f_3 = 1850 \text{ Hz} \tag{3.43}$$

The natural frequency of the beam with all three loads can be determined by substituting Eqs. 3.41, 3.42, and 3.43 into Eq. 3.38

$$\frac{1}{f_n^2} = \frac{1}{(1850)^2} + \frac{1}{(1390)^2} + \frac{1}{(1850)^2}$$

$$f_n = 953 \text{ Hz} \tag{3.44}$$

Comparing the results shown in Fig. 3.7 with the results shown in Fig. 3.4 shows that Dunkerley's method will give a resonant frequency slightly lower than the true value.

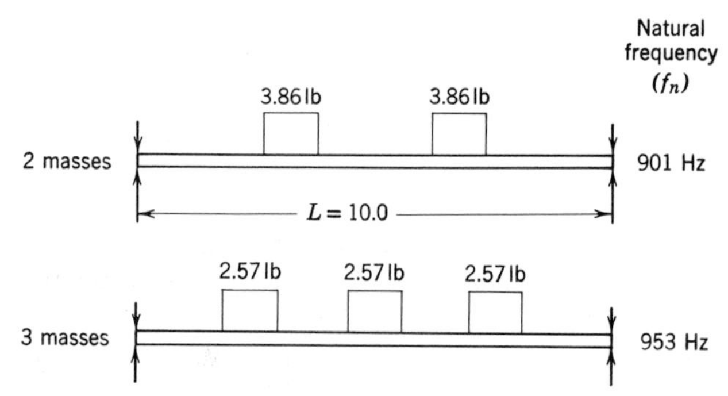

FIGURE 3.7. Natural frequencies for a simply supported beam using Dunkerley's Equation.

3.3. Influence Coefficients and Matrix Iteration

Another method for determining the natural frequencies of a beam with lumped masses is to assume that the dynamic deflections that occur during resonance are proportional to the static deflections at each point along the beam. The dynamic loading can then be used, along with influence coefficients, to determine the dynamic deflections at each point.

Consider a simply supported beam with several concentrated masses vibrating at one of its resonant modes (Fig. 3.8). The inertia force acting on each mass can be expressed as a function of its acceleration using Newton's second law of motion: $F = m\ddot{Y}$. The acceleration term $\ddot{Y}$ is determined from the circular frequency Ω for simple harmonic motion. Each mass; m_i, in the system with a displacement Y_i will be acted upon by the inertia force $-m_i\ddot{Y}_i$. Then using Eq. 2.41

$$-m_i\ddot{Y}_i = m_i\Omega^2 Y_i \tag{3.45}$$

An influence coefficient can be defined as the deflection at point i due to a unit (1-lb) load applied at point j, which is usually written as $\mathscr{A}_{ij}$.

The deflection at each mass point can then be written in terms of the influence coefficient and the inertia force.

$$\begin{aligned}
Y_1 &= \mathscr{A}_{11}(m_1\Omega^2 Y_1) + \mathscr{A}_{12}(m_2\Omega^2 Y_2) + \mathscr{A}_{13}(m_3\Omega^2 Y_3) + \cdots \\
Y_2 &= \mathscr{A}_{21}(m_1\Omega^2 Y_1) + \mathscr{A}_{22}(m_2\Omega^2 Y_2) + \mathscr{A}_{23}(m_3\Omega^2 Y_3) + \cdots \\
Y_3 &= \mathscr{A}_{31}(m_1\Omega^2 Y_1) + \mathscr{A}_{32}(m_2\Omega^2 Y_2) + \mathscr{A}_{33}(m_3\Omega^2 Y_3) + \cdots \\
&\vdots
\end{aligned}$$

These equations can be written in the general matrix form

$$\begin{bmatrix} Y_1 \\ Y_2 \\ Y_3 \\ \cdot \\ \cdot \end{bmatrix} = \frac{\Omega^2}{g} \begin{bmatrix} \mathscr{A}_{11}W_1 & \mathscr{A}_{12}W_2 & \mathscr{A}_{13}W_3 & \cdot \\ \mathscr{A}_{21}W_1 & \mathscr{A}_{22}W_2 & \mathscr{A}_{23}W_3 & \cdot \\ \mathscr{A}_{31}W_1 & \mathscr{A}_{32}W_2 & \mathscr{A}_{33}W_3 & \cdot \\ \cdot & \cdot & \cdot & \cdot \\ \cdot & \cdot & \cdot & \cdot \end{bmatrix} \begin{bmatrix} Y_1 \\ Y_2 \\ Y_3 \\ \cdot \\ \cdot \end{bmatrix} \tag{3.46}$$

The matrix operation may be started by assuming some value for each deflection term in the righthand column. The value of unity (one) is usually

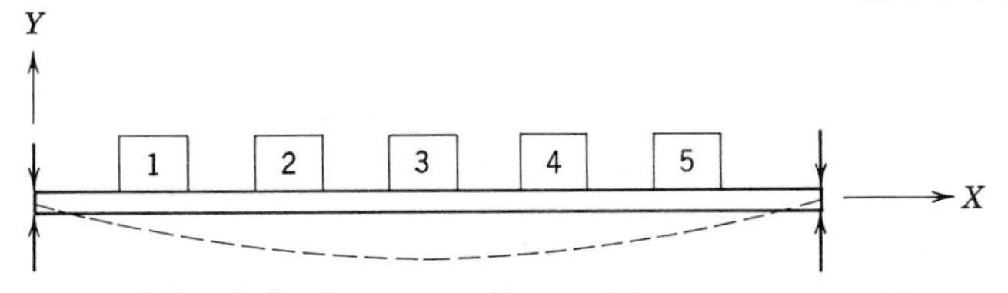

FIGURE 3.8. A simply supported beam with many masses.

convenient. The resulting column can then be normalized by dividing each term in the column by the smallest term of the column. This procedure is called iteration. It is continued until there is no change, or only a very small change between the assumed value and the computed value, which shows the convergence is complete. This may require three or four iterations.

The matrix iteration technique is very convenient to use with beams that have their weight concentrated at several discrete points. The beam itself then is assumed to be weightless, but it still retains its elastic properties characterized by its length, modulus of elasticity, and moment of inertia for its cross-section.

Consider a two-mass system that is used to simulate a uniformly loaded beam with simply supported ends, as shown in Fig. 3.9.

All of the beam weight is concentrated in the two masses, hence the beam is weightless. The beam, however, is assumed to retain its elastic properties.

The deflection equations can be written as

$$Y_1 = \frac{\Omega^2}{g}\left[\mathscr{A}_{11} W_1 Y_1 + \mathscr{A}_{12} W_2 Y_2\right]$$

$$Y_2 = \frac{\Omega^2}{g}\left[\mathscr{A}_{21} W_1 Y_1 + \mathscr{A}_{22} W_2 Y_2\right]$$

The matrix form for the above equations becomes:

$$\begin{bmatrix} Y_1 \\ Y_2 \end{bmatrix} = \frac{\Omega^2}{g} \begin{bmatrix} \mathscr{A}_{11} W_1 & \mathscr{A}_{12} W_2 \\ \mathscr{A}_{21} W_1 & \mathscr{A}_{22} W_2 \end{bmatrix} \begin{bmatrix} Y_1 \\ Y_2 \end{bmatrix} \tag{3.47}$$

The influence coefficient, $\mathscr{A}_{11}$, represents the deflection at point 1 due to a unit (1-lb) load acting at point 1.

The deflection equation as shown in a standard handbook is (see Eq. 3.5):

$$\mathscr{A}_{11} = \frac{Pa^2 b^2}{3\,EIL} \tag{3.48}$$

where $P = 1.0$ lb unit load
$a = 3.33$ in.
$b = 6.67$ in.
$E = 10.5 \times 10^6$ lb/in.^{2} (aluminum)
$I = 1.0$ in.^{4}
$L = 10.0$ in. length

Substituting into Eq. 3.48

$$\mathscr{A}_{11} = \frac{(1.0)(3.33)^2(6.67)^2}{3(10.5 \times 10^6)(1.0)(10.0)} = 1.57 \times 10^{-6} \text{ in./lb} \tag{3.49}$$

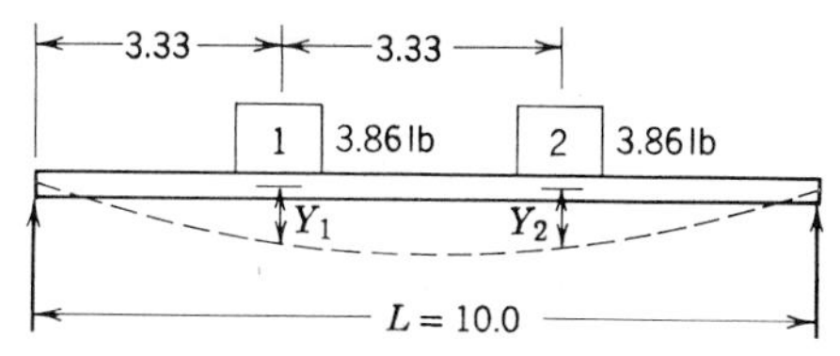

FIGURE 3.9. A simply supported beam with two masses.

The influence coefficient, $\mathscr{A}_{21}$, represents the deflection at point 2 due to a unit (1-lb) load acting at point 1.

Using a standard handbook, the deflection equation is (see Eq. 3.7):

$$\mathscr{A}_{21} = \frac{PaV}{6\,EIL}\,(L^2 - V^2 - a^2) \tag{3.50}$$

where $P = 1.0$ lb. unit load
$a = 3.33$ in.
$V = 3.33$ in.
$E = 10.5 \times 10^6$ lb/in.^{2} (aluminum)
$I = 1.0$ in.^{4}
$L = 10.0$ in.

Substituting into Eq. 3.50

$$\mathscr{A}_{21} = \frac{(1.0)\,(3.33)\,(3.33)}{6(10.5 \times 10^6)\,(1.0)\,(10.0)}\,[(10.0)^2 - (3.33)^2 - (3.33)^2]$$

$$\mathscr{A}_{21} = 1.37 \times 10^{-6} \text{ in./lb} \tag{3.51}$$

From Eqs. 3.47 and 3.49 and Fig. 3.9

$$\mathscr{A}_{11}W_1 = (1.57 \times 10^{-6})\,(3.86) = 6.04 \times 10^{-6} \text{ in.} \tag{3.52}$$

From Eqs. 3.47 and 3.51 and Fig. 3.9

$$\mathscr{A}_{21}W_1 = (1.37 \times 10^{-6})\,(3.86) = 5.27 \times 10^{-6} \text{ in.} \tag{3.53}$$

An examination of Eq. 3.6 shows that it is the same as Eq. 3.52. Also Eq. 3.8 is the same as Eq. 3.53. This is due to the symmetry of this particular problem and it can be used to determine the other terms in Eq. 3.47.

Using Maxwell's theorem of reciprocity, $\mathscr{A}_{12} = \mathscr{A}_{21}$; hence

$$\mathscr{A}_{12}W_2 = (1.37 \times 10^{-6})\,(3.86) = 5.27 \times 10^{-6} \text{ in.} \tag{3.54}$$

Because of symmetry, $\mathscr{A}_{11} = \mathscr{A}_{22}$; hence

$$\mathscr{A}_{22}W_2 = (1.57 \times 10^{-6})\,(3.86) = 6.04 \times 10^{-6} \text{ in.} \tag{3.55}$$

Substituting Eqs. 3.52, 3.53, 3.54, and 3.55 into Eq. 3.47 gives

$$\begin{bmatrix} Y_1 \\ Y_2 \end{bmatrix} = \frac{\Omega^2 \times 10^{-6}}{g} \begin{bmatrix} 6.04 & 5.27 \\ 5.27 & 6.04 \end{bmatrix} \begin{bmatrix} Y_1 \\ Y_2 \end{bmatrix}$$

The iteration procedure can be started by assuming a value of unity (1.0) for each deflection term.

$$\begin{bmatrix} Y_1 \\ Y_2 \end{bmatrix} = \frac{\Omega^2 \times 10^{-6}}{g} \begin{bmatrix} 6.04 & 5.27 \\ 5.27 & 6.04 \end{bmatrix} \begin{bmatrix} 1.0 \\ 1.0 \end{bmatrix}$$

Multiply the 2×2 matrix by the column matrix (row $\times$ column) then add the rows

$$\begin{bmatrix} Y_1 \\ Y_2 \end{bmatrix} = \frac{\Omega^2 \times 10^{-6}}{g} \begin{bmatrix} (6.04 + 5.27) \\ (5.27 + 6.04) \end{bmatrix}$$

$$\begin{bmatrix} Y_1 \\ Y_2 \end{bmatrix} = \frac{\Omega^2 \times 10^{-6}}{g} \begin{bmatrix} 11.31 \\ 11.31 \end{bmatrix}$$

Normalize the column by dividing each term by the smallest value. In this case the values are the same, 11.31

$$\begin{bmatrix} Y_1 \\ Y_2 \end{bmatrix} = \frac{11.31\Omega^2 \times 10^{-6}}{g} \begin{bmatrix} 1.0 \\ 1.0 \end{bmatrix}$$

Since there is no change between the assumed value of 1.0 and the computed value of 1.0 in the column matrix, the convergence is complete with only one iteration

$$\begin{bmatrix} 1.0 \\ 1.0 \end{bmatrix} = \frac{11.31\Omega^2 \times 10^{-6}}{g} \begin{bmatrix} 1.0 \\ 1.0 \end{bmatrix}$$

Solving for Ω

$$\frac{11.31\Omega^2 \times 10^{-6}}{g} = 1$$

$$\Omega = \left(\frac{g}{11.31 \times 10^{-6}} \right)^{1/2}$$

where $g = 386$ in./sec^2 for gravity

$$\Omega = \left(\frac{386}{11.31 \times 10^{-6}} \right)^{1/2} = 5.85 \times 10^3 \text{ rad/sec}$$

Solving for the fundamental resonant frequency

$$f_n = \frac{\Omega}{2\pi} = \frac{5.85 \times 10^3}{2\pi} = 930 \text{ Hz} \tag{3.56}$$

This compares with 930 Hz shown by Eq. 3.16.

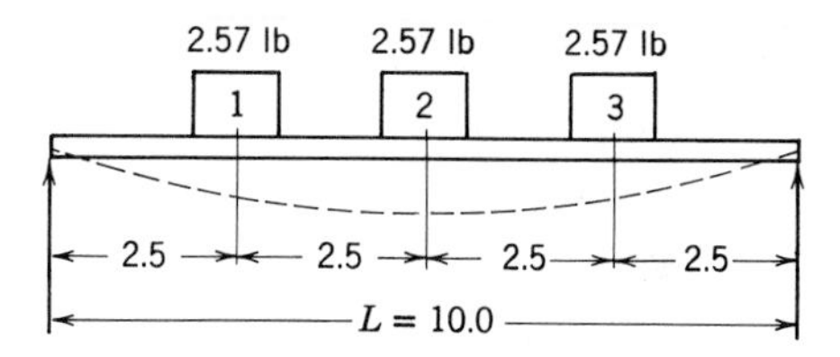

FIGURE 3.10. A simply supported beam with three masses.

Consider a three-mass system that is used to simulate a uniformly loaded beam with simply supported ends, as shown in Fig. 3.10. The general matrix shown by Eq. 3.46 can be written as follows:

$$\begin{bmatrix} Y_1 \\ Y_2 \\ Y_3 \end{bmatrix} = \frac{\Omega^2}{g} \begin{bmatrix} \mathscr{A}_{11}W_1 & \mathscr{A}_{12}W_2 & \mathscr{A}_{13}W_3 \\ \mathscr{A}_{21}W_1 & \mathscr{A}_{22}W_2 & \mathscr{A}_{23}W_3 \\ \mathscr{A}_{31}W_1 & \mathscr{A}_{32}W_2 & \mathscr{A}_{33}W_3 \end{bmatrix} \begin{bmatrix} Y_1 \\ Y_2 \\ Y_3 \end{bmatrix} \tag{3.57}$$

The influence coefficient, $\mathscr{A}_{11}$, represents the deflection at point 1 due to a unit (1-lb) load acting at point 1.

The deflection equation as shown in a standard handbook is (see Eq. 3.5)

$$\mathscr{A}_{11} = \frac{Pa^2b^2}{3\,EIL} \tag{3.58}$$

where $P = 1.0$-lb. unit load
$a = 2.5$ in.
$b = 7.5$ in.
$E = 10.5 \times 10^6$ (aluminum)
$I = 1.0$ in.^{4} moment of inertia
$L = 10.0$ in. length

Substituting into Eq. 3.58

$$\mathscr{A}_{11} = \frac{(1.0)(2.5)^2(7.5)^2}{3(10.5 \times 10^6)(1.0)(10.0)} = 1.11 \times 10^{-6} \text{ in./lb} \tag{3.59}$$

From Fig. 3.10 and Eq. 3.57

$$\mathscr{A}_{11}W_1 = (1.11 \times 10^{-6})(2.57) = 2.86 \times 10^{-6} \text{ in.} \tag{3.60}$$

Due to the symmetry of this particular problem, Eq. 3.60 is the same as Eq. 3.17. This symmetry can be used to determine the influence coefficients in Eq. 3.57

$$\mathscr{A}_{21}W_1 = 3.51 \times 10^{-6} \text{ in.} \quad \text{(see Eq. 3.18)}$$
$$\mathscr{A}_{31}W_1 = 2.23 \times 10^{-6} \text{ in.} \quad \text{(see Eq. 3.19)}$$
$$\mathscr{A}_{12}W_2 = 3.51 \times 10^{-6} \text{ in.} \quad \text{(see Eq. 3.21)}$$

$$\mathscr{A}_{22}W_2 = 5.10 \times 10^{-6} \text{ in.} \quad \text{(see Eq. 3.23)}$$
$$\mathscr{A}_{32}W_2 = 3.51 \times 10^{-6} \text{ in.} \quad \text{(see Eq. 3.25)}$$
$$\mathscr{A}_{13}W_3 = 2.23 \times 10^{-6} \text{ in.} \quad \text{Reciprocity}$$
$$\mathscr{A}_{23}W_3 = 3.51 \times 10^{-6} \text{ in.} \quad \text{Reciprocity}$$
$$\mathscr{A}_{33}W_3 = 2.86 \times 10^{-6} \text{ in.} \quad \text{Symmetry} \tag{3.61}$$

Substituting Eqs. 3.60 and 3.61 into Eq. 3.57

$$\begin{bmatrix} Y_1 \\ Y_2 \\ Y_3 \end{bmatrix} = \frac{\Omega^2 \times 10^{-6}}{g} \begin{bmatrix} 2.86 & 3.51 & 2.23 \\ 3.51 & 5.10 & 3.51 \\ 2.23 & 3.51 & 2.86 \end{bmatrix} \begin{bmatrix} Y_1 \\ Y_2 \\ Y_3 \end{bmatrix}$$

Start the iteration procedure by assuming a value of 1.0 for each deflection term.

$$\begin{bmatrix} Y_1 \\ Y_2 \\ Y_3 \end{bmatrix} = \frac{\Omega^2 \times 10^{-6}}{g} \begin{bmatrix} 2.86 & 3.51 & 2.23 \\ 3.51 & 5.10 & 3.51 \\ 2.23 & 3.51 & 2.86 \end{bmatrix} \begin{bmatrix} 1.0 \\ 1.0 \\ 1.0 \end{bmatrix}$$

Multiply the 3×3 matrix by the column matrix (row $\times$ column) then add the rows.

$$\begin{bmatrix} Y_1 \\ Y_2 \\ Y_3 \end{bmatrix} = \frac{\Omega^2 \times 10^{-6}}{g} \begin{bmatrix} (2.86 + 3.51 + 2.23) \\ (3.51 + 5.10 + 3.51) \\ (2.23 + 3.51 + 2.86) \end{bmatrix} = \frac{\Omega^2 \times 10^{-6}}{g} \begin{bmatrix} 8.60 \\ 12.12 \\ 8.60 \end{bmatrix}$$

Normalize the column matrix by dividing through by the smallest value (8.60).

$$\begin{bmatrix} Y_1 \\ Y_2 \\ Y_3 \end{bmatrix} = \frac{8.60 \times 10^{-6}\Omega^2}{g} \begin{bmatrix} 1.00 \\ 1.41 \\ 1.00 \end{bmatrix}$$

The iteration procedure was started by assuming a value of 1.0 for each deflection term and ended with values of $Y_1 = 1.00$, $Y_2 = 1.41$, and $Y_3 = 1.00$.

Using these new deflections, start the second matrix iteration by assuming $Y_1 = 1.00$, $Y_2 = 1.41$, and $Y_3 = 1.00$.

$$\begin{bmatrix} Y_1 \\ Y_2 \\ Y_3 \end{bmatrix} = \frac{\Omega^2 \times 10^{-6}}{g} \begin{bmatrix} 2.86 & 3.51 & 2.23 \\ 3.51 & 5.10 & 3.51 \\ 2.23 & 3.51 & 2.86 \end{bmatrix} \begin{bmatrix} 1.00 \\ 1.41 \\ 1.00 \end{bmatrix}$$

Multiply the 3×3 matrix by the column matrix (row $\times$ column).

$$\begin{bmatrix} Y_1 \\ Y_2 \\ Y_3 \end{bmatrix} = \frac{\Omega^2 \times 10^{-6}}{g} \begin{bmatrix} (2.86)(1.0) + (3.51)(1.41) + (2.23)(1.0) \\ (3.51)(1.0) + (5.10)(1.41) + (3.51)(1.0) \\ (2.23)(1.0) + (3.51)(1.41) + (2.86)(1.0) \end{bmatrix}$$

Add the rows in the 3×3 matrix.

$$\begin{bmatrix} Y_1 \\ Y_2 \\ Y_3 \end{bmatrix} = \frac{\Omega^2 \times 10^{-6}}{g} \begin{bmatrix} 10.04 \\ 14.21 \\ 10.04 \end{bmatrix}$$

Normalize the column matrix by dividing through by the smallest value (10.04).

$$\begin{bmatrix} Y_1 \\ Y_2 \\ Y_3 \end{bmatrix} = \frac{10.04\,\Omega^2 \times 10^{-6}}{g} \begin{bmatrix} 1.00 \\ 1.415 \\ 1.00 \end{bmatrix}$$

The second iteration was started by assuming a value of $Y_1 = 1.0$, $Y_2 = 1.41$, and $Y_3 = 1.0$ and ended with $Y_1 = 1.0$, $Y_2 = 1.415$, and $Y_3 = 1.0$. These values are close enough to the assumed values to indicate the convergence is complete. The natural frequency can then be determined.

$$\begin{bmatrix} 1.0 \\ 1.41 \\ 1.0 \end{bmatrix} = \frac{10.04\,\Omega^2 \times 10^{-6}}{g} \begin{bmatrix} 1.0 \\ 1.415 \\ 1.0 \end{bmatrix}$$

$$\frac{10.04\,\Omega^2 \times 10^{-6}}{g} = 1$$

$$\Omega = \left(\frac{g}{10.04 \times 10^{-6}}\right)^{1/2} = \left(\frac{386}{10.04 \times 10^{-6}}\right)^{1/2}$$

$$\Omega = 0.62 \times 10^4 \text{ rad/sec}$$

Solving for the fundamental resonant frequency

$$f_n = \frac{\Omega}{2\pi} = \frac{0.62 \times 10^4}{2\pi} = 986 \text{ Hz} \qquad (3.62)$$

This compares with 985 Hz shown by Eq. 3.35.

3.4. Inertia-Force Balance Method

A fourth method for determining the natural frequencies of a lumped-mass system is to write the differential equations of motion in terms of the inertia forces and spring forces. This method is very convenient to use with a digital computer for determining the manner in which a spring–mass system responds to a sinusoidal vibration input.

Consider a system with one spring and one mass with no damping (Fig. 3.11). If the mass is displaced slightly and released, it will vibrate at its natural frequency. At the instant the mass is released from the displaced position, the spring will tend to pull the mass back toward its

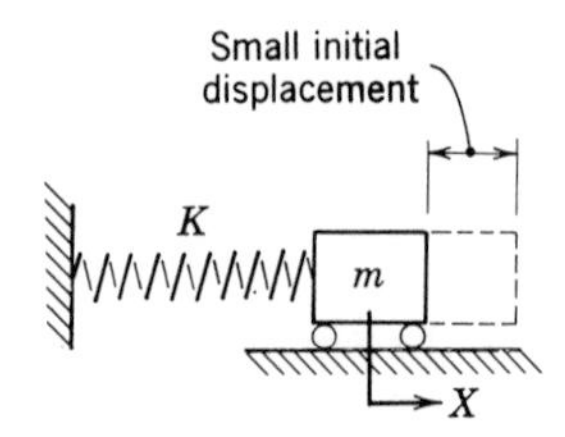

FIGURE 3.11. A single spring mass system without damping.

equilibrium position. The force developed by the spring is KX. The acceleration force acting on the mass as it moves toward its equilibrium position can be determined from Newton's second law of motion $F = m\ddot{X}$.

A free-body force diagram of the system is shown in Fig. 3.12.

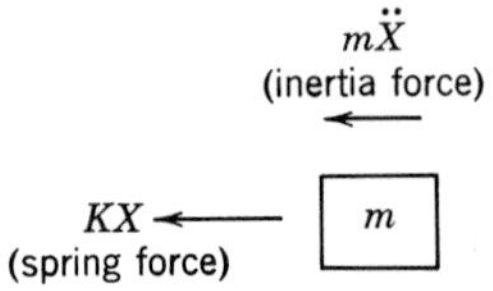

$m\ddot{X}$
(inertia force)

KX
(spring force)

m

FIGURE 3.12. A free body force diagram of a simple system.

Considering the forces in the X direction

$$-m\ddot{X} - KX = 0 \tag{3.63}$$

Assuming harmonic motion takes place, the solution to the above equation can be taken in the following form

$$X = a \sin \Omega t \tag{3.64}$$

then

$$\dot{X} = \frac{dX}{dt} = \Omega a \cos \Omega t$$

and

$$\ddot{X} = \frac{d^2X}{dt^2} = -\Omega^2 a \sin \Omega t \tag{3.65}$$

Substituting Eqs. 3.64 and 3.65 into Eq. 3.63

$$-m(-\Omega^2 a \sin \Omega t) - K(a \sin \Omega t) = 0$$

Dividing through by $a \sin \Omega t$

$$m\Omega^2 = K$$

$$\Omega = \left(\frac{K}{m}\right)^{1/2} \text{ rad/sec} \tag{3.66}$$

The single-spring-and-mass system shown in Fig. 3.11 can also be used to simulate a beam with a single concentrated load, as shown in Fig. 3.1.

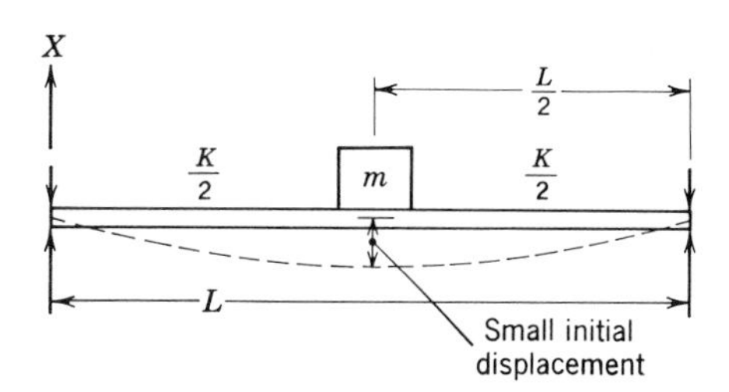

FIGURE 3.13. A simply supported beam with a concentrated mass.

Again assume the beam is weightless but it maintains elastic properties characterized by its length, modulus of elasticity, and moment of inertia for its cross-section.

If the mass is displaced slightly and released, it will vibrate at its natural frequency.

A simply supported beam with a concentrated mass is shown in Fig. 3.13.

A free-body force diagram of the system is shown in Fig. 3.14.

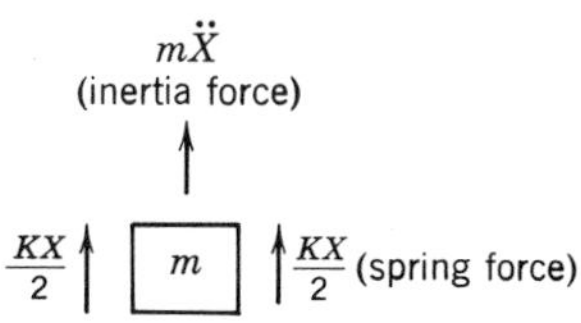

$m\ddot{X}$
(inertia force)

$\frac{KX}{2}$ m $\frac{KX}{2}$ (spring force)

FIGURE 3.14. A free body force diagram of the beam mass.

The complete beam represents the total spring system. Since each half of the beam is considered as a separate spring, each half of the beam must have half of the total spring rate.

Equations 3.63, 3.64, and 3.65 also apply to the beam with the single concentrated mass. The natural frequency can be determined with Eq. 3.66 by changing radians per second to cycles per second.

$$f_n = \frac{\Omega}{2\pi} = \frac{1}{2\pi}\left(\frac{Kg}{W}\right)^{1/2} \tag{3.67}$$

For example, if the same beam geometry as shown in Fig. 3.1 is used

$W = 7.72$ lb weight

$g = 386$ in./sec^2 gravity

$E = 10.5 \times 10^6$ lb/in.^{2} (aluminum)

$I = 1.0$ in.^{4}

$L = 10.0$ in.

The spring rate can be determined from Eq. 3.1.

$$K = \frac{W}{\delta_{st}} = \frac{48EI}{L^3}$$

$$K = \frac{48(10.5 \times 10^6)(1.0)}{(10.0)^3} = 0.504 \times 10^6 \text{ lb/in.}$$

Substituting into Eq. 3.67

$$f_n = \frac{1}{2\pi}\left[\frac{(0.504 \times 10^6)(386)}{7.72}\right]^{1/2}$$

$$f_n = 798 \text{ Hz} \tag{3.68}$$

Comparing Eq. 3.68 with Eq. 3.4 shows the results are the same.

Consider a system with two masses and three springs (Fig. 3.15). If mass 2 is displaced slightly, it will cause mass 1 to displace slightly less, as shown. The instant mass 2 is released from its displaced position, all three springs will tend to return the two masses to their equilibrium positions. The two masses will then tend to vibrate at their natural frequencies. A free-body force diagram of the system is shown in Fig. 3.16.

Considering each mass separately, the forces in the X direction can be written as follows:

for mass 1

$$-m_1\ddot{X}_1 - K_1X_1 + K_2(X_2 - X_1) = 0 \tag{3.69}$$

for mass 2

$$-m_2\ddot{X}_2 - K_2(X_2 - X_1) - K_3X_2 = 0 \tag{3.70}$$

Assuming harmonic motion takes place, the solution to the above equations can be taken in the following form:

$$X_1 = a_1 \sin \Omega t \tag{3.71}$$

$$\ddot{X}_1 = -\Omega^2 a_1 \sin \Omega t \tag{3.72}$$

$$X_2 = a_2 \sin \Omega t \tag{3.73}$$

$$\ddot{X}_2 = -\Omega^2 a_2 \sin \Omega t \tag{3.74}$$

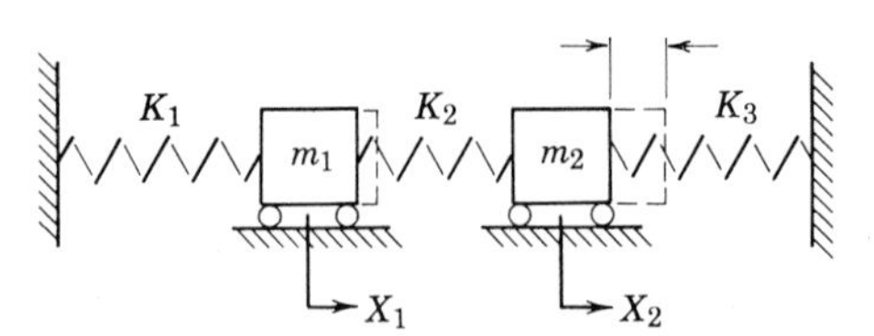

FIGURE 3.15. A three spring-two mass system without damping.

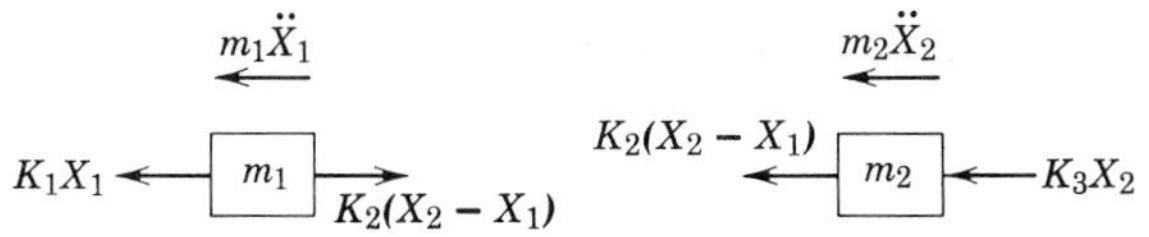

$m_1\ddot{X}_1$ $m_2\ddot{X}_2$

K_1X_1 m_1 $K_2(X_2 - X_1)$ $K_2(X_2 - X_1)$ m_2 K_3X_2

FIGURE 3.16. A free body force diagram for the two mass system.

Substituting Eqs. 3.71 and 3.72 into Eq. 3.69 for mass 1 motion

$$-m_1(-\Omega^2 a_1 \sin \Omega t) - K_1(a_1 \sin \Omega t) + K_2(a_2 \sin \Omega t - a_1 \sin \Omega t) = 0$$

$$(m_1\Omega^2 a_1 - K_1 a_1 + K_2 a_2 - K_2 a_1) \sin \Omega t = 0 \tag{3.75}$$

Substituting Eqs. 3.73 and 3.74 into Eq. 3.70 for mass 2 motion

$$-m_2(-\Omega^2 a_2 \sin \Omega t) - K_2(a_2 \sin \Omega t - a_1 \sin \Omega t) - K_3(a_2 \sin \Omega t) = 0$$

$$(m_2\Omega^2 a_2 - K_2 a_2 + K_2 a_1 - K_3 a_2) \sin \Omega t = 0 \tag{3.76}$$

Dividing Eqs. 3.75 and 3.76 by $\sin \Omega t$ and collecting terms leads to two equations.

$$(m_1\Omega^2 - K_1 - K_2)a_1 + K_2 a_2 = 0 \quad \text{(for mass 1)} \tag{3.77}$$

$$K_2 a_1 + (m_2\Omega^2 - K_2 - K_3)a_2 = 0 \quad \text{(for mass 2)} \tag{3.78}$$

If the determinant of the coefficients a_1 and a_2 is set equal to zero, the frequency equation of the two-mass system can be determined.

$$\begin{bmatrix} m_1\Omega^2 - K_1 - K_2 & K_2 \\ K_2 & m_2\Omega^2 - K_2 - K_3 \end{bmatrix} = 0$$

Notice the matrix is symmetrical about the diagonal that extends from the upper left down to the lower right. This symmetry is characteristic of this type of problem.

The determinant can be expanded by cross-multiplying, then combining terms.

$$(m_1\Omega^2 - K_1 - K_2)(m_2\Omega^2 - K_2 - K_3) - K_2^2 = 0$$

$$\Omega^4(m_1 m_2) - \Omega^2(m_1 K_2 + m_1 K_3 + m_2 K_1 + m_2 K_2) + K_1 K_2 + K_1 K_3 + K_2 K_3 = 0$$

Divide through by $m_1 m_2$ and note that $m = W/g$. This results in the frequency equation.

$$\Omega^4 - \Omega^2 g \left[\frac{K_1 + K_2}{W_1} + \frac{K_2 + K_3}{W_2}\right] + \left[\frac{K_1 K_2 + K_1 K_3 + K_2 K_3}{W_1 W_2}\right] g^2 = 0 \tag{3.79}$$

This equation will yield two resonant frequencies.

The two-mass–three-spring system shown in Fig. 3.15 can also be used to simulate a beam with two concentrated masses, as shown in Fig.

X
K1
m1
K2
m2
K3
L
Small initial deflection

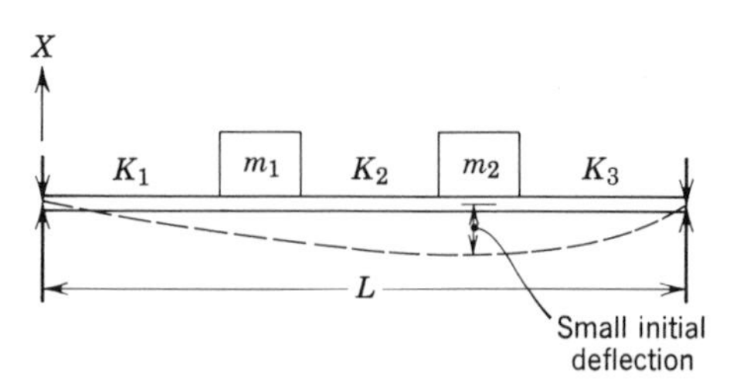

FIGURE 3.17. A simply supported beam with two masses.

3.17. Assume the beam is weightless but that it still maintains its elastic properties.

If mass 2 is displaced slightly, it will cause mass 1 to displace slightly less, as shown. The instant mass 2 is released, all three springs will tend to return the masses to their equilibrium positions.

A free-body force diagram of the system is shown in Fig. 3.18.

Equations 3.69–3.79 also apply to the beam with two concentrated masses and the natural frequencies can be determined with Eq. 3.79.

For example, if the same beam geometry as shown in Fig. 3.2 is used.

$$W_1 = W_2 = 3.86 \text{ lb weight}$$
$$g = 386 \text{ in/sec}^2 \text{ gravity}$$
$$E = 10.5 \times 10^6 \text{ lb/in.}^2 \text{ (aluminum)}$$
$$I = 1.0 \text{ in.}^4 \text{ moment of inertia}$$
$$L = 10.0 \text{ in. length}$$

The spring rate, K_1, can be approximated from the static deflection previously calculated. Since the deflection at the first load was shown by Eq. 3.12 to be 11.31×10^{-6} in.

$$K_1 = \frac{W_1}{\delta_1} = \frac{3.86}{11.31 \times 10^{-6}} = 0.341 \times 10^6 \text{ lb/in.} \tag{3.80}$$

The deflection at the second load, due to symmetry, was the same. This was shown by Eq. 3.14 as 11.31×10^{-6} in.

$$K_3 = \frac{W_2}{\delta_2} = \frac{3.86}{11.31 \times 10^{-6}} = 0.341 \times 10^6 \text{ lb/in.} \tag{3.81}$$

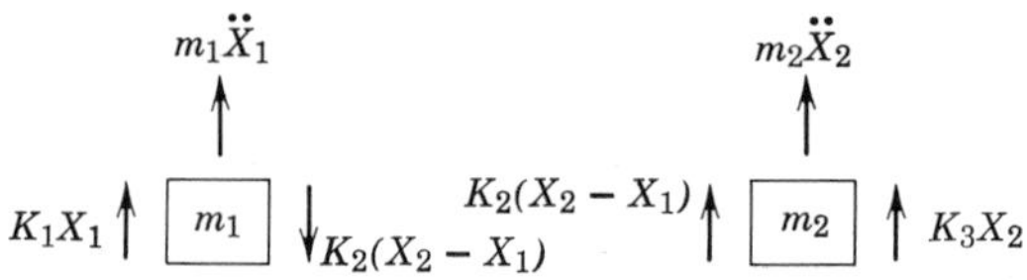

m1Ẍ1
m2Ẍ2
K1X1
m1
K2(X2 − X1)
K2(X2 − X1)
m2
K3X2

FIGURE 3.18. A free body force diagram of the two beam masses.

Spring K_2 has very little effect on the fundamental resonant mode of the symmetrical 2-mass system. This is because the two masses move in the same direction with equal amplitudes, which puts very little strain in spring K_2. During the second resonant mode, the two masses move in opposite direction with equal amplitudes. This places a direct strain on spring K_2, which influences the resonant frequency.

Assume that spring K_2 has the same spring rate as K_1 and K_3.

$$K_2 = 0.341 \times 10^6 \text{ lb/in.}$$

The two resonant frequencies of the two-mass system can be determined from Eq. 3.79:

$$K_1 + K_2 = K_2 + K_3 = 0.341 \times 10^6 + 0.341 \times 10^6 = 0.682 \times 10^6$$
$$K_1K_2 = K_1K_3 = K_2K_3 = (0.341 \times 10^6)(0.341 \times 10^6) = 0.116 \times 10^{12}$$

also $g = 386$ in./sec^2 gravity
$W_1 = W_2 = 3.86$ lb
$W_1W_2 = (3.86)^2$

Substituting into Eq. 3.79

$$\Omega^4 - \Omega^2(386)\left[\frac{2(0.682 \times 10^6)}{3.86}\right] + \left[\frac{3(0.116 \times 10^{12})}{(3.86)^2}\right](386)^2 = 0$$

Let

$$Y = \Omega^2 \quad (3.82)$$

$$Y^2 - 1.364 \times 10^8 + 0.348 \times 10^{16} = 0 \quad (3.83)$$

The standard quadratic equation can be used to solve Eq. 3.83.

$$Y = \frac{-b \pm (b^2 - 4\,ac)^{1/2}}{2\,a} \quad (3.84)$$

where $b = -1.364 \times 10^8$
$a = 1.0$
$c = 0.348 \times 10^{16}$

$$Y = \frac{1.364 \times 10^8 \pm [(-1.364 \times 10^8)^2 - 4(0.348 \times 10^{16})]^{1/2}}{2}$$

$$Y_1 = \frac{1.364 \times 10^8 - 0.684 \times 10^8}{2} = 0.340 \times 10^8 \quad (3.85)$$

$$Y_2 = \frac{1.364 \times 10^8 + 0.684 \times 10^8}{2} = 1.024 \times 10^8 \quad (3.86)$$

From Eq. 3.82

$$f_1 = \frac{(0.340 \times 10^8)^{1/2}}{2\pi} = 930 \text{ Hz} \tag{3.87}$$

$$f_2 = \frac{(1.024 \times 10^8)^{1/2}}{2\pi} = 1610 \text{ Hz} \tag{3.88}$$

If the value of spring K_2 is doubled

$$K_2 = 2(0.341 \times 10^6) = 0.682 \times 10^6 \text{ lb/in.}$$

Then substituting into Eq. 3.79 results in the equation

$$Y^2 - 2.046 \times 10^8\, Y + 0.580 \times 10^{16} = 0$$

and

$$Y = \frac{2.046 \times 10^8 \pm 1.363 \times 10^8}{2}$$

$$f_1 = \frac{(0.341 \times 10^8)^{1/2}}{2\pi} = 930 \text{ Hz} \tag{3.89}$$

$$f_2 = \frac{(1.704 \times 10^8)^{1/2}}{2\pi} = 2080 \text{ Hz} \tag{3.90}$$

Notice that doubling the stiffness of spring K_2 had no effect on the fundamental resonant mode of 930 Hz. Also note that these results check well with two other methods which are shown by Eqs. 3.16 and 3.56.

If a two-spring–two-mass system must be analyzed (Fig. 3.19), Eq. 3.79 can be altered by eliminating terms with K_3.

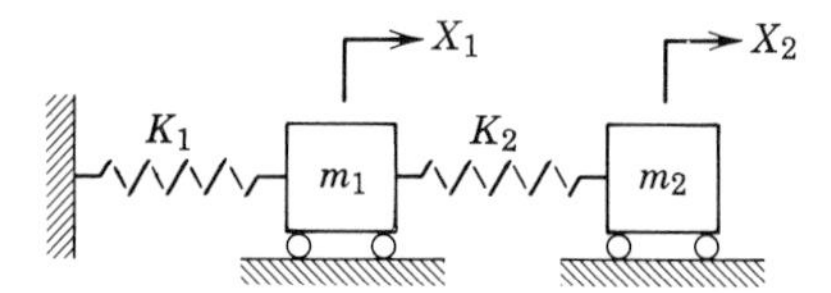

FIGURE 3.19. A two spring-two mass system.

The natural frequency equation becomes

$$\Omega^4 - \Omega^2 g\left(\frac{K_1 + K_2}{W_1} + \frac{K_2}{W_2}\right) + \left(\frac{K_1 K_2}{W_1 W_2}\right) g^2 = 0 \tag{3.91}$$

Consider a system with three masses and four springs (Fig. 3.20). If mass number 2 at the center is displaced slightly, it will cause the other two masses to displace slightly less, as shown. The instant mass 2 is released from its displaced position, all four springs will tend to return

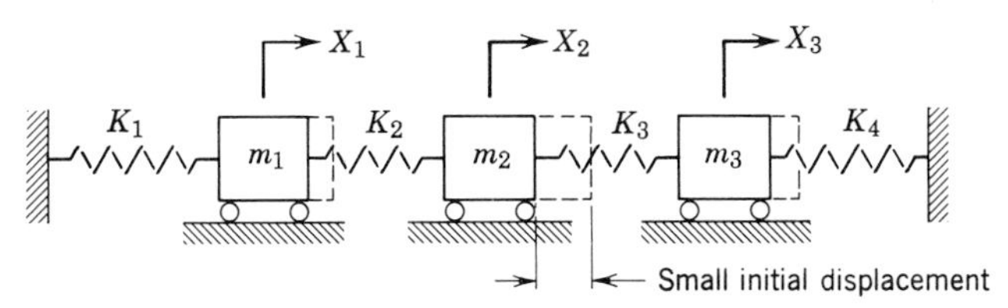

FIGURE 3.20. A four spring-three mass system without damping.

the three masses to their equilibrium positions. The three masses will then tend to vibrate at their natural frequencies.

A free-body force diagram of the system is shown in Fig. 3.21.

Considering each mass separately, the forces in the X direction can be written as follows:

For mass 1

$$-m_1\ddot{X}_1 - K_1X_1 + K_2(X_2 - X_1) = 0 \tag{3.92}$$

For mass 2

$$-m_2\ddot{X}_2 - K_2(X_2 - X_1) - K_3(X_2 - X_3) = 0 \tag{3.93}$$

For mass 3

$$-m_3\ddot{X}_3 + K_3(X_2 - X_3) - K_4X_3 = 0 \tag{3.94}$$

Assuming harmonic motion, the solution to the above equations can be taken in the following form:

$$X_1 = a_1 \sin \Omega t \tag{3.95}$$

$$\ddot{X}_1 = -\Omega^2 a_1 \sin \Omega t \tag{3.96}$$

$$X_2 = a_2 \sin \Omega t \tag{3.97}$$

$$\ddot{X}_2 = -\Omega^2 a_2 \sin \Omega t \tag{3.98}$$

$$X_3 = a_3 \sin \Omega t \tag{3.99}$$

$$\ddot{X}_3 = -\Omega^2 a_3 \sin \Omega t \tag{3.100}$$

Substituting Eqs. 3.95 and 3.96 into Eq. 3.92 for mass 1 motion.

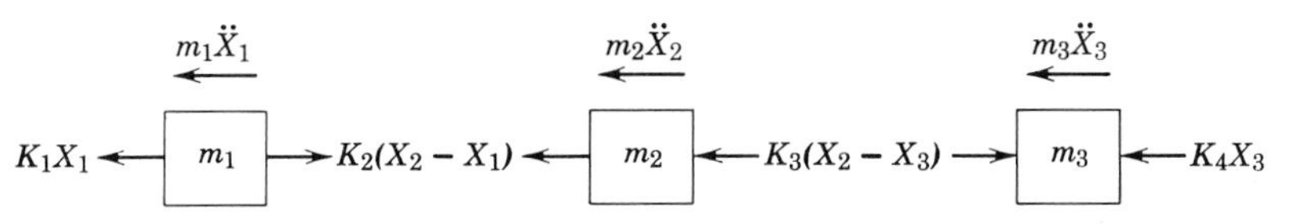

FIGURE 3.21. A free body force diagram for the three mass system.

$$-m_1(-\Omega^2 a_1 \sin \Omega t) - K_1(a_1 \sin \Omega t) + K_2(a_2 \sin \Omega t - a_1 \sin \Omega t) = 0$$

then

$$(m_1\Omega^2 a_1 - K_1 a_1 + K_2 a_2 - K_2 a_1) \sin \Omega t = 0 \tag{3.101}$$

Substituting Eq. 3.97 and 3.98 into Eq. 3.93 for mass 2 motion.

$$-m_2(-\Omega^2 a_2 \sin \Omega t) - K_2(a_2 \sin \Omega t - a_1 \sin \Omega t)$$
$$- K_3(a_2 \sin \Omega t - a_3 \sin \Omega t) = 0$$

then

$$(m_2\Omega^2 a_2 - K_2 a_2 + K_2 a_1 - K_3 a_2 + K_3 a_3) \sin \Omega t = 0 \tag{3.102}$$

Substituting Eqs. 3.99 and 3.100 into Eq. 3.94 for mass 3 motion.

$$-m_3(-\Omega^2 a_3 \sin \Omega t) + K_3(a_2 \sin \Omega t - a_3 \sin \Omega t)$$
$$- K_4 a_3 \sin \Omega t = 0$$

then

$$(m_3\Omega^2 a_3 + K_3 a_2 - K_3 a_3 - K_4 a_3) \sin \Omega t = 0 \tag{3.103}$$

Dividing Eqs. 3.101, 3.102, and 3.103 by $\sin \Omega t$:

$$(m_1\Omega^2 - K_1 - K_2)\, a_1 + K_2 a_2 = 0 \qquad \text{(for mass 1)}$$
$$K_2 a_1 + (m_2\Omega^2 - K_2 - K_3)\, a_2 + K_3 a_3 = 0 \qquad \text{(for mass 2)}$$
$$K_3 a_2 + (m_3\Omega^2 - K_3 - K_4)\, a_3 = 0 \qquad \text{(for mass 3)}$$

If the determinant of the coefficients a_1, a_2, and a_3 is set equal to zero, the frequency equation of the 3-mass system can be determined.

$$\begin{bmatrix} m_1\Omega^2 - K_1 - K_2 & K_2 & 0 \\ K_2 & m_2\Omega^2 - K_2 - K_3 & K_3 \\ 0 & K_3 & m_3\Omega^2 - K_3 - K_4 \end{bmatrix} = 0$$

Notice the matrix is symmetrical about the diagonal that extends from the upper left down to the lower right.

The determinant can be expanded by minors.

$$(m_1\Omega^2 - K_1 - K_2)\,[(m_2\Omega^2 - K_2 - K_3)\,(m_3\Omega^2 - K_3 - K_4) - K_3^2]$$
$$- (K_2)\,[(K_2)\,(m_3\Omega^2 - K_3 - K_4) - 0] = 0$$

$$\Omega^6 m_1 m_2 m_3 - \Omega^4 (m_1 m_3 K_2 + m_1 m_3 K_3 + m_1 m_2 K_3 + m_2 m_3 K_1 + m_2 m_3 K_2$$
$$+ m_1 m_2 K_4) + \Omega^2 (m_1 K_2 K_3 + m_1 K_2 K_4 + m_1 K_3 K_4 + m_3 K_1 K_2 + m_3 K_1 K_3$$
$$+ m_2 K_1 K_3 + m_2 K_1 K_4 + m_3 K_2 K_3 + m_2 K_2 K_3 + m_2 K_2 K_4) - K_1 K_2 K_3$$
$$- K_1 K_2 K_4 - K_1 K_3 K_4 - K_2 K_3 K_4 = 0$$

Divide through by $m_1 m_2 m_3$ and note that $m = W/g$. This results in the frequency equation.

$$\Omega^6-\Omega^4 g\left(\frac{K_1+K_2}{W_1}+\frac{K_2+K_3}{W_2}+\frac{K_3+K_4}{W_3}\right)+\Omega^2 g^2\left(\frac{K_1K_2+K_1K_3+K_2K_3}{W_1W_2}\right.$$
$$\left.+\frac{K_1K_3+K_1K_4+K_2K_3+K_2K_4}{W_1W_3}+\frac{K_2K_3+K_2K_4+K_3K_4}{W_2W_3}\right)$$
$$-\left(\frac{K_1K_2K_3+K_1K_2K_4+K_1K_3K_4+K_2K_3K_4}{W_1W_2W_3}\right)g^3=0 \tag{3.104}$$

This equation will yield three resonant frequencies.

The three-mass–four-spring system shown in Fig. 3.20 can also be used to simulate a beam with three concentrated masses, as in Fig. 3.22. Assume the beam is weightless but that it still maintains its elastic properties. If mass number 2 is displaced slightly, it will cause the other two masses to displace slightly less. The instant mass 2 is released, all four springs will tend to return the three masses to their equilibrium positions. The three masses will then tend to vibrate at their natural frequencies. A free-body force diagram of the system is shown in Fig. 3.23.

Equations 3.92–3.104 also apply to the beam with three concentrated masses and the natural frequencies can be determined with Eq. 3.104. For example, if the same beam geometry as shown in Fig. 3.10 is used

$$\begin{aligned} W_1 = W_2 = W_3 &= 2.57 \text{ lb weight} \\ g &= 386 \text{ in./sec}^2 \text{ gravity} \\ E &= 10.5\times 10^6 \text{ lb/in.}^2 \text{ (aluminum)} \\ I &= 1.0 \text{ in.}^4 \text{ moment of inertia} \\ L &= 10.0 \text{ in. length} \end{aligned}$$

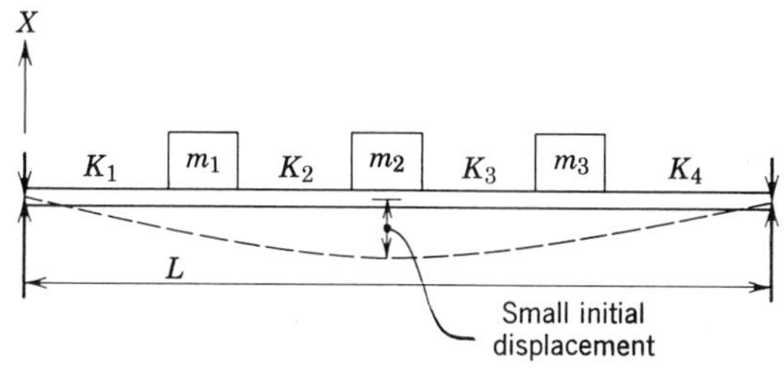

FIGURE 3.22. A simply supported beam with three masses.

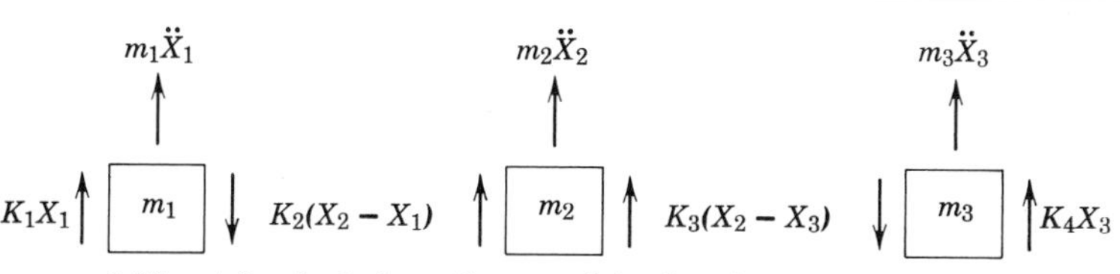

FIGURE 3.23. A free body force diagram of the three beam masses.

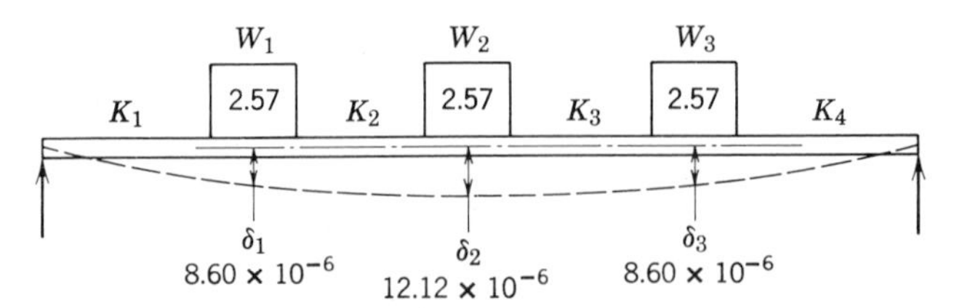

FIGURE 3.24. Deflections for a simply supported beam with three masses.

The spring rates K_1, K_2, K_3, and K_4 can be approximated from the static deflections previously calculated by Eqs. 3.29, 3.31, and 3.33. These deflections are shown in Fig. 3.24.

The approximate spring rates can be determined from the "deflection differences" at each load, as follows:

$$\left.\begin{aligned} K_1 &= \frac{W_1}{\delta_1} = \frac{2.57}{8.60 \times 10^{-6}} = 0.299 \times 10^6 \text{ lb/in.} \\ K_2 &= \frac{W_2}{\delta_2 - \delta_1} = \frac{2.57}{(12.12 - 8.60) \times 10^{-6}} = 0.730 \times 10^6 \text{ lb/in.} \\ K_3 &= \frac{W_2}{\delta_2 - \delta_3} = \frac{2.57}{(12.12 - 8.60) \times 10^{-6}} = 0.730 \times 10^6 \text{ lb/in.} \\ K_4 &= \frac{W_3}{\delta_3} = \frac{2.57}{8.60 \times 10^{-6}} = 0.299 \times 10^6 \text{ lb/in.} \end{aligned}\right\} \tag{3.105}$$

The three resonant frequencies can be determined with the use of Eq. 3.104. The method of solution is quite lengthy since it is equivalent to a cubic equation. Standard handbooks show some convenient methods for solving this type of equation, as follows.

Substitute Eq. 3.105 into Eq. 3.104

$$\Omega^6 - \Omega^4(386)\left(\frac{1.029}{2.57} + \frac{1.460}{2.57} + \frac{1.029}{2.57}\right) \times 10^6 + \Omega^2(386)^2\left(\frac{0.218 + 0.218 + 0.533}{6.60}\right.$$

$$\left. + \frac{0.218 + 0.894 + 0.533 + 0.218}{6.60} + \frac{0.533 + 0.218 + 0.218}{6.60}\right) \times 10^{12}$$

$$-\left(\frac{0.159 + 0.0652 + 0.0652 + 0.159}{16.97}\right) \times 10^{18}(386)^3 = 0$$

Let $Y = \Omega^2$, then

$$Y^3 - 5.284 \times 10^8 Y^2 + 6.763 \times 10^{16} Y - 1.518 \times 10^{24} = 0 \tag{3.106}$$

This can be put into the form

$$Y^3 + PY^2 + qY + r = 0$$

Let

$$Y = \left(X - \frac{P}{3}\right) \tag{3.107}$$

Then Eq. 3.106 simplifies to the following:

$$X^3 + aX + b = 0$$

$$\left.\begin{aligned} \text{where } P &= -5.284 \times 10^8 \\ q &= 6.763 \times 10^{16} \\ r &= -1.518 \times 10^{24} \end{aligned}\right\} \text{ from Eq. 3.106}$$

Then

$$a = \frac{1}{3}(3q - P^2) = \frac{1}{3}[3(6.763 \times 10^{16}) - (-5.284 \times 10^8)^2]$$

$$a = -2.543 \times 10^{16}$$

and

$$b = \frac{1}{27}(2P^3 - 9Pq + 27r)$$

$$b = \frac{1}{27}[2(-5.284 \times 10^8)^3 - 9(-5.284 \times 10^8)(6.763 \times 10^{16}) + 27(-1.518 \times 10^{24})]$$

$$b = \frac{1}{27}(-14.435 \times 10^{24}) = -0.5346 \times 10^{24}$$

Check the type of roots in the equation

$$\frac{b^2}{4} + \frac{a^3}{27} = \frac{(-0.5346 \times 10^{24})^2}{4} + \frac{(-2.543 \times 10^{16})^3}{27} = -0.537 \times 10^{48}$$

Since this value is less than zero, a trigonometric solution is more convenient to use.

$$\cos\phi = \left(\frac{\dfrac{b^2}{4}}{-\dfrac{a^3}{27}}\right)^{1/2} = \left[\frac{\dfrac{(-0.5346 \times 10^{24})^2}{4}}{-\dfrac{(-2.543 \times 10^{16})^3}{27}}\right]^{1/2} = 0.3425$$

$$\phi = 69^\circ\, 58'$$

$$X_1 = 2\left(-\frac{a}{3}\right)^{1/2} \cos\frac{\phi}{3} = 2\left[-\frac{(-2.543 \times 10^{16})}{3}\right]^{1/2} \cos\frac{69.96^\circ}{3}$$

$$X_1 = 1.6908 \times 10^8 \tag{3.108}$$

$$X_2 = 2\left(-\frac{a}{3}\right) \quad \cos\left(\frac{\phi}{3} + 120°\right)$$

$$X_2 = -1.4765 \times 10^8 \tag{3.109}$$

$$X_3 = 2\left(-\frac{a}{3}\right)^{1/2} \cos\left(\frac{\phi}{3} + 240°\right)$$

$$X_3 = -0.2143 \times 10^8 \tag{3.110}$$

From Eqs. 3.107–3.110

$$Y_1 = \left(X_1 - \frac{P}{3}\right) = \left[1.6908 \times 10^8 - \frac{(-5.284 \times 10^8)}{3}\right]$$

$$Y_1 = 3.4521 \times 10^8 \tag{3.111}$$

$$Y_2 = \left(X_2 - \frac{P}{3}\right) = 0.2848 \times 10^8 \tag{3.112}$$

$$Y_3 = \left(X_3 - \frac{P}{3}\right) = 1.547 \times 10^8 \tag{3.113}$$

The fundamental resonant mode is obtained from Eq. 3.112 where $Y = \Omega^2$.

$$f_n = \frac{\Omega}{2\pi} = \frac{\sqrt{Y_2}}{2\pi} = \frac{(0.2848 \times 10^8)^{1/2}}{2\pi}$$

$$f_n = 850 \text{ Hz} \tag{3.114}$$

Comparing these results with the 985-Hz fundamental resonant frequency shown by Eq. 3.35 shows the "deflection difference" method is about 13.7% lower than the true resonance for this beam with simply supported ends.

The very same beam was analyzed with a fixed end condition and the "deflection difference" method showed a resonant frequency about 8.7% lower than the true fundamental resonance.

Many other beam configurations have been examined with the "deflection difference" method and they appear to show fundamental resonances that are about 8–20% lower than the true fundamental resonance.*

The natural frequencies of a 3-spring–3-mass system shown in Fig. 3.25 can be determined from Eq. 3.104 by eliminating terms with K_4. The

*The best results are obtained with an odd number of masses, where the center mass is located near the center of the beam. The poorest results are obtained when all of the masses are clustered near one end of the beam.

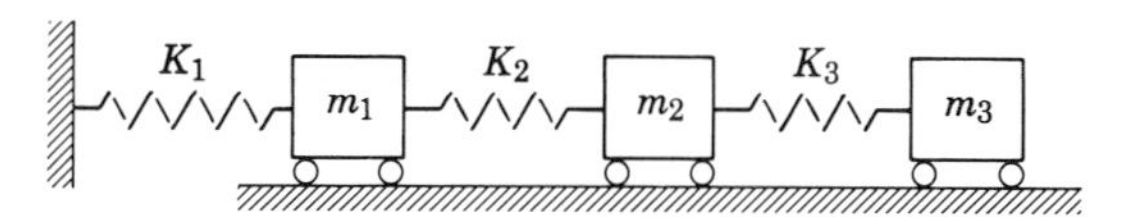

FIGURE 3.25. A three spring-three mass system.

natural frequency equation becomes

$$\Omega^6 - \Omega^4 g\left(\frac{K_1+K_2}{W_1}+\frac{K_2+K_3}{W_2}+\frac{K_3}{W_3}\right)+\Omega^2 g^2\left(\frac{K_1K_2+K_1K_3+K_2K_3}{W_1W_2}\right.$$

$$\left.+\frac{K_1K_3+K_2K_3}{W_1W_2}+\frac{K_2K_3}{W_2W_3}\right)-\left(\frac{K_1K_2K_3}{W_1W_2W_3}\right)g^3=0 \qquad (3.115)$$

3.5. General Equations of Motion

The equations of motion discussed in the previous section of this chapter, were all developed by considering the actions and reactions of the masses and springs when one mass in the system was slightly displaced. This method of analysis was shown in Figs. 3.15, 3.16, 3.20, and 3.21. The models used to demonstrate this method were simple and easy to understand. Most models used in the analysis of lumped mass systems are not quite as easy to analyze. For example, consider the relatively simple model shown in Fig. 3.26. If mass 2 is displaced slightly, it will cause masses 1, 3, and 4 to displace slightly less. However, when the spring forces acting on each mass are analyzed in a free-body force diagram, the direction of the force from spring 4 on masses 1 and 4 cannot

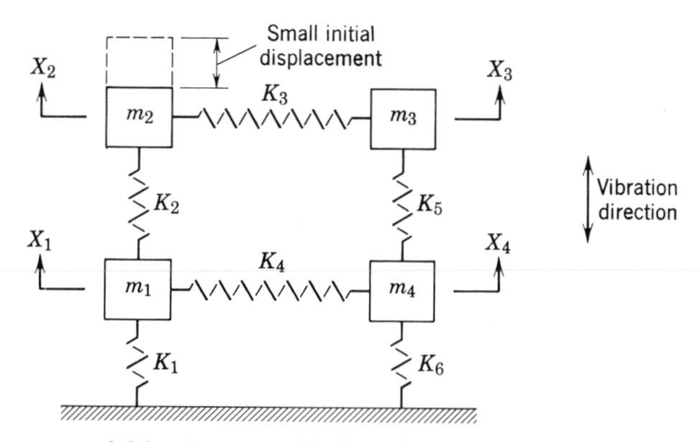

FIGURE 3.26. A system with six springs and four masses.

be determined without knowing the relative stiffness of springs 2, 3, and 5. A free body force diagram of such a system is shown in Fig. 3.27.

If the rule for "general equations of motion" is utilized, all of the forces will automatically be correct, without even using a free-body force diagram. With this rule, all of the forces associated with a specific mass are treated as positive; all of the other forces are treated as negative.

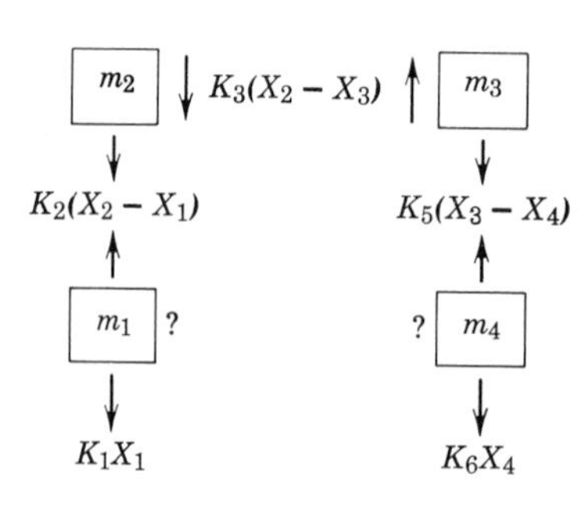

FIGURE 3.27. A free body force diagram for the four mass system.

For example, in Fig. 3.26, write the equations of motion for each mass in the system, starting with mass 1. All forces associated with mass 1 will be positive and all forces associated with the other masses will be negative.

For mass 1: keep X_1 positive.

$$+m_1\ddot{X}_1 + K_1X_1 + K_2(X_1 - X_2) + K_4(X_1 - X_4) = 0$$

For mass 2: keep X_2 positive

$$+m_2\ddot{X}_2 + K_2(X_2 - X_1) + K_3(X_2 - X_3) = 0$$

For mass 3: keep X_3 positive.

$$+m_3\ddot{X}_3 + K_3(X_3 - X_2) + K_5(X_3 - X_4) = 0$$

For mass 4: keep X_4 positive.

$$+m_4\ddot{X}_4 + K_4(X_4 - X_1) + K_5(X_4 - X_3) + K_6X_4 = 0$$

Using the method just described, the equations of motion can be written for the three-mass system shown in Fig. 3.20, to compare with the method previously used.

For mass 1: keep X_1 positive.

$$m_1\ddot{X}_1 + K_1X_1 + K_2(X_1 - X_2) = 0 \tag{3.116}$$

For mass 2: keep X_2 positive.

$$m_2\ddot{X}_2 + K_2(X_2 - X_1) + K_3(X_2 - X_3) = 0 \tag{3.117}$$

For mass 3: keep X_3 positive.

$$m_3\ddot{X}_3 + K_3(X_3 - X_2) + K_4X_3 = 0 \tag{3.118}$$

A comparison of Eqs. 3.116–3.118 shows that they are exactly the same as Eqs. 3.92–3.94.

Equations of motion, similar to those shown above, are often used for complex spring–mass systems because these equations can be solved very quickly with a digital computer. The solution is often made at many different frequency points which provides a transmissibility response curve for every mass in the system (see Chapter 8).

3.6. Forced Vibrations–Equations of Motion

When a lumped-spring–mass system is subjected to a sinusoidal vibration force, the system will initially tend to vibrate at its own natural frequency while it follows the external forcing frequency. Because there is damping present in every real system, all motion not sustained by the external sinusoidal force will gradually die out and the system will vibrate at the frequency of the exciting force. This is generally called steady-state vibration.

A lumped-mass system can be forced to vibrate by shaking individual masses, or groups of masses, in the system. Consider a three-mass system where mass 1 is being acted upon by a sinusoidal force as shown in the Fig. 3.28. The steady-state equations of motion can be determined by considering the forces acting on each mass.

For mass 1: keep X_1 positive

$$m_1\ddot{X}_1 + K_1X_1 + K_2(X_1 - X_2) = F_0 \sin \Omega t$$

For mass 2: keep X_2 positive

$$m_2\ddot{X}_2 + K_2(X_2 - X_1) + K_3(X_2 - X_3) = 0$$

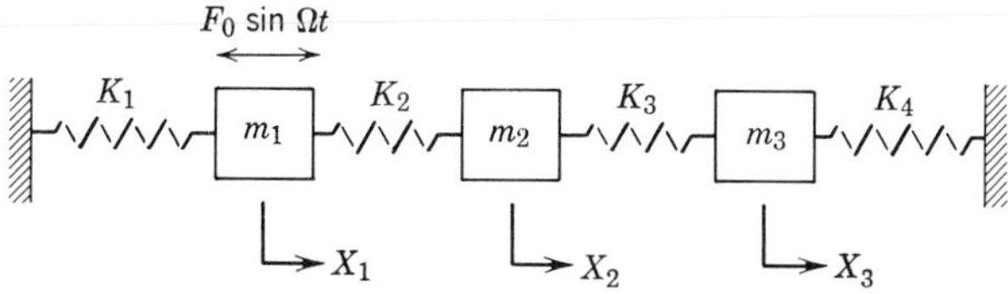

FIGURE 3.28. Forced vibrations for a four spring-three mass system.

For mass 3: keep X_3 positive

$$m_3\ddot{X}_3 + K_3(X_3 - X_2) + K_4X_3 = 0$$

Collecting terms in these equations and rewriting:

For mass 1

$$m_1\ddot{X}_1 + X_1(K_1 + K_2) - X_2K_2 = F_0 \sin \Omega t \tag{3.119}$$

For mass 2

$$m_2\ddot{X}_2 + X_2(K_2 + K_3) - X_1K_2 - X_3K_3 = 0 \tag{3.120}$$

For mass 3

$$m_3\ddot{X}_3 + X_3(K_3 + K_4) - X_2K_3 = 0 \tag{3.121}$$

Notice that the forcing function $F_0 \sin \Omega t$ only appears with the equation for mass 1 since it is acting directly on mass 1.

After substituting values for K_1, K_2, and K_3, the three equations above can be written in the matrix form and solved directly by a digital computer. Programs such as IBM's "LISA" (linear systems analysis) will solve problems such as this at many frequency points and provide a printout of the response curve for each mass (see Chapter 8).

Real systems have damping, which must be included in any dynamic stress analysis or the stresses will become infinite at resonance. Consider a three-mass system with damping where the supports at each end of the system are subjected to a sinusoidal input force in the vertical direction (Fig. 3.29).

The steady-state equations of motion can be determined by considering the inertia forces, the damping forces and the spring forces acting on each mass.

The input force, $F_0 \sin \Omega t$, does not act on any mass directly. Instead, the input force passes through the supports at each end of the beam as shown in Fig. 3.29. This means the end supports will be set into motion and this motion will be transferred to each end mass through a spring and a damper. Damping for this problem was assumed to be proportional to velocity.

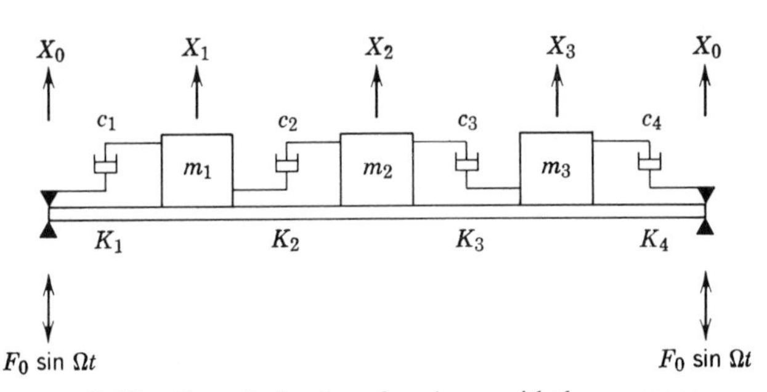

FIGURE 3.29. Forced vibrations for a beam with three masses.

If a computer is to be used to determine the response of each mass at many frequency points, then the computer will be programmed to use a sinusoidal input forcing function. Under these conditions, only the coefficients of the forcing function are required. This will come out automatically if the relative motion of the supports is included in the steady-state equations of motion, as shown below.

For mass 1: keep X_1 positive

$$m_1\ddot{X}_1 + c_1(\dot{X}_1 - \dot{X}_0) + c_2(\dot{X}_1 - \dot{X}_2) + K_1(X_1 - X_0) + K_2(X_1 - X_2) = 0$$

For mass 2: keep X_2 positive

$$m_2\ddot{X}_2 + c_3(\dot{X}_2 - \dot{X}_1) + c_3(\dot{X}_2 - \dot{X}_3) + K_2(X_2 - X_1) + K_3(X_2 - X_3) = 0$$

For mass 3: keep X_3 positive

$$m_3\ddot{X}_3 + c_3(\dot{X}_3 - \dot{X}_2) + c_4(\dot{X}_3 - \dot{X}_0) + K_3(X_3 - X_2) + K_4(X_3 - X_0) = 0$$

Collecting terms in the above equations:

For mass 1

$$m_1\ddot{X}_1 + \dot{X}_1(c_1 + c_2) - \dot{X}_2c_2 + X_1(K_1 + K_2) - X_2K_2 = \dot{X}_0c_1 + X_0K_1 \quad (3.122)$$

For mass 2

$$m_2\ddot{X}_2 - \dot{X}_1c_2 + \dot{X}_2(c_2 + c_3) - \dot{X}_3c_3 - X_1K_2 + X_2(K_2 + K_3) - X_3K_3 = 0 \quad (3.123)$$

For mass 3

$$m_3\ddot{X}_3 - \dot{X}_2c_3 + \dot{X}_3(c_3 + c_4) - X_2K_3 + X_3(K_3 + K_4) = \dot{X}_0c_4 + X_0K_4 \quad (3.124)$$

Equations 3.122–3.124 can be put into a matrix format which will make it easier to program the equations for a computer solution (Fig. 3.30).

	Column 1 Terms with X_1	Column 2 Terms with X_2	Column 3 Terms with X_3		Column 4 Terms with X_0
Row 1 for mass 1	m_1 $c_1 + c_2$ $K_1 + K_2$	$-c_2$ $-K_2$	0		c_1 K_1
Row 2 for mass 2	$-c_2$ $-K_2$	m_2 $c_2 + c_3$ $K_2 + K_3$	$-c_3$ $-K_3$	=	0
Row 3 for mass 3	0	$-c_3$ $-K_3$	m_3 $c_3 + c_4$ $K_3 + K_4$		c_4 K_4

FIGURE 3.30. Matrix for a three mass system with forced vibration.

Notice the matrix is symmetrical about the diagonal, which is characteristic for this type of problem.

All of the values shown in the matrix can be calculated from the physical properties of the system. Sample problems of this type are detailed in Chapter 8.

3.7. Lagrange's Equation

Lagrange's equation can also be used to establish the equations of motion for the three-mass system shown in Fig. 3.29. This equation presents a method for writing the equations of motion in terms of kinetic energy, strain energy, and dissipation energy.

The kinetic energy of a lumped-mass system is the sum of the kinetic energies of each mass in the system.

$$T = \tfrac{1}{2} m_1 \dot{X}_1^2 + \tfrac{1}{2} m_2 \dot{X}_2^2 + \dots \tag{3.125}$$

The strain energy of a lumped-mass system is the sum of the strain energies in each of the springs.

$$U = \tfrac{1}{2} K_1 (X_1 - X_0)^2 + \tfrac{1}{2} K_2 (X_2 - X_1)^2 + \dots \tag{3.126}$$

The dissipation energy lost in a lumped-mass system is the sum of the energies lost in each of the dashpots.

$$D = \tfrac{1}{2} c_1 (\dot{X}_1 - \dot{X}_0)^2 + \tfrac{1}{2} c_2 (\dot{X}_2 - \dot{X}_1)^2 + \dots \tag{3.127}$$

Lagrange's equation for this system is

$$\frac{d}{dt}\left(\frac{\partial T}{\partial \dot{X}_i}\right) - \frac{\partial T}{\partial X_i} + \frac{\partial U}{\partial X_i} + \frac{\partial D}{\partial \dot{X}_i} = 0 \tag{3.128}$$

The total kinetic energy of the lumped-mass system shown in Fig. 3.29 is

$$T = \tfrac{1}{2} m_1 \dot{X}_1^2 + \tfrac{1}{2} m_2 \dot{X}_2^2 + \tfrac{1}{2} m_3 \dot{X}_3^2 \tag{3.129}$$

The total strain energy of the springs in the lumped mass system is

$$U = \tfrac{1}{2} K_1 (X_1 - X_0)^2 + \tfrac{1}{2} K_2 (X_2 - X_1)^2 + \tfrac{1}{2} K_3 (X_3 - X_2)^2 + \tfrac{1}{2} K_4 (X_0 - X_3)^2 \tag{3.130}$$

The total energy dissipated in the dashpots of the lumped-mass system is

$$D = \tfrac{1}{2} c_1 (\dot{X}_1 - \dot{X}_0)^2 + \tfrac{1}{2} c_2 (\dot{X}_2 - \dot{X}_1)^2 + \tfrac{1}{2} c_3 (\dot{X}_3 - \dot{X}_2)^2 + \tfrac{1}{2} c_4 (\dot{X}_0 - \dot{X}_3)^2 \tag{3.131}$$

Follow the operations as shown in Lagrange's equation (3.128) and start with Eq. 3.129

$$\frac{d}{dt}\left(\frac{\partial T}{\partial \dot{X}_1}\right) = m_1\ddot{X}_1 \tag{3.132}$$

$$\frac{d}{dt}\left(\frac{\partial T}{\partial \dot{X}_2}\right) = m_2\ddot{X}_2 \tag{3.133}$$

$$\frac{d}{dt}\left(\frac{\partial T}{\partial \dot{X}_3}\right) = m_3\ddot{X}_3 \tag{3.134}$$

From Eqs. 3.128 and 3.129

$$\frac{\partial T}{\partial X_1} = \frac{\partial T}{\partial X_2} = \frac{\partial T}{\partial X_3} = 0 \tag{3.135}$$

From Eqs. 3.128 and 3.130

$$\frac{\partial U}{\partial X_1} = K_1(X_1 - X_0) - K_2(X_2 - X_1) \tag{3.136}$$

$$\frac{\partial U}{\partial X_2} = K_2(X_2 - X_1) - K_3(X_3 - X_2) \tag{3.137}$$

$$\frac{\partial U}{\partial X_3} = K_3(X_3 - X_2) - K_4(X_0 - X_3) \tag{3.138}$$

From Eqs. 3.128 and 3.131

$$\frac{\partial D}{\partial \dot{X}_1} = c_1(\dot{X}_1 - \dot{X}_0) - c_2(\dot{X}_2 - \dot{X}_1) \tag{3.139}$$

$$\frac{\partial D}{\partial \dot{X}_2} = c_2(\dot{X}_2 - \dot{X}_1) - c_3(\dot{X}_3 - \dot{X}_2) \tag{3.140}$$

$$\frac{\partial D}{\partial \dot{X}_3} = c_3(\dot{X}_3 - \dot{X}_2) - c_4(\dot{X}_0 - \dot{X}_3) \tag{3.141}$$

Substituting Eqs. 3.132, 3.136, and 3.139 into Eq. 3.128 (for $i = 1$).

$$m_1\ddot{X}_1 + K_1(X_1 - X_0) - K_2(X_2 - X_1) + c_1(\dot{X}_1 - \dot{X}_0) - C_2(\dot{X}_2 - \dot{X}_1) = 0$$

$$m_1\ddot{X}_1 + \dot{X}_1(c_1 + c_2) - \dot{X}_2c_2 + X_1(K_1 + K_2) - X_2K_2 = \dot{X}_0c_1 + X_0K_1 \tag{3.142}$$

Substituting Eqs. 3.133, 3.137, and 3.140 into Eq. 3.128 (for $i = 2$).

$$m_2\ddot{X}_2 + K_2(X_2 - X_1) - K_3(X_3 - X_2) + c_2(\dot{X}_2 - \dot{X}_1) - c_3(\dot{X}_3 - \dot{X}_2) = 0$$

$$m_2\ddot{X}_2 - \dot{X}_1c_2 + \dot{X}_2(c_2 + c_3) - \dot{X}_3c_3 - X_1K_2 + X_2(K_2 + K_3) - X_3K_3 = 0 \tag{3.143}$$

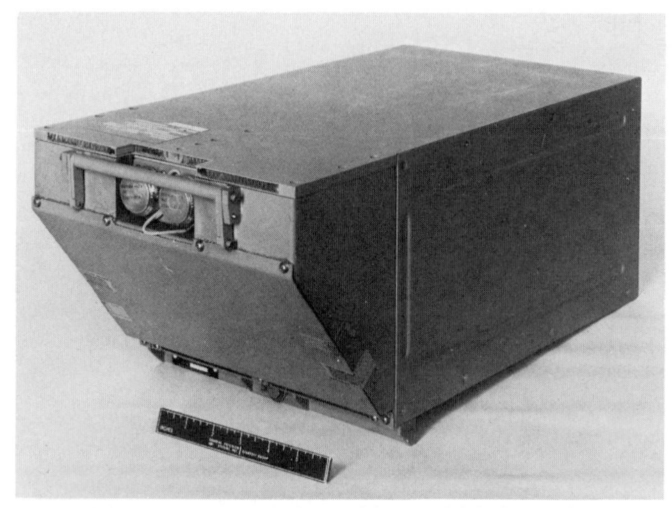

FIGURE 3.31. An electronic box with a multiple-fin air heat exchanger as part of the box structure for the top and bottom surfaces (courtesy Kearfott division, The Singer Company).

Substituting Eqs. 3.134, 3.138, and 3.141 into Eq. 3.128 (for $i = 3$)

$$m_3\ddot{X}_3 + K_3(X_3 - X_2) - K_4(X_0 - X_3) + c_3(\dot{X}_3 - \dot{X}_2) - c_4(\dot{X}_0 - \dot{X}_3) = 0$$

$$m_3\ddot{X}_3 - \dot{X}_2c_3 + \dot{X}_3(c_3 + c_4) - X_2K_3 + X_3(K_3 + K_4) = \dot{X}_0c_4 + X_0K_4 \tag{3.144}$$

Comparing Eqs. 3.122, 3.123, and 3.124 with Eqs. 3.142, 3.143, and 3.144 using Lagrange's equations shows they are exactly the same.

3.8. Forces in the Springs

The general reason for analyzing a multiple-spring–mass–damper system is to determine the forces, stresses, and fatigue life of the system. Since the springs represent the elastic members of the structure, the forces in the springs really represent the forces in different structural elements that were originally used to define the springs.

In order to determine the forces in the springs, it is necessary to determine the relation between the impressed sinusoidal force and the response of a damped-spring–mass system. Consider, for example, the single-degree-of-freedom system shown in Fig. 3.32. A vector diagram can be used to show the force relations in the system during vibration. The projections of the force vectors on the vertical axis represent the instantaneous magnitude of these vectors when they act on the mass. Note that

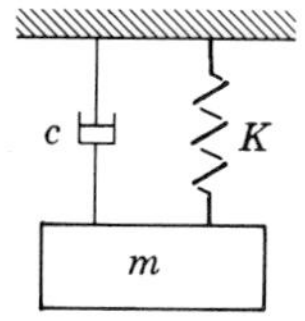

FIGURE 3.32. Forced vibration on a single mass system with damping.

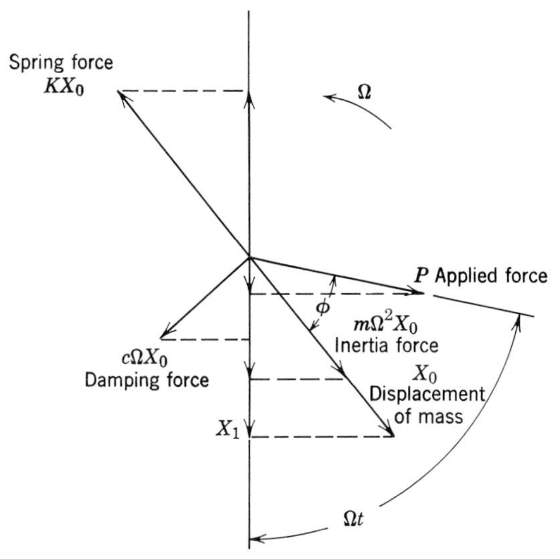

FIGURE 3.33. A vector diagram showing force relations during vibration.

when a differentiation is made, it is equivalent to multiplying the vector by Ω and rotating it through 90° in a forward direction (Fig. 3.33).

The force vector will act ϕ degrees ahead of the displacement vector X_0 so as to balance the dynamic forces in the system. The magnitude of these forces at any instant of time will be their projection on the vertical axis. The magnitude of the displacement at any instant of time will also be the projection on the vertical axis as shown in the following equation:

$$X_1 = X_0 \cos(\Omega t - \phi) \tag{3.145}$$

Consider a two-degrees-of-freedom system excited through the support (Fig. 3.34). A vector diagram can be used to show the displacement relations of the support along with the phase for each mass (Fig. 3.35).

The maximum displacement of the support is shown by the vector X_0. The displacement of the support at any instant of time is shown by the projection X_s on the vertical axis. The displacement of mass 1 will be a

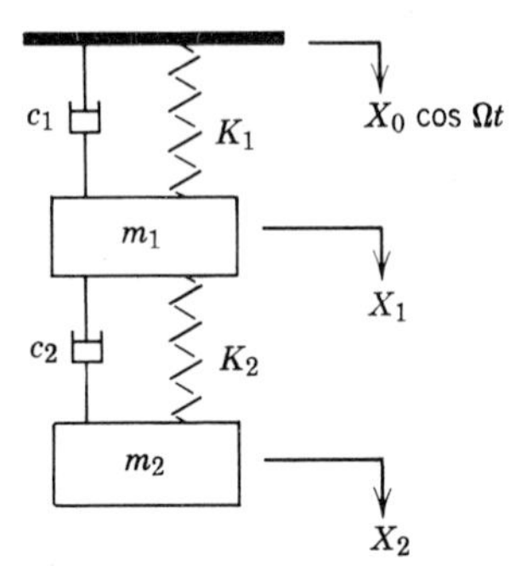

FIGURE 3.34. A two degree-of-freedom system excited at the base.

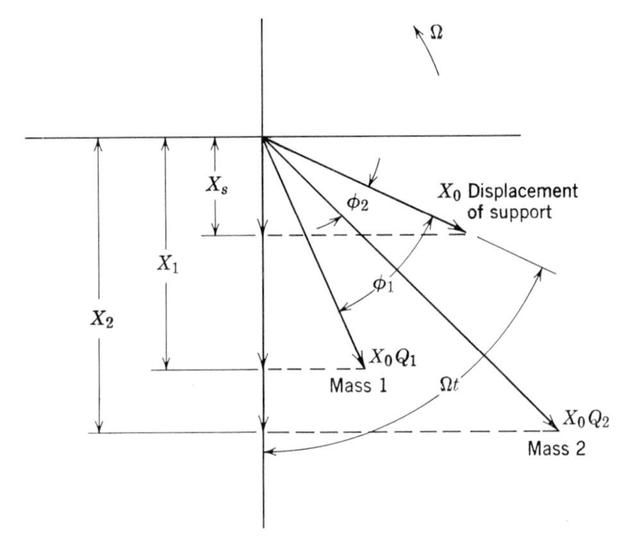

FIGURE 3.35. A vector diagram showing displacement relations for a two-degree-of-freedom system.

function of the transmissibility, Q_1, for mass 1 which can be written as X_0Q_1. The displacement of mass 1 at any instant of time will be the projection X_1 on the vertical axis. The same is also true for mass 2 which will have a displacement that is a function of its transmissibility, Q_2, which results in X_0Q_2. The displacement of mass 2 at any instant of time will be the projection X_2 on the vertical axis. These displacement equations are as follows:

$$X_s = X_0 \cos \Omega t \tag{3.146}$$

$$X_1 = X_0Q_1 \cos (\Omega t - \phi_1) \tag{3.147}$$

$$X_2 = X_0 Q_2 \cos(\Omega t - \phi_2) \tag{3.148}$$

The force in spring 1 will be a function of the relative displacement between mass 1 and the support as shown in the following equation.

$$P_1 = K_1(X_1 - X_s) \tag{3.149}$$

Substituting Eqs. 3.146 and 3.147 into 3.149

$$P_1 = K_1[X_0 Q_1 \cos(\Omega t - \phi_1) - X_0 \cos \Omega t]$$

Expanding the equation and regrouping

$$P_1 = K_1 X_0[(Q_1 \cos \phi_1 - 1) \cos \Omega t + (Q_1 \sin \phi_1) \sin \Omega t] \tag{3.150}$$

In Chapter 2 it was shown that the displacement of a spring–mass system could be related to the G force and frequency as (see Eq. 2.42)

$$X_0 = \frac{9.8 G_{\text{in}}}{f^2} \tag{3.151}$$

Substituting Eq. 3.151 into Eq. 3.150

$$P_1 = \frac{9.8 G_{\text{in}} K_1}{f^2} [(Q_1 \cos \phi_1 - 1) \cos \Omega t + (Q_1 \sin \phi_1) \sin \Omega t]$$

For convenience let

$$\left.\begin{aligned} \theta &= \Omega t \\ A &= Q_1 \cos \phi_1 - 1 \\ B &= Q_1 \sin \phi_1 \\ b &= \frac{9.8 G_{\text{in}} K_1}{f^2} \end{aligned}\right\} \tag{3.152}$$

then

$$P_1 = b(A \cos \theta + B \sin \theta) \tag{3.153}$$

The maximum force in spring K_1 can be determined by taking the first derivative of the above equation and setting it equal to zero to find the maximum point.

$$\frac{dP_1}{d\theta} = b(-A \sin \theta + B \cos \theta) = 0 \text{ for maximum}$$

In order for the above expression to be true

$$A \sin \theta = B \cos \theta$$

$$\tan \theta = \frac{B}{A} \tag{3.154}$$

This can be represented by a right triangle as shown in Fig. 3.36.

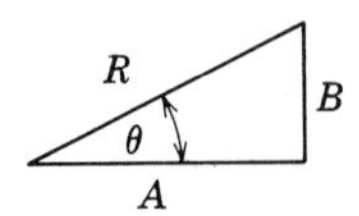

FIGURE 3.36. A right triangle for trigonometric functions.

$$\sin\theta = \frac{B}{R} \qquad \text{and} \qquad \cos\theta = \frac{A}{R} \tag{3.155}$$

Substitute these values into Eq. 3.153

$$P_{1\max} = b\left[A\left(\frac{A}{R}\right) + B\left(\frac{B}{R}\right)\right] = b\left(\frac{A^2+B^2}{R}\right) = b\left(\frac{R^2}{R}\right)$$

$$P_{1\max} = b\sqrt{A^2+B^2} \tag{3.156}$$

Substituting Eq. 3.152 into 3.156

$$P_{1\max} = \frac{9.8G_{\text{in}}K_1}{f^2}\,[(Q_1\cos\phi_1 - 1)^2 + (Q_1\sin\phi_1)^2]^{1/2}$$

$$P_{1\max} = \frac{9.8G_{\text{in}}K_1}{f^2}\,(Q_1{}^2 - 2Q_1\cos\phi_1 + 1)^{1/2} \tag{3.157}$$

At low frequencies the transmissibility of mass 1 will be very close to 1.0 and the phase angle, ϕ_1 will be almost zero, so cos ϕ_1 will be 1.0. Equation (3.157) is then

$$P_{1\max} = \frac{9.8G_{\text{in}}K_1}{f^2}\,[(1) - 2(1) + 1]^{1/2} = 0$$

The force in spring 1 would be zero.

At frequencies near the resonant frequency of mass 1, the phase angle ϕ_1 will be close to 90°, hence cos ϕ_1 will be close to zero. Equation 3.157 then becomes

$$P_{1\max} = \frac{9.8G_{\text{in}}K_1}{f^2}\,(Q_1{}^2 + 1)^{1/2}$$

For low damping where the transmissibility is greater than about 10, the expression is approximately

$$P_{1\max} = \frac{9.8G_{\text{in}}K_1Q_1}{f^2} \tag{3.158}$$

The force in spring 2 is a function of the relative displacement between mass 1 and mass 2.

$$P_2 = K_2(X_2 - X_1) \tag{3.159}$$

Substituting Eq. 3.147 and 3.148 into 3.159

$$P_2 = K_2[X_0Q_2 \cos(\Omega t - \phi_2) - X_0Q_1 \cos(\Omega t - \phi_1)]$$

Expanding the equation and regrouping

$$P_2 = K_2X_0[(Q_2 \cos\phi_2 - Q_1 \cos\phi_1) \cos\Omega t + (Q_2 \sin\phi_2 - Q_1 \sin\phi_1) \sin\Omega t]$$

Using Eq. (3.151) and for convenience let

$$\left.\begin{aligned} \theta &= \Omega t \\ A &= Q_2 \cos\phi_2 - Q_1 \cos\phi_1 \\ B &= Q_2 \sin\phi_2 - Q_1 \sin\phi_1 \\ b &= \frac{9.8G_{\text{in}}K_2}{f^2} \end{aligned}\right\} \tag{3.160}$$

then

$$P_2 = b(A \cos\theta + B \sin\theta) \tag{3.161}$$

This is exactly the same as Eq. 3.153 so the maximum force in spring K_2 will be the same as Eq. 3.156.

$$P_{2\max} = b(A^2 + B^2)^{1/2} \tag{3.162}$$

Substituting Eq. 3.160 into 3.162

$$P_{2\max} = \frac{9.8G_{\text{in}}K_2}{f^2}[(Q_2 \cos\phi_2 - Q_1 \cos\phi_1)^2 + (Q_2 \sin\phi_2 - Q_1 \sin\phi_1)^2]^{1/2}$$

This can be simplified to the following:

$$P_{2\max} = \frac{9.8G_{\text{in}}K_2}{f^2}[Q_2^2 - 2Q_1Q_2(\cos\phi_2 \cos\phi_1 + \sin\phi_2 \sin\phi_1) + Q_1^2] \tag{3.163}$$

At low frequencies, the transmissibility Q_1 of mass 1 and Q_2 of mass 2 is very close to 1.0 and the phase angles ϕ_1 and ϕ_2 are very close to zero, so $\cos\phi_1$ and $\cos\phi_2$ are 1.0.

$$P_{2\max} = \frac{9.8G_{\text{in}}K_2}{f^2}(1 - 2 + 1)^{1/2} = 0$$

The force in spring 2 would then be zero.

Since this is a two-degrees-of-freedom system, there will normally be two different resonances. The greatest load in spring 2 would occur when there is 180° phase difference between mass 1 and mass 2. For example, if the phase angle ϕ_1 for mass 1 is 90° and the phase angle ϕ_2 for mass 2 is

270°, then Eq. 3.163 is

$$P_{2\max} = \frac{9.8G_{\text{in}}K_2}{f^2}[Q_2^2 - 2Q_1Q_2(-1) + Q_1^2]^{1/2}$$

$$P_{2\max} = \frac{9.8G_{\text{in}}K_2}{f^2}(Q_1 + Q_2) \tag{3.164}$$

In order to determine the forces in the springs it is necessary to know the transmissibility and the phase angle of each mass during vibration. In any multiple-spring–mass system, with damping, these calculations can be very long. However, with the use of a computer, solutions to complex problems can be obtained very quickly.

Many companies manufacture different types of computers, both analog and digital. IBM is, of course, the biggest computer manufacturer and more people are using IBM computers than any other make today. IBM has a computer program called "LISA", an acronym for linear systems analysis. This program was written for electrical engineers, but it does an excellent job of solving problems in mechanical vibrations and shock. The input format is very simple and the program provides a print-out of the results.

3.9. Sample Problem

Consider, for example, the two-degrees-of-freedom system, shown in Fig. 3.37, excited through the base.

The equations of motion can be written for this system by following the rules previously outlined (see Figs. 3.26 and 3.27).

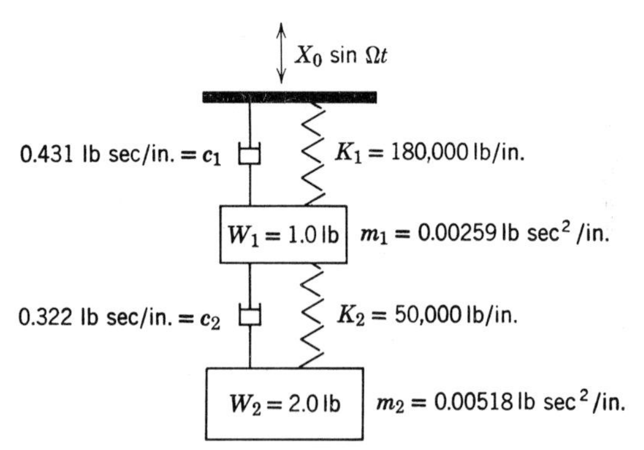

FIGURE 3.37. Forced vibrations of a two mass system with damping.

For mass 1

$$m_1\ddot{X}_1 + c_1(\dot{X}_1 - \dot{X}_0) + K_1(X_1 - X_0) + c_2(\dot{X}_1 - \dot{X}_2) + K_2(X_1 - X_2) = 0$$

$$m_1\ddot{X}_1 + \dot{X}_1(c_1 + c_2) - c_2\dot{X}_2 + X_1(K_1 + K_2) - K_2X_2 = c_1\dot{X}_0 + K_1X_0 \quad (3.165)$$

For mass 2

$$m_2\ddot{X}_2 + c_2(\dot{X}_2 - \dot{X}_1) + K_2(X_2 - X_1) = 0$$

$$m_2\ddot{X}_2 + c_2\dot{X}_2 - c_2\dot{X}_1 + K_2X_2 - K_2X_1 = 0 \quad (3.166)$$

Substitute the values shown in Fig. 3.37.

For mass 1 using Eq. 3.165

$$2.59 \times 10^{-3}\ddot{X}_1 + 0.753\dot{X}_1 - 0.322\dot{X}_2 + 2.30 \times 10^5 X_1 - 0.50 \times 10^5 X_2 = 0.431\dot{X}_0 + 1.80 \times 10^5 X_0 \quad (3.167)$$

For mass 2 using Eq. 3.166

$$5.18 \times 10^{-3}\ddot{X}_2 + 0.322\dot{X}_2 - 0.322\dot{X}_1 + 0.50 \times 10^5 X_2 - 0.50 \times 10^5 X_1 = 0 \quad (3.168)$$

These two equations can be put into a matrix form similar to that shown in Fig. 3.30; this makes it easier to set up the data for the computer. Only the numerical coefficients are entered, as shown in Fig. 3.38.

	Column 1 Terms with X_1	Column 2 Terms with X_2		Column 3 Terms with X_0
Row 1 for mass 1	2.59×10^{-3} 0.753 2.30×10^5	-0.322 -0.50×10^5	=	0.431 1.80×10^5
for mass 2	-0.322 -0.50×10^5	5.18×10^{-3} 0.322 0.50×10^5		0

FIGURE 3.38. Matrix of a two mass system with forced vibration.

For the IBM LISA computer program, computer cards would be punched as follows (each line represents data on one card):

TITLE, (up to 66 spaces can be used for title)
READ, MATRIX
1,1,2,2.59E−3,0.753,2.30E5

```
1,2,1,S,–0.322,–0.50E5
1,3,1,0.431,1.80E5
2,2,2,5.18E–3,0.322,0.50E5
DEFINE, V1 = V(1)
DEFINE, V2 = V(2)
COMPUTE, POLY, V1, V2
DATA, FREQUENCY = 0,2000,20,CPS,LIN
COMPUTE, BODE, V1, V2
LABEL, (up to 66 spaces can be used for a label)
PPLOT
EXIT
```

This is all of the input that is required to run the program, which will print out the transmissibility and phase angle and also produce a plot of these functions as well. The computer will make these calculations at frequencies of 0, 20, 40, 60, 80, and so on up to 2000 Hz in linear increments.

The three numbers preceeding the numerical data input, such as 1,1,2, represent the row of the matrix, the column of the matrix, then the highest order of the differential, respectively. For the highest order of the differential, an $\ddot{X}$ would be 2 since it represents the second derivative and an $\dot{X}$ would be 1 since it represents the first derivative.

The letter E is used for indicating the exponent so a value of 50,000 is represented by 0.50E5. Also the letter S is used for indicating symmetry in the matrix, which simplifies the data input.

A plot of the transmissibility versus frequency for each mass, and a plot of the phase angle versus frequency for each mass, as determined by the computer, is shown in Figure 3.39. (See Chapter 8 for more details.) These figures were plotted from data taken at 20-cycle increments. For a more accurate plot of the transmissibilities in the resonant regions, the frequency data should be taken in smaller increments.

Figure 3.39 shows there are two resonances for the spring mass system shown in Fig. 3.37. Forces in springs K_1 and K_2 can be determined at any frequency shown in the computer run, which will have the transmissibility and phase angle listed. Consider, for example, frequencies of 440 and 1520 Hz, which have the following data listed by the computer, for an input force of $1.0G$.

FREQUENCY (Hz)		TRANSMISSIBILITY (Q)	PHASE ANGLE (ϕ)
mass 1	440	5.61	153.8°
mass 2	440	26.86	157.6°
mass 1	1520	22.56	91.1°
mass 2	1520	2.67	267.2°

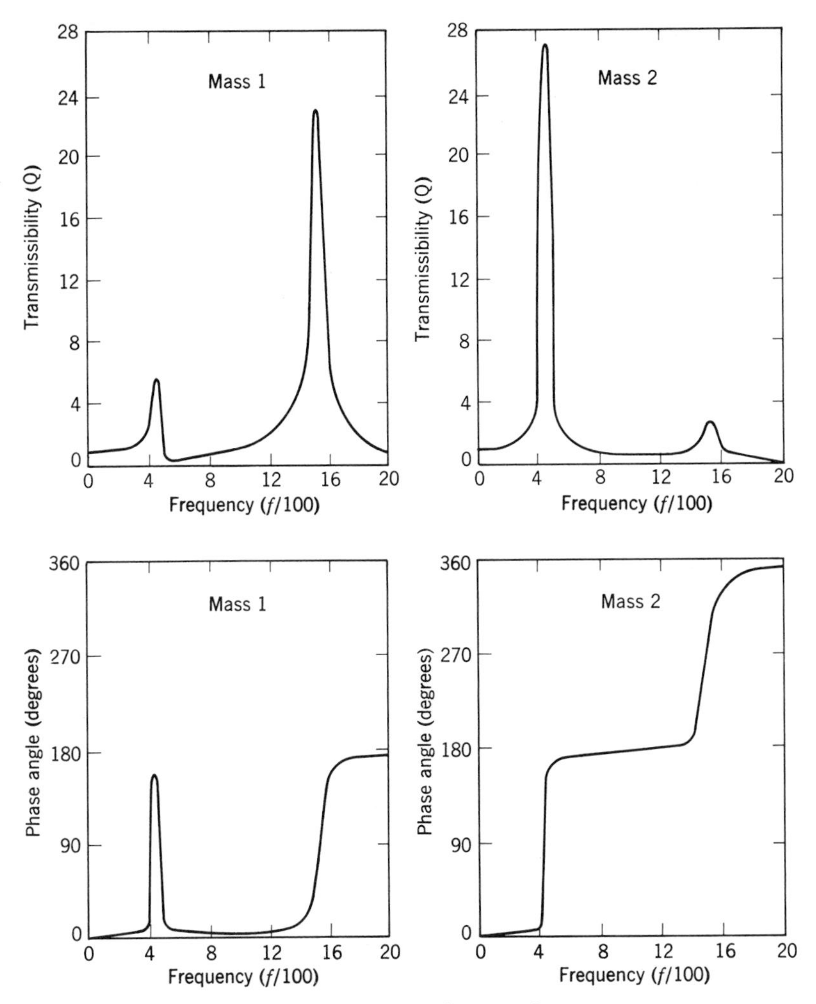

FIGURE 3.39. Transmissibility and phase angle curves for a two mass system.

The force in spring 1 ($K_1 = 180{,}000$ lb/in.) is

$$P_{1\,\max} = \frac{9.8 G_{\text{in}} K_1}{f^2} (Q_1^2 - 2Q_1 \cos \phi_1 + 1)^{1/2} \quad \text{(see Eq. 3.157)}$$

At 440 Hz, $Q_1 = 5.61$ and $\phi_1 = 153.8°$ so

$$P_{1\,\max} = \frac{9.8(1.0)(1.8 \times 10^5)}{(440)^2} [(5.61)^2 - 2(5.61) \cos 153.8 + 1]^{1/2}$$

$$P_{1\,\max} = 59.6 \text{ lb} \quad (440 \text{ Hz}) \tag{3.169}$$

At 1520 Hz, $Q_1 = 22.56$ and $\phi_1 = 91.1°$ so

$$P_{1\,max} = \frac{(9.8)(1.0)(1.8 \times 10^5)}{(1520)^2}[(22.56)^2 - 2(22.56)\cos 91.1 + 1]^{1/2}$$

$$P_{1\,max} = 17.3 \text{ lb} \quad (1520 \text{ Hz}) \tag{3.170}$$

The force in spring 2 ($K_2 = 50{,}000$ lb/in.) is

$$P_{2\,max} = \frac{9.8 G_{in} K_2}{f^2}[Q_2{}^2 - 2Q_1Q_2(\cos\phi_2 \cos\phi_1 + \sin\phi_2 \sin\phi_1) + Q_1{}^2]^{1/2}$$

(see Eq. 3.163)

At 440 Hz, $Q_2 = 26.86$, $Q_1 = 5.61$, $\phi_2 = 157.6°$, $\phi_1 = 153.8°$ so

$$P_{2\,max} = \frac{(9.8)(1.0)(0.50 \times 10^5)}{(440)^2}[(26.86)^2 - 2(5.61)(26.86)$$

$$\times (\cos 157.6° \cos 153.8° + \sin 157.6° \sin 153.8°) + (5.61)^2]^{1/2}$$

$$P_{2\,max} = 53.8 \text{ lb} \quad (440 \text{ Hz}) \tag{3.171}$$

At 1520 Hz, $Q_2 = 2.67$, $Q_1 = 22.56$, $\phi_2 = 267.2°$, $\phi_1 = 91.1°$

$$P_{2\,max} = \frac{(9.8)(1.0)(0.50 \times 10^5)}{(1520)^2}[(2.67)^2 - 2(22.56)(2.67)$$

$$\times (\cos 267.2° \cos 91.1° + \sin 267.2° \sin 91.1°) + (22.56)^2]^{1/2}$$

$$P_{2\,max} = 5.35 \text{ lb} \quad (1520 \text{ Hz}) \tag{3.172}$$

Since the force in each spring is known, the stress in each spring can be determined if the geometry of the structure is defined. (See Chapter 8 for more details.)

CHAPTER 4

BEAMS AND SUSPENDED ELECTRONIC COMPONENTS

4.1. SUSPENDED ELECTRONIC COMPONENTS

Electronic component parts such as resistors, capacitors, and diodes are often suspended between solder terminals, which are mounted to circuit boards or structural members. If the component body is not cemented, tied, or supported in any manner, resonances may develop in the suspended component when it is subjected to a sinusoidal vibration environment. Under these conditions, the body of the component acts like a concentrated mass and the electrical lead wires act like a spring (Fig. 4.1).

If the solder terminals are mounted on a very rigid resonance-free structure, the suspended component can be analyzed by approximating it as a beam with a concentrated load. If the component body contacts the support structure during vibration, it will substantially reduce the transmissibility of the system and the following analysis will no longer be valid.

Another type of system that uses suspended component parts is the cordwood module. The name "cordwood" comes from the stacking of components such as resistors, capacitors, and diodes like a cord of wood,

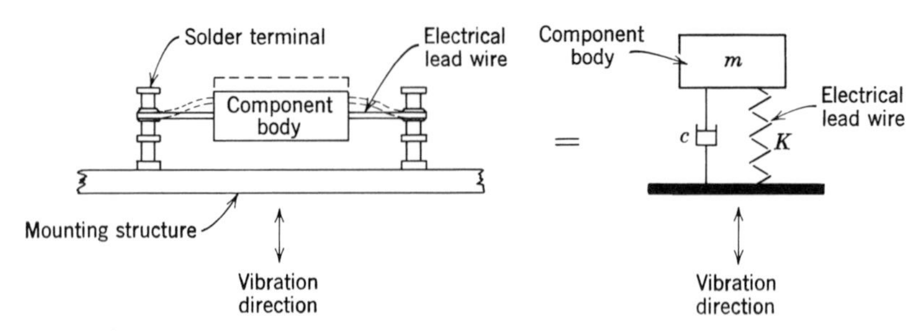

FIGURE 4.1. An electronic component part suspended between solder terminals.

where the longitudinal axes of all the components are parallel. The electronic component parts are usually suspended between two small printed-circuit boards, sometimes called jig wafers. Modules of this type may contain anywhere from six to 60 components, depending upon the application and cost. Large modules are difficult to repair unless the defective component is accessible at the outer perimeter. Therefore some modules may be discarded because it may be cheaper to replace an entire module than it is to repair it.

Several cordwood modules are often mounted on a large plug-in type of printed-circuit board that acts as a master interconnecting circuit-board (Fig. 4.2).

Lead wires must be good electrical conductors, so they are made of metal. The metals most commonly used in commercial and military applications are conveniently shown in the military standard publication *MIL-STD-1267B*, March 6, 1968, under the title: "Leads, Weldable, For Electronic Component Parts."

The most common lead materials are nickel, iron, and copper, although silver is sometimes used on some types of high-powered diodes. Electrical lead-wire metals will exhibit similar fatigue characteristics associated

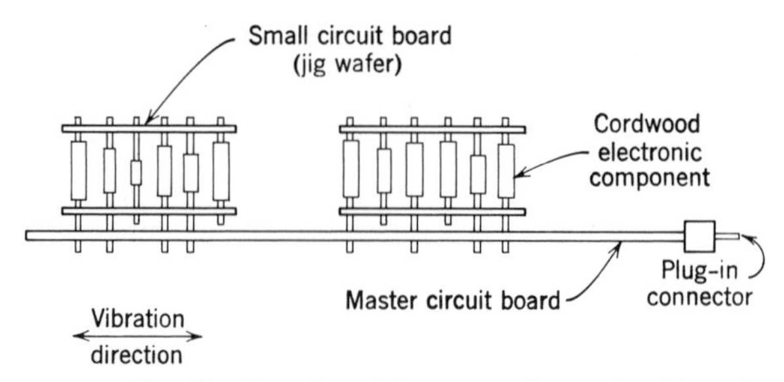

FIGURE 4.2. Cordwood modules mounted on a circuit board.

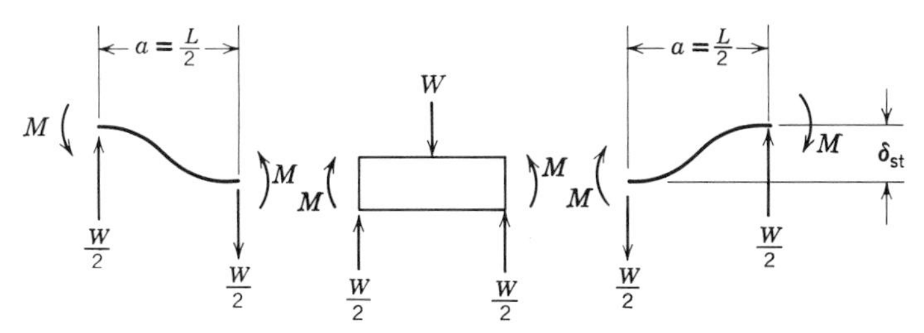

FIGURE 4.3. Free body force diagram of a suspended component.

with these metals when they are used as major structural elements subjected to alternating stress loads.

Lunney and Crede reported on resonant fatigue tests of resistors and capacitors suspended by their electrical lead wires [3]. These same two authors were involved in vibration and fatigue analysis of electronic equipment based on test data taken in the actual operating environments [7].

The bodies of most component parts such as resistors, capacitors, and diodes are much larger in diameter than their electrical lead wires. Therefore, under a dynamic load perpendicular to the longitudinal axis of the component, most of the deflection will be due to bending in the electrical lead wires. A very good approximation of the resonant frequency for a suspended component can be obtained by considering the electrical lead wire as a weightless beam, fixed at both ends, with a concentrated mass at the center. The static deflection of the suspended component can be determined from an examination of the free-body diagram shown in Fig. 4.3.

$$\delta_{st} = \frac{(W/2)a^3}{12EI} = \frac{Wa^3}{24EI}$$

Let L = total length, so $a = L/2$

$$\delta_{st} = \frac{W}{24EI}\left(\frac{L}{2}\right)^3 = \frac{WL^3}{192EI} \tag{4.1}$$

This is the same as the equation for a concentrated load in the center of a beam with fixed ends (Fig. 4.4).

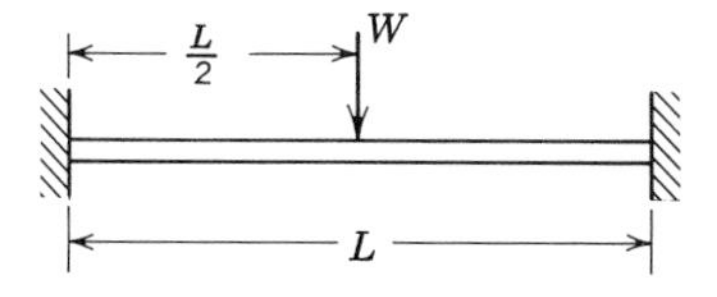

FIGURE 4.4. A beam with fixed ends and a concentrated load.

The natural frequency of the suspended electronic component part can then be approximated from the static deflection equation for a single-degree-of-freedom system (see Chapter 2, Eq. 2.10):

$$f_n = \frac{1}{2\pi}\left(\frac{g}{\delta_{st}}\right)^{1/2} \tag{4.2}$$

Substituting Eq. 4.1 into Eq. 4.2:

$$f_n = \frac{1}{2\pi}\left(\frac{192EIg}{WL^3}\right)^{1/2} \tag{4.3}$$

4.2. Sample Problem

Consider, for example, a typical tantalytic capacitor type CS 12-A (A case size), as shown in Fig. 4.5. Determine the natural frequency, transmissibility, and dynamic stresses in the electrical lead wires.

The physical properties of the capacitor are as follows:

$W = 0.86 \times 10^{-3}$ lb weight

$E = 30 \times 10^6$ psi nickel lead wire

$I = \frac{\pi d^4}{64} = \frac{\pi}{64}(0.020)^4 = 0.785 \times 10^{-8}$ in.^{4} moment of inertia

$L = 2(0.445) = 0.890$ in. length of lead wire

$g = 386$ in/sec^2 gravity

Substituting into Eq. 4.3 for the natural frequency

$$f_n = \frac{1}{2\pi}\left[\frac{(1.92 \times 10^2)(30 \times 10^6)(0.785 \times 10^{-8})(386)}{(0.86 \times 10^{-3})(0.89)^3}\right]^{1/2}$$

$$f_n = 855 \text{ Hz} \tag{4.4}$$

A series of vibration tests were run on many different types of electronic component parts with different suspension lengths ranging from 0.75 to

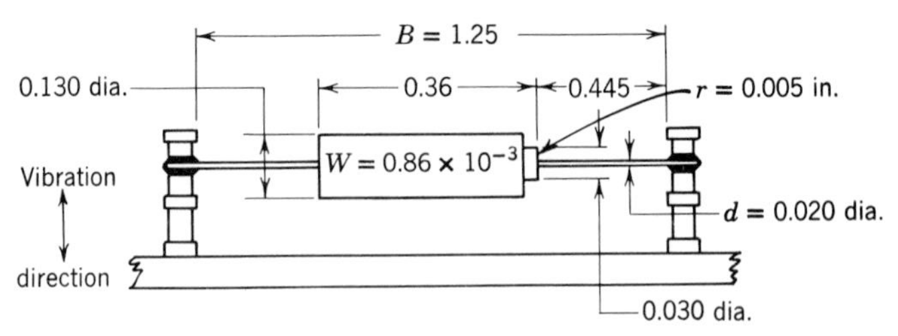

FIGURE 4.5. A tantalytic capacitor suspended between solder terminals.

1.625 in., in several increments. A strobe light was used with a calibrated microscope, to determine the resonant frequency and maximum displacement amplitude for each component during a 2-G, 5-G, and 10-G sinusoidal sweep to 2000 Hz. For the tantalytic capacitor in Fig. 4.5, a comparison of the calculated resonance and the test resonance is shown in Table 4.1.

TABLE 4.1. CS 12-A CAPACITOR

Suspended Length (B in.)	Resonant Frequency (Hz)	
	Analysis	Test
0.75	2940	> 2000
0.85	2090	1700
1.00	1400	1175
1.25	855	800
1.50	590	585
1.62	507	525

In addition to the resonant frequency of the suspended component part, it was very desirable to see if there was any way of approximating the transmissibility at resonance. All the test data were gathered, and the various physical properties were arranged and rearranged to see if some sort of a pattern could be established. Although there was a great deal of scatter in the data, a large number of the components seemed to follow a pattern related to

$$C = \frac{GEI}{WL} \tag{4.5}$$

where E = modulus of elasticity of wire, lb/in.^{2}
G = peak input acceleration G force
I = moment of inertia of lead wire, in.^{4}
L = length of electrical lead wire, in.
W = weight of component body, lb
f_n = natural frequency, Hz
C = dynamic constant

The resulting curve was plotted in Fig. 4.7 as C versus A. The transmissibility for the suspended component can then be determined from

$$Q = A(f_n)^{1/2} \tag{4.6}$$

The value of A is obtained from Fig. 4.7 after the value of C has been determined from Eq. 4.5. The curve in Fig. 4.7 can be used for accelera-

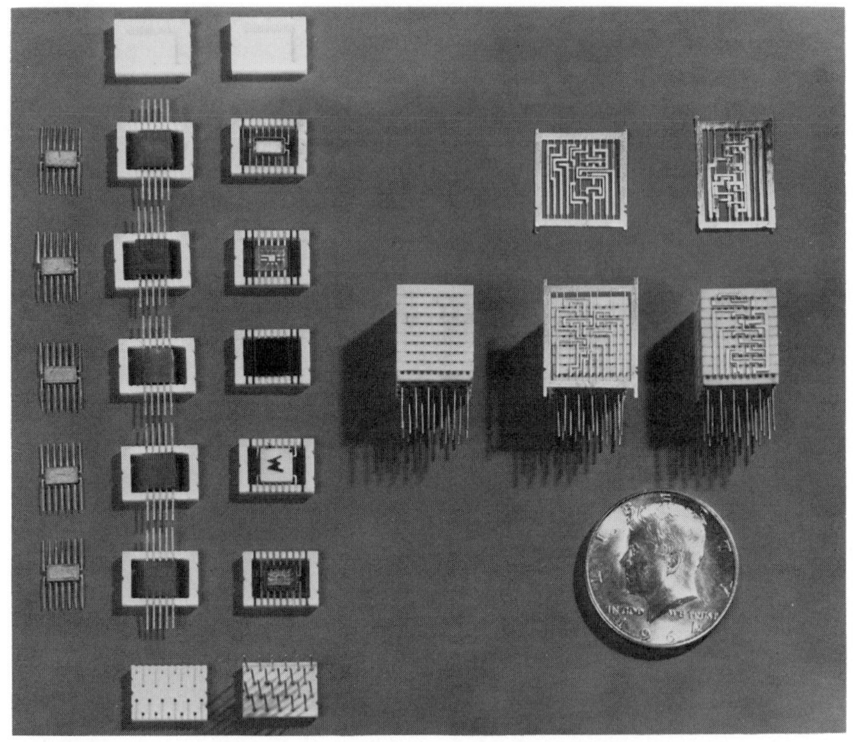

FIGURE 4.6. Building up a small module with the use of flatpack integrated circuits (courtesy Norden division of United Aircraft).

tion input forces from about 2Gs peak to about 10Gs peak, since this was the range of the test data.

The transmissibility for a 5-G peak input to the capacitor shown in Fig. 4.5 can now be approximated using the following data:

$$E = 30 \times 10^6 \text{ lb/in.}^2 \text{ modulus of elasticity for nickel wire}$$

$$I = \frac{\pi d^4}{64} = 0.785 \times 10^{-8} \text{ in.}^4 \text{ moment of inertia}$$

$$W = 0.86 \times 10^{-3} \text{ lb weight}$$

$$L = 2(0.445) = 0.890 \text{ in. length}$$

Substituting into Eq. 4.5

$$C = \frac{(5.0)(30 \times 10^6)(0.785 \times 10^{-8})}{(0.86 \times 10^{-3})(0.89)} = 1540 \tag{4.7}$$

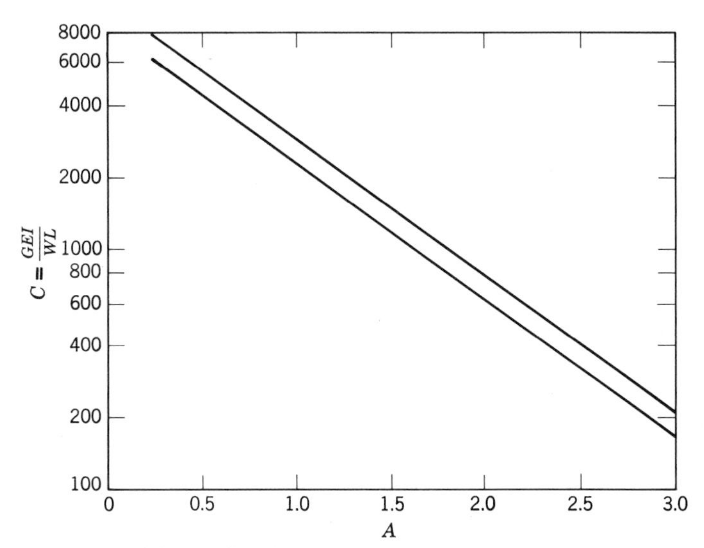

FIGURE 4.7. Curve for determining transmissibilities of suspended components.

From the curve in Fig. 4.7, the average value for A is about 1.35. Then substituting Eq. 4.4 into Eq. 4.6

$$Q = 1.35(855)^{1/2}$$

$$Q = 39.5 \tag{4.8}$$

Test data showed a Q value of 37.5 for this particular component, with a 5-G peak input.

Test data was not always consistent. There were fluctuations in the transmissibilities of similar electronic component parts. A great many of these fluctuations were believed to be due to some slipping of the electrical lead wires in the vibration fixture, different electronic component body sizes for similar parts, different diameters on the electrical lead wires, and nonuniform solder coatings on the electrical lead wires.

If the input G force and the transmissibility of a suspended electronic component part at resonance is known, the dynamic bending stresses can be calculated and the fatigue life of the part can be estimated.

Again consider the tantalytic capacitor shown in Fig. 4.5. If the capacitor is mounted on a rigid bulkhead within an electronic box that has a transmissibility of 2.0 at 855 Hz, and if the electronic box will be subjected to a 5-G peak sinusoidal vibration input, the capacitor will receive a 10-G peak vibration input at its resonance of 855 Hz, shown by Eq. 4.4.

The maximum dynamic load, P_d, acting on the body of the capacitor can be determined from the equation

$$P_d = G_{\text{in}} Q W \tag{4.9}$$

where $G_{\text{in}} = 10.0G$ peak input acceleration

$$C = \frac{GEI}{WL} = \frac{(10)(30 \times 10^6)(0.785 \times 10^{-8})}{(0.86 \times 10^{-3})(0.890)} = 3080$$

$$A = 0.9 \text{ (from Fig. 4.7)}$$

$$Q = A(f_n)^{1/2} = 0.9(855)^{1/2} = 26.3$$

Substituting into Eq. 4.9

$$P_d = (10.0)(26.3)(0.86 \times 10^{-3}) = 0.226 \text{ lb} \tag{4.10}$$

The maximum dynamic bending moment in the electrical lead wires on the capacitor can be determined from Figs. 4.3 and 4.5.

$$M = \frac{P_d a}{4} = \frac{(0.226)(0.445)}{4} = 0.0252 \text{ in. lb} \tag{4.11}$$

The dynamic bending stress in the wire can be determined from the standard bending stress equation, which includes a theoretical stress concentration factor, K_t (see Chapter 10).

$$S_b = K_t \frac{Mc}{I} \tag{4.12}$$

where $M = 0.0252$ in. lb (see Eq. 4.11)

$$c = \frac{d}{2} = \frac{0.020}{2} = 0.010 \text{ in. wire radius}$$

$$I = \frac{\pi d^4}{64} = 0.785 \times 10^{-8} \text{ in.}^4 \text{ moment of inertia}$$

K_t = theoretical or geometric stress-concentration factor where the electrical lead wire joins the component body (from Peterson[18] and Fig. 4.5)

$$\frac{r}{d} = \frac{0.005}{0.020} = 0.25 \quad \text{and} \quad \frac{D}{d} = \frac{0.030}{0.020} = 1.5$$

$$K_t = 1.34 \text{ (bending)}$$

Substituting into Eq. 4.12

$$S_b = \frac{(1.34)(0.0252)(0.010)}{0.785 \times 10^{-8}} = 43{,}000 \text{ lb/in.}^2 \tag{4.13}$$

An examination of the fatigue S–N curve for the electrical lead wire (shown in Chapter 10, Fig. 10.10*a*), shows the expected fatigue life would probably be about 8×10^6 cycles if there were no nicks or scratches in the wire to create additional stress concentrations.

If a qualification test plan requires a resonant dwell period of 30 min, as many tests do, and if by chance the dwell happens to be at 855 Hz, the expected fatigue life of the capacitor lead wires can be approximated as follows:

$$t = \frac{N}{f} \tag{4.14}$$

where t = time to fail

$N = 8 \times 10^6$ fatigue cycles to fail

$f_n = 855$ Hz resonant frequency

$$t = \frac{8 \times 10^6 \text{ cycles to fail}}{855 \text{ cycles/sec} \times 60 \text{ sec/min}}$$

$$t = 156 \text{ min to fail} \tag{4.15}$$

It would appear that this capacitor will pass the qualification test, if the test does not have to be repeated too many times.

If the qualification test program requires only sinusoidal sweeps from 10 to 2000 Hz and back to 10 Hz several times, then it is possible to approximate the time it takes to sweep through the capacitor resonance by considering the half-power points. These are the points where the power that can be absorbed by a damper is proportional to the square of the amplitude at a given frequency. For a system that has a transmissibility greater than about 10, the curve in the area of the resonance is approximately symmetrical. The half-power points are often considered to be the bandwidth of the system (Fig. 4.8).

The time it takes to sweep through the half power points can be determined from the expression

$$t = \frac{\log_e \dfrac{1 + \dfrac{1}{2Q}}{1 - \dfrac{1}{2Q}}}{R \log_e 2} \tag{4.16}$$

where t = time in minutes

R = sweep rate in octaves/min
(10 to 20 Hz in 1 min; 20 to 40 Hz in 1 min, etc.)

$Q = 26.3$ transmissibility (see Eq. 4.9)

Assuming a sweep rate of 0.5 octaves/min, the time it takes to sweep

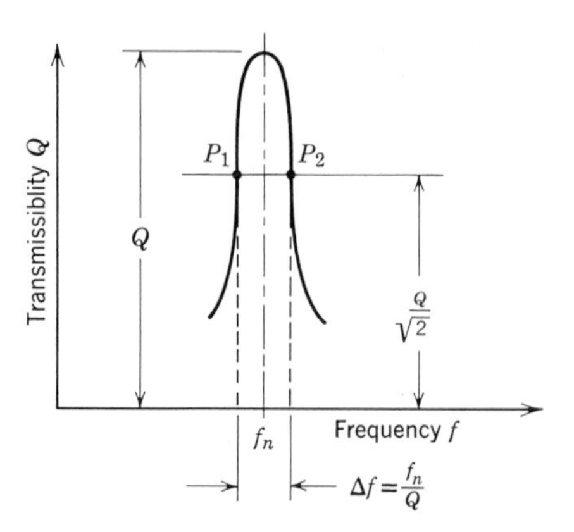

FIGURE 4.8. Half power points for a transmissibility curve.

through the half power points is as follows:

$$t = \frac{\log_e \dfrac{1+\dfrac{1}{52.6}}{1-\dfrac{1}{52.6}}}{0.5 \log_e 2} = 0.11 \text{ min} = 6.6 \text{ sec} \tag{4.17}$$

The number of fatigue cycles that will be accumulated in 6.6 sec or one single sweep from 10 to 2000 Hz is

$$n = f_n t = 855 \text{ cycles/sec} \times 6.6 \text{ sec} = 5640 \text{ cycles}$$

The number of single sweeps (NS) required for a failure is

$$\text{NS} = \frac{N}{n} = \frac{8 \times 10^6 \text{ cycles to fail}}{5.64 \times 10^3 \text{ cycles/sweep}}$$

$$\text{NS} = 1420 \text{ single sweeps to fail} \tag{4.18}$$

This capacitor should be capable of passing the qualification test.

4.3. Natural Frequency of a Uniform Beam

The natural frequency of a uniform beam can usually be determined by considering the strain energy and the kinetic energy of the vibrating beam. There are several different ways of expressing the strain energy and the kinetic energy associated with the beam. Some methods are approximate while others are exact. The exact methods usually require more work than the approximate methods and the approximate methods are often satisfactory for engineering solutions, so the approximate methods will be given more consideration.

Boundary conditions are very important in exact methods as well as in approximate methods. For approximate methods, if the geometric boundary conditions are satisfied, the resulting natural-frequency equations will usually be quite accurate for the fundamental resonant frequency. The geometric boundary conditions consist only of the slope and the deflection. However these boundary conditions must be met by the type of deflection curve that is used to describe the vibrating beam.

The technique of assuming a deflection curve for a vibrating system, from which the natural-frequency equation is determined, is called the Rayleigh method after Lord Rayleigh [28] who first proposed this method. If the deflection curve happens to be the exact curve for the boundary conditions then the frequency equation will also be exact.

A uniform beam with simply supported ends is shown in Fig. 4.9. The bending moment at any point X can be determined by taking moments at X:

$$M_x = \frac{wL}{2}X - \frac{wX^2}{2} \tag{4.19}$$

The strain energy of the beam can be expressed as

$$U = \int_0^L \frac{M_x^2 dX}{2EI} \tag{4.20}$$

Substituting Eq. 4.19 into Eq. 4.20

$$U = \frac{1}{2EI}\int_0^L \left(\frac{w^2L^2X^2}{4} - \frac{w^2LX^3}{2} + \frac{w^2X^4}{4}\right) dX$$

$$U = \frac{w^2L^5}{240EI} \tag{4.21}$$

The kinetic energy of the vibrating beam can be expressed as

$$T = \frac{w\Omega^2}{2g}\int_0^L Y^2 \mathrm{d}X \tag{4.22}$$

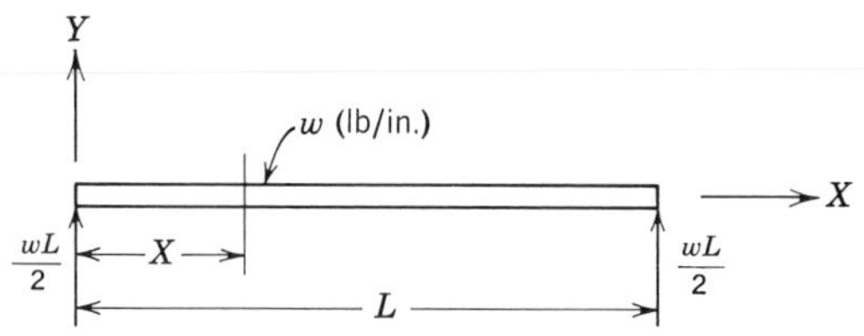

FIGURE 4.9. A uniform beam with simply supported ends.

The deflection at any point on the beam can be obtained from a structural handbook as

$$Y=\frac{w}{24EI}\,(L^3X-2LX^3+X^4) \tag{4.23}$$

Substituting Eq. 4.23 into Eq. 4.22:

$$T=\frac{w\Omega^2}{2g}\left(\frac{w^2}{576E^2I^2}\right)\int_0^L (L^6X^2-4L^4X^4+2L^3X^5+4L^2X^6-4LX^7+X^8)\,dX$$

$$T=\frac{w^3\Omega^2}{2(576)gE^2I^2}\left(\frac{L^9}{20.329}\right) \tag{4.24}$$

If there is no energy lost in the system, at resonance the strain energy must equal the kinetic energy so Eq. 4.21 must equal Eq. 4.24.

$$\frac{w^2L^5}{240EI}=\frac{w^3\Omega^2}{2(576)gE^2I^2}\left(\frac{L^9}{20.329}\right)$$

$$\Omega^2=\frac{97.58EIg}{wL^4} \tag{4.25}$$

$$f_n=\frac{\Omega}{2\pi}$$

$$f_n=\frac{9.87}{2\pi}\left(\frac{EIg}{wL^4}\right)^{1/2} \tag{4.26}$$

A trigonometric function could have been used to describe the deflection curve of the simply supported beam shown in Fig. 4.10.

Consider the deflection equation

$$Y=Y_0\sin\frac{\pi X}{L} \tag{4.27}$$

This curve satisfies the geometric boundary conditions of slope and deflection as follows:

when

$$\left.\begin{array}{ll} X=0, & Y=0 \\ X=L, & Y=0 \\ X=L/2, & Y=Y_0 \end{array}\right\}\text{deflections}$$

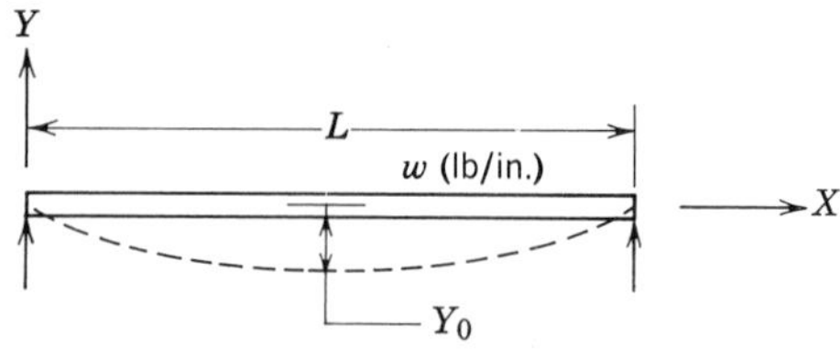

FIGURE 4.10. Deflection curve for a simply supported beam.

For the slope

$$\theta = \frac{dY}{dX} = \frac{\pi}{L} Y_0 \cos \frac{\pi X}{L}$$

$$\left.\begin{aligned} \text{when } X = 0,\ \theta &= \frac{\pi}{L} Y_0 \\ X = L,\ \theta &= -\frac{\pi}{L} Y_0 \\ X = \frac{L}{2},\ \theta &= 0 \end{aligned}\right\} \text{slope}$$

The strain energy Eq. 4.20 can be written in a slightly different form by making use of the general bending moment equation

$$EI \frac{d^2Y}{dX^2} = M \tag{4.28}$$

Substituting Eq. 4.28 into Eq. 4.20

$$U = \frac{EI}{2} \int_0^L \left(\frac{d^2Y}{dX^2}\right)^2 dX \tag{4.29}$$

From Eq. 4.27

$$\frac{dY}{dX} = \frac{\pi}{L} Y_0 \cos \frac{\pi X}{L}$$

$$\frac{d^2Y}{dX^2} = -\frac{\pi^2}{L^2} Y_0 \sin \frac{\pi X}{L}$$

$$\left(\frac{d^2Y}{dX^2}\right)^2 = \frac{\pi^4}{L^4} Y_0^2 \sin^2 \frac{\pi X}{L} \tag{4.30}$$

Substituting Eq. 4.30 into Eq. 4.29 for the strain energy of the beam

$$U = \frac{EI}{2} \int_0^L \left(\frac{\pi^4}{L_4} Y_0^2 \sin^2 \frac{\pi X}{L}\right) dX$$

$$U = \frac{EI\pi^4 Y_0^2}{2L^4} \left(\frac{L}{2}\right) = \frac{EI\pi^4 Y_0^2}{4L^3} \tag{4.31}$$

The kinetic energy of the vibrating beam can be obtained from Eq. 4.22 along with Eq. 4.27.

$$T = \frac{w\Omega^2}{2g} \int_0^L Y_0^2 \sin^2 \frac{\pi X}{L} dX$$

$$T = \frac{w\Omega^2 Y_0^2}{2g} \left(\frac{L}{2}\right) \tag{4.32}$$

At resonance, the kinetic energy is equal to the strain energy so Eq. 4.32 is equal to Eq. 4.31.

$$\frac{EI\pi^4 Y_0^2}{4L^3} = \frac{w\Omega^2 Y_0^2 L}{4g}$$

$$\Omega^2 = \frac{\pi^4 EIg}{wL^4}$$

$$f_n = \frac{\pi^2}{2\pi}\left(\frac{EIg}{wL^4}\right)^{1/2} \tag{4.33}$$

This results in the exact frequency equation (as shown by Eq. 4.26) because the exact deflection curve was used in Eq. 4.27. A polynomial can also be used to describe the deflection curve for a simply supported beam, shown in Fig. 4.11.

Consider the deflection equation

$$Y = Y_0\left(1 - \frac{X^2}{a^2}\right) \tag{4.34}$$

This curve satisfies the geometric boundary conditions of slope and deflection as follows:

$$\left.\begin{array}{ll} \text{when } X = 0, & Y = Y_0 \\ X = a, & Y = 0 \\ X = -a, & Y = 0 \end{array}\right\} \text{deflection}$$

For the slope

$$\theta = \frac{dY}{dX} = -\frac{2Y_0 X}{a^2}$$

$$\left.\begin{array}{ll} \text{when } X = 0, & \theta = 0 \\ X = a, & \theta = -\dfrac{2Y_0}{a} \\ X = -a, & \theta = \dfrac{2Y_0}{a} \end{array}\right\} \text{slope}$$

From Eq. 4.34

$$\frac{dY}{dX} = -\frac{2Y_0 X}{a^2}$$

$$\frac{d^2Y}{dX^2} = -\frac{2Y_0}{a^2}$$

$$\left(\frac{d^2Y}{dX^2}\right)^2 = \frac{4Y_0^2}{a^4} \tag{4.35}$$

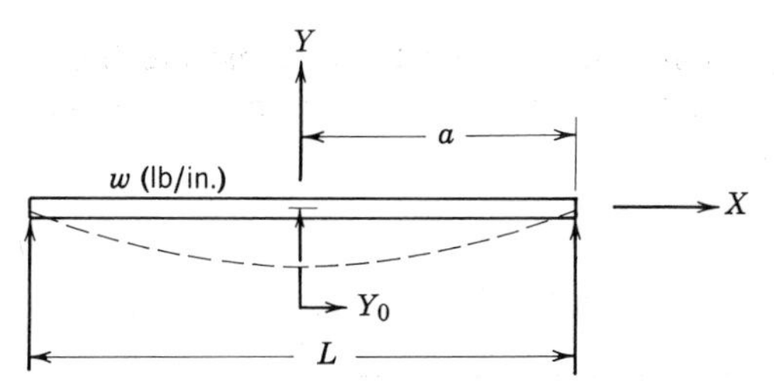

FIGURE 4.11. A simply supported beam with the Y axis at the center.

For the strain energy, substituting Eq. 4.35 into Eq. 4.29

$$U = \frac{EI}{2} \int_0^a \frac{4Y_0^2}{a^4} \, dX$$

$$U = \frac{2EIY_0^2}{a^4}(a) = \frac{2EIY_0^2}{a^3} \tag{4.36}$$

The kinetic energy can be obtained by substituting Eq. 4.34 into Eq. 4.22.

$$T = \frac{w\Omega^2 Y_0^2}{2g} \int_0^a \left(1 - \frac{2X^2}{a^2} + \frac{X^4}{a^4}\right) dX$$

$$T = \frac{w\Omega^2 Y_0^2 a}{2g}\left(\frac{8}{15}\right) \tag{4.37}$$

At resonance, the kinetic energy must equal the strain energy so Eq. 4.37 must equal Eq. 4.36.

$$\frac{2EIY_0^2}{a^3} = \frac{4w\Omega^2 Y_0^2 a}{15g}$$

$$\Omega^2 = \frac{EIg}{wa^4}\left(\frac{15}{2}\right)$$

$$f_n = \frac{\Omega}{2\pi} = \frac{2.74}{2\pi}\left(\frac{EIg}{wa^4}\right)^{1/2}$$

since $a = L/2$,

$$f_n = \frac{10.96}{2\pi}\left(\frac{EIg}{wL^4}\right)^{1/2} \tag{4.38}$$

Comparing these results with the exact-frequency equation of Eq. 4.26 shows the polynomial results in a frequency that is about 11.1% too high. In fact, the Rayleigh method always results in a frequency value that will be greater than the true value, unless the exact deflection curve is used.

The last equation can be used to demonstrate the natural frequency calculation of a simply supported beam with a uniform load. Consider the beam shown in Fig. 4.12.

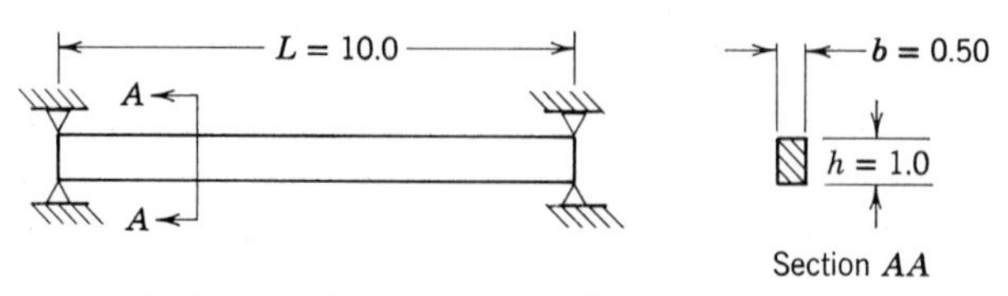

FIGURE 4.12. A uniform beam with simply supported ends.

If the beam is made of aluminum:

$E = 10.5 \times 10^6$ lb/in.^{2} modulus of elasticity

$b = 0.50$ in. beam width

$h = 1.0$ in. height of beam

$I = \frac{bh^3}{12} = \frac{(0.50)(1.0)^3}{12} = 0.0417$ in.^{4} moment of inertia

$g = 386$ in. sec^2 gravity

$L = 10.0$ in. length

$\rho = 0.10$ lb/in.^{3} density of aluminum

$W = bhL\rho = (0.50)(1.0)(10.0)(0.10) = 0.50$ lb weight of beam

$w = \frac{W}{L} = \frac{0.50}{10.0} = 0.05$ lb/in.

Substituting into Eq. 4.38

$$f_n = \frac{10.96}{2\pi}\left[\frac{(10.5 \times 10^6)(0.0417)(386)}{(0.05)(10.0)^4}\right]^{1/2}$$

$$f_n = 1013 \text{ Hz} \tag{4.39}$$

Now assume the beam is made of steel

$E = 29 \times 10^6$ lb/in.^{2} modulus of elasticity

$\rho = 0.283$ lb/in.^{3} density

$W = (0.50)(1.0)(10.0)(0.283) = 1.415$ lb weight

$w = \frac{W}{L} = \frac{1.415}{10.0} = 0.1415$ lb/in.

Substituting into Eq. 4.38

$$f_n = \frac{10.96}{2\pi}\left[\frac{(29 \times 10^6)(0.0417)(386)}{(0.141)(10.0)^4}\right]^{1/2}$$

$$f_n = 1005 \text{ Hz} \tag{4.40}$$

Comparing Eqs. 4.40 and 4.39 shows the natural frequency of the steel beam is almost the same as the natural frequency of the aluminum beam, even when the steel beam is almost three times heavier. A closer examination of the natural frequency equation shows that it depends upon the ratio of the modulus of elasticity and the density. These two factors can be examined for the beam shown in Fig. 4.12 considering several different materials.

Magnesium: $\frac{E}{\rho} = \frac{6.5 \times 10^6}{0.065} = 100 \times 10^6$ in.

Aluminum: $\frac{E}{\rho} = \frac{10.5 \times 10^6}{0.10} = 105 \times 10^6$ in.

Steel: $\frac{E}{\rho} = \frac{29.0 \times 10^6}{0.283} = 102 \times 10^6$ in.

Beryllium: $\frac{E}{\rho} = \frac{42.0 \times 10^6}{0.068} = 619 \times 10^6$ in.

The only two items that will change in the natural-frequency equation are the modulus of elasticity and the density. An examination of the ratios shown above indicates that the natural frequency of the beam shown in Fig. 4.12 will be about the same for magnesium, aluminum, and steel. A beryllium beam, however, would have a much higher natural frequency.

The Rayleigh method is very convenient for determining the resonant frequency of a uniform beam with a concentrated load (Fig. 4.13).

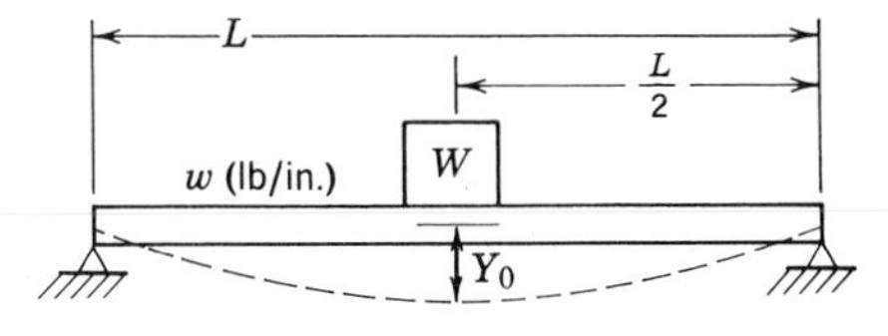

FIGURE 4.13. A uniform beam with a concentrated load at the center.

Consider the trigonometric function shown by Eq. 4.27. This same expression can be used to approximate the deflection curve for the uniform beam with a concentrated mass in the center. This results in the same strain energy as shown by Eq. 4.31

$$U = \frac{EI\pi^4 Y_0^2}{4L^3} \tag{4.41}$$

The kinetic energy must now include the effects of the concentrated load in addition to the uniform load. The total kinetic energy then will be the same as Eq. 4.22 with the extra kinetic energy of the concentrated load.

$$T = \frac{w\Omega^2}{2g} \int_0^L Y^2\, dX + \tfrac{1}{2} m V^2 \tag{4.42}$$

where

$$V = Y_0 \Omega$$

The kinetic energy of the uniform beam will be the same as Eq. 4.32 with the additional kinetic energy of the concentrated load.

$$T = \frac{w\Omega^2 Y_0^2 L}{4g} + \frac{W Y_0^2 \Omega^2}{2g} \tag{4.43}$$

Since the kinetic energy must equal the strain energy at resonance, Eq. 4.43 must be equal to Eq. 4.41.

$$\frac{w\Omega^2 Y_0^2 L}{4g} + \frac{W Y_0^2 \Omega^2}{2g} = \frac{EI\pi^4 Y_0^2}{4L^3}$$

$$\Omega^2 = \frac{EI\pi^4 g}{L^3(wL + 2W)}$$

$$f_n = \frac{\Omega}{2\pi} = \frac{\pi}{2}\left[\frac{EIg}{L^3(wL + 2W)}\right]^{1/2} \tag{4.44}$$

If a concentrated load on a weightless beam is desired, wL will be zero and the equation becomes

$$f_n = 1.11\left(\frac{EIg}{WL^3}\right)^{1/2}$$

Comparing this to the exact equation for a concentrated mass at the center of a weightless beam shows there is an error of about 0.6%.

If the concentrated load is zero, the natural frequency equation becomes

$$f_n = \frac{\pi}{2}\left(\frac{EIg}{wL^4}\right)^{1/2}$$

This is the exact equation for a simply supported beam with a uniform load.

The Rayleigh method can be used to determine the fundamental resonant frequency for beams with various end conditions. This method can be used with a trigonometric or a polynomial expression as long as the geometric boundary conditions of slope and deflection are satisfied. Some typical deflections for different boundary conditions are shown along with the resulting natural frequency equations (Fig. 4.14).

CASE 1 Cantilever Beam – Uniform Load

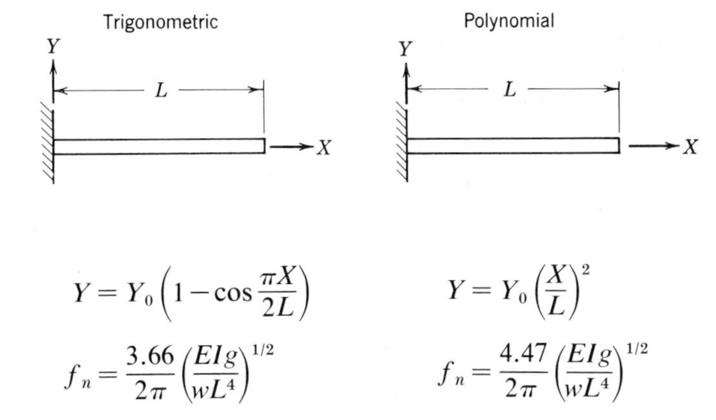

$$Y = Y_0\left(1 - \cos\frac{\pi X}{2L}\right) \qquad Y = Y_0\left(\frac{X}{L}\right)^2$$

$$f_n = \frac{3.66}{2\pi}\left(\frac{EIg}{wL^4}\right)^{1/2} \qquad f_n = \frac{4.47}{2\pi}\left(\frac{EIg}{wL^4}\right)^{1/2}$$

CASE 2 Fixed–Fixed Beam – Uniform Load

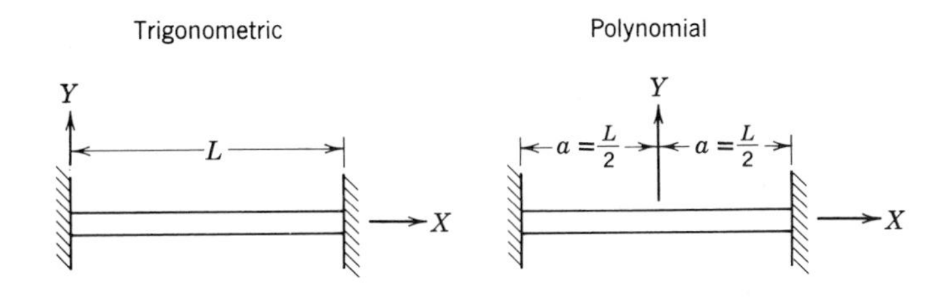

$$Y = Y_0\left(1 - \cos\frac{2\pi X}{L}\right) \qquad Y = Y_0\left[1 - \left(\frac{X}{a}\right)^2\right]^2$$

$$f_n = \frac{22.7}{2\pi}\left(\frac{EIg}{wL^4}\right)^{1/2} \qquad f_n = \frac{22.4}{2\pi}\left(\frac{EIg}{wL^4}\right)^{1/2}$$

FIGURE 4.14. Displacement curves and natural frequency equations.

4.4. Free–Free Beam with End Masses

The resonant frequency of a free–free beam with end masses can be determined with the use of the Rayleigh method and a trigonometric function. A free–free beam is usually considered to be a floating beam, similar perhaps to a piece of wood floating in water. This type of structure may also be found in the vibration laboratory in the form of a vibration fixture. In this case the nodal points of the beam would be bolted to the head of the vibration machine. The beam becomes a vibration fixture that might support an optical system which requires the accurate alignment of two electronic boxes at each end of the beam (Fig. 4.15). The system shown in Fig. 4.15 can be approximated as a free–free beam with end masses, as shown in Fig. 4.16.

Consider the trigonometric expression that satisfies the boundary conditions

$$Y = Y_0 \sin \frac{\pi X}{L} - A \tag{4.45}$$

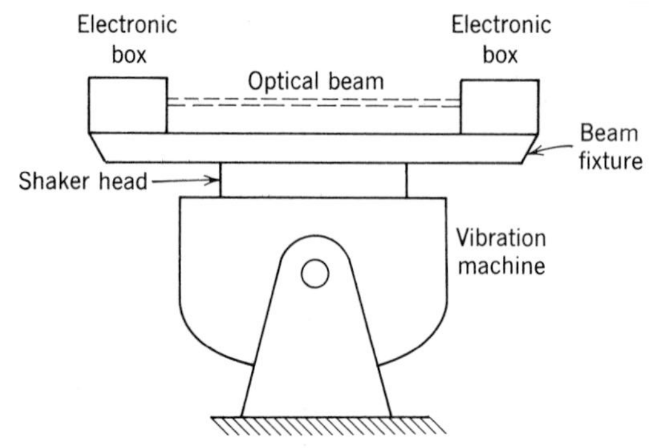

FIGURE 4.15. A vibration fixture system with two electronic boxes.

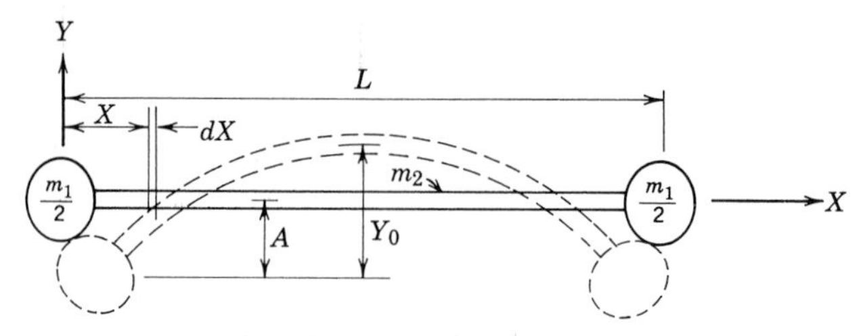

FIGURE 4.16. A free–free beam with end masses.

At the fundamental resonant mode, the total momentum Mo_1 of the end masses m_1 is

$$\mathrm{Mo}_1 = 2\left(\frac{m_1}{2}\right) V$$

where velocity, $V = A\Omega$

$$\mathrm{Mo}_1 = m_1 A\Omega \tag{4.46}$$

The momentum Mo_2 of a small section of the uniform beam mass m_2 is

$$d\mathrm{Mo}_2 = dm_2 V$$

where $dm_2 = (w/g)\ dX$ and $V = Y\Omega$

$$d\mathrm{Mo}_2 = \frac{w}{g} \Omega Y dX$$

The momentum of the beam can be determined by integrating over the length of the beam

$$\mathrm{Mo}_2 = \frac{w}{g} \Omega \int_0^L Y\ dX \tag{4.47}$$

Substituting Eq. 4.45 into Eq. 4.47

$$\mathrm{Mo}_2 = \frac{w}{g} \Omega \int_0^L \left(Y_0 \sin \frac{\pi X}{L} - A\right) dX$$

$$\mathrm{Mo}_2 = \frac{w}{g} \Omega \int_0^L Y_0 \sin \frac{\pi X}{L}\ dX - \frac{w}{g} \Omega \int_0^L A\ dX \tag{4.48}$$

The total vertical momentum of the system during vibration must be zero.

$$\mathrm{Mo}_2 - \mathrm{Mo}_1 = 0 \tag{4.49}$$

Substituting Eq. 4.46 and 4.48 into Eq. 4.49

$$\frac{w}{g} \Omega \int_0^L Y_0 \sin \frac{\pi X}{L}\ dX - \frac{w}{g} \Omega \int_0^L A\ dX - m_1 A \Omega = 0$$

$$\frac{w}{g} Y_0 \left(-\frac{L}{\pi} \cos \frac{\pi X}{L}\right)_0^L - \frac{w}{g} (AX)_0^L - m_1 A = 0$$

$$\frac{w}{g} Y_0 \left(\frac{2L}{\pi}\right) = A\left(\frac{w}{g} L + m_1\right)$$

since

$$\frac{w}{g} L = m_2 \tag{4.50}$$

$$\frac{2m_2Y_0}{\pi} = A(m_2 + m_1)$$

$$A = \frac{2Y_0}{\pi}\left(\frac{m_2}{m_2 + m_1}\right) \tag{4.51}$$

From Eq. 4.45

$$\frac{dY}{dX} = Y_0 \frac{\pi}{L} \cos \frac{\pi X}{L}$$

$$\frac{d^2Y}{dX^2} = -Y_0 \frac{\pi^2}{L^2} \sin \frac{\pi X}{L}$$

$$\left(\frac{d^2Y}{dX^2}\right)^2 = Y_0^2 \frac{\pi^4}{L^4} \sin^2 \frac{\pi X}{L} \tag{4.52}$$

The strain energy of the uniform beam during vibration will be

$$U = \frac{EI}{2} \int_0^L \left(\frac{d^2Y}{dX^2}\right)^2 dX \qquad \text{(see Eq. 4.29)}$$

Substituting Eq. 4.52 into the above expression

$$U = \frac{EIY_0^2\pi^4}{2L^4} \int_0^L \sin^2 \frac{\pi X}{L}\, dX$$

$$U = \frac{EIY_0^2\pi^4}{2L^4}\left(\frac{L}{2}\right) \tag{4.53}$$

The kinetic energy of the vibrating beam will be

$$T_2 = \frac{w\Omega^2}{2g} \int_0^L Y^2\, dX \qquad \text{(see Eq. 4.22)}$$

Substituting Eq. 4.45 into above

$$T_2 = \frac{w\Omega^2}{2g} \int_0^L \left(Y_0^2 \sin^2 \frac{\pi X}{L} - 2AY_0 \sin \frac{\pi X}{L} + A^2\right) dX$$

Since

$$\int_0^L \sin^2 \frac{\pi X}{L}\, dX = \frac{L}{2}$$

$$\int_0^L \sin \frac{\pi X}{L}\, dX = \frac{2L}{\pi}$$

$$T_2 = \frac{w\Omega^2}{2g}\left[Y_0^2\left(\frac{L}{2}\right) - \frac{4LAY_0}{\pi} + A^2L\right] \tag{4.54}$$

The kinetic energy of the vibrating end masses is

$$T_1 = \tfrac{1}{2}(2)\left(\frac{m_1}{2}\right)V^2 = \tfrac{1}{2}m_1A^2\Omega^2 \tag{4.55}$$

The total kinetic energy of the vibrating system is

$$T = T_1 + T_2 = \frac{w\Omega^2}{2g}\left[Y_0{}^2\frac{L}{2} - \frac{4LAY_0}{\pi} + A^2L\right] + \tfrac{1}{2}m_1A^2\Omega^2 \tag{4.56}$$

Substitute Eqs. 4.50 and 4.51 into Eq. 4.56 for the total kinetic energy of the vibrating system. Then

$$T = \frac{w\Omega^2}{2g}\left[\frac{Y_0{}^2L}{2} - \frac{8Y_0{}^2L}{\pi^2}\left(\frac{m_2}{m_2+m_1}\right) + \frac{4Y_0{}^2L}{\pi^2}\left(\frac{m_2}{m_2+m_1}\right)^2\right]$$

$$+\frac{m_1\Omega^2}{2}\left(\frac{4Y_0{}^2}{\pi^2}\right)\left(\frac{m_2}{m_2+m_1}\right)^2$$

$$T = \frac{\Omega^2Y_0{}^2}{2}\left\{m_2\left[\frac{1}{2} - \frac{8}{\pi^2}\left(\frac{m_2}{m_2+m_1}\right) + \frac{4}{\pi^2}\left(\frac{m_2}{m_2+m_1}\right)^2\right] + \frac{4}{\pi^2}m_1\left(\frac{m_2}{m_2+m_1}\right)^2\right\}$$

$$T = \frac{\Omega^2Y_0{}^2}{2}\left[\frac{m_2}{2} - \frac{8}{\pi^2}\left(\frac{m_2{}^2}{m_2+m_1}\right) + \frac{4}{\pi^2}\left(\frac{m_2{}^2}{m_2+m_1}\right)\right]$$

$$T = \Omega^2Y_0{}^2\left[\frac{m_2}{4} - \frac{2}{\pi^2}\left(\frac{m_2{}^2}{m_2+m_1}\right)\right] \tag{4.57}$$

At resonance, the kinetic energy must equal the strain energy so Eq. 4.57 will equal Eq. 4.53.

$$\Omega^2Y_0{}^2\left[\frac{m_2}{4} - \frac{2}{\pi^2}\left(\frac{m_2{}^2}{m_2+m_1}\right)\right] = \frac{EIY_0{}^2\pi^4}{4L^3}$$

$$\Omega^2 = \frac{EI\pi^4}{4L^3\left(\dfrac{m_2}{m_2+m_1}\right)[0.25(m_2+m_1) - 0.203m_2]}$$

$$\Omega^2 = \frac{EI\pi^4}{\left(\dfrac{m_2}{m_2+m_1}\right)(m_1+0.188m_2)L^3}$$

$$f_n = \frac{\Omega}{2\pi} = \frac{\pi}{2}\left[\frac{EI(m_2+m_1)}{L^3m_2(m_1+0.188m_2)}\right]^{1/2}$$

If weights instead of masses are used, the equation becomes

$$f_n = \frac{\pi}{2}\left[\frac{EI(W_2+W_1)g}{L^3W_2(W_1+0.188W_2)}\right]^{1/2} \tag{4.58}$$

If there are no end masses on the free–free beam, $W_1 = 0$ and the equation becomes

$$f_n = \frac{22.8}{2\pi}\left(\frac{EIg}{W_2L^3}\right)^{1/2} \tag{4.59}$$

This is the same as the equation for a uniform beam with fixed ends (see Fig. 4.14, Case 2). One of the characteristics of a free–free beam is that it does happen to have the same natural frequency as a fixed–fixed beam.

4.5. Sample Problem

Consider the situation shown in Fig. 4.15 where two separate electronic boxes are tested which are optically linked to each other and require a precise separation distance. A large amount of relative motion between the two electronic boxes in the vertical direction is undesirable. The vibration test consists of sinusoidal sweeps from 5 to 500 Hz and back to 5 Hz several times. Therefore, to be conservative, a design goal of 750 Hz was established for the vibration fixture so the natural frequency would be 50% higher than the highest forcing frequency, to avoid exciting the fixture resonance.

Care must be used in designing a fixture of this type because balance is very important, so provisions should be made for balance weights. Dynamic shifts in the center of gravity may occur if one or both electronic boxes develops a severe resonance below 500 Hz. A dynamic shift in the center of gravity (CG) might result in a rocking mode in the shaker head. (See Chapter 9 for more information on vibration testing.)

A preliminary vibration-fixture design considered the use of a 2-in.-thick aluminum plate 5 in. wide and 32 in. long. Is this satisfactory or is a stiffer fixture plate required? A sketch of the proposed system is shown in Fig. 4.17.

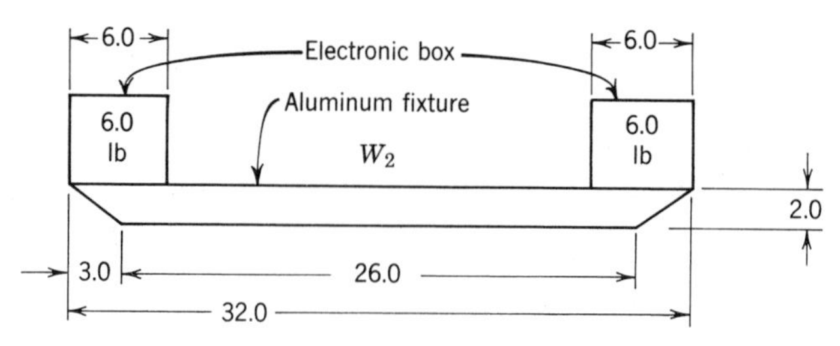

FIGURE 4.17. A vibration fixture with electronic boxes on opposite ends.

The natural frequency for this system as a free–free beam with end masses can be determined from Eq. 4.58

$$f_n = \frac{\pi}{2}\left[\frac{EI(W_2+W_1)g}{L^3W_2(W_1+0.188W_2)}\right]^{1/2}$$

where $E = 10.5 \times 10^6$ lb/in.^{2} aluminum

$$I = \frac{bh^3}{12} = \frac{(5.0)(2.0)^3}{12} = 3.33 \text{ in.}^4 \text{ (5-in. wide plate)}$$

W_1 = weight of 2 electronic boxes plus 3-in. tips of vibration fixture, 5 in. wide

$= 2(6) + 2(5)(3)(0.1) = 12 + 3.0 = 15.0$ lb

W_2 = weight of fixture up to center of electronic boxes

$= (26)(5.0)(2.0)(0.1) = 26.0$ lb

$L = 26.0$-in. length to center of boxes

$g = 386$ in./sec^2 gravity

$$f_n = \frac{\pi}{2}\left\{\frac{10.5\times 10^6(3.33)(26+15)(3.86\times 10^2)}{(26.0)^3(26)[15+(0.188)(26)]}\right\}^{1/2}$$

$$f_n = \frac{\pi}{2}(6.08\times 10^4)^{1/2} = 387 \text{ Hz} \tag{4.60}$$

Since the resonant frequency is below 750 Hz, the fixture design will not be satisfactory.

If a 3-in.-thick aluminum plate is used for the vibration fixture, $W_1 =$ 16.5 lb, $W_2 = 39$ lb, $I = 11.25$ in.^{4}

The natural frequency will be 619 Hz, which is still unsatisfactory. Therefore, a thicker plate might be used, or ribs could be added to the 2-in. plate to increase the stiffness.

4.6. Nonuniform Cross-Sections

Electronic equipment must make use of every cubic inch of volume there is because there usually is not very much volume available. When there is a little extra space, it is usually located in an area that is difficult to reach. Often structural modifications must be made to accommodate equipment that just seems to grow larger and larger with every redesign. This usually requires notching, reinforcing, or moving major load-carrying members. The net result is a chassis or a structure that supports all of the electronics at the expense of a cross-section with notches and cutouts along its length.

In order to determine the resonant frequency of this type of structure, it is necessary to consider the effects of nonuniform cross-sections. If there are many different cross-sections that must be considered, most methods of analysis can become very long and time consuming. However there is a trade-off that can be made which reduces the amount of work at the expense of some accuracy. This trade-off involves the calculation of an equivalent moment of inertia for a uniform structure that will have approximately the same stiffness as the nonuniform structure. The loss of accuracy is only about 5–10% in the natural frequency for many cases, which is well worth the amount of work that can be saved.

Consider a cantilever beam with two different cross-sections that form two steps (Fig. 4.18). Castigliano's strain energy theorem can be used to determine the deflection under the load at the free end of the beam. The general deflection equation is (see Chapter 5, Section 5.20).

$$\delta = \frac{1}{EI} \int_1^2 M_1 \frac{\partial M_1}{\partial P} \, dX$$

The deflection for the step beam shown in Fig. 4.18 is

$$\delta = \frac{1}{EI_1} \int_o^a M_1 \frac{\partial M_1}{\partial P} \, dX_1 + \frac{1}{EI_2} \int_o^b M_2 \frac{\partial M_2}{\partial P} \, dX_2 \tag{4.61}$$

The bending moment at point 1 is

$$M_1 = PX_1$$

$$\frac{\partial M_1}{\partial P} = X_1$$

The bending moment at point 2 is

$$M_2 = P(a + X_2)$$

$$\frac{\partial M_2}{\partial P} = (a + X_2)$$

Substituting into Eq. 4.61

$$\delta = \frac{P}{EI_1} \int_0^a X_1^2 \, dX_1 + \frac{P}{EI_2} \int_0^b (a^2 + 2aX_2 + X_2^2) \, dX_2$$

$$\delta = \frac{Pa^3}{3EI_1} + \frac{P}{EI_2} \left(a^2 b + ab^2 + \frac{b^3}{3} \right) \tag{4.62}$$

Let

$$a = b = \frac{L}{2} \tag{4.63}$$

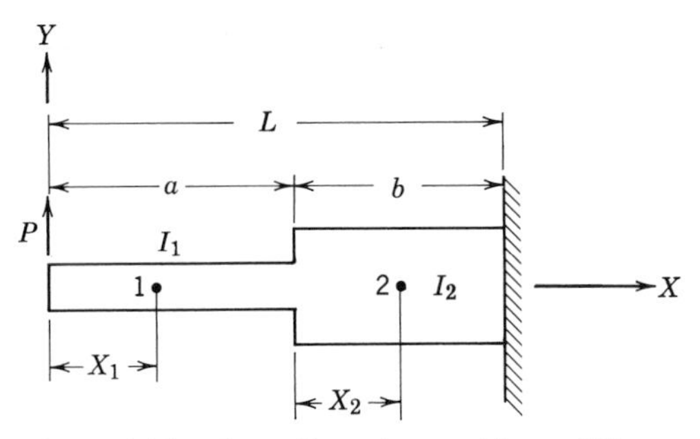

FIGURE 4.18. A cantilever beam with two different cross sections.

$$\delta = \frac{PL^3}{24}\left(\frac{1}{EI_1} + \frac{7}{EI_2}\right) \tag{4.64}$$

The deflection of a *uniform* cantilever beam with a concentrated load at the end can be obtained from any structural handbook

$$\delta = \frac{PL^3}{3EI_{av}} \tag{4.65}$$

An average moment of inertia I_{av} is used in the above expression so that the average moment of inertia can be determined by forcing the deflections in Eq. 4.64 and Eq. 4.65 to be equal.

$$\frac{PL^3}{3EI_{av}} = \frac{PL^3}{24}\left(\frac{1}{EI_1} + \frac{7}{EI_2}\right)$$

Solving for the average moment of inertia

$$I_{av} = \frac{8I_1I_2}{7I_1 + I_2} \tag{4.66}$$

The average moment of inertia for the two-step beam in Fig. 4.18 can be determined for two different moments of inertia.

Let

$$I_1 = 2 \text{ in.}^4 \qquad \text{and} \qquad I_2 = 4 \text{ in.}^4 \tag{4.67}$$

Substituting Eq. 4.67 into Eq. 4.66

$$I_{av} = \frac{8(2)(4)}{7(2) + 4} = 3.55 \text{ in.}^4 \tag{4.68}$$

Now let us see if there is some other way that an average moment of inertia might be determined, so as to give results similar to Eq. 4.68 without going through so much work.

Consider a method for averaging the moment of inertia as a function of the length of each section for the stepped beam shown in Fig. 4.18.

$$I'_{av} = \frac{aI_1 + bI_2}{a+b} \tag{4.69}$$

Substituting Eqs. 4.63 and 4.67 into Eq. 4.69

$$I'_{av} = \frac{\left(\frac{L}{2}\right)(2) + \left(\frac{L}{2}\right)(4)}{\frac{L}{2} + \frac{L}{2}} = 3.0 \text{ in.}^4 \tag{4.70}$$

Comparing Eq. 4.70 with Eq. 4.68 indicates the approximate averaging method shown in Eq. 4.69 is only about 15.5% lower than the true value. Since the natural frequency is a function of the square root of the moment of inertia, the error in the natural frequency will be only about 8.1% too low.

Some other proportions can be examined. For example, consider the proportion

$$a = 2b$$

so

$$a = \tfrac{2}{3}L \qquad \text{and} \qquad b = \tfrac{1}{3}L \tag{4.71}$$

Substituting Eq. 4.71 into Eq. 4.62

$$\delta = \frac{PL^3}{81E}\left(\frac{19I_1 + 8I_2}{I_1 I_2}\right) \tag{4.72}$$

To get the same deflection as a uniform cantilever beam with a concentrated load, force Eq. 4.72 to equal Eq. 4.65.

$$\frac{PL^3}{3EI_{av}} = \frac{PL^3}{81E}\left(\frac{19I_1 + 8I_2}{I_1 I_2}\right)$$

$$I_{av} = 27\left(\frac{I_1 I_2}{19I_1 + 8I_2}\right)$$

Substituting Eq. 4.67 into above

$$I_{av} = 27\left(\frac{8}{38+32}\right) = 3.08 \text{ in.}^4 \tag{4.73}$$

This is the correct value for the average moment of inertia.

Compute the approximate average value with the method shown by Eq. 4.69. This can be done with Eqs. 4.71 and 4.67 as follows:

$$I'_{av} = \frac{(\frac{2}{3}L)(2) + (\frac{1}{3}L)(4)}{\frac{2}{3}L + \frac{1}{3}L} = 2.66 \text{ in.}^4 \tag{4.74}$$

Comparing Eq. 4.74 with Eq. 4.73 indicates the approximate averaging method shown by Eq. 4.69 is only about 13.6% lower than the true value.

There is a second method of averaging the moment of inertia as a function of the length of each section for the stepped beam shown in Fig. 4.18. Consider the following

$$I''_{\mathrm{av}} = \frac{a+b}{\dfrac{a}{I_1}+\dfrac{b}{I_2}} \tag{4.75}$$

Substituting Eqs. 4.63 and 4.67 into Eq. 4.75

$$I''_{\mathrm{av}} = \frac{\dfrac{L}{2}+\dfrac{L}{2}}{\dfrac{\left(\frac{L}{2}\right)}{2}+\dfrac{\left(\frac{L}{2}\right)}{4}} = 2.66 \text{ in.}^4 \tag{4.76}$$

Comparing this with the correct value of 3.55 in.^{4} shown by Eq. 4.68 indicates the approximate averaging method shown by Eq. 4.75 is about 25% lower than the true value when $a = b$. The resonant frequency would be 13.4% lower.

For some other proportions, when $a = 2b$ (as shown by Eq. 4.71 along with Eq. 4.67), Eq. 4.75 becomes

$$I''_{\mathrm{AV}} = \frac{(\frac{2}{3}L)+(\frac{1}{3}L)}{\dfrac{(\frac{2}{3}L)}{2}+\dfrac{(\frac{1}{3}L)}{4}} = 2.40 \text{ in.}^4 \tag{4.77}$$

Comparing this with the correct value of 3.08 in.^{4} shown by Eq. 4.73 indicates the approximate averaging method shown by Eq. 4.75 is about 22% lower than the true value, when $a = 2b$. The resonant frequency would be about 11.7% lower.

The same type of analysis can be made for a cantilever beam with three different cross-sections that form three steps (Fig. 4.19). The deflection for the stepped beam is

$$\delta = \frac{1}{EI_1}\int_0^a M_1 \frac{\partial M_1}{\partial P}\,dX_1 + \frac{1}{EI_2}\int_0^b M_2 \frac{\partial M_2}{\partial P}\,dX_2 + \frac{1}{EI_3}\int_0^c M_3 \frac{\partial M_3}{\partial P}\,dX_3$$

Substituting the proper bending moments results in the following deflection equation

$$\delta = \frac{Pa^3}{3EI_1} + \frac{P}{EI_2}\left(a^2b + ab^2 + \frac{b^3}{3}\right) + \frac{P}{EI_3}\left(A^2c + Ac^2 + \frac{c^3}{3}\right) \tag{4.78}$$

Let $a = b = c = L/3$; then

$$A = \tfrac{2}{3}L \tag{4.79}$$

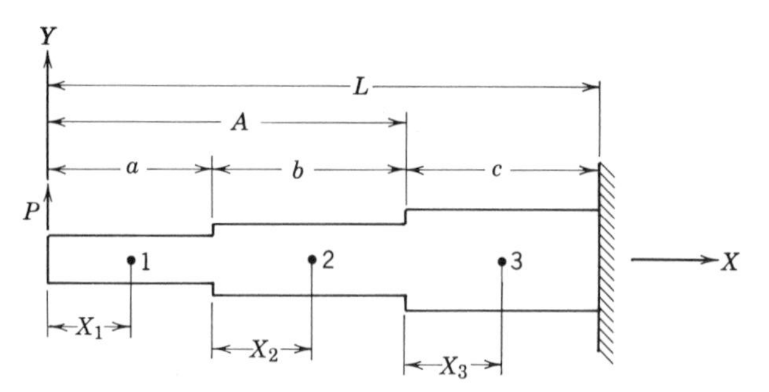

FIGURE 4.19. A cantilever beam with three different cross sections.

$$\delta = \frac{PL^3}{81E}\left(\frac{1}{I_1} + \frac{7}{I_2} + \frac{19}{I_3}\right) \tag{4.80}$$

Set Eq. 4.80 equal to the cantilever beam deflection shown by Eq. 4.65 and solve for the average moment of inertia.

$$I_{\text{av}} = \frac{27 I_1 I_2 I_3}{I_2 I_3 + 7 I_1 I_3 + 19 I_1 I_2}$$

Let $I_1 = 2$ in.^{4}, $I_2 = 4$ in.^{4}, and $I_3 = 6$ in.^{4} (4.81)

$$I_{\text{av}} = \frac{27(48)}{24 + 84 + 152} = 4.98 \text{ in.}^4 \tag{4.82}$$

Compute an approximate average moment of inertia as a function of the length of each section for the stepped beam shown in Fig. 4.19, using the first method shown by Eq. 4.69.

$$I'_{\text{av}} = \frac{aI_1 + bI_2 + cI_3}{a + b + c} \tag{4.83}$$

Substituting Eqs. 4.79 and 4.81 into Eq. 4.83

$$I' = \frac{\left(\frac{L}{3}\right)(2) + \left(\frac{L}{3}\right)(4) + \left(\frac{L}{3}\right)(6)}{\frac{L}{3} + \frac{L}{3} + \frac{L}{3}} = 4.00 \text{ in.}^4 \tag{4.84}$$

Comparing the correct average of 4.98 in.^{4} shown by Eq. 4.82 with the approximate average of 4.00 in.^{4} shown by Eq. 4.84 indicates the approximate averaging method in this case is about 19.7% lower than the true value. The resonant frequency would be about 10.3% lower than the true value.

Using the second method of averaging, the moment of inertia as shown by Eq. 4.75 is

$$I''_{av} = \frac{a+b+c}{\dfrac{a}{I_1}+\dfrac{b}{I_2}+\dfrac{c}{I_3}} \tag{4.85}$$

Substituting Eqs. 4.79 and 4.81 into Eq. 4.85

$$I''_{av} = \frac{\dfrac{L}{3}+\dfrac{L}{3}+\dfrac{L}{3}}{\dfrac{\left(\dfrac{L}{3}\right)}{2}+\dfrac{\left(\dfrac{L}{3}\right)}{4}+\dfrac{\left(\dfrac{L}{3}\right)}{6}} = 3.28 \text{ in.}^4 \tag{4.86}$$

Comparing the correct average of 4.98 in.^{4} from Eq. 4.82 with the value above from the second method of averaging the moment of inertia, shows the second method is about 34% lower than the true value. The resonant frequency would be about 18.9% lower than the true value.

If the relative length of the step sections on the beam are changed so that $a = L/2$, $b = L/3$, and $c = L/6$, then the correct average moment of inertia for the stepped beam is

$$I_{av} = 3.94 \text{ in.}^4$$

Using the first method for approximating the average moment of inertia as shown by Eq. 4.83

$$I'_{av} = 3.33 \text{ in.}^4$$

The error for the first method is 15.5% too low, so the resonant frequency would be about 8% too low.

Using the second method for approximating the average moment of inertia as shown by Eq. 4.85

$$I''_{av} = 2.77 \text{ in.}^4$$

The error for the second method is 29.7% too low, so the resonant frequency would be about 16% too low.

There are, of course, many other methods that can be used to determine an average moment of inertia for a stepped beam. For any of these methods it is generally desirable to choose one that is conservative, so that it will produce a moment of inertia that is slightly lower than the true value. This will then result in a calculated natural frequency that is slightly lower than the true value. This is preferred because most electronic structures have many bolted and riveted joints which tend to reduce the stiffness and, therefore, the natural frequency.

Consider, for example, a third method for averaging the moment of inertia of a stepped beam as follows:

$$I'''_{av} = \frac{aI_1^2 + bI_2^2 + cI_3^2}{aI_1 + bI_2 + cI_3}$$

This third method will produce more accurate results than the two previous methods, shown by Eqs. 4.83 and 4.85, when the step beam segments are approximately equal in length. However, when the beam segment with the smallest moment of inertia is considerably longer than the other beam segments, the average moment of inertia becomes greater than the true value. This can be demonstrated with the three-step beam shown in Fig. 4.19. Let $a = \frac{3}{4}L$, $b = \frac{1}{8}L$, $c = \frac{1}{8}L$, and $I_1 = 2\text{ in.}^4$, $I_2 = 4\text{ in.}^4$, and $I_3 = 6\text{ in.}^4$ The correct average moment of inertia for the stepped beam can be determined from Eqs. 4.65 and 4.78. This value is

$$I_{av} = 3.05\text{ in.}^4$$

The first approximate method for calculating the average moment of inertia for the three-segment beam is shown by Eq. 4.83. This is

$$I'_{av} = 2.66\text{ in.}^4$$

Using the second approximate method, as shown by Eq. 4.85

$$I''_{av} = 2.34\text{ in.}^4$$

Now if the third method is used to approximate the moment of inertia for the same beam, the average value is

$$I'''_{av} = 3.57\text{ in.}^4$$

This is greater than the correct value for the average moment of inertia, which makes the third approximate method undesirable for the conditions indicated.

4.7. Sample Problem

Consider, for example, an electronic box with mounting flanges at each end and a cross-section that has four different moments of inertia along its length (Fig. 4.20). It is required to determine the approximate resonant frequency during vibration in the vertical direction (along the Y axis).

$$I_a = 3.5\text{ in.}^4,\ I_b = 2.5\text{ in.}^4,\ I_c = 4.2\text{ in.}^4,\ I_d = 5.5\text{ in.}^4$$

The approximate average moment of inertia for an equivalent uniform beam can be determined with an equation similar to Eqs. 4.69 and 4.83.

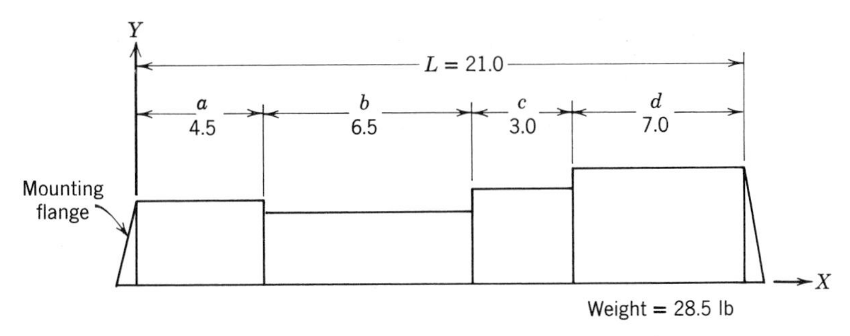

FIGURE 4.20. An electronic chassis with four different cross sections.

$$I'_{av} = \frac{aI_a + bI_b + cI_c + dI_d}{a+b+c+d} \tag{4.87}$$

$$I'_{av} = \frac{(4.5)(3.5) + (6.5)(2.5) + (3.0)(4.2) + (7.0)(5.5)}{4.5+6.5+3.0+7.0}$$

$$I'_{av} = 3.96 \text{ in.}^4$$

Approximating the chassis as a simply supported uniform beam, the natural frequency is (see Eq. 4.33)

$$f_n = \frac{\pi}{2}\left(\frac{EIg}{WL^3}\right)^{1/2} \tag{4.88}$$

where $E = 10.5 \times 10^6$ lb/in.^{2} aluminum

$I = I'_{av} = 3.96$ in.^{4}

$g = 386$ in./sec^2 gravity

$W = wL = 28.5$ lb weight

$L = 21.0$ in. length

$$f_n = \frac{\pi}{2}\left[\frac{(10.5\times 10^6)(3.96)(3.86\times 10^2)}{(28.5)(21.0)^3}\right]^{1/2}$$

$$f_n = 387 \text{ Hz} \tag{4.89}$$

If there is some question concerning the integrity of several structural members that are part of a riveted assembly and there is a fear of some relative motion in this assembly, there could be a reduction in the stiffness of the electronic box which would reduce the natural frequency. A conservative approximation of the natural frequency could then be made by using an approximate average moment of inertia that is known to be relatively low. Under these circumstances it might be desirable to use an expression similar to Eqs. 4.75 and 4.85.

$$I''_{\text{av}} = \frac{a+b+c+d}{\dfrac{a}{I_a}+\dfrac{b}{I_b}+\dfrac{c}{I_c}+\dfrac{d}{I_d}} \tag{4.90}$$

$$I''_{\text{av}} = \frac{4.5+6.5+3.0+7.0}{\dfrac{4.5}{3.5}+\dfrac{6.5}{2.5}+\dfrac{3.0}{4.2}+\dfrac{7.0}{5.5}}$$

$$I''_{\text{av}} = 3.57 \text{ in.}^4$$

Substituting this value into Eq. 4.88

$$f_n = 370 \text{ Hz} \tag{4.91}$$

For this particular electronic box, there is not much difference in the resonant frequency for the two different methods used in determining the approximate average moment of inertia.

4.8. Composite Beams

Composite laminations are often used in electronic boxes because of electrical, thermal, and vibration requirements. Sometimes a printed-circuit board is mounted so close to a metal bulkhead that it is possible for bare metal lead wires on the circuit board to contact the metal bulkhead during vibration. Short circuits can result, and they may damage the electronic equipment. In order to prevent possible short circuits, a thin strip of epoxy fiberglass, 0.005 in. thick, may be cemented to the bulkhead. Epoxy fiberglass is hard, tough, and can resist the pounding of many sharp points during a resonant condition. The bulkhead, of course, then becomes a composite lamination.

Printed-circuit boards often use metal-strip laminations of aluminum or copper to conduct away the heat. In some cases, thin copper strips are bonded to a circuit board and electronic component parts such as resistors, diodes, flatpacks, and transistors are cemented to these copper strips and their electrical lead wires are soldered to the printed-circuit board.

Aluminum plates have been laminated to printed-circuit boards in order to conduct away the heat. In some cases, the components are mounted directly on the aluminum which has clearance holes for the electrical lead wires to permit soldering to the circuit board on the back side of the aluminum plate. In other cases, the components may be mounted on a circuit board only 0.005 in. thick. The heat will then flow right through the epoxy fiberglass circuit board, with a moderate temperature rise, and be conducted away by the aluminum plate.

Sometimes it is necessary to thermally isolate an electronic component, such as a gyro, which must be maintained at a constant temperature to provide the required accuracy. A closely controlled heating system may be used to maintain a constant gyro temperature over a wide external temperature range. Since most structural members are made of metal, and are good heat conductors, the thermal isolation system will usually consist of a metal laminated to a plastic or a ceramic.

Two very common materials that are often used in electronic box structures are aluminum and epoxy fiberglass. The natural frequency of a composite beam of these materials can be determined by considering a combination of the physical properties of both materials.

Consider the case of a simply supported laminated beam that has a uniform load distribution along its length (Fig. 4.21). If the lamination of aluminum and epoxy fiberglass is side by side, as shown in the cross-section view of Fig. 4.22, then an equivalent beam of one material can be used to find the natural frequency.

The width of the epoxy fiberglass lamination can be reduced to make it equivalent in its stiffness to an aluminum beam, using the *EI* stiffness factor. The subscripts a and e refer to aluminum and epoxy respectively.

$$E_a I_a = E_e I_e$$

$$\frac{E_a b_a h_a^3}{12} = \frac{E_e b_e h_e^3}{12}$$

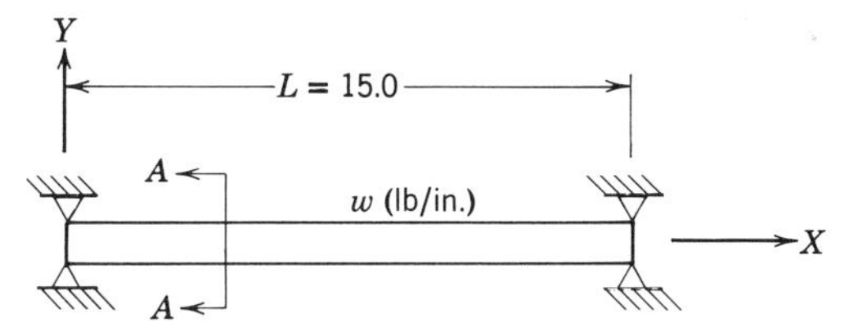

FIGURE 4.21. A simply supported laminated beam.

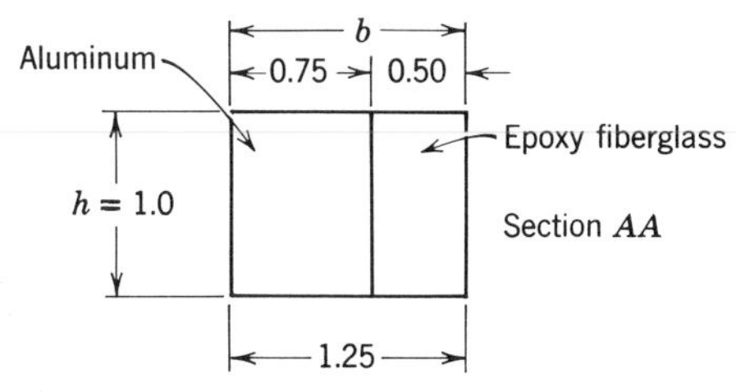

FIGURE 4.22. Cross section of a laminated beam.

Since the height of the aluminum section h_a is the same as the height of the epoxy section h_e, they cancel. The aluminum equivalent of the epoxy section is

$$b_a = b_e \frac{E_e}{E_a} \tag{4.92}$$

where b_a = width of aluminum equivalent to epoxy

$b_e = 0.50$-in. width of epoxy

$E_e = 2.0 \times 10^6$ lb/in.^{2} epoxy fiberglass

$E_a = 10.5 \times 10^6$ lb/in.^{2} aluminum

$$b_a = (0.50)\left(\frac{2.0 \times 10^6}{10.5 \times 10^6}\right) = 0.0952 \text{ in.} \tag{4.93}$$

The equivalent width of a solid aluminum beam can now be determined by adding the increment shown above, to the width of the aluminum section.

$$b_{eq} = 0.75 + 0.0952 = 0.8452 \text{ in. aluminum}$$

The natural frequency of this beam can be determined from the equation of a uniform beam as follows:

$$f_n = \frac{\pi}{2}\left(\frac{EIg}{WL^3}\right)^{1/2} \qquad \text{(see Eq. 4.33)}$$

where $E = 10.5 \times 10^6$ lb/in.^{2} aluminum

$$I = \frac{b_{eq}h^3}{12} = \frac{(0.8452)(1.0)^3}{12} = 0.0704 \text{ in.}^4$$

$g = 386$ in./sec^2 gravity

$L = 15.0$ in. length

$W_a = (15)(1.0)(0.75)(0.10 \text{ lb/in.}^3) =$ 1.125 lb aluminum

$W_e = (15)(1.0)(0.50)(0.065 \text{ lb/in.}^3) =$ $\underline{0.487}$ lb epoxy

W = total beam weight = 1.612 lb

$$f_n = \frac{\pi}{2}\left[\frac{(10.5 \times 10^6)(0.0704)(386)}{(1.612)(15.0)^3}\right]^{1/2}$$

$$f_n = 360 \text{ Hz} \tag{4.94}$$

If the lamination of aluminum and epoxy is stacked in the vertical direction as shown in the cross-section view of Fig. 4.23, then a composite moment of inertia can be computed for the two materials. The same cross-sectional area for the aluminum and epoxy fiberglass is used to keep the same weights.

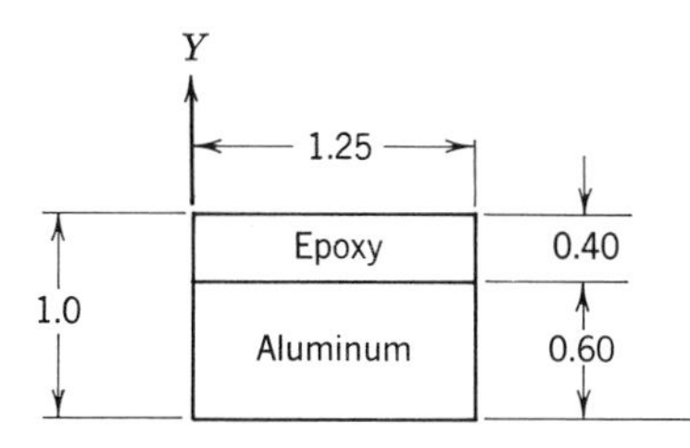

FIGURE 4.23. Cross section of a laminated beam.

The following tabulations can be set up to simplify the composite moment of inertia calculations for the beam stiffness factor EI.

SECTION	AREA	E_0	AE	Y	AEY
Epoxy	0.50	2×10^6	1.00×10^6	0.80	0.80×10^6
Aluminum	0.75	10.5×10^6	7.87×10^6	0.30	2.36×10^6
			8.87×10^6		3.16×10^6

$$\bar{Y} = \frac{\sum AEY}{\sum AE} = \frac{3.16 \times 10^6}{8.87 \times 10^6} = 0.356\text{-in. centroid}$$

SECTION	c	c^2	AEc^2	I_0	$E_0 I_0$
Epoxy	0.444	0.196	0.196×10^6	0.0066	0.0132×10^6
Aluminum	0.056	0.0031	0.024×10^6	0.0225	0.2360×10^6
			0.220×10^6		0.2492×10^6

The composite beam stiffness factor is

$$EI = \sum AEc^2 + \sum E_0 I_0 = 0.220 \times 10^6 + 0.249 \times 10^6$$

$$EI = 0.469 \times 10^6 \text{ lb in.}^2 \tag{4.95}$$

Substituting into Eq. 4.33 for the natural frequency

$$f_n = \frac{\pi}{2}\left[\frac{(0.469 \times 10^6)(386)}{(1.612)(15.0)^3}\right]^{1/2} = 286 \text{ Hz} \tag{4.96}$$

Eq. 4.92 can also be used with the composite cross-section shown in Fig. 4.23. The result will then be a T-section with only one material involved.

The equivalent width of an aluminum section is

$$b_a = 1.25\left(\frac{2.0 \times 10^6}{10.5 \times 10^6}\right) = 0.238 \text{ in.}$$

This results in the aluminum T-section shown in Fig. 4.24.

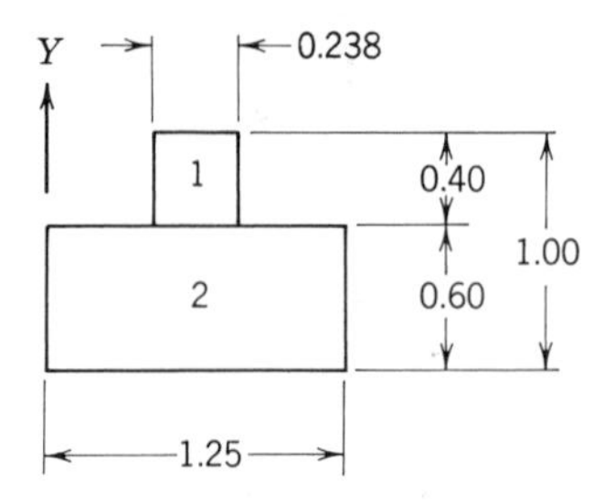

FIGURE 4.24. A "T" section used to simulate a laminated beam.

Table 4.2 can be set up to simplify the calculation for the moment of inertia of the cross-section.

TABLE 4.2

Item	Area	Y	AY	c	c^2	Ac^2	I_0
1	0.095	0.80	0.076	0.444	0.196	0.0186	0.00127
2	0.750	0.30	0.225	0.056	0.0031	0.0023	0.02250
Total	0.845		0.301			0.0209	0.02377

$$\bar{Y} = \frac{\sum AY}{\sum A} = \frac{0.301}{0.845} = 0.356\text{-in. centroid}$$

$$I = \sum Ac^2 + \sum I_0 = 0.0209 + 0.02377 = 0.04467 \text{ in.}^4$$

The beam stiffness factor becomes

$$EI = 10.5 \times 10^6\ (0.04467) = 0.469 \times 10^6 \text{ lb in.}^2 \qquad (4.97)$$

Comparing Eq. 4.97 with Eq. 4.95 shows the beam stiffness factors are exactly the same for both methods.

If the composite lamination of aluminum and epoxy fiberglass is stacked in three layers like a sandwich, as shown in the cross-section view of Fig. 4.25, then the beam stiffness factor, EI, can be calculated more

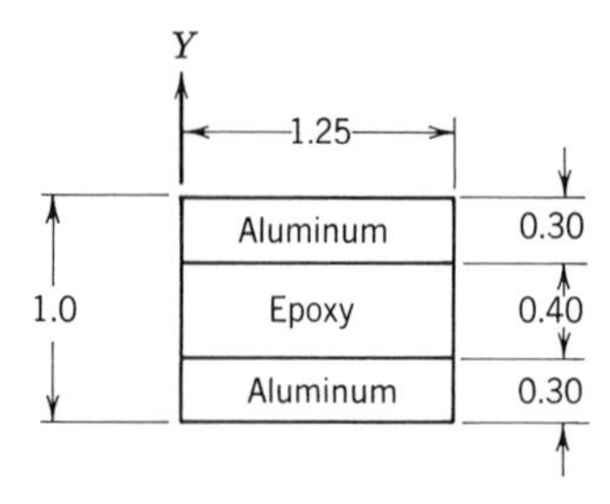

FIGURE 4.25. Cross section of a beam with three laminations.

easily. The same cross-section area for the aluminum and epoxy fiberglass is used to keep the same weights.

For the aluminum section

$$E_a I_a = 10.5 \times 10^6 \left(\frac{1.25}{12}\right)[(1.0)^3 - (0.40)^3] = 1.025 \times 10^6 \text{ lb in.}^2$$

For the epoxy section

$$E_e I_e = 2.0 \times 10^6 \left(\frac{1.25}{12}\right)(0.40)^3 = 0.0133 \times 10^6 \text{ lb in.}^2$$

The composite beam stiffness factor is

$$EI = E_a I_a + E_e I_e = 1.025 \times 10^6 + 0.0133 \times 10^6$$

$$EI = 1.0383 \times 10^6 \text{ lb in.}^2 \tag{4.98}$$

Use Eq. 4.33 to determine the natural frequency of the sandwich cross-section composite beam.

$$f_n = \frac{\pi}{2}\left[\frac{(1.0383 \times 10^6)(386)}{(1.612)(15.0)^3}\right]^{1/2} = 425 \text{ Hz} \tag{4.99}$$

The composite beams shown in this section must be cemented together in order to eliminate relative motion at the interface between the aluminum and the epoxy. If bolted joints are used, relative motion will usually occur between the aluminum and the epoxy, which will reduce the stiffness and the natural frequency. It takes a large number of large bolts to prevent relative motion between two members at high frequencies and high G forces.

CHAPTER 5

ELECTRONIC COMPONENTS, FRAMES, AND RINGS

5.1. Electronic Components Mounted on Circuit Boards

Electronic boxes are being required to occupy less space while providing more functions, so the emphasis has been put on reducing the size of electronic component parts. The development of solid-state electronic parts, such as integrated circuits, sharply reduced the physical size of the parts and permitted more functions to be included in a smaller volume. Even small electronic component parts, however, must be mounted to provide the proper heat removal, accessibility, and structural integrity, depending upon the environment.

Because space is very limited in most aircraft, spacecraft, submarines, and even automobiles, electronic equipment must be supported by many different types of structures that can be adapted to the geometry of the system. Also, the physical size and shape of many electronic component parts themselves may permit them to be analyzed as structural members.

Consider, for example, some typical electronic component parts such as integrated circuits, resistors, capacitors, and diodes mounted on printed-circuit boards by their electrical lead wires. This is a common practice in the electronics industry because it permits low-cost production and easy maintenance. These component parts are generally soldered to the printed-circuit board by dip-soldering, wave-soldering, and even hand-soldering. The printed-circuit boards are usually of the plug-in type which are guided along the edges to permit easy connector engagement. Under these conditions, the edges of the printed-circuit board can usually be considered as simply supported.

A typical installation might be like that shown in Fig. 5.1. During vibration in an axis perpendicular to the plane of the printed-circuit board, the acceleration forces will produce deflections in the circuit board. As the circuit board bends back and forth, bending stresses are developed in the electrical lead wires that fasten the electronic component parts to the circuit board (Fig. 5.2).

Electronic component parts can be mounted on printed-circuit boards in many different ways. The ability of these components to survive a

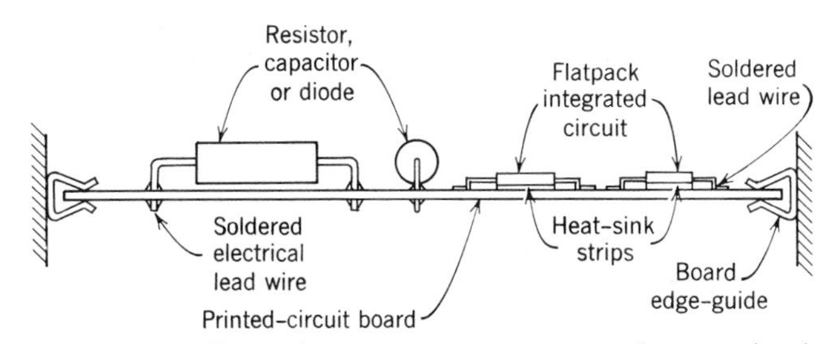

FIGURE 5.1. Electronic component parts mounted on a printed circuit board.

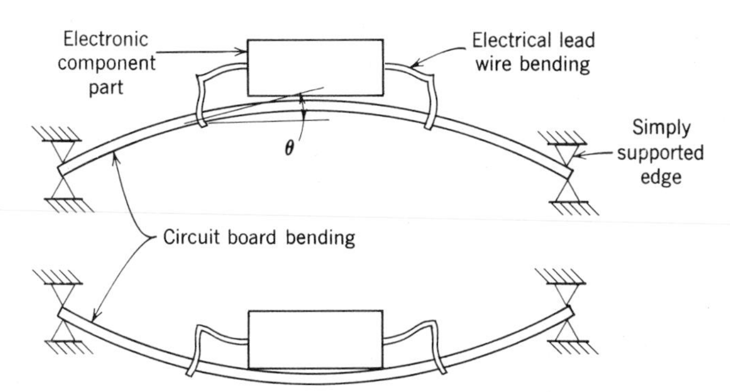

FIGURE 5.2. Bending in the component lead wires on a vibrating circuit board.

severe vibration environment will depend upon many different factors such as component size, resonant frequency of the circuit board, acceleration G forces, method of mounting components, type of strain relief in the electrical lead wires, location of the component, and duration of the vibration environment.

Most component failures in a severe-vibration environment will be due to cracked solder joints, cracked seals, or broken electrical lead wires. These failures are usually due to dynamic stresses that develop because of relative motion between the electronic component body, the electrical lead wires, and the printed-circuit board. This relative motion is generally most severe during resonant conditions that can develop in the electronic component part or in the printed-circuit board.

Resonances may develop in the component part when the body of the component acts as the mass and the electrical lead wires act as the springs. These resonances are usually not too severe if the body of the component is in contact with the printed-circuit board, since this contact will sharply reduce the relative motion of the component. If resonances of this type do develop, it is an easy task to tie or cement the component part to the circuit board.

If resonances develop in the circuit board, large displacements can force the electrical lead wires to bend back and forth as the circuit board vibrates up and down (Fig. 5.2). If the stress levels are high enough and if the number of fatigue cycles is great enough, then fatigue failures can be expected in the solder joints and the electrical lead wires.

The most severe stress condition will be for a component part mounted at the center of a printed-circuit board. For a rectangular board, the most severe condition will occur when the body of the component part is parallel to the short side of the circuit board. This is due to the more rapid change of curvature for the short side of the board compared to the long side of the board, for the same displacement. Tying the electronic component part to the circuit board with lacing cord, at the center of the component body, will generally have very little affect on the relative motion between the component body and the circuit board when the circuit board is in resonance. Cementing the component body to the circuit board or tying both ends of the component part to the circuit board will greatly improve the fatigue life in the solder joints and in the electrical lead wires.

Conformal coatings are quite often used on printed-circuit boards to protect the circuitry from moisture. Although most conformal coatings are quite thin, from about 0.0005 in. to about 0.010 in. thick, they will still act as a cement so that they will also improve the fatigue life of the electronic component parts.

An examination of the physical proportions of a typical resistor, capacitor, diode, or integrated circuit shows that the component body is much larger than the component electrical lead wire. The stiffness of the component body is therefore much greater than the stiffness of the lead wire. Because of this, a good approximation of the mounting geometry can be obtained by assuming that all of the deflection between the component body and the circuit board is due to bending in the electrical lead wires. Under these conditions, the loads, deflections and stresses in the electri-

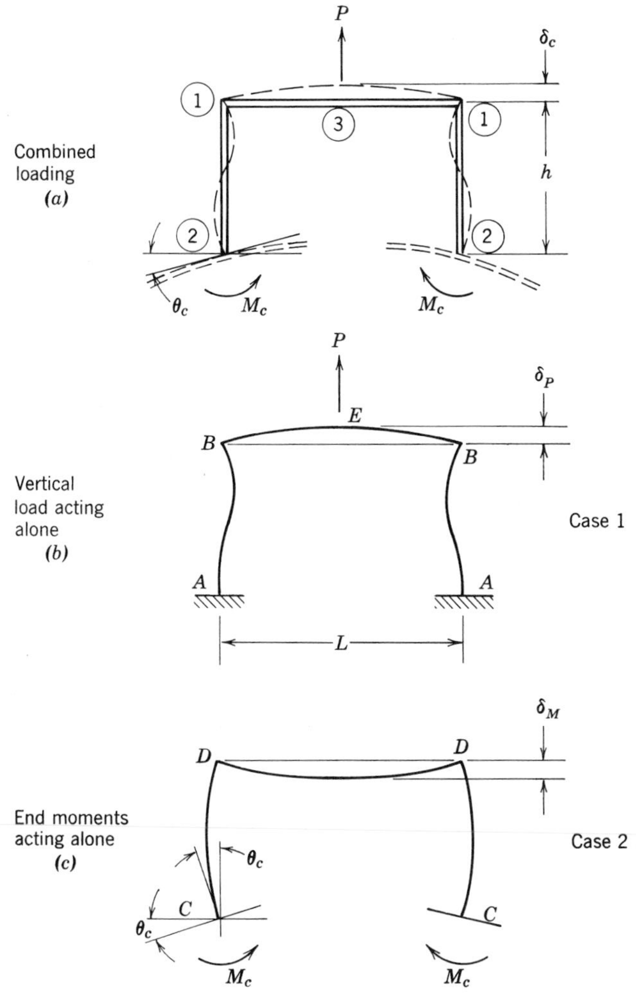

FIGURE 5.3. Bending modes in the electrical lead wires of components mounted on circuit boards.

cal lead wires can be approximated by considering the lead wires as rectangular bents.

The most severe stress conditions will usually occur at the fundamental resonant mode of the circuit board, because displacements are usually the greatest then. The points where the electrical lead wires are soldered to the circuit board will tend to remain perpendicular to the circuit board as it bends back and forth. The points where the electrical lead wires are joined to the body of the component will also tend to remain fixed, because the component body is so much stiffer than the electrical lead wires. This results in a complex bending mode in the electrical lead wires that can be approximated as a combination of two separate loading patterns on rectangular bents, as shown in Fig. 5.3*a*.

CASE 1

A vertical load acting alone, due to the vertical displacement (Fig. 5.3*b*).

CASE 2

End moments acting alone, due to the angular rotation of the circuit board (Fig. 5.3*c*.)

5.2. Bent with a Vertical Load and Fixed Ends

CASE 1

Although the electrical lead wires will usually have the same cross-section, consider a more general case where the horizontal member has a different cross-section than the two vertical members, as shown in Fig. 5.4.

Superposition can be used to determine the various loads, moments, and deflections. The method of analysis is to assume that deflections are

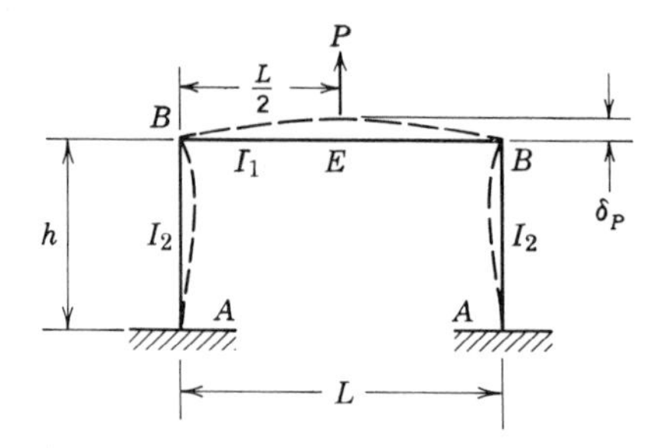

FIGURE 5.4. Bent with a vertical load and fixed ends.

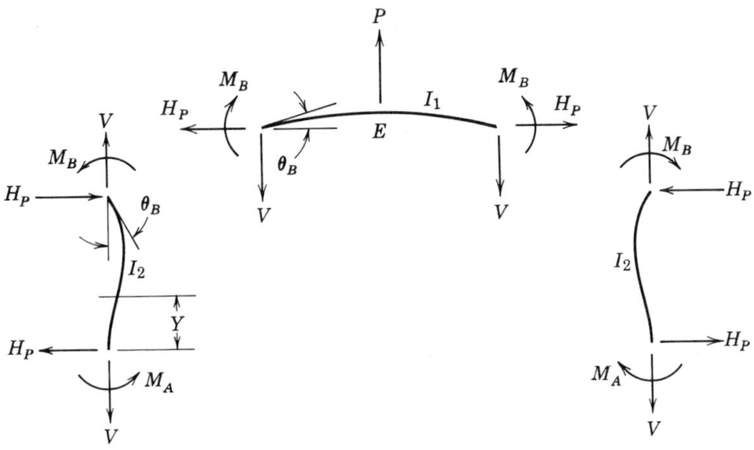

FIGURE 5.5. Free body force diagram of a bent with fixed ends and a concentrated load.

small so right angles remain right angles after rotation and stresses do not exceed the elastic limit of the material.

A free-body force diagram of the rectangular bent is shown in Fig. 5.5.

Consider one vertical leg AB. Superposition can be used to determine the angular rotation. The effect of the vertical force V is small (Fig. 5.6).

The angles at the bottom of the vertical leg can be determined from a standard handbook.

$$\theta_1 = \frac{M_B h}{6EI_2} \quad \text{and} \quad \theta_2 = \frac{M_A h}{3EI_2} \tag{5.1}$$

$$\theta_A = \theta_1 - \theta_2 = 0 \tag{5.2}$$

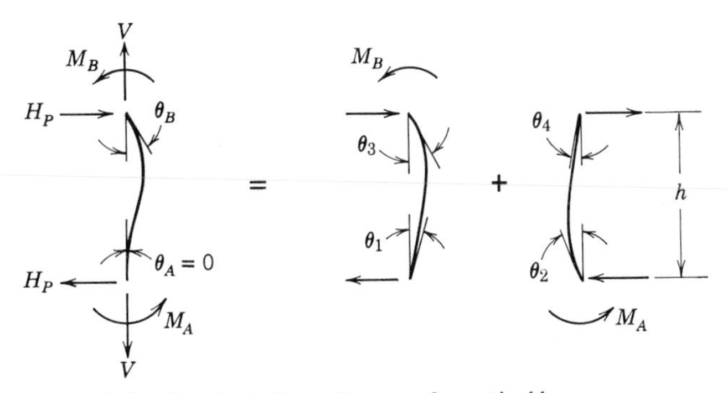

FIGURE 5.6. Free body force diagram of a vertical leg.

Substituting Eq. 5.1 into 5.2

$$\frac{M_B h}{6EI_2} - \frac{M_A h}{3EI_2} = 0$$

$$M_A = \tfrac{1}{2} M_B \tag{5.3}$$

The angles at the top of the vertical leg are

$$\theta_3 = \frac{M_B h}{3EI_2} \quad \text{and} \quad \theta_4 = \frac{M_A h}{6EI_2} \tag{5.4}$$

$$\theta_B = \theta_3 - \theta_4 \tag{5.5}$$

Substituting Eqs. 5.3 and 5.4 into Eq. 5.5

$$\theta_B = \frac{M_B h}{3EI_2} - \frac{(\frac{1}{2} M_B h)}{6EI_2} = \frac{M_B h}{4EI_2} \tag{5.6}$$

Since the angle is small, the change in the angle due to the action of the vertical force V is small and therefore can be ignored with very little effect on the accuracy.

Superposition can be used to determine the angular rotation on the horizontal leg. Angular changes due to H_P will also be small (Fig. 5.7).

Angles can be determined from a standard handbook.

$$\theta_5 = \frac{PL^2}{16EI_1}; \quad \theta_6 = \frac{M_B L}{3EI_1}; \quad \theta_7 = \frac{M_B L}{6EI_1} \tag{5.7}$$

From Figs. 5.5 and 5.7

$$\theta_B = \theta_5 - (\theta_6 + \theta_7) \tag{5.8}$$

Substituting Eqs. 5.6 and 5.7 into Eq. 5.8

$$\frac{M_B h}{4EI_2} = \frac{PL^2}{16EI_1} - \left(\frac{M_B L}{3EI_1} + \frac{M_B L}{6EI_1}\right)$$

$$M_B = \frac{PL^2}{\dfrac{4hI_1}{I_2} + 8L} \tag{5.9}$$

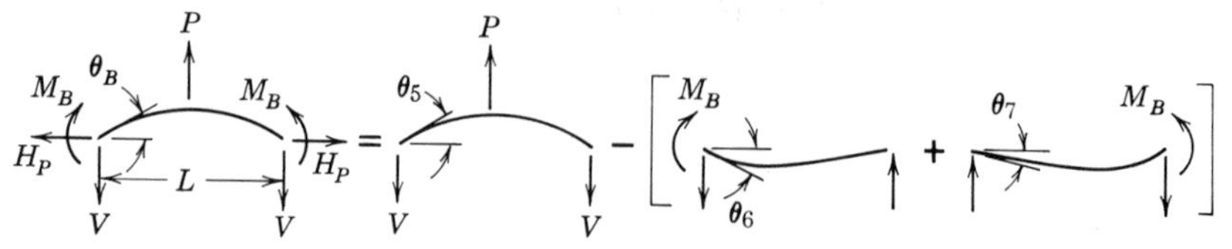

FIGURE 5.7. Free body force diagram of a horizontal leg.

Let

$$K = \frac{hI_1}{LI_2} \tag{5.10}$$

Substituting Eq. 5.10 into 5.9

$$M_B = \frac{PL}{4K+8} \tag{5.11}$$

From Eq. 5.3

$$M_A = \frac{PL}{8K+16} \tag{5.12}$$

The horizontal force H_p can be determined by taking moments about point A on the vertical leg, using Fig. 5.6

$$\sum M_A = 0$$

$$M_A + M_B - H_p h = 0 \tag{5.13}$$

Substituting Eqs. 5.11 and 5.12 into Eq. 5.13 and solving for H_p

$$H_p = \frac{3PL}{2h(4K+8)} \tag{5.14}$$

The vertical force V can be determined by considering forces in the vertical direction

$$V = \frac{P}{2} \tag{5.15}$$

The maximum deflection under the load can be determined by superposition (Fig. 5.8).

Using standard handbook equations

$$\delta_1 = \frac{PL^3}{48EI_1}; \quad \delta_2 = \delta_3 = \frac{M_B L^2}{16EI_1} \tag{5.16}$$

From Figs. 5.4 and 5.8

$$\delta_P = \delta_1 - (\delta_2 + \delta_3) \tag{5.17}$$

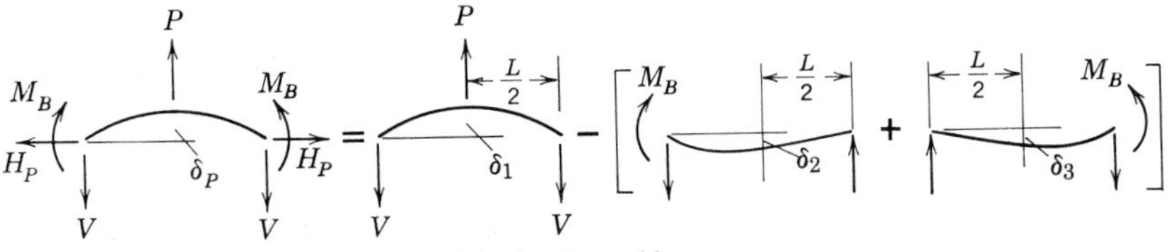

FIGURE 5.8. Deflection diagram of the horizontal leg.

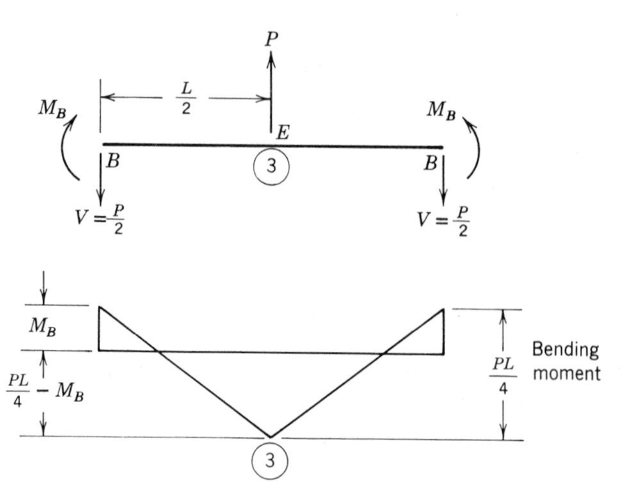

FIGURE 5.9. Bending moment diagram of the horizontal leg.

Substituting Eqs. 5.11 and 5.16 into Eq. 5.17

$$\delta_P = \frac{PL^3}{48EI_1} - \frac{PL}{4K+8}\left(\frac{L^2}{8EI_1}\right)$$

$$\delta_P = \frac{PL^3}{48EI_1}\left(1 - \frac{3}{2K+4}\right) \tag{5.18}$$

The bending moment at the center of the horizontal leg, under the vertical load, can be determined from the bending moment diagram (Fig. 5.9).

Taking moments about point E

$$M_E = \frac{P}{2}\left(\frac{L}{2}\right) - M_B \tag{5.19}$$

Substituting Eq. 5.11 into 5.19

$$M_E = \frac{PL}{4K+8}(K+1) \tag{5.20}$$

5.3. Bent with Hinged Ends and End Moments

CASE 2

Superposition can be used to determine the loads, moments and deflections. (Fig. 5.10). Again assume that deflections are small and stresses do

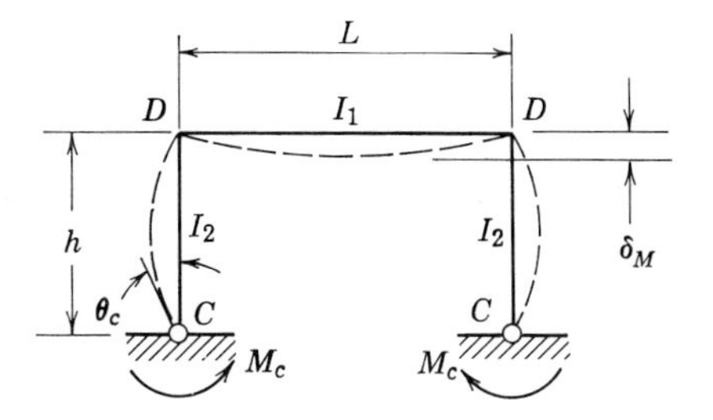

FIGURE 5.10. Bent with hinged ends and end moments.

not exceed the elastic limit of the material. A free-body diagram of the rectangular bent appears as shown in Fig. 5.11.

Consider the vertical leg CD. Superposition can be used to determine the angular rotation (Fig. 5.12).

Using standard handbook equations

$$\theta_1 = \frac{M_D h}{6EI_2}; \quad \theta_2 = \frac{M_c h}{3EI_2} \tag{5.21}$$

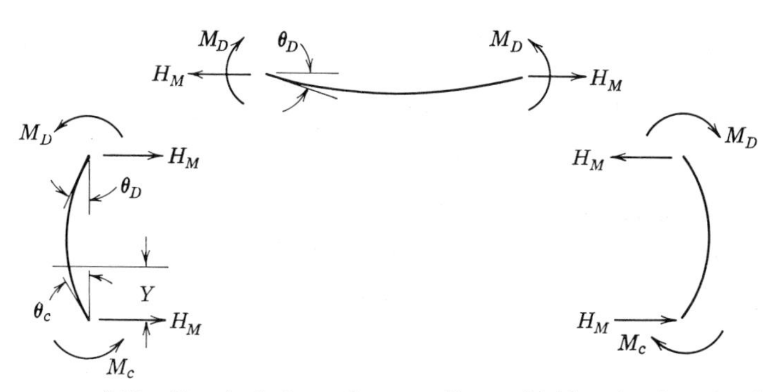

FIGURE 5.11. Free body force diagram of bent with hinged ends and end moments.

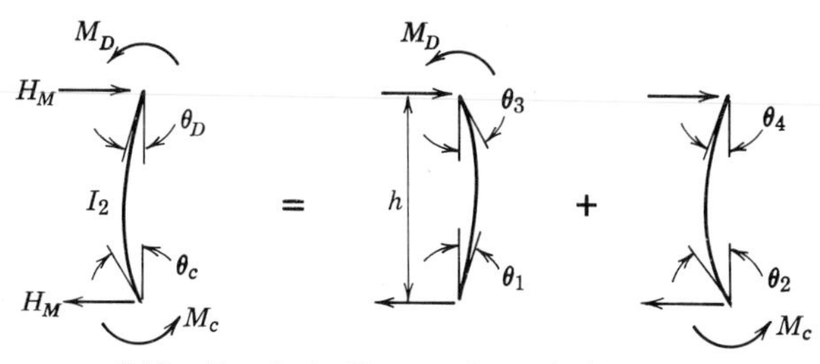

FIGURE 5.12. Free body diagram of a vertical leg.

$$\theta_c = \theta_2 - \theta_1 \tag{5.22}$$

Substituting Eq. 5.21 into 5.22

$$\theta_c = \frac{M_c h}{3EI_2} - \frac{M_D h}{6EI_2} \tag{5.23}$$

Solving for M_D from above equation

$$M_D = 2M_c - \frac{6EI_2}{h}\theta_c \tag{5.24}$$

Using standard handbook equations

$$\theta_3 = \frac{M_D h}{3EI_2}; \quad \theta_4 = \frac{M_c h}{6EI_2} \tag{5.25}$$

$$\theta_D = \theta_4 - \theta_3 \tag{5.26}$$

Substituting Eqs. 5.25 and 5.24 into Eq. 5.26

$$\theta_D = \frac{M_c h}{6EI_2} - \frac{h}{3EI_2}\left[2M_c - \frac{6EI_2}{h}\theta_c\right] = 2\theta_c - \frac{M_c h}{2EI_2} \tag{5.27}$$

Consider the horizontal leg. Superposition can be used to determine the angular rotation (Fig. 5.13).

Using standard handbook equations

$$\theta_5 = \frac{M_D L}{3EI_1}; \quad \theta_6 = \frac{M_D L}{6EI_1} \tag{5.28}$$

$$\theta_D = \theta_5 + \theta_6 \tag{5.29}$$

Substituting Eqs. 5.27 and 5.28 into Eq. 5.29 and solving for M_D as follows

$$2\theta_c - \frac{M_c h}{2EI_2} = \frac{M_D L}{3EI_1} + \frac{M_D L}{6EI_1}$$

$$M_D = \frac{2EI_1}{L}\left(2\theta_c - \frac{M_c h}{2EI_2}\right)$$

$$M_D = \frac{4EI_1\theta_c}{L} - KM_c \tag{5.30}$$

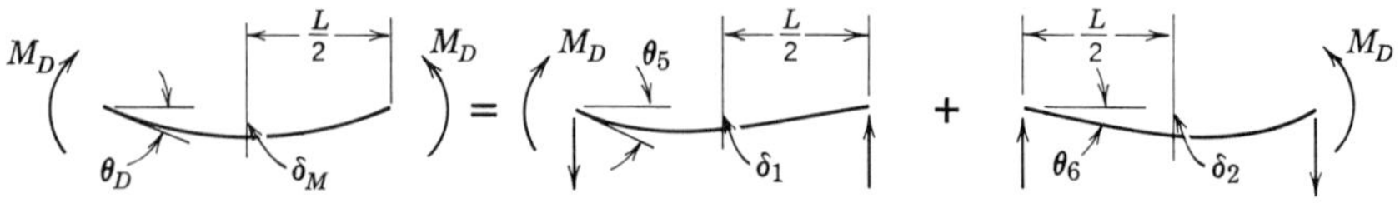

FIGURE 5.13. Free body deflection of the horizontal leg.

Substituting Eq. 5.30 into Eq. 5.23

$$\theta_c = \frac{M_c h}{3EI_2} - \frac{h}{6EI_2}\left(\frac{4EI_1\theta_c}{L} - KM_c\right)$$

$$\theta_c = \frac{M_c h}{3EI_2}\left(1 + \frac{K}{2}\right) - \frac{4\theta_c}{6}\left(\frac{hI_1}{LI_2}\right)$$

$$M_c = \frac{\theta_c\left(1 + \frac{2K}{3}\right)3EI_2}{h\left(1 + \frac{K}{2}\right)} = \frac{2\theta_c(3 + 2K)EI_2}{h(2 + K)} \tag{5.31}$$

The maximum vertical deflection of the horizontal leg, due to the end moments, can be determined with the use of superposition and a standard handbook for deflection equations. The deflections as shown in Fig. 5.13 are as follows:

$$\delta_1 = \delta_2 = \frac{M_D L^2}{16EI_1}$$

$$\delta_M = 2\delta_1 = \frac{M_D L^2}{8EI_1} \tag{5.32}$$

Substituting Eq. 5.30 into Eq. 5.32

$$\delta_M = \frac{\theta_c L}{2} - \frac{hLM_c}{8EI_2}$$

Substituting Eq. 5.31 into above

$$\delta_M = \frac{\theta_c L}{2} - \frac{hL}{8EI_2}\left(\frac{2\theta_c(3 + 2K)EI_2}{h(2 + K)}\right)$$

$$\delta_M = \frac{\theta_c L}{4}\left[2 - \left(\frac{3 + 2K}{2 + K}\right)\right] \tag{5.33}$$

The horizontal force H_M can be determined from Fig. 5.12 by taking moments about point c:

$$M_c + M_D = H_M h$$

Substituting Eq. 5.30 into the above equation

$$M_c + \frac{4EI_1\theta_c}{L} - KM_c = H_M h$$

$$H_M = \frac{M_c}{h}(1 - K) + \frac{4EI_1\theta_c}{hL} \tag{5.34}$$

5.4. Combining Stresses and Deflections due to Case 1 and Case 2

An examination of Figs. 5.3, 5.5, and 5.11 shows the maximum deflections in the top horizontal legs are in the opposite directions for case 1 and case 2. The combined deflection at the top leg must therefore be the difference between these two deflections.

$$\delta_c = \delta_P - \delta_M$$

Substituting Eqs. 5.18 and 5.33 into the above equation

$$\delta_c = \frac{PL^3}{48EI_1}\left(1 - \frac{3}{2K+4}\right) - \frac{\theta_c L}{4}\left[2 - \left(\frac{3+2K}{2+K}\right)\right] \tag{5.35}$$

Also from Figs. 5.3, 5.5, and 5.11, the moments at the intersection of the vertical and horizontal legs, M_B for case 1 and M_D for case 2, act in the same direction. The total moment at point 1 is

$$M_1 = M_B + M_D$$

Substituting Eqs. 5.11 and 5.30 into the above equation

$$M_1 = \frac{PL}{4K+8} + \left(\frac{4EI_1\theta_c}{L} - KM_c\right) \tag{5.36}$$

The moments at the ends of the bents, M_A for case 1 and M_c for case 2, also act in the same direction. The total moment at point 2 in Fig. 5.3*a* is

$$M_2 = M_A + M_c$$

Substituting Eqs. 5.12 and 5.31 into the above equation

$$M_2 = \frac{PL}{8K+16} + \frac{2EI_2\theta_c}{h}\left(\frac{3+2K}{2+K}\right) \tag{5.37}$$

Also, the horizontal force H in the horizontal leg for both cases 1 and 2 acts in the same direction. The total force is

$$H_T = H_P + H_M$$

Substituting Eqs. 5.14 and 5.34 into the above

$$H_T = \frac{3PL}{2h(4K+8)} + \frac{M_c}{h}(1-K) + \frac{4EI_1\theta_c}{hL} \tag{5.38}$$

An examination of Figs. 5.9 and 5.11 shows the bending moments at the center of the top leg are in opposite directions for case 1 and case 2. The total bending moment is

$$M_3 = M_E - M_D$$

Substituting Eqs. 5.20 and 5.30 into the above

$$M_3 = \frac{PL}{4K+8}(K+1) - \left(\frac{4EI_1\theta_c}{L} - KM_c\right) \tag{5.39}$$

5.5. Sample Problem

Consider a printed-circuit board with flatpack integrated circuits mounted on one side of the circuit board, shown in Fig. 5.14. This printed-circuit board is mounted in an electronic box that will be vibrated in a direction perpendicular to the plane of the circuit board. The natural frequency for the circuit board in the problem, as shown by test data, is as follows:

$$f_n = 160 \text{ Hz} \tag{5.40}$$

This particular circuit board is mounted in a rack that supports the three sides of the board. Vibration tests show that the plug-in connector acts like a simply supported side so that the circuit board can be approximated as a flat plate with four simply supported edges.

Vibration tests on this circuit board also show that the transmissibility at resonance happens to be relatively close to the relation (see Chapter 6)

$$Q = (f_n)^{1/2} \tag{5.41}$$

The transmissibilities of many different types of circuit boards can be approximated with this relation. However, since there are a wide variety

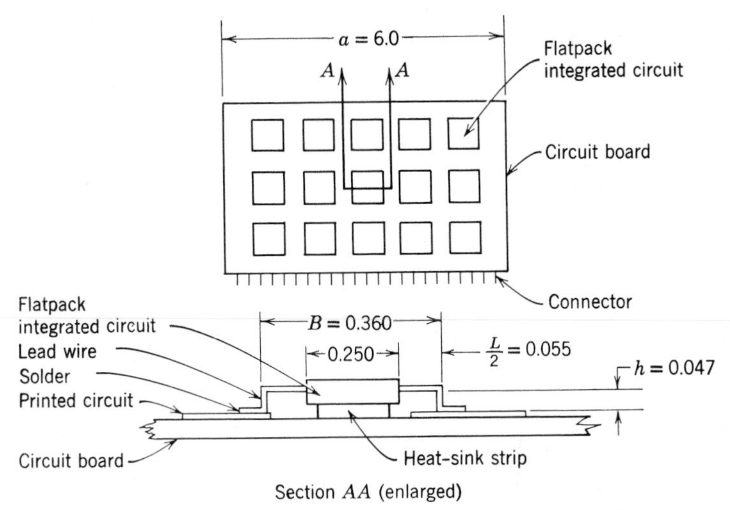

FIGURE 5.14. Printed circuit board with flat pack integrated circuits.

of circuit boards with a great variation in the mechanical joints that dissipate energy, there will also be a great variation in the resulting transmissibilities. If there is no test data available on a new group of circuit boards, for example, Eq. 5.41 is a good place to start when a dynamic analysis must be made.

Substituting Eq. 5.40 into Eq. 5.41

$$Q = (160)^{1/2} = 12.6 \tag{5.42}$$

The maximum displacement at the center of the circuit board can be approximated by assuming that the circuit board acts like a single-degree-of-freedom system during resonance. The single-amplitude displacement can then be determined from Chapter 2, Eq. 2.42.

$$\delta = \frac{9.8G}{f_n^2} = \frac{9.8G_{\text{in}}Q}{f_n^2} \tag{5.43}$$

The electronic box will be subjected to a 5.0-G-peak sinusoidal vibration input. The box is very rigid so that it will not amplify the input to the circuit board at a frequency of 160 Hz.

The displacement at the center of the circuit board is

$$\delta = \frac{9.8(5.0)(12.6)}{(160)^2} = 0.0242 \text{ in.} \tag{5.44}$$

Since the circuit board is simply supported on all four sides, the deflection curve at its fundamental resonant mode can be approximated as a sine wave. As the board vibrates up and down, the electrical lead wires on the integrated circuits will be forced to bend back and forth as shown in Fig. 5.15, when the flatpack is mounted at the center of the board.

The relative deflection of the electrical lead wires on the flatpack can be approximated from the sine curve as follows:

$$\delta_c = \delta\left(1 - \sin\frac{\pi X}{a}\right) \tag{5.45}$$

where $\delta = 0.0242$ in.

$X = 2.82$ in.

$a = 6.0$ in. circuit board length

$$\delta_c = 0.0242\left[1 - \sin\frac{\pi(2.82)}{6.0}\right]$$

$$\delta_c = 0.000121 \text{ in.} \tag{5.46}$$

This represents the total deflection due to the combined loading, as shown in Fig. 5.3a, when only the lead wires are considered.

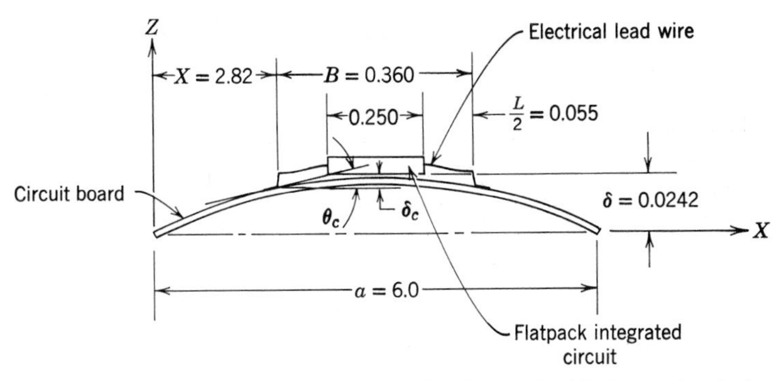

FIGURE 5.15. Deflection of a printed circuit board with flat packs during vibration.

The slope θ_c of the bending circuit board, at the lead wires, can be determined from the first derivative of the deflection curve, which was approximated as a sine curve.

$$Z = \delta \sin \frac{\pi X}{a} \tag{5.47}$$

$$\theta_c = \frac{dZ}{dX} = \delta \frac{\pi}{a} \cos \frac{\pi X}{a} \tag{5.48}$$

where $\delta = 0.0242$ in
$a = 6.0$ in.
$X = 2.82$ in.

$$\theta_c = \frac{(0.0242)\pi}{6.0} \cos \frac{\pi(2.82)}{6.0}$$

$$\theta_c = 0.00121 \text{ rad} \tag{5.49}$$

Since the stiffness of the flatpack body is so much greater than the stiffness of the electrical lead wires, all of the deflection will be due to the deflection in the wires. The wires will then tend to deform, as shown in Fig. 5.16, where the physical dimensions of the system are shown.

The relative displacement of the electrical lead wires, for the combined loading condition shown in Fig. 5.16, can be used to determine the equivalent dynamic load, P, acting on the wires, with the use of Eq. 5.35,

$$\delta_c = \frac{PL^3}{48EI_1}\left(1 - \frac{3}{2K+4}\right) - \frac{\theta_c L}{4}\left[2 - \left(\frac{3+2K}{2+K}\right)\right]$$

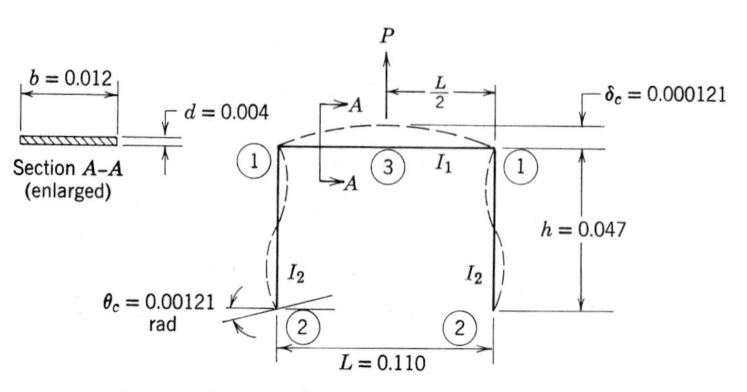

FIGURE 5.16. Bending mode in flat pack electrical lead wires.

where $\delta_c = 0.000121$ in.

$$L = 0.110 \text{ in.}$$

$$E = 30 \times 10^6 \text{ lb/in.}^2 \text{ nickel wire}$$

$$I_1 = I_2 = \frac{bd^3}{12} = \frac{(0.012)(0.004)^3}{12} = 64 \times 10^{-12} \text{ in.}^4$$

$$h = 0.047 \text{ in.}$$

$$K = \frac{h}{L} = \frac{0.047}{0.110} = 0.427 \text{ dimensionless}$$

$$\theta_c = 0.00121 \text{ rad}$$

$$EI = (30 \times 10^6)\ (64 \times 10^{-12}) = 0.192 \times 10^{-2} \text{ lb/in.}^2$$

Substituting into Eq. 5.35

$$1.21 \times 10^{-4} = \frac{P(0.110)^3}{48(0.192 \times 10^{-2})}\left(1 - \frac{3}{4.854}\right) - \frac{1.21 \times 10^{-3}(0.110)}{4}\left(2 - \frac{3.854}{2.427}\right)$$

$$0.121 \times 10^{-3} = 5.46 \times 10^{-3} P - 0.0138 \times 10^{-3}$$

$$P = 0.0247 \text{ lb} \tag{5.50}$$

The bending moment, M_c, developed in the electrical lead wires due to the bending of the circuit board, as shown in Fig. 5.3c, can be determined from Eq. 5.31.

$$M_c = \frac{2\theta_c EI_2}{h}\left(\frac{3 + 2K}{2 + K}\right)$$

where

$$\theta_c = 0.00121 \text{ rad (see Eq. 5.49)}$$

$$E = 30 \times 10^6 \text{ lb/in.}^2 \text{ nickel wire}$$

$$I_1 = I_2 = 64 \times 10^{-12} \text{ in.}^4$$

$$K = \frac{h}{L} = \frac{0.047}{0.110} = 0.427$$

Substituting into Eq. 5.31

$$M_c = \frac{2(1.21 \times 10^{-3})(30 \times 10^6)(64 \times 10^{-12})}{0.047}\left(\frac{3.854}{2.427}\right)$$

$$M_c = 0.157 \times 10^{-3} \text{ lb in.} \tag{5.51}$$

The total bending moment in the electrical lead wire at point 2, shown in Fig. 5.16, can be determined from Eq. 5.37.

$$M_2 = \frac{PL}{8K+16} + \frac{2\theta_c E I_2}{h}\left(\frac{3+2K}{2+K}\right)$$

Equation 5.51 represents the second half of the above equation, thus

$$M_2 = \frac{(0.0247)(0.110)}{8(0.427)+16} + 0.157 \times 10^{-3}$$

$$M_2 = 0.140 \times 10^{-3} + 0.157 \times 10^{-3} = 0.297 \times 10^{-3} \text{ lb in.} \tag{5.52}$$

The total bending moment in the electrical lead wires at point 1, shown in Fig. 5.16, can be determined from Eq. 5.36.

$$M_1 = \frac{PL}{4K+8} + \left(\frac{4EI_1\theta_c}{L} - KM_c\right)$$

$$M_1 = \frac{(0.0247)(0.110)}{9.708} + \left[\frac{4(0.192 \times 10^{-2})(1.21 \times 10^{-3})}{0.110} - 0.427(0.157 \times 10^{-3})\right]$$

$$M_1 = 0.297 \times 10^{-3} \text{ lb in.} \tag{5.53}$$

The total bending moment in the electrical lead wires at point 3, shown in Fig. 5.16, can be determined from Eq. 5.39.

$$M_3 = \frac{PL}{4K+8}(K+1) - \left(\frac{4EI_1\theta_c}{L} - KM_c\right)$$

$$M_3 = \frac{(0.0247)(0.110)(1.427)}{9.708} - \left[\frac{4(0.192 \times 10^{-2})(1.21 \times 10^{-3})}{0.110} - 0.427(0.157 \times 10^{-3})\right]$$

$$M_3 = 0.382 \times 10^{-3} \text{ lb in.} \tag{5.54}$$

The total horizontal force, H, in the electrical lead wires, as shown in Figs. 5.5 and 5.11, can be determined from Eq. 5.38.

$$H_T = \frac{3PL}{2h(4K+8)} + \frac{M_c}{h}(1-K) + \frac{4EI_1\theta_c}{hL}$$

$$H_T = \frac{3(0.0247)(0.110)}{2(0.047)(9.708)} + \frac{0.157 \times 10^{-3}}{0.047}(0.573) + \frac{4(0.192 \times 10^{-2})(1.21 \times 10^{-3})}{(0.047)(0.110)}$$

$$H_T = 0.0126 \text{ lb} \tag{5.55}$$

The vertical load, V, in the electrical lead wires, as shown in Fig. 5.5, can be determined from Eq. 5.15.

$$V = \frac{P}{2} = \frac{0.0247}{2} = 0.0123 \text{ lb} \tag{5.56}$$

The maximum bending moment in the electrical lead wires was at point 3 as shown by Eq. 5.54. The bending stress in the wire at this point can be determined from the standard bending stress equation (neglecting stress concentration factors)

$$S_b = \frac{M_3 c}{I} \tag{5.57}$$

where $M_3 = 0.382 \times 10^{-3}$ lb in.

$$c = \frac{d}{2} = \frac{0.004}{2} = 0.002 \text{ in.}$$

$I = 64 \times 10^{-12}$ in.^{4} (see Fig. 5.16)

$$S_b = \frac{(0.382 \times 10^{-3})(0.002)}{64 \times 10^{-12}} = 11{,}900 \text{ lb/in.}^2 \tag{5.58}$$

The direct tensile stress in the lead wire at point 3 can be determined from the horizontal load H_T as shown by Eq. (5.55), and the standard tensile stress equation

$$S_t = \frac{H_T}{A} \tag{5.59}$$

where $A = (0.012)(0.004) = 48 \times 10^{-6}$ in.^{2} area of cross-section

$$S_t = \frac{126 \times 10^{-4}}{48 \times 10^{-6}} = 263 \text{ lb/in.}^2 \text{ (small)} \tag{5.60}$$

An examination of the S–N fatigue curve for nickel wire, as shown in Chapter 10, Fig. 10.10a, indicates that an infinite fatigue life can be expected for bending stress levels below about

$$S_b = 42{,}500 \text{ lb/in.}^2 \tag{5.61}$$

Since the bending stress in the flatpack integrated-circuit lead wire is only 11,900 psi, as shown by Eq. 5.58, no failures are expected.

5.6. Bent with a Lateral Load–Fixed Ends

When the vibration direction is in the same plane as the printed circuit board and parallel to the axis of the resistor, the inertia force on the resistor can result in another type of bending in the component lead wires, as shown in Fig. 5.17.

Since the body of the resistor is much stiffer than the electrical lead wires, almost all of the bending deflection will occur in the lead wires.

If it is assumed that deflections are small and stresses do not exceed the elastic limit, then a good approximation of the system can be obtained by considering only the deflection of the electrical lead wires. The inertia force acting on the resistor body can then be treated as a concentrated load.

There must be enough clearance between the body of the resistor and the printed-circuit board, so that the resistor does not contact the board during vibration, in order to make the following analysis valid.

This type of mounting is not recommended for electronic component parts subjected to vibration. However, since some manufacturers still mount components this way, it is desirable to know the characteristics (Fig. 5.18).

Superposition can be used to determine the loads, moments, and deflections. Assume that deflections are small and that stresses do not exceed the elastic limit. A free-body force diagram of the rectangular bent will appear as shown in Fig. 5.19.

Consider the vertical leg AB. Using superposition, the concentrated load and the moment at the top of the vertical leg can be considered separately, as shown in Fig. 5.20.

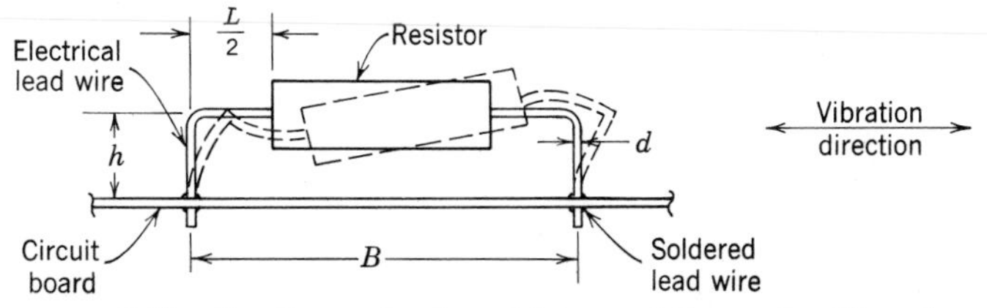

FIGURE 5.17. Bending mode of a resistor mounted on a circuit board.

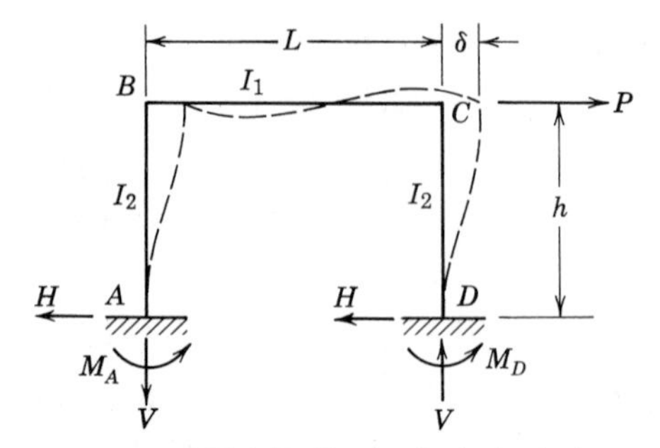

FIGURE 5.18. Deflection in the lead wires on a vibrating resistor.

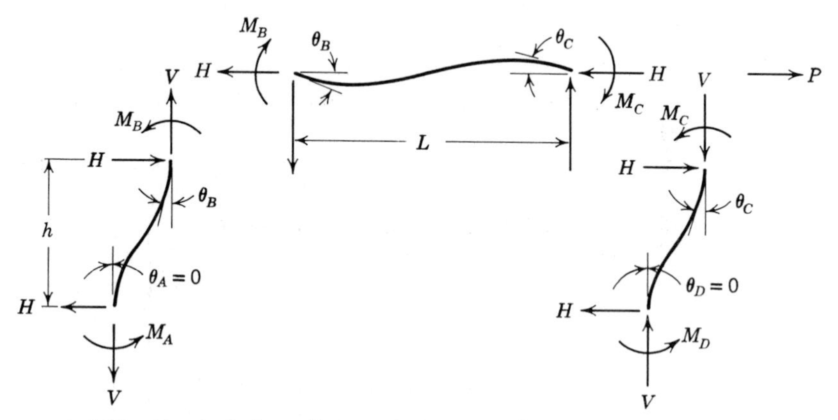

FIGURE 5.19. Free body force diagram of a bent in the lateral direction.

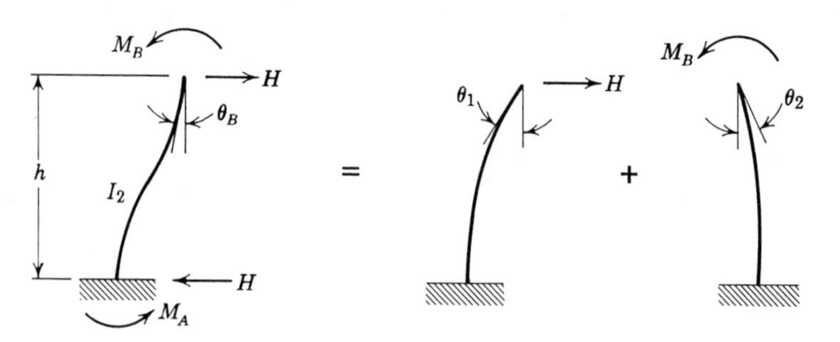

FIGURE 5.20. Free body force diagram of the vertical leg.

Using standard handbook equations

$$\theta_1 = \frac{Hh^2}{2EI_2} \quad \text{and} \quad \theta_2 = \frac{M_B h}{EI_2}$$

$$\theta_B = \theta_1 - \theta_2$$

Substituting into the above equation

$$\theta_B = \frac{Hh^2}{2EI_2} - \frac{M_B h}{EI_2} \tag{5.62}$$

Consider the horizontal leg BC. Superposition can be used to determine the angular rotation at point B (Fig. 5.21).

Using standard handbook equations

$$\theta_3 = \frac{M_B L}{3EI_1} \quad \text{and} \quad \theta_4 = \frac{M_c L}{6EI_1}$$

$$\theta_B = \theta_3 - \theta_4$$

Substitute into the above equation and note that $M_B = M_c$ because of symmetry.

$$\theta_B = \frac{M_B L}{6EI_1} \tag{5.63}$$

Substituting Eq. 5.63 into Eq. 5.62 and solving for M_B

$$\frac{M_B L}{6EI_1} = \frac{Hh^2}{2EI_2} - \frac{M_B h}{EI_2}$$

$$M_B\left(\frac{L}{6I_1} + \frac{h}{I_2}\right) = \frac{Hh^2}{2I_2}$$

Let

$$K = \frac{hI_1}{LI_2} \tag{5.64}$$

$$M_B\left(\frac{h}{6KI_2} + \frac{h}{I_2}\right) = \frac{Hh^2}{2I_2}$$

$$M_B = \frac{Hh}{2\left(\frac{1}{6K} + 1\right)} = \frac{Hh}{2}\left(1 - \frac{1}{6K+1}\right) \tag{5.65}$$

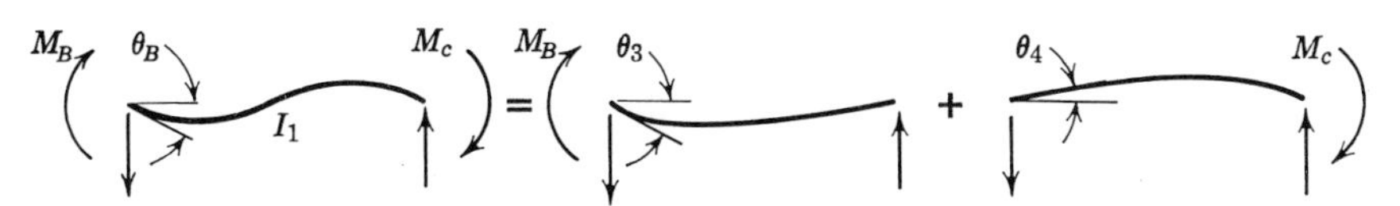

FIGURE 5.21. Free body force diagram of the horizontal leg.

Considering the forces in the horizontal direction and because of the symmetry of the system

$$H = \frac{P}{2} \tag{5.66}$$

Substituting Eq. 5.66 into Eq. 5.65

$$M_B = \frac{Ph}{4}\left(1 - \frac{1}{6K+1}\right) \tag{5.67}$$

The moment at the fixed end of the vertical leg can be determined by considering the sum of all moments (Fig. 5.20).

$$M_A + M_B = Hh \tag{5.68}$$

Substituting Eq. 5.66 and 5.67 into the above

$$M_A + \frac{Ph}{4}\left(1 - \frac{1}{6K+1}\right) = \frac{Ph}{2}$$

$$M_A = \frac{Ph}{4}\left(1 + \frac{1}{6K+1}\right) \tag{5.69}$$

The vertical force, V, can be determined by considering the sum of all the external moments shown in Figs. 5.18 and 5.19.

$$M_A + M_D + VL = PH \tag{5.70}$$

Substitute Eq. 5.69 into Eq. 5.70 and note that $M_A = M_D$ due to symmetry.

$$\frac{Ph}{2}\left(1 + \frac{1}{6K+1}\right) + VL = Ph$$

$$V = \frac{Ph}{2L}\left(1 - \frac{1}{6K+1}\right) = \frac{3PhK}{L(1+6K)} \tag{5.71}$$

Superposition can be used to determine the deflection at the top of the bent by considering the action of the force and moment separately (Fig. 5.22).

Using standard handbook equations

$$\delta_1 = \frac{Ph^3}{6EI_2} \quad \text{and} \quad \delta_2 = \frac{M_B h^2}{2EI_2} \tag{5.72}$$

$$\delta = \delta_1 - \delta_2 \tag{5.73}$$

Substituting Eqs. 5.67 and 5.72 into the above

$$\delta = \frac{Ph^3}{6EI_2} - \frac{h^2}{2EI_2}\left[\frac{Ph}{4}\left(1 - \frac{1}{6K+1}\right)\right]$$

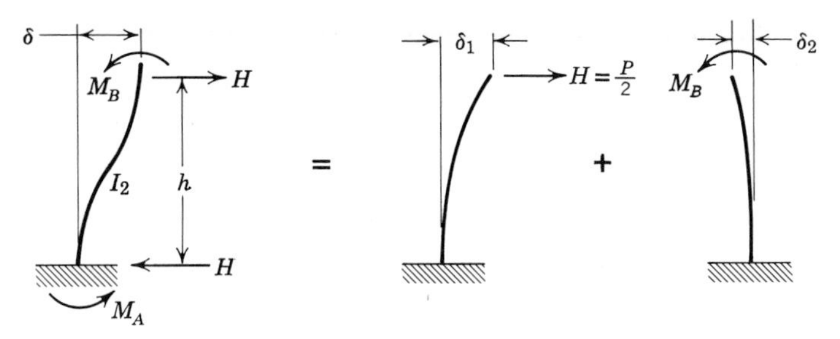

FIGURE 5.22. Deflection diagram of the vertical leg.

$$\delta = \frac{Ph^3}{24EI_2}\left(1 + \frac{3}{6K+1}\right) \tag{5.74}$$

If the resistor shown in Fig. 5.17 is a $\frac{1}{2}$-watt MIL-style RN 70 metal-film type, the mounting dimensions will be as shown in Chapter 6, Fig. 6.37. The natural frequency can be determined from the static deflection equation, using Eq. 2.10 along with Eq. 5.74 where the force P then becomes the static resistor weight. If the transmissibility of the resistor at resonance is approximated by $A\,(f_n)^{1/2}$ then the dynamic load acting on the resistor can be determined from the input G force. Assuming this dynamic load turns out to be one pound, bending moments, forces, and stresses in the electrical lead wire can be determined. Starting with the bending moment at point A, the equation is as follows:

$$M_A = \frac{Ph}{4}\left(1 + \frac{1}{6K+1}\right) \text{ (see Eq. 5.69)}$$

where $P = 1.0$ lb dynamic load on the resistor
$h = 0.174$ in.
$L = 0.226$ in.
$K = \frac{hI_1}{LI_2} = 0.77$
$I_1 = I_2 = 4.52 \times 10^{-8}$ in.^{4}
$d = 0.031$ in. electrical lead wire diameter
$B = 1.10$ in.
$E = 15 \times 10^6$ lb/in.^{2} (copper wire)

Substituting into Eq. 5.69

$$M_A = \frac{(1.0)(0.174)}{4}\left[1 + \frac{1}{6(0.77)+1}\right] = 0.0516 \text{ lb in.} \tag{5.75}$$

The bending moment at point B can be determined from Eq. 5.67.

$$M_B = \frac{Ph}{4}\left(1 - \frac{1}{6K+1}\right) = \frac{(1.0)(0.174)}{4}\left[1 - \frac{1}{6(0.77)+1}\right]$$

$$M_B = 0.0357 \text{ lb in.} \tag{5.76}$$

The vertical force can be determined from Eq. 5.71.

$$V = \frac{Ph}{2L}\left(1 - \frac{1}{6K+1}\right) = \frac{(1.0)(0.174)}{2(0.226)}\left[1 - \frac{1}{6(0.77)+1}\right]$$

$$V = 0.317 \text{ lb} \tag{5.77}$$

From Eq. 5.66

$$H = \frac{P}{2} = \frac{1.0}{2} = 0.50 \text{ lb.} \tag{5.78}$$

The maximum bending stress at point A is (neglecting stress concentration factors):

$$S_A = \frac{M_A c}{I} = \frac{(0.0516)\left(\frac{0.031}{2}\right)}{4.52 \times 10^{-8}} = 17{,}700 \text{ lb/in.}^2 \tag{5.79}$$

The bending stress at point B is

$$S_B = \frac{M_B c}{I} = \frac{(0.0357)\left(\frac{0.031}{2}\right)}{4.52 \times 10^{-8}} = 12{,}200 \text{ lb/in.}^2 \tag{5.80}$$

The axial (tension or compression) stress in the vertical leg AB is

$$S_t = \frac{V}{A} \tag{5.81}$$

where $V = 0.317$ lb (see Eq. 5.77)

$$A = \frac{\pi}{4} d^2 = \frac{\pi}{4}(0.031)^2 = 7.54 \times 10^{-4} \text{ in.}^2 \text{ area}$$

$$S_t = \frac{0.317}{7.54 \times 10^{-4}} = 421 \text{ lb/in.}^2 \tag{5.82}$$

The bending stress at point A and the axial stress in the vertical leg AB act in the same direction so they can be added together to get the maximum tensile or compressive stress:

$$S_T = S_A + S_t = 17{,}700 + 421 = 18{,}121 \text{ lb/in.}^2 \tag{5.83}$$

5.7. Bent with Transverse Load–Fixed Ends

The third axis of vibration that must be considered for electronic components mounted on printed-circuit boards is in the plane of the circuit board, but perpendicular to the axis of the electronic component part (Fig. 5.23). There must be enough clearance between the body of the component and the printed-circuit board to ensure that they do not make contact during vibration in order to make the following analysis valid.

Again, this type of mounting is not recommended for electronic component parts subjected to severe vibration conditions. It is much better to have the component cemented, tied, or fastened more securely to the circuit board. However, since some manufacturers still mount components in this fashion, it is desirable to know the characteristics of the structure.

Since the body of most resistors, capacitors, diodes, and flatpack integrated circuits is much stiffer then the electrical lead wires, the body of the component can be ignored when the geometry of the structure is being analyzed. The weight of the component can then be used to determine the magnitude of the dynamic load acting on the system (Fig. 5.24).

Using superposition, loads, moments, and deflections can be determined by considering a free-body force diagram, as shown in Fig. 5.25.

Consider the vertical leg AB. When the top horizontal leg BC bends, a torsional moment is imposed on the vertical leg. This torsional moment can be shown by the righthand rule where the thumb points in the

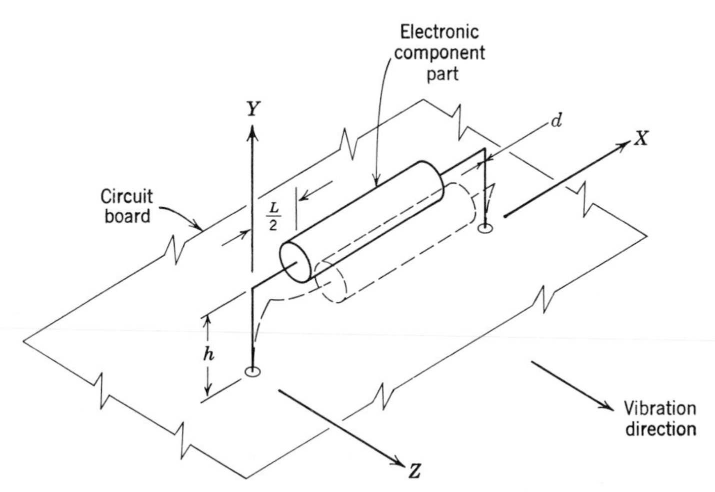

FIGURE 5.23. Transverse vibration of a component mounted on a circuit board.

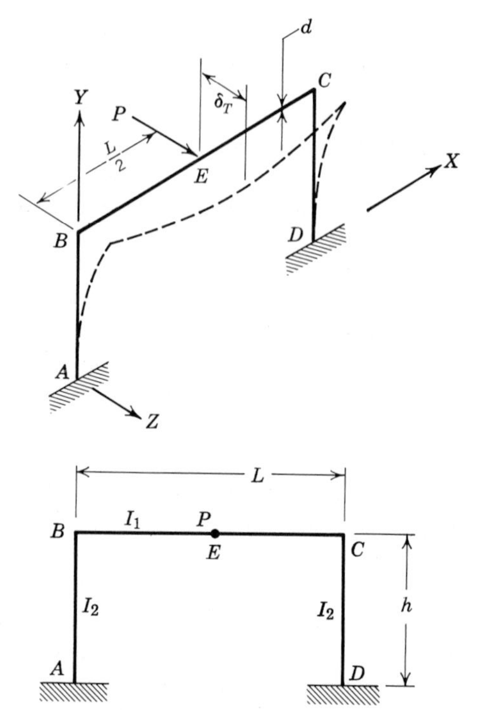

FIGURE 5.24. Transverse deflection mode of a bent with a concentrated load.

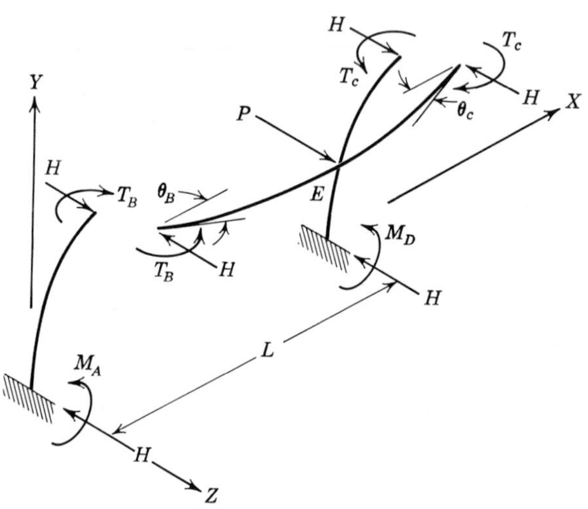

FIGURE 5.25. Free body force diagram of a bent with a transverse load.

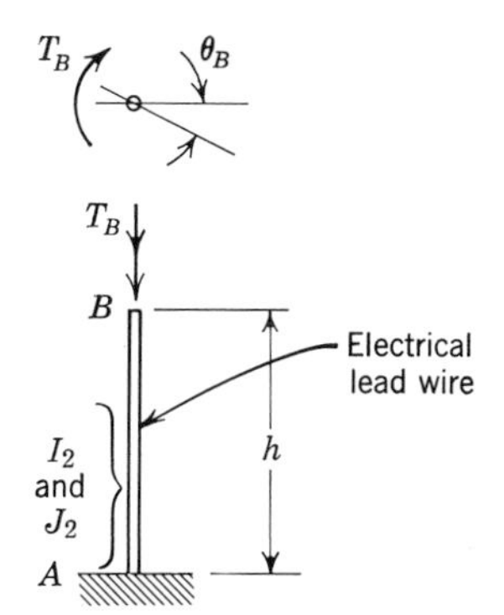

FIGURE 5.26. Torsional deflection in the vertical leg.

direction of the double arrow and the fingers show the direction of the moment (Fig. 5.26).

The torsional displacement of the vertical leg will be approximately the same as the angular displacement of the horizontal leg, at their junction, if the displacements are small. A standard handbook can be used to determine this angle, as shown in Fig. 5.26.

$$\theta_B = \frac{T_B h}{GJ_2} \tag{5.84}$$

Consider the horizontal leg BC. Superposition can be used to determine the angular rotation at point B due to bending in the horizontal leg BC and torsion in the vertical leg AB, as shown in Fig. 5.27.

Using standard handbook equations

$$\theta_1 = \frac{PL^2}{16EI_1}; \quad \theta_2 = \frac{T_B L}{3EI_1}; \quad \theta_3 = \frac{T_c L}{6EI_1} \tag{5.85}$$

$$\theta_B = \theta_1 - (\theta_2 + \theta_3) \tag{5.86}$$

Substitute Eq. 5.86 into Eq. 5.85 and note that $T_B = T_c$ because of symmetry.

$$\theta_B = \frac{PL^2}{16EI_1} - \frac{T_B L}{2EI_1} \tag{5.87}$$

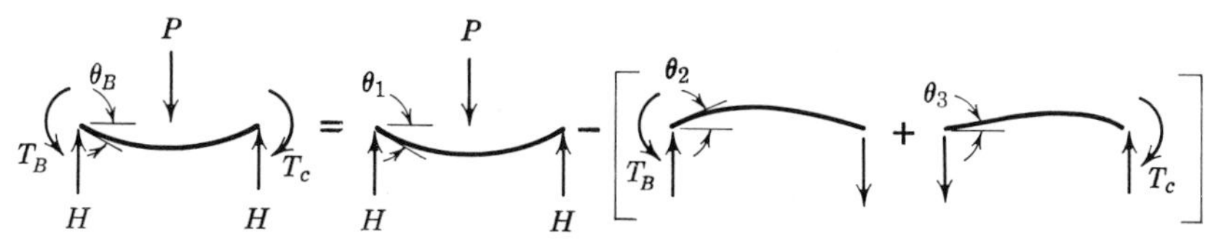

FIGURE 5.27. Free body force diagram of the horizontal leg.

Substituting Eq. 5.84 into Eq. 5.87

$$\frac{T_B h}{GJ_2} = \frac{PL^2}{16EI_1} - \frac{T_B L}{2EI_1}$$

$$T_B = \frac{PL^2}{16EI_1\left(\frac{h}{GJ_2} + \frac{L}{2EI_1}\right)} = \frac{PL^2 GJ_2}{8(2hEI_1 + LGJ_2)} \tag{5.88}$$

The fixed end moments M_A and M_D and force H can be determined by analyzing the external forces acting on the system (Fig. 5.25).

$$M_A + M_D = Ph \tag{5.89}$$

From symmetry $M_A = M_D$

$$M_A = \frac{Ph}{2} \tag{5.90}$$

Also

$$2H = P$$

$$H = \frac{P}{2} \tag{5.91}$$

The bending moment at point E, under the load, can be determined from a consideration of the moments on the horizontal leg BC as shown in Fig. 5.25.

$$M_E = H\frac{L}{2} - T_B \tag{5.92}$$

Substituting Eqs. 5.88 and 5.91 into the above

$$M_E = \frac{PL}{4} - \frac{PL^2}{16EI_1\left(\frac{h}{GJ_2} + \frac{L}{2EI_1}\right)}$$

$$M_E = \frac{PL}{4}\left[1 - \frac{L}{4EI_1\left(\frac{h}{GJ_2} + \frac{L}{2EI_1}\right)}\right] \tag{5.93}$$

The maximum deflection will occur in the horizontal leg, at the concentrated load. Superposition can be used by considering the action of the force on the horizontal leg and vertical leg separately. Consider the horizontal leg first as shown in Fig. 5.28.

Using equations from a standard handbook

$$\delta_1 = \frac{PL^3}{48EI_1} \tag{5.94}$$

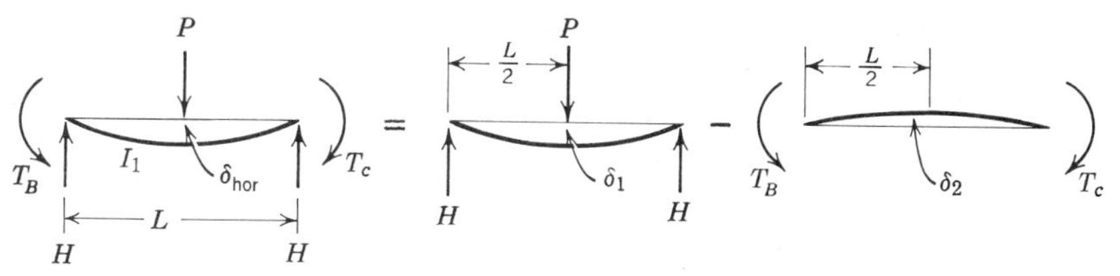

FIGURE 5.28. Deflection diagram of the horizontal leg.

$$\delta_2 = \frac{T_B L^2}{8EI_1} \tag{5.95}$$

$$\delta_{hor} = \delta_1 - \delta_2 \tag{5.96}$$

Substituting Eqs. 5.94 and 5.95 into the above

$$\delta_{hor} = \frac{PL^3}{48EI_1} - \frac{T_B L^2}{8EI_1} \tag{5.97}$$

Next consider the vertical leg, as shown in Fig. 5.29.

$$\delta_{vert} = \frac{\left(\frac{P}{2}\right) h^3}{3EI_2} = \frac{Ph^3}{6EI_2} \tag{5.98}$$

The total deflection is

$$\delta_T = \delta_{hor} + \delta_{vert} \tag{5.99}$$

Substituting Eqs. 5.97 and 5.98 into the above

$$\delta_T = \frac{PL^3}{48EI_1} - \frac{T_B L^2}{8EI_1} + \frac{Ph^3}{6EI_2} \tag{5.100}$$

Equation 5.88 can be used to determine the torsional moment in the vertical leg of the bent, if the load P is known.

Assume Fig. 5.23 represents a $\frac{1}{2}$-watt RN 70 resistor and that the dynamic loading is equal to one pound, based on the input vibration G

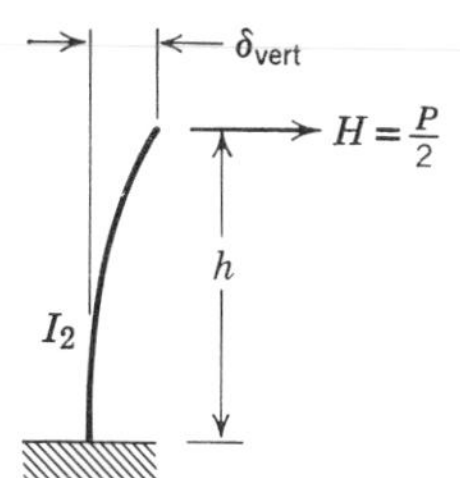

FIGURE 5.29. Bending deflection of the vertical leg.

level, the dynamic response of the resistor, and the weight. Starting with the torsional moment developed in the vertical leg at point B

$$T_B = \frac{PL^2}{16EI_1\left(\dfrac{h}{GJ_2}+\dfrac{L}{2EI_1}\right)} \qquad \text{(see Eq. 5.88)}$$

where $P = 1.0$ lb

$E = 15 \times 10^6$ lb/in.^{2} copper wire

$G = 5.8 \times 10^6$ lb/in.^{2} shear modulus

$h = 0.174$ in.

$L = 0.226$ in.

$$I_1 = I_2 = \frac{\pi d^4}{64} = \frac{\pi}{64}(0.031)^4 = 4.52 \times 10^{-8} \text{ in.}^4$$

$$J_2 = \frac{\pi d^4}{32} = \frac{\pi}{32}(0.031)^4 = 9.04 \times 10^{-8} \text{ in.}^4$$

$$T_B = \frac{(1.0)(0.226)^2}{16(15\times 10^6)(4.52\times 10^{-8})\left[\dfrac{0.174}{(5.8\times 10^6)(9.04\times 10^{-8})}+\dfrac{0.226}{2(15\times 10^6)(4.52\times 10^{-8})}\right]}$$

$$T_B = \frac{0.051}{10.85(0.332+0.166)} = 0.00944 \text{ lb in.} \tag{5.101}$$

The torsional shear stress in vertical leg AB will be uniform along the entire leg:

$$S_s = \frac{T_B R}{J_2} \tag{5.102}$$

where $T_B = 0.00944$ lb in.

$$R = \frac{d}{2} = \frac{0.031}{2} = 0.0155 \text{ in.}$$

$J_2 = 9.04 \times 10^{-8}$ in.^{4}

$$S_s = \frac{(9.44\times 10^{-3})(1.55\times 10^{-2})}{9.04\times 10^{-8}} = 1620 \text{ lb/in.}^2 \tag{5.103}$$

The bending moment at point E on the horizontal leg BC can be determined from Eq. 5.93.

$$M_E = \frac{PL}{4}\left[1 - \frac{L}{4EI_1\left(\dfrac{h}{GJ_2}+\dfrac{L}{2EI_1}\right)}\right]$$

$$M_E = \frac{(1.0)(0.226)}{4}\left[1 - \frac{0.226}{(2.71)(0.332+0.1667)}\right] = 0.0470 \text{ lb in.} \quad (5.104)$$

The bending stress at point E then becomes (neglecting stress concentration factors)

$$S_E = \frac{M_E R}{I_1} = \frac{(0.0470)(0.0155)}{4.52 \times 10^{-8}} = 16{,}100 \text{ lb/in.}^2 \quad (5.105)$$

The maximum bending moment is shown by Eq. 5.90 at points A and D.

$$M_A = M_D = \frac{Ph}{2} = \frac{(1.0)(0.174)}{2} = 0.0870 \text{ lb in.}$$

The maximum bending stress in the electrical lead wire is

$$S_A = \frac{M_A R}{I_1} = \frac{(0.0870)(0.0155)}{4.52 \times 10^{-8}} = 29{,}800 \text{ lb/in.}^2$$

The maximum deflection can be determined from Eqs. 5.88 and 5.100.

$$\delta_T = \frac{PL^3}{48EI_1} + \frac{Ph^3}{6EI_2} - \frac{PL^2GJ_2}{8(2hEI_1 + LGJ_2)}\left(\frac{L^2}{8EI_1}\right)$$

$$\delta_T = \frac{P}{2}\left[\frac{L^3}{24EI_1} + \frac{h^3}{3EI_2} - \frac{L^4GJ_2}{32EI_1(2hEI_1 + LGJ_2)}\right] \quad (5.106)$$

where $EI_1 = 15 \times 10^6(4.52 \times 10^{-8}) = 0.678 \text{ lb in.}^2$
$GJ_2 = 5.8 \times 10^6(9.04 \times 10^{-8}) = 0.524 \text{ lb in.}^2$

Substituting into Eq. 5.106

$$\delta_T = \frac{1.0}{2}\left\{\frac{(0.226)^3}{24(0.678)} + \frac{(0.174)^3}{3(0.678)} - \frac{(0.226)^4(0.524)}{32(0.678)[2(0.174)(0.678) + (0.226)(0.524)]}\right\}$$

$$\delta_T = \tfrac{1}{2}(0.707 \times 10^{-3} + 2.59 \times 10^{-3} - 0.178 \times 10^{-3})$$

$$\delta_T = 1.56 \times 10^{-3} \text{ in.} \quad (5.107)$$

5.8. END CONDITIONS DURING HIGH-FREQUENCY RESONANCES

Many different types of bents are used as structural members in electronic packages. These bents must usually be analyzed to determine bending stresses and deflections. Since the bending stresses and deflections, to a great extent, depend upon the end conditions, it is important to estimate accurately the end conditions of the bent. For example, if the ends of a bent were welded to a rigid plate that had a cross-section thickness much greater than the thickness of the bent, the ends of this bent would be considered as fixed. If this bent had its ends bolted to the plate instead of welded, then there is a good possibility the ends of the bent might be much closer to a simply supported condition than a fixed condition.

The coefficient of fixity for bolted ends will depend upon the number and size of the bolts used to fasten the ends of the bent. In general, it might be stated that the stiffer the bent the more difficult it is to achieve fixed ends with bolts.

For example, vibration tests were run on a rectangular bent 10 in. high by 10 in. long using four bolts with a 0.50-in. diameter in each leg, or a total of eight bolts. Since the bent was to be used as part of a vibration fixture, it was desirable to keep the resonant frequency above 1000 Hz, so the design goal was 1500 Hz. The vibration tests showed a much lower resonant frequency than expected. The analysis showed the results were much closer to a hinged end condition than fixed ends for vibration in the vertical direction, as shown in Fig. 5.30.

A strobe light and a microscope were used to examine the base of the bent to see if relative motion at the base could be observed to account for the reduced stiffness. A microscope was required to see clearly the small deflections at the base. No attempt was made to measure these deflections. Even when the bolts were torqued extra tight the deflections could not be eliminated.

Experiences such as this have shown that it is very difficult to prevent relative motion at bolted interfaces when high-frequency resonances

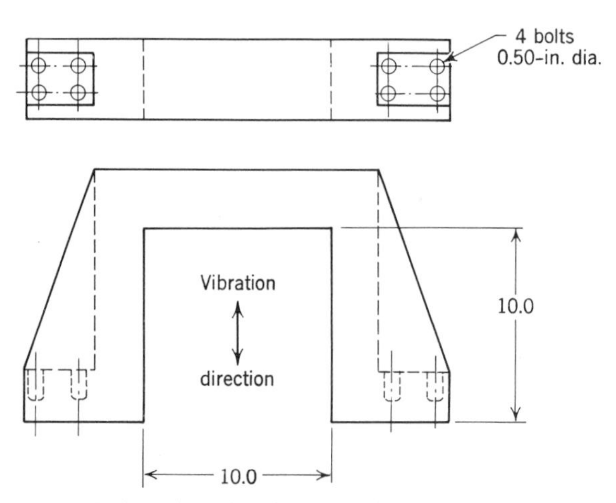

FIGURE 5.30. Rigid vibration fixture in the shape of a bent.

occur. Therefore, when an analysis of a rigid structure with bolted interfaces is being made, it is better to be conservative by estimating the end fixity to be closer to a simply supported condition than to a fixed condition, unless test data show otherwise.

5.9. Bent with a Vertical Load–Hinged Ends

The bending moments, loads, and deflections of a rectangular bent with hinged ends can be determined with the use of superposition, considering a concentrated load in the vertical direction as shown in Fig. 5.31. A free-body force diagram of the rectangular bent will appear as shown in Fig. 5.32:

Consider the vertical leg AB. The angle at the top of the leg, considering small displacements is:

$$\theta_B = \frac{M_B h}{3EI_2} \tag{5.108}$$

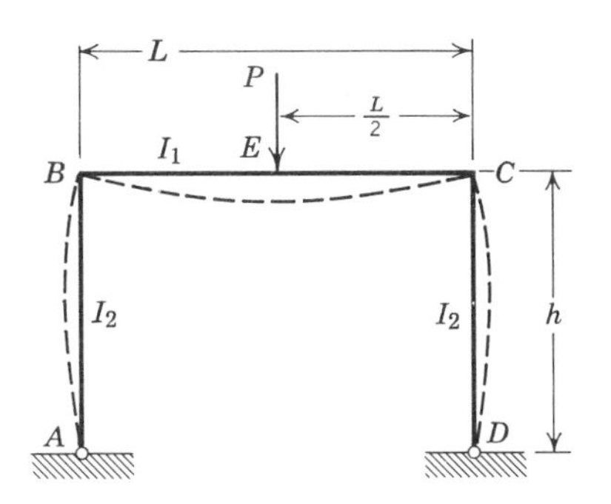

FIGURE 5.31. Bent with a vertical load and hinged ends.

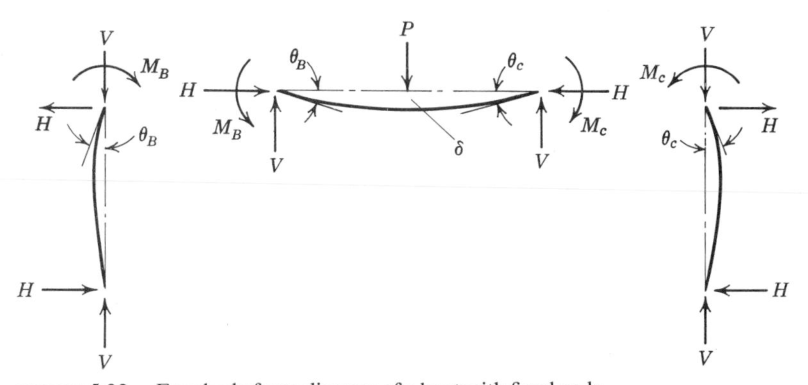

FIGURE 5.32. Free body force diagram of a bent with fixed ends.

Consider the horizontal leg BC

$$\theta_1 = \frac{PL^2}{16EI_1}; \quad \theta_2 = \frac{M_B L}{3EI_1}; \quad \theta_3 = \frac{M_c L}{6EI_1} \tag{5.109}$$

$$\theta_B = \theta_1 - (\theta_2 + \theta_3) \tag{5.110}$$

Substitute Eq. 5.109 into Eq. 5.110 and note that $M_B = M_c$ because of symmetry.

$$\theta_B = \frac{PL^2}{16EI_1} - \frac{M_B L}{2EI_1} \tag{5.111}$$

Substituting Eq. 5.108 into Eq. 5.111

$$\frac{M_B h}{3EI_2} = \frac{PL^2}{16EI_1} - \frac{M_B L}{2EI_1}$$

Let $K = hI_1/LI_2$ and solve for M_B.

$$M_B = \frac{3PL}{8(2K+3)} \tag{5.112}$$

Considering forces in the vertical direction

$$V = \frac{P}{2}$$

Considering moments acting on the vertical leg AB

$$Hh = M_B = \frac{3PL}{8(2K+3)}$$

$$H = \frac{3PL}{8h(2K+3)} \tag{5.113}$$

Using Fig. 5.33 to find the deflection at the center of the horizontal leg

$$\delta_1 = \frac{PL^3}{48EI_1}; \quad \delta_2 = \delta_3 = \frac{M_B L^2}{16EI_1}$$

$$\delta = \delta_1 - (\delta_2 + \delta_3)$$

$$\delta = \frac{PL^3}{48EI_1} - \frac{M_B L^2}{8EI_1}$$

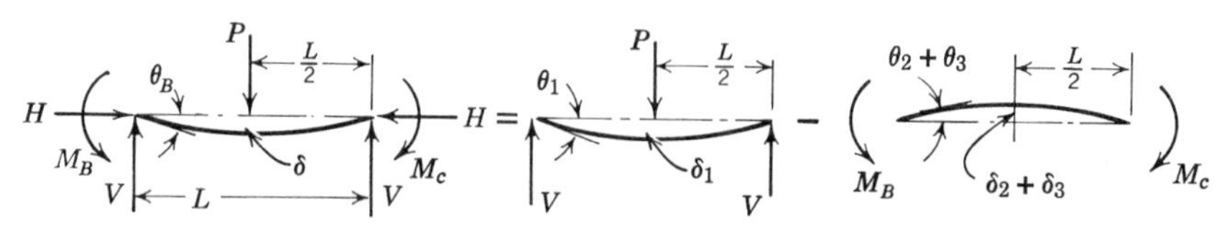

FIGURE 5.33. Deflection diagram of the horizontal leg.

Substituting from Eq. 5.112

$$\delta = \frac{PL^3}{48EI_1}\left[1 - \frac{9}{4(2K+3)}\right] \tag{5.114}$$

5.10. Bent with a Lateral Load—Hinged Ends

Consider a bent with hinged ends and a concentrated load acting in the lateral direction. The deflection can be broken up into two parts. The first part considers bending of the vertical legs AB and CD only, with no bending in the horizontal leg BC. The second part considers bending of the horizontal leg BC only, with no bending in the vertical legs. When the horizontal leg bends, the angular displacement causes an additional displacement of the vertical leg through pure rotation (Fig. 5.34).

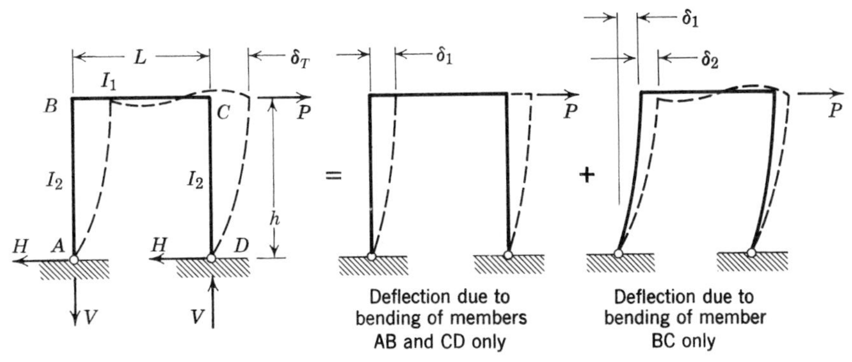

FIGURE 5.34. Bent with a lateral load and hinged ends.

The vertical leg AB is shown in Fig. 5.35.

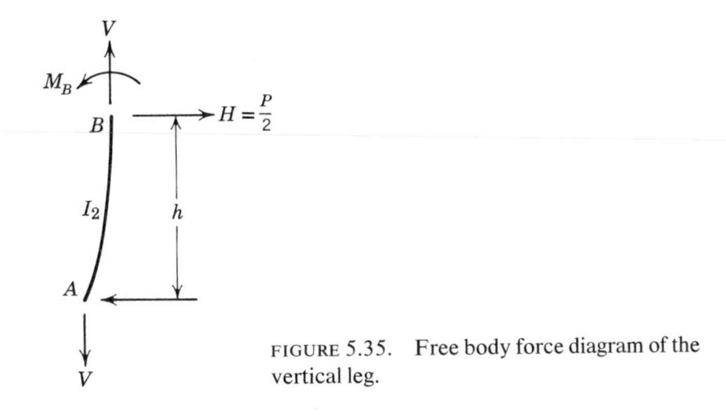

FIGURE 5.35. Free body force diagram of the vertical leg.

Considering forces in the horizontal direction

$$H = \frac{P}{2} \tag{5.115}$$

Considering moments about point B

$$M_B = Hh = \frac{Ph}{2} \tag{5.116}$$

Considering leg AB as a cantilevered beam with a concentrated end load

$$\delta_1 = \frac{Hh^3}{3EI_2} = \frac{Ph^3}{6EI_2} \tag{5.117}$$

The horizontal leg BC is shown in Fig. 5.36.

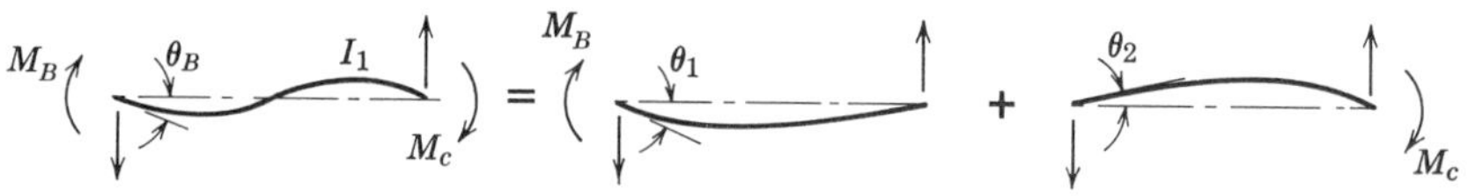

FIGURE 5.36. Free body force diagram of the horizontal leg.

The angular displacements are

$$\theta_1 = \frac{M_B L}{3EI_1} \quad \text{and} \quad \theta_2 = \frac{M_c L}{6EI_1} \tag{5.118}$$

$$\theta_B = \theta_1 - \theta_2 \tag{5.119}$$

Substitute Eq. 5.118 into Eq. 5.119 and note that $M_B = M_c$ due to symmetry.

$$\theta_B = \frac{M_B L}{6EI_1} \tag{5.120}$$

Substituting Eq. 5.116 into Eq. 5.120

$$\theta_B = \frac{PhL}{12EI_1} \tag{5.121}$$

The vertical leg will rotate through the angle θ_B due to bending of the horizontal leg shown in Fig. 5.34.

$$\delta_2 = h\theta_B = \frac{Ph^2 L}{12EI_1} \tag{5.122}$$

The total deflection of the bent will then be the sum of δ_1 and δ_2.

$$\delta_T = \delta_1 + \delta_2 = \frac{Ph^3}{6EI_2} + \frac{Ph^2L}{12EI_1}$$

Let $K = hI_1/LI_2$

$$\delta_T = \frac{Ph^3}{6EI_2}\left(1 + \frac{1}{2K}\right) \tag{5.123}$$

5.11. Enclosed Frame–Hinged Ends

A completely enclosed frame can be evaluated using the same superposition methods. The simply supported frame with a concentrated load in the lateral direction is shown in Fig. 5.37.

The vertical leg AB is shown in Fig. 5.38.

At the top of the vertical leg AB

$$\theta_1 = \frac{Ph^2}{4EI_2}; \quad \theta_2 = \theta_A; \quad \theta_3 = \frac{M_B h}{EI_2}$$

$$\theta_B = \theta_1 + \theta_2 - \theta_3$$

$$\theta_B = \frac{Ph^2}{4EI_2} + \theta_A - \frac{M_B h}{EI_2} \tag{5.124}$$

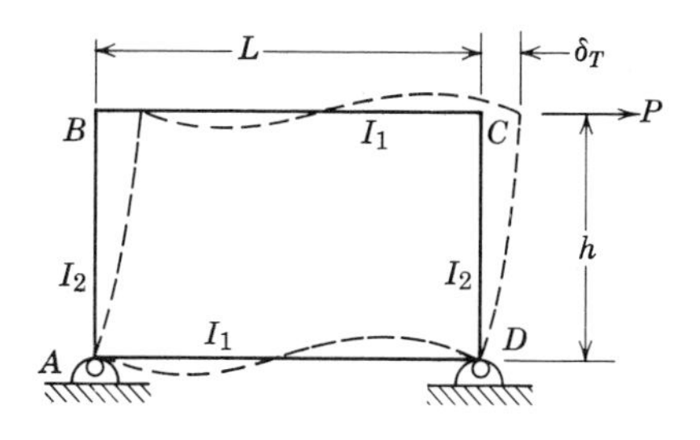

FIGURE 5.37. Enclosed frame with hinged supports and a lateral load.

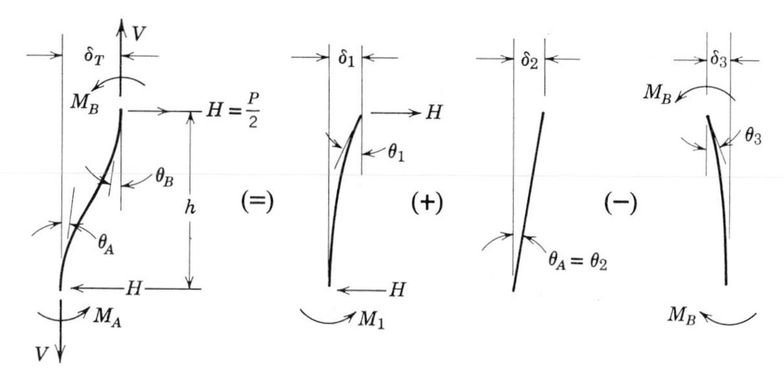

FIGURE 5.38. Free body force diagram of the vertical leg.

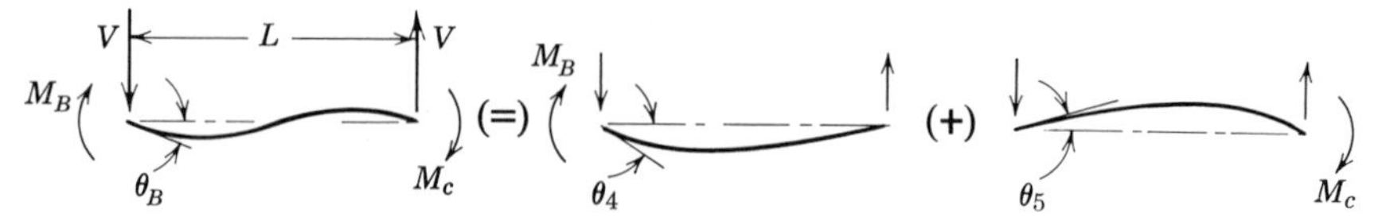

FIGURE 5.39. Free body force diagram of the top horizontal leg.

At the bottom of the vertical leg AB

$$M_A = M_1 - M_B = \frac{Ph}{2} - M_B \tag{5.125}$$

The top horizontal leg BC is shown in Fig. 5.39.

$$\theta_4 = \frac{M_B L}{3EI_1}; \quad \theta_5 = \frac{M_c L}{6EI_1}; \quad M_B = M_c \text{ by symmetry}$$

$$\theta_B = \theta_4 - \theta_5 = \frac{M_B L}{3EI_1} - \frac{M_B L}{6EI_1} = \frac{M_B L}{6EI_1} \tag{5.126}$$

Substituting Eq. 5.126 into Eq. 5.124

$$\frac{M_B L}{6EI_1} = \frac{Ph^2}{4EI_2} + \theta_A - \frac{M_B h}{EI_2} \tag{5.127}$$

The bottom horizontal leg AD is shown in Fig. 5.40.

$$\theta_6 = \frac{M_A L}{3EI_1}; \quad \theta_7 = \frac{M_D L}{6EI_1}; \quad M_A = M_D \text{ by symmetry}$$

$$\theta_A = \theta_6 - \theta_7 = \frac{M_A L}{6EI_1} \tag{5.128}$$

Substituting Eq. 5.128 into Eq. 5.127

$$\frac{M_B L}{6EI_1} + \frac{M_B h}{EI_2} = \frac{Ph^2}{4EI_2} + \frac{M_A L}{6EI_1} \tag{5.129}$$

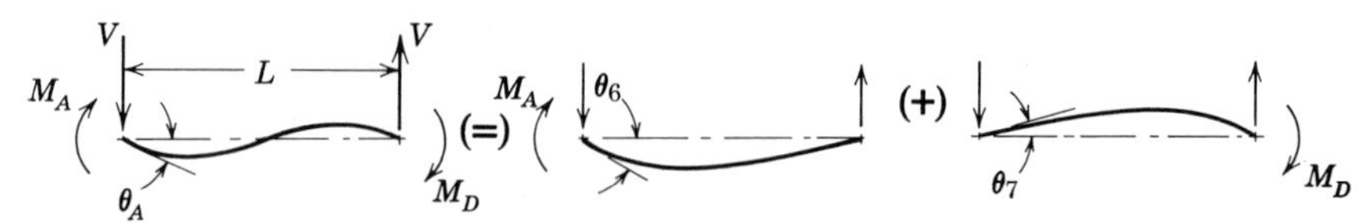

FIGURE 5.40. Free body force diagram of the bottom horizontal leg.

Substituting Eq. 5.125 into Eq. 5.129

$$\frac{M_B L}{6I_1}+\frac{M_B h}{I_2}=\frac{Ph^2}{4I_2}+\frac{L}{6I_1}\left(\frac{Ph}{2}-M_B\right)$$

$$M_B\left(\frac{h}{I_2}+\frac{L}{3I_1}\right)=\frac{Ph}{4}\left(\frac{h}{I_2}+\frac{L}{3I_1}\right)$$

$$M_B=\frac{Ph}{4} \tag{5.130}$$

The deflection can be determined as follows:

$$\delta_1=\frac{Hh^3}{3EI_2}=\frac{Ph^3}{6EI_2}$$

$$\delta_2=h\theta_A=\frac{M_A hL}{6EI_1}=\frac{Ph}{4}\left(\frac{hL}{6EI_1}\right)=\frac{Ph^2L}{24EI_1}$$

$$\delta_3=\frac{M_B h^2}{2EI_2}=\frac{Ph}{4}\left(\frac{h^2}{2EI_2}\right)=\frac{Ph^3}{8EI_2}$$

The total deflection then becomes

$$\delta_T=\delta_1+\delta_2-\delta_3$$

$$\delta_T=\frac{Ph^3}{6EI_2}+\frac{Ph^2L}{24EI_1}-\frac{Ph^3}{8EI_2}$$

$$\delta_T=\frac{Ph^2}{2EI_2}\left(\frac{h}{3}+\frac{LI_2}{12I_1}-\frac{h}{4}\right)$$

$$\delta_T=\frac{Ph^2}{24EI_2}\left(h+L\frac{I_2}{I_1}\right) \tag{5.131}$$

5.12. Cordwood Modules

A common type of electronic component mounting is the cordwood module. This module is usually made up of component parts such as resistors, capacitors, and diodes standing up between two small printed-circuit boards that are sometimes called jig wafers (see Fig. 5.41).

The cordwood module subassembly is usually dip-soldered to a master circuit board which can be plugged into an electronic box for easy service and maintenance (see also Fig. 4.2).

During vibration in an axis perpendicular to the plane of the master circuit board, dynamic deflections in the master circuit board will cause bending in the electrical lead wires and bending in the small circuit-board jig wafers. This type of bending deflection will very often lead to high

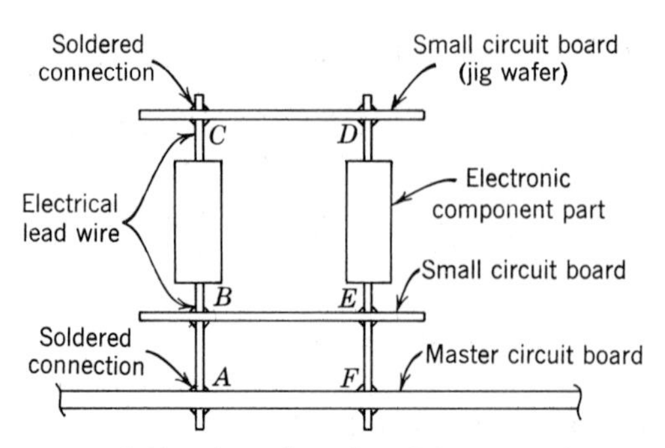

FIGURE 5.41. A cordwood module mounted on a circuit board.

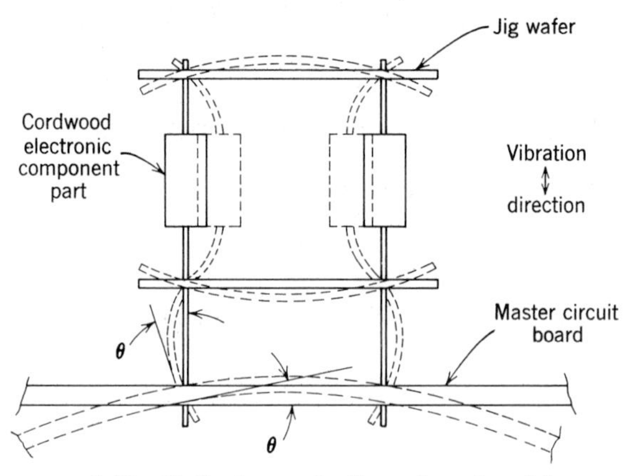

FIGURE 5.42. Deflection mode of a cordwood module.

bending stresses in the solder joints. If the number of fatigue cycles is great enough, fatigue failures may occur in the solder joints.

Consider the case of a small cordwood module with the electronic component parts in a symmetrical arrangement. The angular displacement of the master circuit board will induce the same angular displacement in the electrical lead wires, where the lead wires are soldered to the master circuit board (Fig. 5.42).

The bending moments in the electrical lead wires can be approximated by ignoring the body of the electronic component. Since the body is so much stiffer than the leads, most of the bending will be in the electrical leads and printed-circuit boards, as shown in Fig. 5.43. The horizontal force due to the kinetic energy of the component body can be ignored here

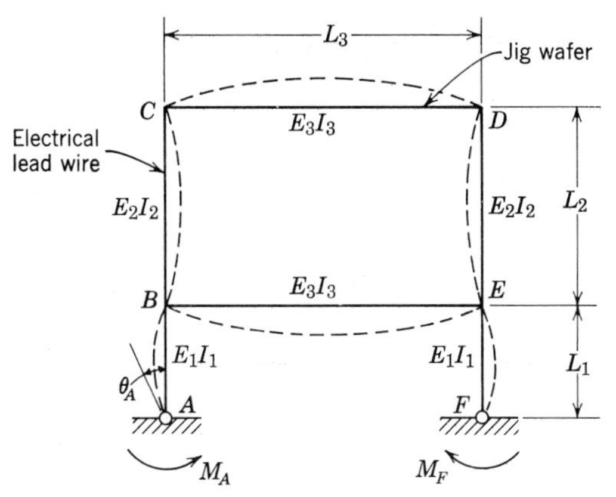

FIGURE 5.43. Deflection mode of the electrical lead wires in a cordwood module.

if the natural frequency of the component body, suspended on its leads, is not excited (see Chapter 4, Section 4.1). A free-body force diagram of the bent appears as shown in Fig. 5.44.

The vertical leg AB is shown in Fig. 5.45.

$$\theta_1 = \frac{M_A L_1}{3E_1 I_1}; \qquad \theta_3 = \frac{M_{B1} L_1}{6E_1 I_1}$$

$$\theta_A = \theta_1 - \theta_3$$

$$\theta_A = \frac{M_A L_1}{3E_1 I_1} - \frac{M_{B1} L_1}{6E_1 I_1} \tag{5.132}$$

$$\theta_2 = \frac{M_A L_1}{6E_1 I_1}; \qquad \theta_4 = \frac{M_{B1} L_1}{3E_1 I_1}$$

$$\theta_{B1} = \theta_2 - \theta_4$$

$$\theta_{B1} = \frac{M_A L_1}{6E_1 I_1} - \frac{M_{B1} L_1}{3E_1 I_1} \tag{5.133}$$

The horizontal leg BE is shown in Fig. 5.46.

$$\theta_5 = \frac{M_{B2} L_3}{3E_3 I_3}; \quad \theta_7 = \frac{M_{B2} L_3}{6E_3 I_3} \qquad (\text{Note: } M_{B2} = M_{E2})$$

$$\theta_{B2} = \theta_5 + \theta_7 = \frac{M_{B2} L_3}{2E_3 I_3} \tag{5.134}$$

From continuity, the angles at point B must all be equal.

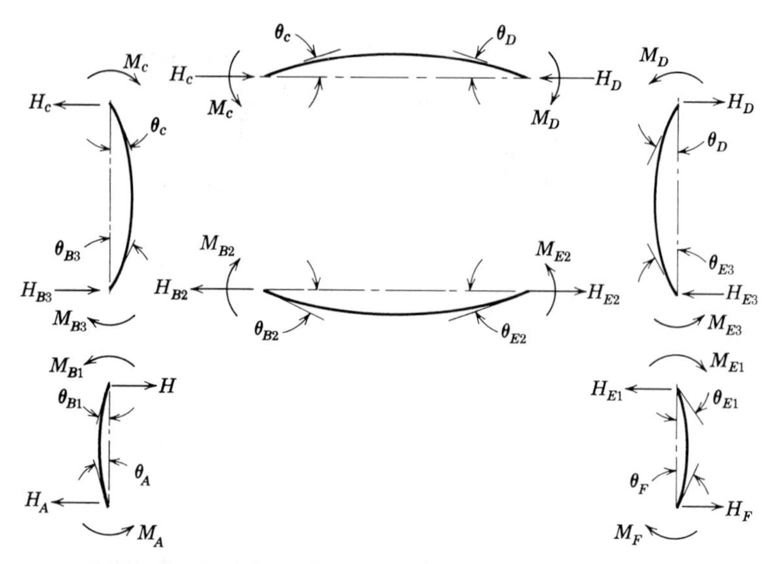

FIGURE 5.44. Free body force diagram of the cordwood module.

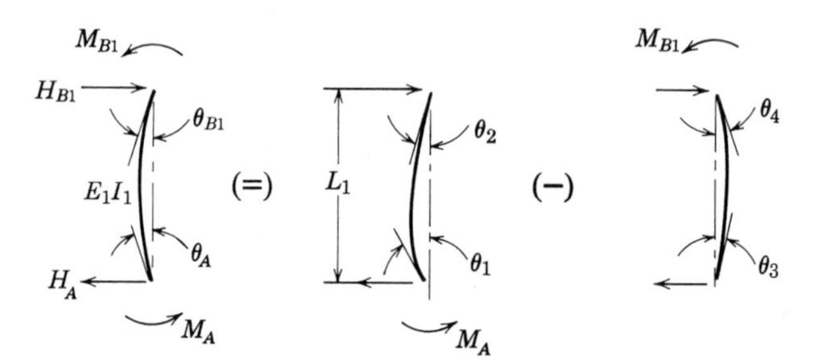

FIGURE 5.45. Free body force diagram of the lower vertical leg.

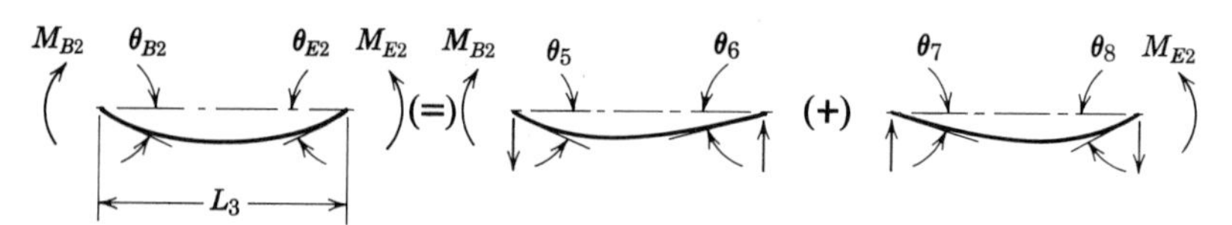

FIGURE 5.46. Free body force diagram of the lower horizontal leg.

$$\theta_{B1} = \theta_{B2} = \theta_{B3} \tag{5.135}$$

Substituting Eq. 5.134 into Eq. 5.133

$$\frac{M_{B2}L_3}{2E_3I_3} = \frac{M_AL_1}{6E_1I_1} - \frac{M_{B1}L_1}{3E_1I_1} \tag{5.136}$$

Let

$$a_1 = \frac{L_1}{E_1I_1},\ a_2 = \frac{L_2}{E_2I_2},\ a_3 = \frac{L_3}{E_3I_3} \tag{5.137}$$

then

$$\frac{M_{B2}a_3}{2} = \frac{M_Aa_1}{6} - \frac{M_{B1}a_1}{3}$$

$$M_{B1} = \frac{M_A}{2} - \frac{3M_{B2}a_3}{2a_1} \tag{5.138}$$

The vertical leg BC is shown in Fig. 5.47.

$$\theta_9 = \frac{M_{B3}L_2}{3E_2I_2} \qquad \text{and} \qquad \theta_{12} = \frac{M_cL_2}{6E_2I_2}$$

$$\theta_{B3} = \theta_9 - \theta_{12}$$

$$\theta_{B3} = \frac{M_{B3}a_2}{3} - \frac{M_ca_2}{6} \tag{5.139}$$

From Eqs. 5.134, 5.135, and 5.139

$$\frac{M_{B2}a_3}{2} = \frac{M_{B3}a_2}{3} - \frac{M_ca_2}{6} \tag{5.140}$$

$$\theta_{10} = \frac{M_{B3}L_2}{6E_2I_2} \qquad \text{and} \qquad \theta_{11} = \frac{M_cL_2}{3E_2I_2}$$

$$\theta_c = \theta_{10} - \theta_{11}$$

$$\theta_c = \frac{M_{B3}a_2}{6} - \frac{M_ca_2}{3} \tag{5.141}$$

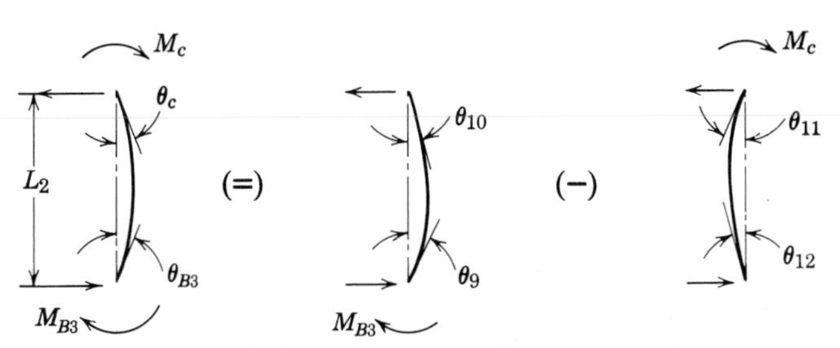

FIGURE 5.47. Free body force diagram of the upper vertical leg.

The horizontal leg CD is shown in Fig. 5.48.

$$\theta_{13} = \frac{M_c L_3}{3E_3 I_3} \quad \text{and} \quad \theta_{14} = \frac{M_c L_3}{6E_3 I_3} \quad (\text{Note: } M_c = M_D)$$

$$\theta_c = \theta_{13} + \theta_{14} = \frac{3M_c L_3}{6E_3 I_3} = \frac{M_c a_3}{2} \tag{5.142}$$

Substituting Eq. 5.142 into Eq. 5.141 and solving for M_c

$$M_c = \frac{M_{B3} a_2}{6\left(\frac{a_2}{3} + \frac{a_3}{2}\right)}$$

Let

$$b = \frac{a_2}{3} + \frac{a_3}{2} \tag{5.143}$$

$$M_c = \frac{M_{B3} a_2}{6b} \tag{5.144}$$

An examination of Fig. 5.44 shows

$$M_{B1} = M_{B2} + M_{B3}$$

Then

$$M_{B3} = M_{B1} - M_{B2} \tag{5.145}$$

Substituting Eqs. 5.138, 5.144, and 5.145 into Eq. 5.140

$$\frac{M_{B2} a_3}{2} = \frac{(M_{B1} - M_{B2}) a_2}{3} - \frac{(M_{B1} - M_{B2}) a_2^2}{36b} \tag{5.146}$$

Substituting Eq. 5.138 into Eq. 5.146

$$M_{B2}\left(\frac{a_3}{2} + \frac{a_2}{3} - \frac{a_2^2}{36b} + \frac{a_3 a_2}{2a_1} - \frac{a_3 a_2^2}{24 a_1 b}\right) = M_A\left(\frac{a_2}{6} - \frac{a_2^2}{72b}\right) \tag{5.147}$$

It is necessary to get M_A in terms of M_{B2}.
Substituting Eq. 5.132 into Eq. 5.138 and solving for M_A

$$M_A = \frac{4\theta_A}{a_1} - \frac{M_{B2} a_3}{a_1} \tag{5.148}$$

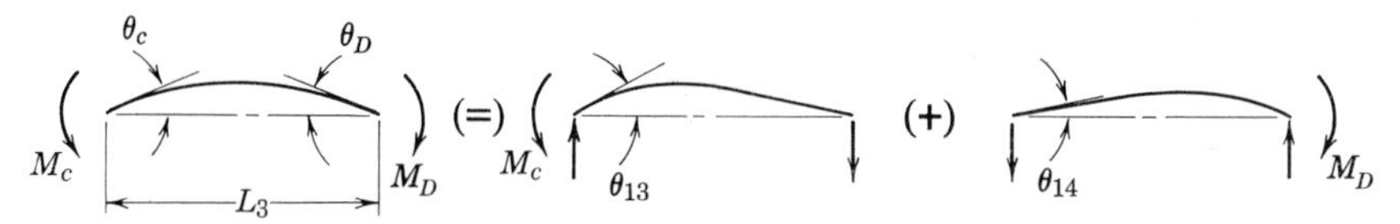

FIGURE 5.48. Free body force diagram of the upper horizontal leg.

Substituting Eq. 5.148 into Eq. 5.147 and solving for M_{B2}

$$M_{B2}=\frac{\dfrac{\theta_A a_2}{3a_1}\left(2-\dfrac{a_2}{6b}\right)}{\dfrac{a_3}{2}+\dfrac{a_2}{3}\left(1-\dfrac{a_2}{12b}\right)+\dfrac{a_3 a_2}{3a_1}\left(2-\dfrac{a_2}{6b}\right)} \tag{5.149}$$

Equation 5.149 can be used to determine the bending moment in horizontal member BE, if the angle θ_A is known. Once the moment M_{B2} is known, all of the other bending moments can be determined.

The angle θ_A can be determined from the size of the master circuit board, the geometry of the cordwood module, the resonant frequency of the master circuit board, and the G force.

5.13 Sample Problem

For example, consider a master printed-circuit board simply supported on four sides with several cordwood modules soldered to the circuit board.

At the fundamental resonant mode of the circuit board, the deflection curve can be approximated by a sine curve. A cordwood module at the center of the circuit board will be forced to deflect as shown in Fig. 5.49.

The master board deflection curve can be approximated by a trigonometric function as follows:

$$Z=\delta \sin \frac{\pi X}{a} \tag{5.150}$$

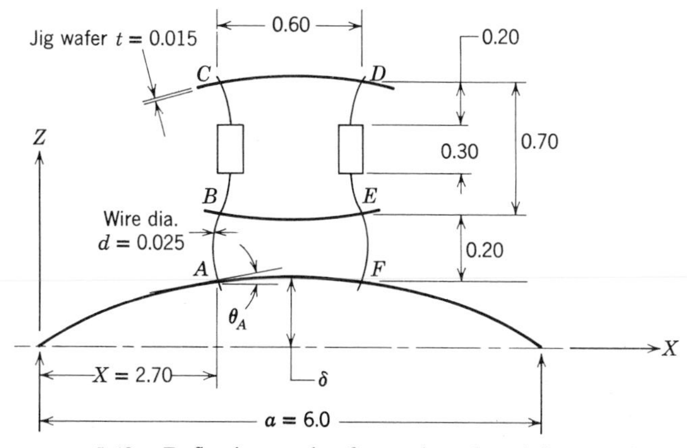

FIGURE 5.49. Deflection mode of a cordwood module on a circuit board.

The slope at any point along the circuit board is

$$\theta_A = \frac{dZ}{dX} = \delta \frac{\pi}{a} \cos \frac{\pi X}{a} \tag{5.151}$$

The master board deflection, δ, can be determined from the natural frequency of the circuit board. Assuming a uniform load distribution, the natural frequency of the circuit board can be approximated as a rectangular plate simply supported on four sides as shown in Fig. 5.51 (see Chapter 6).

$$f_n = \frac{\pi}{2}\left(\frac{D}{\rho}\right)^{1/2}\left(\frac{1}{a^2} + \frac{1}{b^2}\right) \tag{5.152}$$

where $E = 2.0 \times 10^6$ lb/in.^{2} epoxy fiberglass board
$h = 0.062$ in. board thickness
$\mu = 0.12$ Poisson's ratio
$a = 6.0$ in. length
$b = 5.0$ in. width
$W = 0.60$ lb weight

$$D = \frac{Eh^3}{12(1-\mu^2)} = \frac{(2 \times 10^6)(0.062)^3}{12[1-(0.12)^2]} = 40.2 \text{ lb in.}$$

$$\rho = \frac{\text{mass}}{\text{area}} = \frac{W}{gab} = \frac{0.60}{(386)(6.0)(5.0)} = 0.518 \times 10^{-4} \text{ lb sec}^2/\text{in.}^3$$

Substituting into Eq. 5.152

$$f_n = \frac{\pi}{2}\left(\frac{40.2}{0.518 \times 10^{-4}}\right)^{1/2}\left[\frac{1}{(6.0)^2} + \frac{1}{(5.0)^2}\right]$$

$$f_n = 93.7 \text{ Hz} \tag{5.153}$$

The maximum deflection at the center of the master circuit board can be approximated by considering the system to be equivalent to a single-degree-of-freedom system. The deflection can be determined from Eq. 2.42 as follows:

$$\delta = \frac{9.8G}{f_n^2} \quad \text{single amplitude} \tag{5.154}$$

Assume a sinusoidal vibration input of $5G$'s peak to an electronic box that supports the circuit board. Further assume the electronic box will have a transmissibility of 2 at a frequency of 94 Hz. The transmissibility of the circuit board can be approximated by considering the energy dissipation in the circuit board guides, the connector, and the components

FIGURE 5.50. Cordwood modules mounted on a printed-circuit board that has stiffening ribs fastened with screws (courtesy Norden division of United Aircraft).

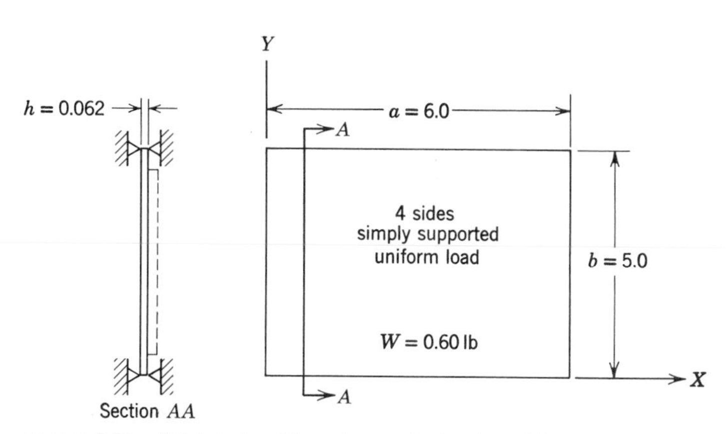

FIGURE 5.51. Printed circuit board supported on four sides.

mounted on the circuit board. If there is no other test information available, a good approximation for an epoxy fiberglass circuit board full of components is about (see Chapter 6 for more details)

$$Q = (f_n)^{1/2} \tag{5.155}$$

For this circuit board, the transmissibility for a 10-G sinusoidal force will probably be about

$$Q = (94)^{1/2} = 9.7; \text{ assume } Q = 10 \tag{5.156}$$

Substituting into Eq. 5.154

$$\delta = \frac{9.8(5.0)(2)(10)}{(93.7)^2} = 0.111 \text{ in.} \tag{5.157}$$

The slope at point A, where the cordwood module is soldered to the master printed circuit board, is shown in Fig. 5.49 and can be determined by Eqs. 5.151 and 5.157.

$$\theta_A = \delta \frac{\pi}{a} \cos \frac{\pi X}{a} = (0.111) \frac{\pi}{6.0} \cos \frac{\pi(2.70)}{6.0}$$

$$\theta_A = 0.00907 \text{ rad} \tag{5.158}$$

Since the component part body is so much stiffer than the lead wires, most of the deflection will be due to bending in the electrical lead wires. If the component body is ignored and only the lead wires are considered, the geometry of the cordwood module will appear as shown in Fig. 5.52. The horizontal force due to the kinetic energy of the component body is ignored here.

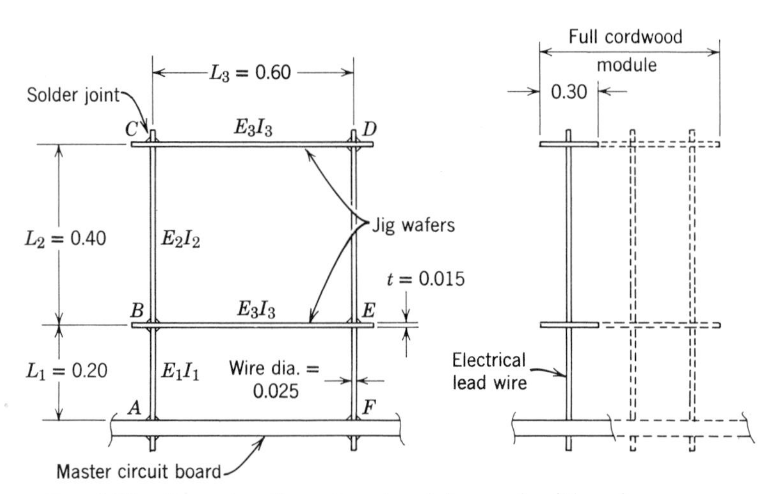

FIGURE 5.52. Geometry of a cordwood module on a circuit board.

Rewriting Eq. 5.149 for convenience

$$M_{B2} = \frac{\dfrac{\theta_A a_2}{3a_1}\left(2 - \dfrac{a_2}{6b}\right)}{\dfrac{a_3}{2} + \dfrac{a_2}{3}\left(1 - \dfrac{a_2}{12b}\right) + \dfrac{a_3 a_2}{3a_1}\left(2 - \dfrac{a_2}{6b}\right)}$$

where $\theta_A = 0.00907$ rad

$L_1 = 0.20$ in., $L_2 = 0.40$ in., $L_3 = 0.60$ in.

$E_1 = E_2 = 15 \times 10^6$ lb/in.^{2} (copper wire)

$E_3 = 2.0 \times 10^6$ lb/in.^{2} (epoxy fiberglass)

$$I_1 = I_2 = \frac{\pi d^4}{64} = \frac{\pi}{64}(0.025)^4 = 1.92 \times 10^{-8} \text{ in.}^4 \text{ (wire)}$$

$$I_3 = \frac{bt^3}{12} = \frac{(0.30)(0.015)^3}{12} = 8.45 \times 10^{-8} \text{ in.}^4 \text{ (board)}$$

$$a_1 = \frac{L_1}{E_1 I_1} = \frac{0.20}{(15 \times 10^6)(1.92 \times 10^{-8})} = 0.695 \text{ lb}^{-1} \text{ in.}^{-1}$$

$$a_2 = \frac{L_2}{E_2 I_2} = \frac{0.40}{(15 \times 10^6)(1.92 \times 10^{-8})} = 1.390 \text{ lb}^{-1} \text{ in.}^{-1}$$

$$a_3 = \frac{L_3}{E_3 I_3} = \frac{0.60}{(2.0 \times 10^6)(8.45 \times 10^{-8})} = 3.55 \text{ lb}^{-1} \text{ in.}^{-1}$$

$$b = \frac{a_2}{3} + \frac{a_3}{2} = \frac{1.39}{3} + \frac{3.55}{2} = 2.238 \text{ lb}^{-1}/\text{in.}^{-1}$$

Substituting into Eq. 5.149

$$M_{B2} = \frac{\dfrac{(9.07 \times 10^{-3})(1.39)}{3(0.695)}\left[2 - \dfrac{1.390}{6(2.238)}\right]}{\dfrac{3.55}{2} + \dfrac{1.39}{3}\left[1 - \dfrac{1.39}{12(2.238)}\right] + \dfrac{(3.55)(1.39)}{3(0.695)}\left[2 - \dfrac{1.39}{6(2.238)}\right]}$$

$$M_{B2} = \frac{6.05 \times 10^{-3}(1.896)}{1.775 + 0.439 + 4.488} = 1.72 \times 10^{-3} \text{ lb in.} \tag{5.159}$$

Equation 5.148 can be used to find M_A as follows:

$$M_A = \frac{4\theta_A}{a_1} - \frac{M_{B2} a_3}{a_1}$$

$$M_A = \frac{4(9.07 \times 10^{-3})}{0.695} - \frac{(1.72 \times 10^{-3})(3.55)}{0.695}$$

$$M_A = 52.2 \times 10^{-3} - 8.78 \times 10^{-3} = 43.42 \times 10^{-3} \text{ lb in.} \tag{5.160}$$

Equation 5.138 can be used to find M_{B1} as follows:

$$M_{B1}=\frac{M_A}{2}-\frac{3M_{B2}a_3}{2a_1}=\frac{43.42\times 10^{-3}}{2}-\frac{3(1.72\times 10^{-3})(3.55)}{2(0.695)}$$

$$M_{B1}=8.53\times 10^{-3}\ \text{lb in.} \tag{5.161}$$

Equation 5.145 can be used to find M_{B3} as follows:

$$M_{B3}=M_{B1}-M_{B2}=8.53\times 10^{-3}-1.72\times 10^{-3}$$

$$M_{B3}=6.81\times 10^{-3}\ \text{lb in.} \tag{5.162}$$

Equation 5.144 can be used to find M_c as follows:

$$M_c=\frac{M_{B3}a_2}{6b}=\frac{(6.81\times 10^{-3})(1.39)}{6(2.238)}$$

$$M_c=0.705\times 10^{-3}\ \text{lb in.} \tag{5.163}$$

The maximum bending moment in the electrical lead wires will occur at point A, which is shown by Eq. 5.160 as 43.4×10^{-3} lb in. The bending stress in the electrical lead wire can be determined from the standard bending stress equation.

$$S_A=\frac{M_A c}{I_1} \tag{5.164}$$

where $M_A=43.42\times 10^{-3}$ lb in.

$$c=\frac{0.025}{2}=0.0125\ \text{in. wire radius}$$

$$I_1=1.92\times 10^{-8}\ \text{in.}^4\ \text{wire moment of inertia}$$

$$S_A=\frac{(43.42\times 10^{-3})(1.25\times 10^{-2})}{1.92\times 10^{-8}}=28{,}270\ \text{lb/in.}^2 \tag{5.165}$$

The bending stress in the electrical lead wire due to M_{B1}, shown in Fig. 5.44, can be determined from Eq. (5.161) as follows:

$$S_{B1}=\frac{M_{B1}c}{I_1}=\frac{(8.53\times 10^{-3})(1.25\times 10^{-2})}{1.92\times 10^{-8}}=5550\ \text{lb/in.}^2 \tag{5.166}$$

The bending stress in the electrical lead wire at point c, in Fig. 5.52, can be determined from Eq. 5.163 as follows:

$$S_c=\frac{M_c c}{I_2}=\frac{(0.705\times 10^{-3})(1.25\times 10^{-2})}{1.92\times 10^{-8}}=459\ \text{lb/in.}^2 \tag{5.167}$$

An examination of the fatigue curves for copper wire (see Chapter 10, Fig. 10.10*b*) shows that fatigue failures might be expected in the electrical lead wires at points A and F if the number of fatigue cycles is high enough.

An examination of the fatigue curves for solder (see Chapter 10, Fig. 10.11) shows that fatigue failures might be expected in the solder itself at points A, F, B, and E shown in Figs. 5.49 and 5.52.

High stresses in the electrical lead wires are developed because large displacement amplitudes occur during circuit-board resonances. Increasing the resonant frequency will reduce the displacements and stresses very rapidly as shown by Eqs. 5.151 and 5.154.

5.14. Solder-Joint Stresses

The phrase "cold solder joint" is very well known to most production engineers. These engineers know how difficult it is to ensure the reliability of every solder joint. They also know that if a solder joint is not properly made, or if a solder joint should fail, it may result in an intermittent electrical connection that can be very difficult to locate.

Even if a solder joint is properly made, high bending or shear stresses coupled with many stress-reversal cycles can lead to fatigue failures.

An examination of many solder-joint fatigue failures indicates that a large percentage of these failures appear to be due to a "shear tear-out" type of failure induced by bending of the electrical lead wires. Since there are many different ways in which a solder joint can be formed, it is rather difficult to generalize on the size and shape of these solder joints. Also, many printed-circuit boards have printed circuits on only one side of the board. Many circuit boards with circuits on both sides, or even multiple-layer printed-circuit boards, may use plated-through holes or eyelets to improve the integrity of the soldered joint.

Printed-circuit boards with circuits on one side of the board only will usually have the weakest solder joints. These solder joints are more likely to fail because of "shear tear-out" in the solder, due to bending in the electrical lead wires as shown in Fig. 5.53.

A bending moment acting on the single-sided circuit board will have to fracture only one solder joint. A bending moment acting on the double-sided circuit board will have to fracture a double solder joint on the electrical lead wire, thus this joint is much stronger (also see Chapter 6, Section 6.8).

The shear-out type of failure in the single-sided circuit board appears, in many cases, to occur at a diameter in the solder fillet that is approximately 1.5 times the diameter of the electrical lead wire. Many of these fatigue failures also seem to occur about halfway up the height of the solder joint, at the knee of the fillet, where there is a rapid change in the

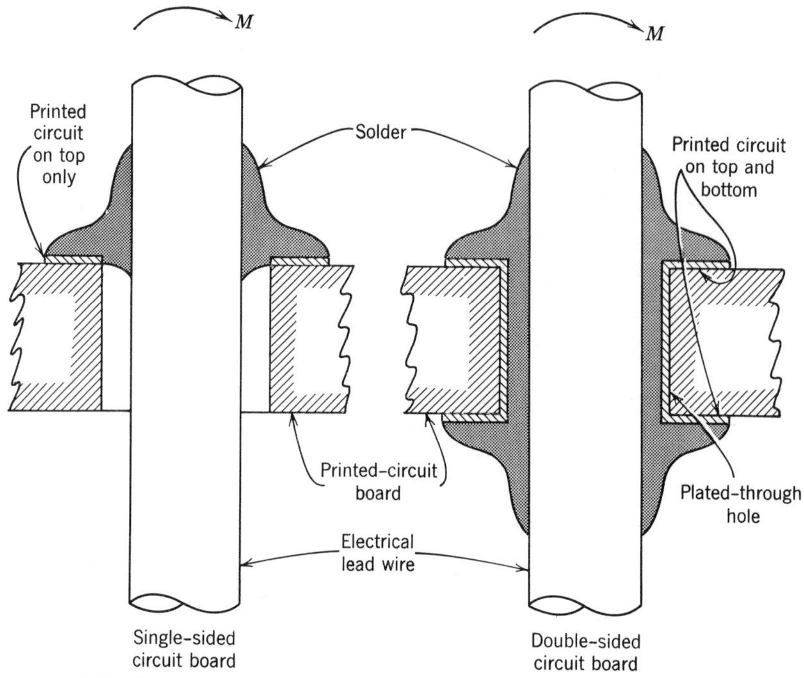

FIGURE 5.53. Solder joints on single and double sided circuit boards.

fillet radius. Some typical measured dimensions are shown in Fig. 5.54.

The maximum shear tear-out stress in the solder joints on the cordwood module will occur at points A and F, as shown in Figs. 5.49 and 5.54. This stress can be determined from the following equation [12]

$$S_s = \frac{M_A}{hA} \tag{5.168}$$

where $M_A = 43.4 \times 10^{-3}$ lb in. (see Eq. 5.160)

$h = 0.010$ in. minimum solder thickness in the fillet radius

$D = 0.032$ in. minimum shear-out diameter

$A = \frac{\pi}{4} D^2 = \frac{\pi}{4} (0.032)^2 = 8.00 \times 10^{-4}$ in.^{2} area at shear-out diameter

$$S_s = \frac{43.4 \times 10^{-3}}{(1.0 \times 10^{-2})(8.00 \times 10^{-4})} = 5420 \text{ lb/in.}^2 \tag{5.169}$$

The minimum shear tear-out stress at point A in Fig. 5.54 can be determined as follows:

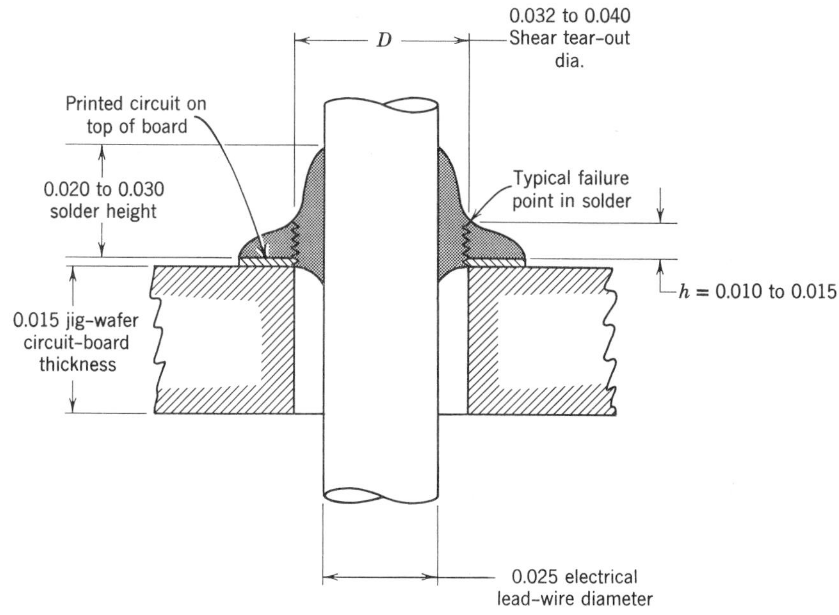

FIGURE 5.54. Fracture point in solder on a single sided circuit board.

$$h = 0.015 \text{ in. maximum solder thickness}$$

$$D = 0.040 \text{ in. maximum shear-out diameter}$$

$$A = \frac{\pi}{4}(0.040)^2 = 12.58 \times 10^{-4} \text{ in.}^2 \text{ area}$$

$$S_s = \frac{43.4 \times 10^{-3}}{(1.5 \times 10^{-2})(12.58 \times 10^{-4})} = 2290 \text{ lb/in.}^2 \tag{5.170}$$

An examination of the fatigue curves for solder (see Chapter 10, Fig. 10.11) shows that fatigue failures can be expected at the solder joint in the cordwood module at points A and F if the number of fatigue cycles is high enough.

Solder-joint stresses can be reduced in several different ways:

1. Increasing the resonant frequency of the printed circuit board will reduce the displacement and solder stresses or resonance.
2. Increasing the damping in the circuit board will decrease the transmissibility at resonance. This might be done with the use of laminations, to introduce shear damping, when the circuit board bends.
3. Using double-sided printed circuits increases the strength of the solder joint, (Fig. 5.53).

4. Increasing the diameter of the copper pads to which the electrical lead wires are soldered will prevent lifting the pad off of the circuit board, which is another form of vibration failure.
5. Bending the end tips of the electrical lead wires 90° and laying them along the printed-circuit runs increases the length of the solder joint.

5.15. Increasing the Cordwood Jig-Wafer Thickness

When vibration fatigue failures have occurred in cordwood-module solder joints, suggestions have often been made to increase the thickness of the jig-wafer printed-circuit boards to decrease the solder-joint stresses. This possibility can be examined by assuming the jig-wafer thickness in the previous problem is doubled, to 0.030 in., and recalculating the bending moments in the electrical lead wires.

Increasing the thickness of the cordwood-module jig-wafers from 0.015 to 0.030 in. will have little or no effect on the natural frequency of the master printed-circuit board. A thicker jig wafer will increase the stiffness of the master circuit board in only a small local area. Since this stiffness is not carried across the entire board, as it would be in the case of a rib, the resonant frequency can be assumed to stay about the same as that shown by Eq. 5.153.

The physical properties of the cordwood module shown in Fig. 5.52 will be the same, except for the thicker jig wafers, as shown by I_3.

$$I_3 = \frac{bt^3}{12} = \frac{(0.30)(0.030)^3}{12} = 67.5 \times 10^{-8} \text{ in.}^4$$

$$a_3 = \frac{L_3}{E_3 I_3} = \frac{0.60}{(2.0 \times 10^6)(67.5 \times 10^{-8})} = 0.445 \text{ lb}^{-1} \text{ in.}^{-1}$$

$$b = \frac{a_2}{3} + \frac{a_3}{2} = \frac{1.390}{3} + \frac{0.445}{2} = 0.686 \text{ lb}^{-1} \text{ in.}^{-1}$$

The bending moment M_{B2} shown in Fig. 5.44 can be determined from Eq. 5.149.

$$M_{B2} = \frac{\dfrac{(9.07 \times 10^{-3})(1.39)}{3(0.695)}\left[2 - \dfrac{1.390}{6(0.686)}\right]}{\dfrac{0.445}{2} + \dfrac{1.390}{3}\left[1 - \dfrac{1.390}{12(0.686)}\right] + \dfrac{(0.445)(1.39)}{3(0.695)}\left[2 - \dfrac{1.390}{6(0.686)}\right]}$$

$$M_{B2} = \frac{6.05 \times 10^{-3}(1.662)}{0.222 + 0.385 + 0.494} = 9.14 \times 10^{-3} \text{ lb in.} \tag{5.171}$$

From Eq. 5.148

$$M_A = \frac{4(9.07\times10^{-3})}{0.695} - \frac{(9.14\times10^{-3})(0.445)}{0.695}$$

$$M_A = 52.2\times10^{-3} - 5.85\times10^{-3} = 46.4\times10^{-3} \text{ lb in.} \qquad (5.172)$$

From Eq. 5.138

$$M_{B1} = \frac{46.4\times10^{-3}}{2} - \frac{3(9.14\times10^{-3})(0.445)}{2(0.695)}$$

$$M_{B1} = 23.2\times10^{-3} - 8.77\times10^{-3} = 14.4\times10^{-3} \text{ lb in.} \qquad (5.173)$$

From Eq. 5.145

$$M_{B3} = 14.4\times10^{-3} - 9.14\times10^{-3} = 5.26\times10^{-3} \text{ lb in.} \qquad (5.174)$$

From Eq. 5.144

$$M_c = \frac{(5.26\times10^{-3})(1.39)}{6(0.686)} = 1.78\times10^{-3} \text{ lb in.} \qquad (5.175)$$

Comparing Eqs. 5.171–5.175 for jig wafers 0.030 in. thick, with Eqs. 5.159–5.163 for jig wafers 0.015 in. thick shows that increasing the jig-wafer thickness from 0.015 to 0.030 in. will tend to increase, not decrease, the bending moments. Therefore the bending stresses in the cordwood-module electrical-lead-wire solder joints will also increase, if the jig-wafer thickness is increased.

The resonant frequency of the master circuit board should be increased by adding ribs, for example, to decrease the dynamic displacements and solder-joint stresses in the cordwood modules.

5.16. Bent with Uniform Lateral Load and Hinged Ends

Superposition can be used to determine the loads, moments, and deflections of a rectangular bent with hinged ends subjected to a uniform lateral load. The deflection can be broken up into two parts. The first part will consist of bending of the vertical legs with no bending of the horizontal leg. The second part will consist of the bending of the horizontal leg with no bending of the vertical legs. However, when the horizontal leg bends, each end will rotate through an angle θ_B which will permit the vertical legs to rotate (not bend) to that same angle as shown (Fig. 5.55). A free-body force diagram of the bent is shown in Fig. 5.56.

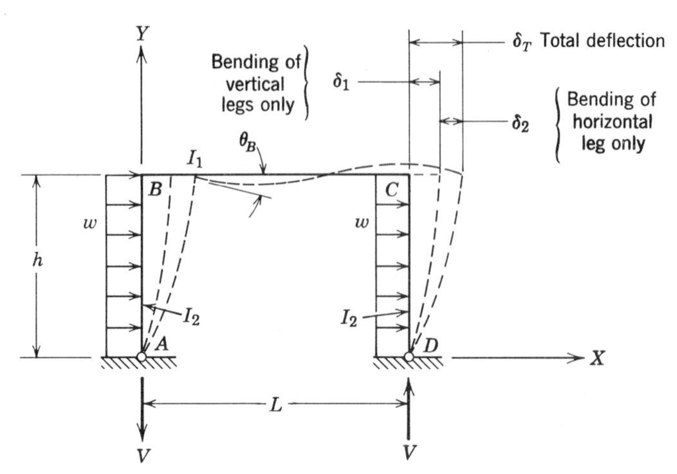

FIGURE 5.55. Bent with uniform lateral load and hinged ends.

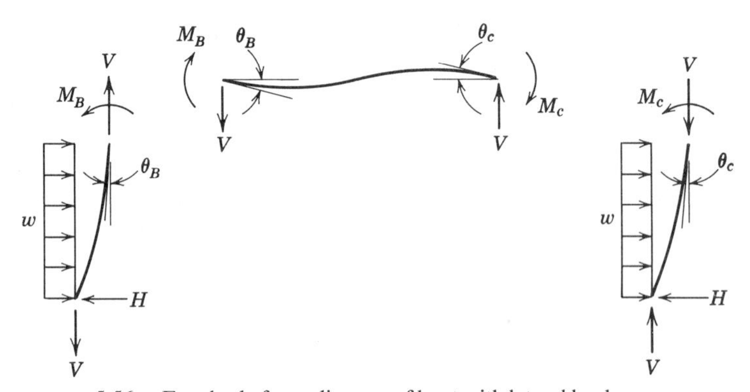

FIGURE 5.56. Free body force diagram of bent with lateral load.

Consider the sum of the forces in the X direction for vertical leg AB

$$H = wh \tag{5.176}$$

Consider moments about point B. For small displacements, the vertical force V will result in small moments so this force can be ignored with very little error.

$$M_B + \frac{wh^2}{2} - Hh = 0$$

From Eq. 5.176

$$M_B = \frac{wh^2}{2} \tag{5.177}$$

The total deflection can be determined by first considering bending of the vertical legs (Fig. 5.57), with no bending in the horizontal leg BC (note: $E_1 = E_2 = E$).

Using standard handbook equations

$$\delta_a = \frac{Hh^3}{3EI_2} \quad \text{and} \quad \delta_b = \frac{wh^4}{8EI_2} \tag{5.178}$$

$$\delta_1 = \delta_a - \delta_b = \frac{Hh^3}{3EI_2} - \frac{wh^4}{8EI_2} = \frac{5wh^4}{24EI_2} \tag{5.179}$$

The bending of the horizontal leg BC is shown in Fig. 5.58.

$$\theta_1 = \frac{M_B L}{3EI_1} \quad \text{and} \quad \theta_2 = \frac{M_B L}{6EI_1} \quad \text{(symmetrical)}$$

$$\theta_B = \theta_1 - \theta_2 = \frac{M_B L}{6EI_1} \tag{5.180}$$

When the horizontal leg BC bends, it rotates through an angle θ_B and the vertical leg must rotate, without bending, through the same angle (Fig. 5.59).

$$\delta_2 = h\theta_B = \frac{M_B Lh}{6EI_1}$$

Using Eq. 5.177

$$\delta_2 = \frac{wh^3 L}{12EI_1} \tag{5.181}$$

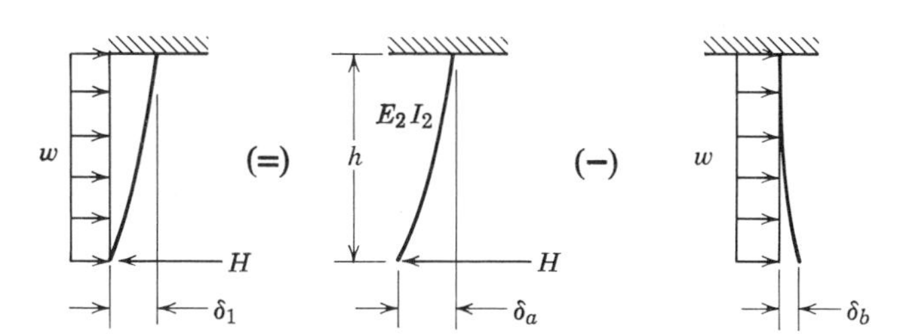

FIGURE 5.57. Free body deflection diagram of vertical leg.

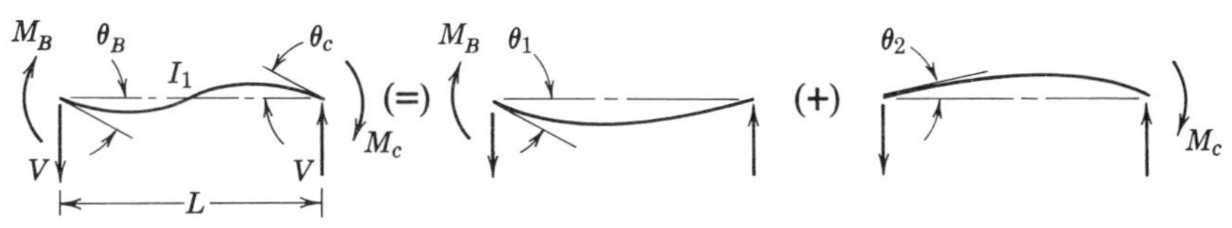

FIGURE 5.58. Free body force diagram of horizontal leg.

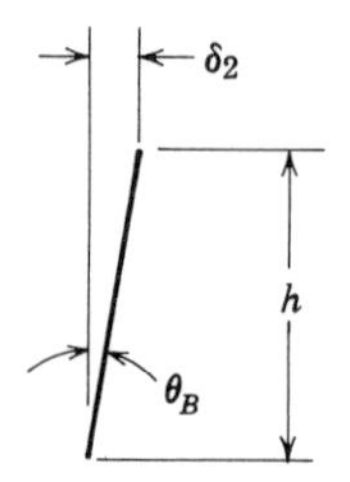

FIGURE 5.59. Rotation of vertical leg without bending.

The total deflection then becomes

$$\delta_T = \delta_1 + \delta_2 = \frac{5wh^4}{24EI_2} + \frac{wh^3L}{12EI_1} = \frac{wh^4}{12EI_2}\left(\frac{5}{2} + \frac{I_2L}{I_1h}\right)$$

Let

$$K = hI_1/LI_2$$

$$\delta_T = \frac{wh^4}{12EI_2}\left(\frac{5}{2} + \frac{1}{K}\right) \tag{5.182}$$

The vertical reaction V can be determined by considering the moments of the external forces about point A as shown in Fig. 5.55.

$$2(wh)\left(\frac{h}{2}\right) - VL = 0$$

$$V = \frac{wh^2}{L} \tag{5.183}$$

5.17. Bent with Uniform Lateral Load and Fixed Ends

A rectangular bent with a uniform lateral load and fixed ends can be analyzed as outlined in Fig. 5.60. A free-body force diagram is shown in Fig. 5.61. The vertical leg AB is shown in Fig. 5.62.

$$\theta_1 = \frac{wh^3}{6EI_2} \quad \text{and} \quad \theta_2 = \frac{M_Bh}{EI_2}$$

$$\theta_B = \theta_1 - \theta_2 = \frac{wh^3}{6EI_2} - \frac{M_Bh}{EI_2} \tag{5.184}$$

The horizontal leg BC is shown in Fig. 5.63.

$$\theta_3 = \frac{M_BL}{3EI_1} \quad \text{and} \quad \theta_4 = \frac{M_cL}{6EI_1} \quad \text{from symmetry, } M_B = M_c$$

$$\theta_B = \theta_3 - \theta_4 = \frac{M_BL}{6EI_1} \tag{5.185}$$

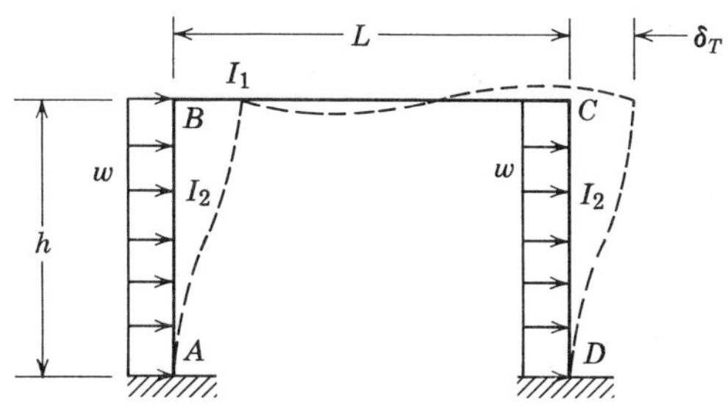

FIGURE 5.60. Bent with uniform lateral load and fixed ends.

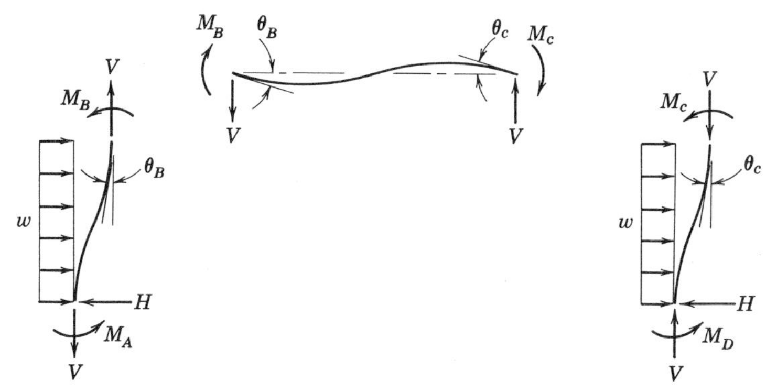

FIGURE 5.61. Free body force diagram of bent with fixed ends.

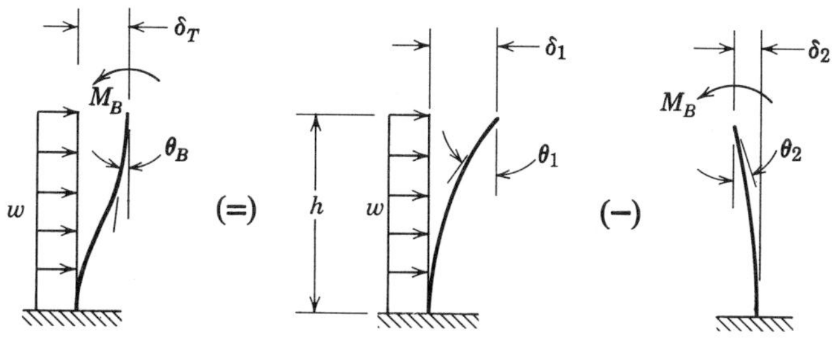

FIGURE 5.62. Free body force diagram of vertical leg.

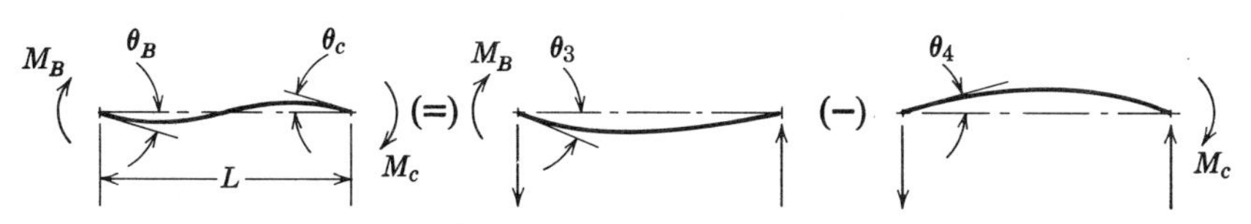

FIGURE 5.63. Free body force diagram of horizontal leg.

Substituting Eq. 5.185 into Eq. 5.184 and solving for M_B

$$M_B\left(\frac{L}{6I_1}+\frac{h}{I_2}\right)=\frac{wh^3}{6I_2}$$

Let $K = hI_1/LI_2$:

$$M_B = wh^2\left(\frac{K}{1+6K}\right) \tag{5.186}$$

The deflections can be determined from Fig. 5.62.

$$\delta_1=\frac{wh^4}{8EI_2} \qquad \text{and} \qquad \delta_2=\frac{M_B h^2}{2EI_2}$$

The total deflection is then as follows:

$$\delta_T=\delta_1-\delta_2=\frac{wh^4}{8EI_2}-\frac{M_B h^2}{2EI_2}$$

Using Eq. 5.186, the total deflection becomes

$$\delta_T=\frac{wh^4}{8EI_2}\left(1-\frac{4K}{1+6K}\right) \tag{5.187}$$

The moment at the fixed end A then becomes

$$M_A=\frac{wh^2}{3}\left(1+\frac{1}{12K+2}\right) \tag{5.188}$$

The vertical force is then

$$V=\frac{wh^2}{3L}\left(1-\frac{1}{6K+1}\right) \tag{5.189}$$

The horizontal force H is

$$H = wh \tag{5.190}$$

5.18. BENT WITH UNIFORM VERTICAL LOAD AND HINGED ENDS

A rectangular bent with a uniform vertical load and hinged ends is shown in Fig. 5.64. A free-body force diagram will appear as in Fig. 5.65.

Using superposition techniques as outlined in previous problems, the bending moment at point B is

$$M_B=\frac{wL^2}{4(2K+3)} \tag{5.191}$$

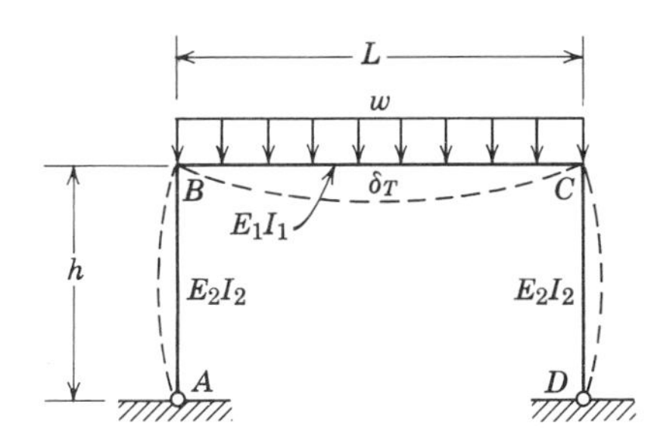

FIGURE 5.64. Bent with uniform vertical load and hinged ends.

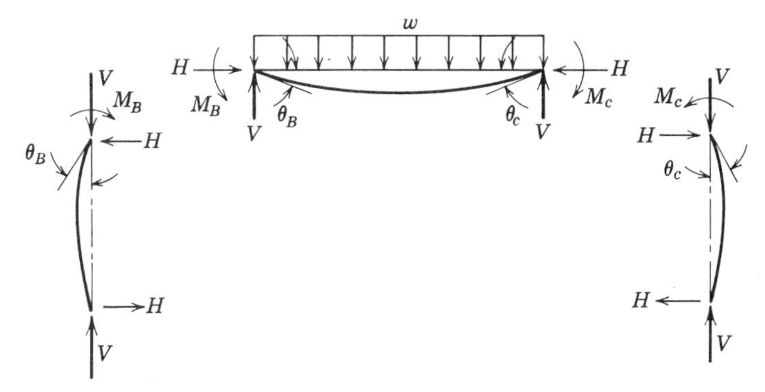

FIGURE 5.65. Free body force diagram of bent with hinged ends.

$$V = \frac{wL}{2} \tag{5.192}$$

$$H = \frac{wL^2}{4h(2K+3)} \tag{5.193}$$

$$\delta_T = \frac{wL^4}{64EI_1}\left(\frac{5}{6} - \frac{2}{2K+3}\right) \tag{5.194}$$

5.19. Bent with Uniform Vertical Load and Fixed Ends

A rectangular bent with a uniform vertical load and fixed ends will appear as shown in Fig. 5.66. A free-body force diagram is shown in Fig. 5.67.

Using superposition techniques as outlined in previous problems results in the following:

$$M_A = \frac{wL^2}{12(K+2)} \tag{5.195}$$

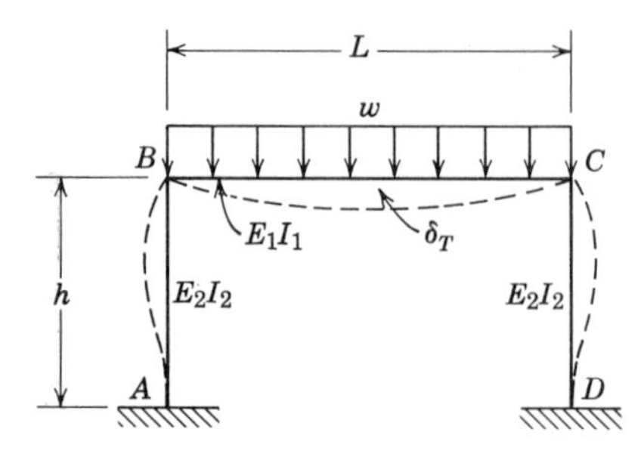

FIGURE 5.66. Bent with uniform vertical load and fixed ends.

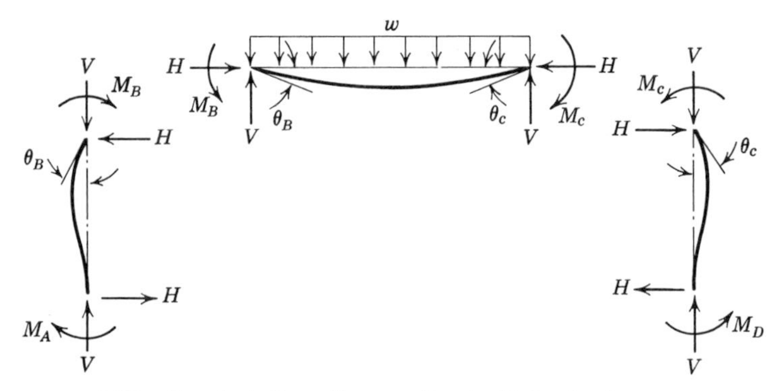

FIGURE 5.67. Free body force diagram of bent with fixed ends.

$$M_B = \frac{wL^2}{6(K+2)} \tag{5.196}$$

$$V = \frac{wL}{2} \quad \text{and} \quad H = \frac{wL^2}{4h(K+2)} \tag{5.197}$$

$$\delta_T = \frac{wL^4}{48EI_1}\left(\frac{5}{8} - \frac{1}{K+2}\right) \tag{5.198}$$

5.20. Strain Energy—Bent with Hinged Ends

Castigliano's strain energy theorem can be used to determine the loads, moments, and deflections in rectangular bents. This theorem states that the partial derivative of the strain energy, with respect to an applied force, will give the deflection produced by that force in the direction of the force.

The total strain energy in a system can usually be accounted for by considering tension and compression, bending, torsion, and shear. Considering each one individually, they are as follows [13]:

$$\text{Tension and compression:} \quad U_t = \int_a^b \frac{P^2\,dX}{2EA} \tag{5.199}$$

$$\text{Bending:} \quad U_B = \int_a^b \frac{M^2\,dX}{2EI} \tag{5.200}$$

$$\text{Torsion:} \quad U_T = \int_a^b \frac{T^2\,dX}{2GJ} \tag{5.201}$$

$$\text{Shear:} \quad U_S = \int_a^b \frac{V^2\,dX}{2GA} \tag{5.202}$$

For example, the lateral displacement of a rectangular bent with hinged ends and a concentrated lateral load can be determined by considering only the strain energy of bending.

Consider the rectangular bent shown in Fig. 5.68. The bending deflections are so much greater than all of the other deflections, such as tension and compression, torsion and shear, that only the bending need be considered unless the beams are very short. A short beam is one whose length is less than about three times the depth. If stresses are within the elastic limit and deflections are small, there is very little error involved in ignoring the other deflection sources.

The strain energy of bending must be considered for two vertical legs and one horizontal leg. Since the bent is symmetrical, only one half of the system can be considered (Fig. 5.69).

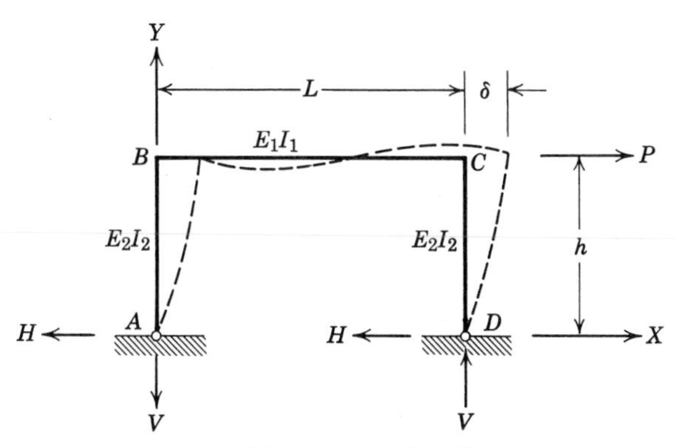

FIGURE 5.68. Bent with a lateral load and hinged ends.

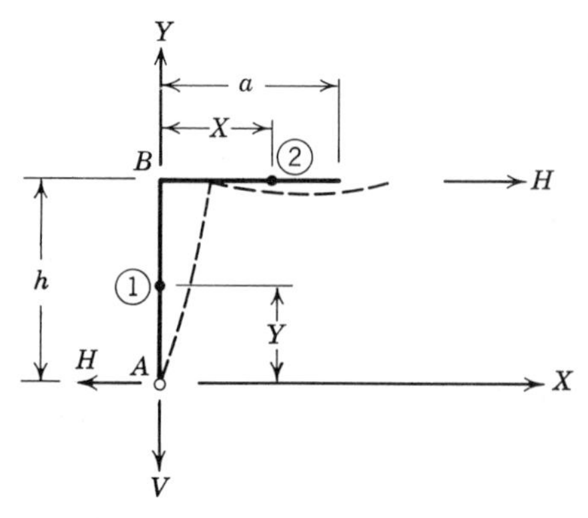

FIGURE 5.69. Half of a bent with a lateral load and hinged ends.

The strain energy of bending was shown to be

$$U_B = \int_a^b \frac{M^2\,dX}{2EI} \qquad \text{(see Eq. 5.200)}$$

Using Castigliano's theorem to find the deflection

$$\delta = \frac{\partial U_B}{\partial P} = \int_a^b \frac{2M \dfrac{\partial M}{\partial P}\,dX}{2EI} = \frac{1}{EI}\int_a^b M \frac{\partial M}{\partial P}\,dX \tag{5.203}$$

With the X and Y axes as shown, the bending moment at point 1 on the vertical leg is

$$M_1 = HY \tag{5.204}$$

Considering all the forces in the X direction

$$H = \frac{P}{2} \tag{5.205}$$

Substituting into Eq. 5.204

$$M_1 = \frac{PY}{2} \tag{5.206}$$

$$\frac{\partial M_1}{\partial P} = \frac{Y}{2} \tag{5.207}$$

With the X and Y axes as shown, the bending moment at point 2 on the horizontal leg is

$$M_2 = Hh - VX \tag{5.208}$$

The force V can be determined by taking moments about point A in Fig. 5.68.

$$Ph = VL$$

$$V = \frac{Ph}{L} \tag{5.209}$$

Substituting Eqs. 5.205 and 5.209 into Eq. 5.208

$$M_2 = \frac{Ph}{2} - \frac{Ph}{L}X = P\left(\frac{h}{2} - \frac{h}{L}X\right) \tag{5.210}$$

$$\frac{\partial M_2}{\partial P} = \left(\frac{h}{2} - \frac{h}{L}X\right) \tag{5.211}$$

Considering the strain energy in two vertical legs and two halves of the horizontal leg as shown in Fig. 5.66, the deflection equation 5.203 can be written as follows (note: $E_1 = E_2 = E$)

$$\delta = \frac{2}{EI_2}\int_0^h M_1 \frac{\partial M_1}{\partial P}\,dY + \frac{2}{EI_1}\int_0^a M_2 \frac{\partial M_2}{\partial P}\,dX$$

Substituting Eqs. 5.206, 5.207, 5.210, and 5.211 into the above

$$\delta = \frac{2}{EI_2}\int_0^h \left(\frac{PY}{2}\right)\left(\frac{Y}{2}\right) dY + \frac{2}{EI_1}\int_0^a P\left(\frac{h}{2} - \frac{h}{L}X\right)\left(\frac{h}{2} - \frac{h}{L}X\right) dX$$

$$\delta = \frac{2P}{EI_2}\int_0^h \frac{Y^2}{4}\,dY + \frac{2P}{EI_1}\int_0^a \left(\frac{h^2}{4} - \frac{h^2X}{L} + \frac{h^2X^2}{L^2}\right) dX$$

$$\delta = \frac{2P}{EI_2}\left(\frac{Y^3}{12}\right)_0^h + \frac{2P}{EI_1}\left(\frac{h^2X}{4} - \frac{h^2X^2}{2L} + \frac{h^2X^3}{3L^3}\right)_0^a$$

$$\delta = \frac{2Ph^3}{12EI_2} + \frac{2P}{EI_1}\left(\frac{h^2a}{4} - \frac{h^2a^2}{2L} + \frac{h^2a^3}{3L^2}\right)$$

Since

$$a = L/2$$

$$\delta = \frac{Ph^3}{6EI_2} + \frac{Ph^2L}{12EI_1}$$

Let $K = hI_1/LI_2$ and solve for δ.

$$\delta = \frac{Ph^3}{6EI_2}\left(1 + \frac{1}{2K}\right) \tag{5.212}$$

An examination of Eq. 5.212 shows it is exactly the same as Eq. 5.123 which was obtained by using superposition.

5.21. Strain Energy—Bent With Fixed Ends

If a rectangular bent has fixed ends, as in Fig. 5.70, it becomes statically indeterminate. Castigliano's theorem can then be utilized by considering various physical characteristics of the structure. For example, it is obvious from Fig. 5.70 that the angular rotation at points A and D will be zero, because these ends are fixed. Also, since the structure is symmetrical, only one half of the structure may be examined as shown in Fig. 5.71. What may not be obvious from the Fig. 5.71 is that the center of the horizontal leg BC becomes a point of inflection, at point E, so the moment M_E must be zero. Also, point E will move only in the horizontal direction and not in the vertical direction. If the X and Y axes are taken as shown in Fig. 5.71, Castigliano's theorem can be used by knowing that the vertical deflection at point E, due to the force V, must be equal to zero. This can be written as follows:

$$\delta_{\text{vert}} = \frac{\partial U}{\partial V} = 0 \tag{5.213}$$

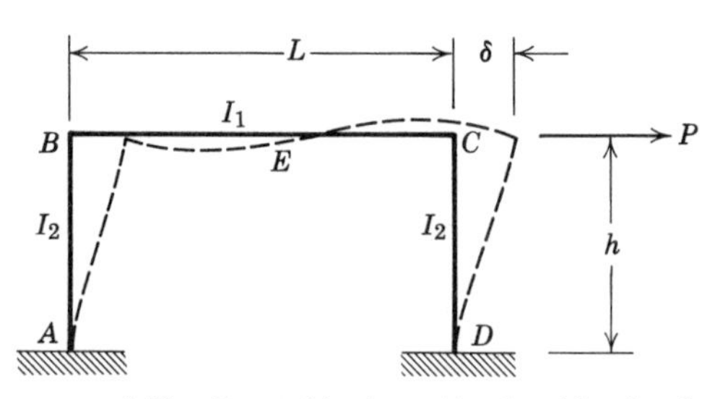

FIGURE 5.70. Bent with a lateral load and fixed ends.

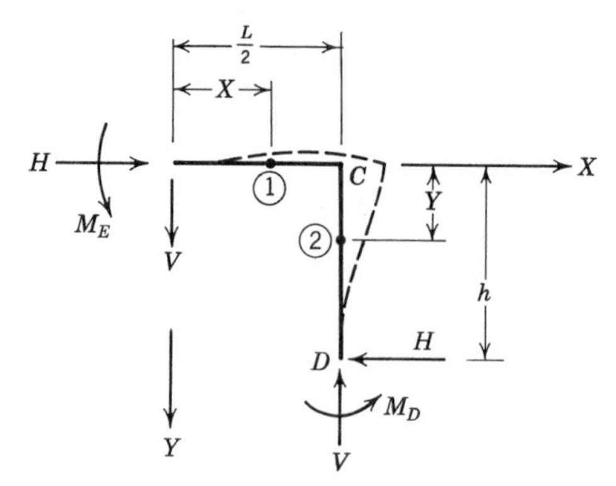

FIGURE 5.71. Half of a bent with a lateral load and fixed ends.

Since only bending deflections will be considered here and since there are bending moments to be considered at points 1 and 2 in Fig. 5.71, the deflection equation can be established using Eqs. 5.200 and 5.203.

$$\delta_{\text{vert}} = \frac{1}{EI_1}\int_0^{L/2} M_1 \frac{\partial M_1}{\partial V}\,dX + \frac{1}{EI_2}\int_0^{h} M_2 \frac{\partial M_2}{\partial V}\,dY = 0 \qquad (5.214)$$

The bending moment at point 1 on the horizontal leg is

$$M_1 = VX \qquad (5.215)$$

$$\frac{\partial M_1}{\partial V} = X \qquad (5.216)$$

The bending moment at point 2 on the vertical leg will be

$$M_2 = V\left(\frac{L}{2}\right) - HY \qquad (5.217)$$

The horizontal force H can be determined by considering all of the forces in the X direction. This results in

$$H = \frac{P}{2} \qquad (5.218)$$

Substituting Eq. 5.218 into Eq. 5.217

$$M_2 = \frac{VL}{2} - \frac{PY}{2} \qquad (5.219)$$

$$\frac{\partial M_2}{\partial V} = \frac{L}{2} \qquad (5.220)$$

Substituting Eqs. 5.215, 5.216, 5.219, and 5.220 into Eq. 5.214.

$$\frac{1}{EI_1}\int_0^{L/2} (VX)\,(X)\,dX + \frac{1}{EI_2}\int_0^{h} \left(\frac{VL}{2} - \frac{PY}{2}\right)\left(\frac{L}{2}\right) dY = 0$$

$$\frac{1}{EI_1}\left(V\frac{X^3}{3}\right)_0^{L/2} + \frac{1}{EI_2}\left[\frac{VL^2}{4}Y - \left(\frac{PL}{4}\right)\left(\frac{Y^2}{2}\right)\right]_0^h = 0$$

$$\frac{1}{EI_1}\left(\frac{VL^3}{24}\right) + \frac{1}{EI_2}\left(\frac{VL^2h}{4} - \frac{PLh^2}{8}\right) = 0$$

$$\frac{VL^2}{6I_1} + \frac{VLh}{I_2} = \frac{Ph^2}{2I_2}$$

Let $K = hI_1/LI_2$ and solve for V.

$$V = \frac{3PhK}{L(1+6K)} \tag{5.221}$$

The moment M_D can be determined from Fig. 5.71 by taking moments about point D.

$$\sum M_D = 0$$

$$M_D + V\left(\frac{L}{2}\right) - Hh = 0 \tag{5.222}$$

Substituting Eqs. 5.221 and 5.218 into the above

$$M_D + \left[\frac{3PhK}{L(1+6K)}\right]\left(\frac{L}{2}\right) - \frac{Ph}{2} = 0$$

With a little algebra this can be written as

$$M_D = \frac{Ph}{4}\left(1 + \frac{1}{6K+1}\right) \tag{5.223}$$

An examination of Eqs. 5.221 and 5.223 shows they are exactly the same as Eqs. 5.71 and 5.69, which were obtained by superposition.

The deflection can be determined using the same method, making sure to include both vertical legs and both halves of the top horizontal leg.

$$\delta = \frac{2}{EI_1}\int_0^{L/2} M_1 \frac{\partial M_1}{\partial P}\, dX + \frac{2}{EI_2}\int_0^h M_2 \frac{\partial M_2}{\partial P}\, dY \tag{5.224}$$

The moment at point 1 is

$$M_1 = VX = \frac{3PhK}{L(1+6K)} X \tag{5.225}$$

$$\frac{\partial M_1}{\partial P} = \frac{3hK}{L(1+6K)} X \tag{5.226}$$

The moment at point 2 is

$$M_2 = \frac{VL}{2} - \frac{PY}{2} = \frac{3PhK}{L(1+6K)}\left(\frac{L}{2}\right) - \frac{PY}{2} \tag{5.227}$$

$$\frac{\partial M_2}{\partial P} = \frac{1}{2}\left[\frac{3hK}{(1+6K)} - Y\right] \tag{5.228}$$

Substituting Eqs. 5.225–5.228 into Eq. 5.224

$$\delta = \frac{2}{EI_1}\int_0^{L/2}\left[\frac{3PhKX}{L(1+6K)}\right]\left[\frac{3hKX}{L(1+6K)}\right] dX$$

$$+ \frac{2}{EI_2}\int_0^h \frac{P}{2}\left[\frac{3hK}{(1+6K)} - Y\right]\left(\frac{1}{2}\right)\left[\frac{3hK}{(1+6K)} - Y\right] dY$$

$$\delta=\frac{18Ph^2K^2}{EI_1L^2(1+6K)^2}\left[\frac{X^3}{3}\right]_0^{L/2}+\frac{P}{2EI_2}\left[\frac{9h^2K^2Y}{(1+6K)^2}-\frac{6hK}{(1+6K)}\left(\frac{Y^2}{2}\right)+\frac{Y^3}{3}\right]_0^h$$

$$\delta=\frac{3Ph^2K^2L}{4EI_1(1+6K)^2}+\frac{Ph^3}{2EI_2}\left[\frac{27K^2-9K(1+6K)+(1+6K)^2}{3(1+6K)^2}\right]$$

$$\delta=\frac{Ph^3}{24EI_2}\left(\frac{36K^2+30K+4}{36K^2+12K+1}\right)$$

This can be written as

$$\delta=\frac{Ph^3}{24EI_2}\left[1+\frac{3(6K+1)}{(6K+1)(6K+1)}\right]$$

$$\delta=\frac{Ph^3}{24EI_2}\left(1+\frac{3}{6K+1}\right) \tag{5.229}$$

An examination of Eq. 5.229 shows it is exactly the same as Eq. 5.74, which was obtained by superposition.

The previous bent with fixed ends could also have been analyzed by knowing that, for small angular displacements, the rotation angle on the vertical leg shown by θ_{B2} will be the same as the rotation angle on the horizontal leg shown by θ_{B1} and indicated in the free-body force diagram of Fig. 5.72 (see also Fig. 5.19). A close inspection of the figure shows that the slope is positive at θ_{B2}, but negative at θ_{B1}, thus

$$\theta_{B2}=-\theta_{B1} \tag{5.230}$$

Castigliano's theorem can be applied here since the partial derivative of the strain energy with respect to an applied moment will give the angular rotation produced by that moment in the direction of the moment. Since the strain energy was due only to bending, Eq. 5.200 is used again.

$$U_B=\int_a^b\frac{M^2\,dX}{2EI} \qquad \text{(see Eq. 5.200)}$$

Using Castigliano's theorem to find the angular rotation due to an applied moment

$$\theta=\frac{\partial U_B}{\partial M_A}=\int_a^b\frac{2M\dfrac{\partial M}{\partial M_A}}{2EI}\,dX=\frac{1}{EI}\int_a^b M\frac{\partial M}{\partial M_A}\,dX \tag{5.231}$$

In Fig. 5.72 the angular rotation due to moment M_B acting at the top of the vertical leg will be approximately the same as the angular rotation due to moment M_B acting on the horizontal leg. Then using Eqs. 5.230 and 5.231, this can be written as follows:

$$\frac{1}{EI_2}\int_0^h M_1\frac{\partial M_1}{\partial M_B}\,dY=-\frac{1}{EI_1}\int_0^{L/2} M_2\frac{\partial M_2}{\partial M_B}\,dX \tag{5.232}$$

If the X and Y axes are taken as shown in Fig. 5.72, the bending moment at point 1 on the vertical leg is

$$M_1 = M_B - HY = M_B - \frac{PY}{2} \tag{5.233}$$

$$\frac{\partial M_1}{\partial M_B} = 1 \tag{5.234}$$

The bending moment at point 2 on the horizontal leg is

$$M_2 = -M_B + VX \tag{5.235}$$

The vertical force can be expressed in terms of M_B by considering the sum of the moments about the left end, at point B, on the horizontal leg (note that from symmetry $M_c = M_D$)

$$V = \frac{2M_B}{L} \tag{5.236}$$

Substituting Eq. 5.236 into Eq. 5.235

$$M_2 = -M_B + \frac{2M_B}{L}X = M_B\left(\frac{2X}{L} - 1\right) \tag{5.237}$$

$$\frac{\partial M_2}{\partial M_B} = \left(\frac{2X}{L} - 1\right) \tag{5.238}$$

Substituting Eqs. 5.233, 5.234, 5.237, and 5.238 into Eq. 5.232

$$\frac{1}{EI_2}\int_0^h \left(M_B - \frac{PY}{2}\right) dY = -\frac{1}{EI_1}\int_0^{L/2} M_B\left(\frac{2X}{L} - 1\right)\left(\frac{2X}{L} - 1\right) dX$$

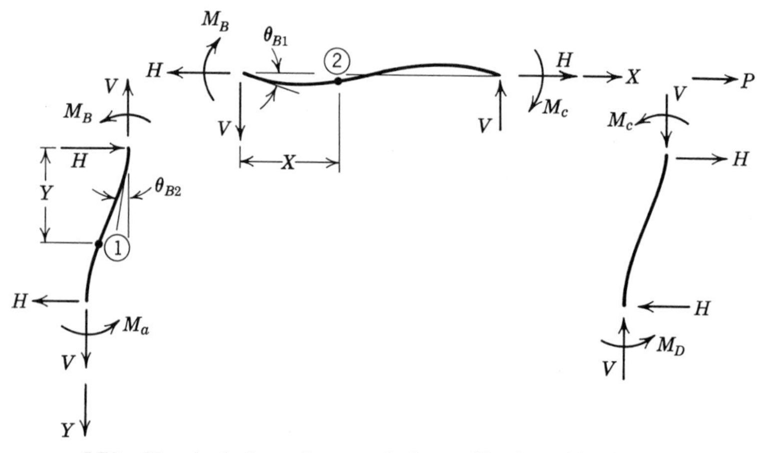

FIGURE 5.72. Free body force diagram of a bent with a lateral load.

$$\frac{1}{EI_2}\left(M_B Y - \frac{PY^2}{4}\right)_0^h = -\frac{1}{EI_1}\left[M_B\left(\frac{4X^3}{3L^2} - \frac{4X^2}{2L} + X\right)\right]_0^{L/2}$$

$$\frac{M_B h}{I_2} + \frac{M_B L}{6I_1} = \frac{Ph^2}{4I_2}$$

Let $K = hI_1/LI_2$ and solve for M_B.

$$M_B = \frac{Ph}{4\left(1 + \frac{1}{6K}\right)}$$

With a little algebra this can be written as

$$M_B = \frac{Ph}{4}\left(1 - \frac{1}{6K+1}\right) \tag{5.239}$$

Substituting Eq. 5.236 into the above

$$V = \frac{Ph}{2L}\left(1 - \frac{1}{6K+1}\right)$$

With a little algebra this can be written as

$$V = \frac{3PhK}{L(6K+1)} \tag{5.240}$$

An examination of Eq. 5.240 shows it is exactly the same as Eq. 5.221. The deflection can be obtained in the same manner as shown in the previous problem.

5.22. Strain Energy – Bent with Circular Arc Section

A modified form of the rectangular bent is very often found in gyros and inertial measuring units where a platform position can be maintained with the use of a gimbal system. The gimbal system, in many cases, can be approximated as a modified bent that has some straight members and some curved members.

Consider the case where the bent has straight sides with fixed ends and a curved top with a concentrated load in the center (Fig. 5.73). Since the bent is symmetrical, it is possible to simplify the problem by considering only half of the system. The vertical deflection produced by the force P acting on the full bent will then be the same as the vertical deflection produced by the force V acting on half of the system.

Since the system is statically indeterminate, it is necessary to find the forces and moments before the deflection can be determined. Castigliano's

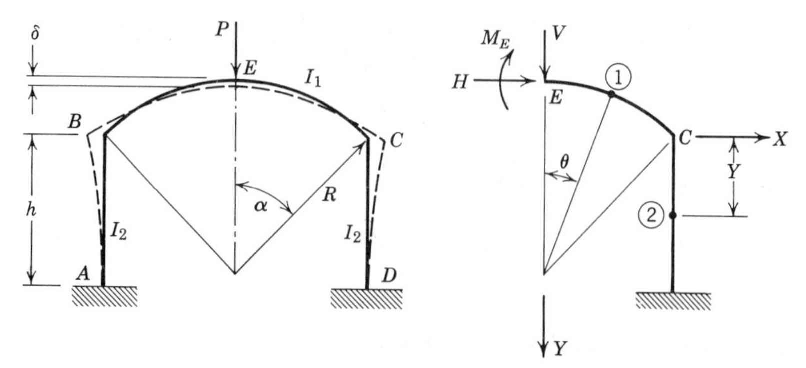

FIGURE 5.73. Bent with fixed ends and an arched top with a vertical load.

theorem is very convenient for this application, so the physical structure can be examined to make use of some obvious characteristics.

There are two unknowns; H and M_E, so two equations must be developed to find these two unknowns. The force V is known since it can be determined easily by considering all of the forces in the vertical direction. This results in

$$V = \frac{P}{2} \tag{5.241}$$

An examination of Fig. 5.73 shows that the relative horizontal deflection at point E, due to the load H, must be zero; this will determine the first equation. Using Castigliano's theorem, this can be written as

$$\delta_{EX} = \frac{\partial U}{\partial H} = 0 \tag{5.242}$$

An examination of Fig. 5.73 also shows that the relative slope at point E, due to the moment M_E, must be zero so this will determine the second equation. Using Castigliano's theorem, this can be written as

$$\theta_E = \frac{\partial U}{\partial M_E} = 0 \tag{5.243}$$

The bending moment at point 1 on the curved top leg is

For $0 \leq \theta \leq \alpha$

$$M_1 = HR(1 - \cos\theta) + M_E - VR\sin\theta \tag{5.244}$$

For the first equation, the horizontal deflection due to force H is zero. From Eq. 5.244

$$\frac{\partial M_1}{\partial H} = R(1 - \cos\theta) \tag{5.245}$$

The bending moment at point 2 on the vertical leg is

$$M_2 = H[R(1-\cos\alpha)+Y]+M_E - VR\sin\alpha \tag{5.246}$$

$$\frac{\partial M_2}{\partial H} = R(1-\cos\alpha)+Y \tag{5.247}$$

Only bending deflections will be considered here since they result in the largest amplitudes. Then, using Eqs. 5.200 and 5.242, the horizontal deflection at point E can be written as

$$\delta_{EX} = \frac{1}{EI_1}\int_0^\alpha M_1 \frac{\partial M_1}{\partial H} R\,d\theta + \frac{1}{EI_2}\int_0^h M_2 \frac{\partial M_2}{\partial H}\,dY = 0 \tag{5.248}$$

Substituting Eqs. 5.244–5.247 into the above

$$\frac{1}{EI_1}\int_0^\alpha [HR(1-\cos\theta)+M_E - VR\sin\theta](R)(1-\cos\theta)R\,d\theta$$

$$+\frac{1}{EI_2}\int_0^h \{H[R(1-\cos\alpha)+Y]+M_E - VR\sin\alpha\}[R(1-\cos\alpha)+Y]\,dy = 0$$

Integrating the above equation and collecting terms

$$H\left\{\frac{R^3}{EI_1}\left[\frac{3\alpha}{2}-2\sin\alpha+\frac{\sin 2\alpha}{4}\right]+\frac{1}{EI_2}\left[R^2h(1-\cos\alpha)^2\right.\right.$$

$$\left.\left.+Rh^2(1-\cos\alpha)+\frac{h^3}{3}\right]\right\}+M_E\left\{\frac{R^2}{EI_1}[\alpha-\sin\alpha]\right.$$

$$\left.+\frac{1}{EI_2}\left[Rh(1-\cos\alpha)+\frac{h^2}{2}\right]\right\}+V\left\{\frac{R^3}{EI_1}[\cos\alpha-1+\tfrac{1}{2}\sin^2\alpha]\right.$$

$$\left.-\frac{R\sin\alpha}{EI_2}\left[Rh(1-\cos\alpha)+\frac{h^2}{2}\right]\right\}=0 \tag{5.249}$$

Let

$$A_1 = \frac{R^3}{EI_1}\left(\frac{3\alpha}{2}-2\sin\alpha+\frac{\sin 2\alpha}{4}\right)$$

$$+\frac{1}{EI_2}\left[R^2h(1-\cos\alpha)^2+Rh^2(1-\cos\alpha)+\frac{h^3}{3}\right]$$

$$A_2 = \frac{R^2}{EI_1}(\alpha-\sin\alpha)+\frac{1}{EI_2}\left[Rh(1-\cos\alpha)+\frac{h^2}{2}\right]$$

$$A_3 = -\left\{\frac{R^3}{EI_1}(\cos\alpha-1+\tfrac{1}{2}\sin^2\alpha)-\frac{R\sin\alpha}{EI_2}\left[Rh(1-\cos\alpha)+\frac{h^2}{2}\right]\right\}$$

Then the above equation can be written as

$$A_1H + A_2M_E = A_3V \tag{5.250}$$

For the second equation, the relative slope at point E due to moment M_E will be zero. This means the partial derivative must be taken with respect to moment M_E as follows:

From Eq. 5.244

$$\frac{\partial M_1}{\partial M_E} = 1 \tag{5.251}$$

From Eq. 5.246

$$\frac{\partial M_2}{\partial M_E} = 1 \tag{5.252}$$

The relative slope at point E can be written as

$$\theta_E = \frac{1}{EI_1}\int_0^{\alpha} M_1 \frac{\partial M_1}{\partial M_E} R\, d\theta + \frac{1}{EI_2}\int_0^{h} M_2 \frac{\partial M_2}{\partial M_E}\, dY = 0 \tag{5.253}$$

Substituting Eqs. 5.244, 5.246, 5.251, and 5.252 into the above

$$\frac{1}{EI_1}\int_0^{\alpha} [HR(1-\cos\theta) + M_E - VR\sin\theta](1)R\, d\theta$$

$$+\frac{1}{EI_2}\int_0^{h} \{H[R(1-\cos\alpha)+Y] + M_E - VR\sin\alpha\}(1)\, dY = 0$$

Integrating the above equation and collecting terms

$$H\left\{\frac{R^2}{EI_1}(\alpha-\sin\alpha) + \frac{1}{EI_2}\left[Rh(1-\cos\alpha)+\frac{h^2}{2}\right]\right\}$$

$$+M_E\left\{\frac{R^2}{EI_1}\left(\frac{\alpha}{R}\right) + \frac{1}{EI_2}(h)\right]$$

$$+V\left[\frac{R^2}{EI_1}(\cos\alpha-1) - \frac{1}{EI_2}(hR\sin\alpha)\right] = 0$$

Let

$$A_4 = \frac{R^2}{EI_1}(\alpha-\sin\alpha) + \frac{1}{EI_2}\left[Rh(1-\cos\alpha)+\frac{h^2}{2}\right]$$

$$A_5 = \frac{R\alpha}{EI_1} + \frac{h}{EI_2}$$

$$A_6 = -\left[\frac{R^2}{EI_1}(\cos\alpha-1) - \frac{1}{EI_2}(hR\sin\alpha)\right]$$

Then the above equation can be written as

$$A_4H + A_5M_E = A_6V \tag{5.254}$$

Equations 5.250 and 5.254 now represent two equations with two unknowns which can be solved simultaneously to give H and M_E for any bent similar to the one shown in Fig. 5.73. Determinants can be used conveniently here to show the solution in terms of V which is a known quantity, as shown by Eq. 5.241.

Repeating the two equations for convenience

$$A_1H + A_2M_E = A_3V$$

$$A_4H + A_5M_E = A_6V$$

The solution by determinants will then be

$$H = \frac{\begin{bmatrix} A_3 & A_2 \\ A_6 & A_5 \end{bmatrix} V}{\begin{bmatrix} A_1 & A_2 \\ A_4 & A_5 \end{bmatrix}} = \frac{A_3A_5 - A_6A_2}{A_1A_5 - A_4A_2} V \tag{5.255}$$

$$M_E = \frac{\begin{bmatrix} A_1 & A_3 \\ A_4 & A_6 \end{bmatrix} V}{\begin{bmatrix} A_1 & A_2 \\ A_4 & A_5 \end{bmatrix}} = \frac{A_1A_6 - A_4A_3}{A_1A_5 - A_4A_2} V \tag{5.256}$$

Once the values for H and M_E are known, the deflection due to the load P can be determined. Taking advantage of the symmetry, the vertical deflection δ can also be determined by considering the force V acting on half of the bent. The deflection at point E can be obtained from the following equation

$$\delta = \frac{1}{EI_1}\int_0^{\alpha} M_1 \frac{\partial M_1}{\partial V} R\, d\theta + \frac{1}{EI_2}\int_0^{h} M_2 \frac{\partial M_2}{\partial V}\, dY \tag{5.257}$$

The partial derivatives of the moment equations must now be taken with respect to V. From Eq. 5.244

$$\frac{\partial M_1}{\partial V} = -R \sin\theta \tag{5.258}$$

From Eq. 5.246

$$\frac{\partial M_2}{\partial V} = -R \sin\alpha \tag{5.259}$$

Substituting Eqs. 5.244, 5.246, 5.258, and 5.259 into Eq. 5.257

$$\delta = \frac{1}{EI_1}\int_0^{\alpha} [HR(1-\cos\theta) + M_E - VR\sin\theta](-R\sin\theta)R\, d\theta$$

$$+ \frac{1}{EI_2}\int_0^{h} \{H[R(1-\cos\alpha) + Y] + M_E - VR\sin\alpha\}(-R\sin\alpha)\, dY$$

Integrating the above equation and collecting terms:

$$\delta = H\left\{\frac{R^3}{EI_1}\left(\frac{\sin^2\alpha}{2} - 1 + \cos\alpha\right) + \frac{R}{EI_2}\left[Rh\sin\alpha\,(\cos\alpha - 1) - \frac{h^2}{2}\sin\alpha\right]\right\}$$
$$+ M_E\left[\frac{R^2}{EI_1}(\cos\alpha - 1) - \frac{Rh\sin\alpha}{EI_2}\right] + V\left[\frac{R^3}{EI_1}\left(\frac{\alpha}{2} - \frac{\sin 2\alpha}{4}\right)\right.$$
$$\left. + \frac{R^2h\sin^2\alpha}{EI_2}\right]$$

Let

$$A_7 = \left\{\frac{R^3}{EI_1}\left(\frac{\sin^2\alpha}{2} - 1 + \cos\alpha\right) + \frac{R}{EI_2}\left[Rh\sin\alpha\,(\cos\alpha - 1) - \frac{h^2}{2}\sin\alpha\right]\right\}$$

$$A_8 = \left[\frac{R^2}{EI_1}(\cos\alpha - 1) - \frac{Rh\sin\alpha}{EI_2}\right]$$

$$A_9 = \left[\frac{R^3}{EI_1}\left(\frac{\alpha}{2} - \frac{\sin 2\alpha}{4}\right) + \frac{R^2h\sin^2\alpha}{EI_2}\right]$$

Then the deflection equation can be written as follows:

$$\delta = A_7H + A_8M_E + A_9V \tag{5.260}$$

5.23. Strain Energy—Circular Arc with Hinged Ends

Castigliano's theorem is very convenient for determining forces, moments, and deflections in circular arcs. For example, consider the semicircular arc with hinged ends and a concentrated load (Fig. 5.74). Since the figure is symmetrical, it is possible to simplify the problem by considering only half of the system. The vertical deflection produced by

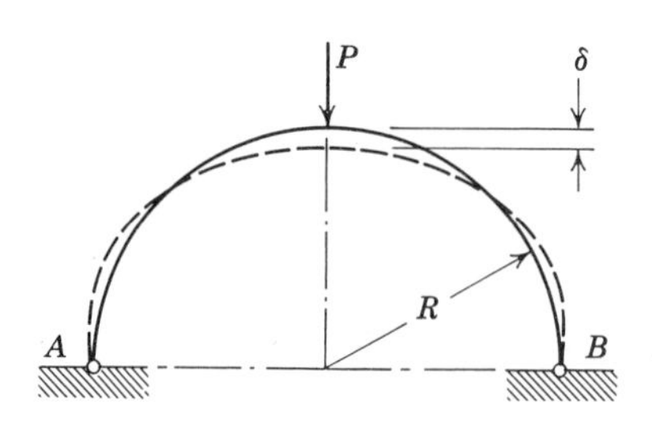

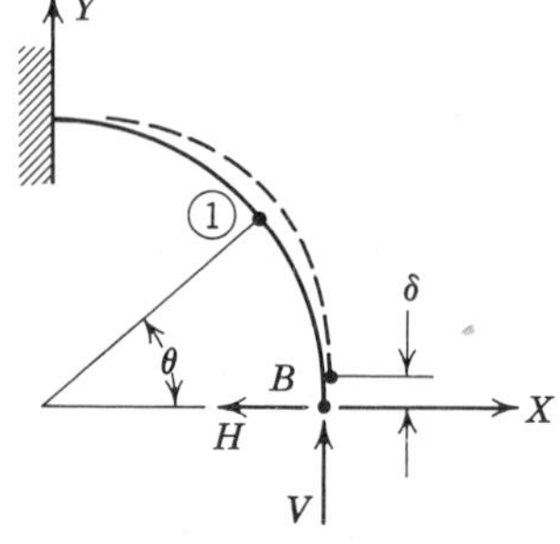

FIGURE 5.74. Circular arc with hinged ends and a vertical load.

the force P acting on the full system, will then be the same as the vertical deflection produced by the force V acting on half of the system.

The vertical force V can be determined by considering the sum of all the forces acting in the vertical direction. This leads to

$$V = \frac{P}{2} \tag{5.261}$$

The horizontal force H is not quite as easy to determine. An examination of the structure shows that the horizontal deflection at point B, due to force H, will be zero. Using Castigliano's theorem, this can be written as follows:

$$\delta_{BX} = \frac{\partial U}{\partial H} = 0 \tag{5.262}$$

When the angle θ is between 0 and 90°, the bending moment at point 1 will be

$$M_1 = VR(1 - \cos\theta) - HR\sin\theta \tag{5.263}$$

$$\frac{\partial M_1}{\partial H} = -R\sin\theta \tag{5.264}$$

The horizontal deflection at point B can be determined from Eqs. 5.200 and 5.262.

$$\delta_{BX} = \frac{1}{EI}\int_0^{\pi/2} M_1 \frac{\partial M_1}{\partial H} R\, d\theta = 0 \tag{5.265}$$

Substituting Eqs. 5.263 and 5.264 into the equation above

$$\delta_{BX} = \frac{1}{EI}\int_0^{\pi/2} [VR(1 - \cos\theta) - HR\sin\theta](-R\sin\theta)R\, d\theta = 0$$

Integrating the above equation and collecting terms

$$\frac{R^3}{EI}\left(-\frac{V}{2} + H\frac{\pi}{4}\right) = 0$$

Only the terms in the brackets can be zero, so this leads to

$$H = \frac{2V}{\pi} \tag{5.266}$$

The vertical deflection due to force V can be determined from the following equation:

$$\delta = \frac{1}{EI}\int_0^{\pi/2} M_1 \frac{\partial M_1}{\partial V} R\, d\theta \tag{5.267}$$

From Eq. 5.263

$$\frac{\partial M_1}{\partial V} = R(1-\cos\theta) \tag{5.268}$$

Substituting Eqs. 5.263 and 5.268 into Eq. 5.267

$$\delta = \frac{1}{EI}\int_0^{\pi/2} [VR(1-\cos\theta) - HR\sin\theta][R(1-\cos\theta)]R\,d\theta$$

Substituting Eq. 5.266 into the above equation and integrating

$$\delta = \frac{R^3V}{EI}\left(\frac{3}{4}\pi - 2 - \frac{1}{\pi}\right) \tag{5.269}$$

Substituting Eq. 5.261 into Eq. 5.269

$$\delta = \frac{PR^3}{52.6EI} \tag{5.270}$$

5.24. Strain Energy—Circular Arc with Fixed Ends

If the circular arc has fixed ends, it becomes statically indeterminate and requires more work to solve for the redundant forces. A semicircular arc with fixed ends and a concentrated load is shown in Fig. 5.75.

Since the figure is symmetrical, the problem can be simplified by considering only half of the system.

There are two unknowns, H and M_c, so that two equations must be developed to find these forces. The force V is known, since an examination of all the forces in the vertical direction will show the following:

$$V = \frac{P}{2} \tag{5.271}$$

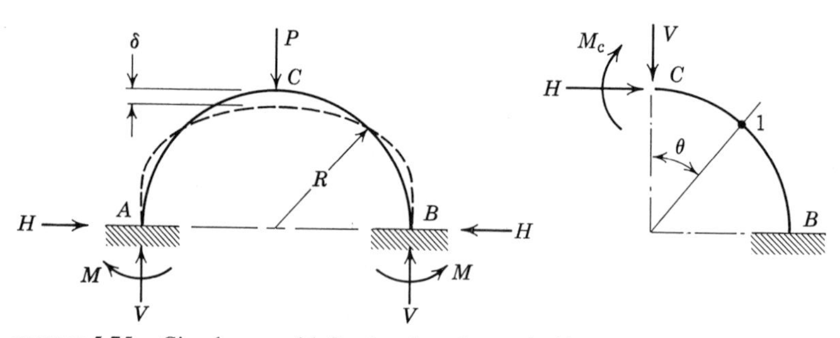

FIGURE 5.75. Circular arc with fixed ends and a vertical load.

The two physical properties that can be used to find the unknown forces are that the horizontal deflection at point C due to force H is zero, and that the relative slope at point C due to the moment M_c is zero.

Using Castigliano's theorem, these two equations will be very similar to Eqs. 5.242 and 5.243.

The bending moment at point 1 is

For $0 \leqslant \theta \leqslant \pi/2$

$$M_1 = HR(1-\cos\theta) + M_c - VR\sin\theta \tag{5.272}$$

$$\frac{\partial M_1}{\partial H} = R(1-\cos\theta) \tag{5.273}$$

The horizontal deflection at point C can be written as

$$\delta_{CX} = \frac{1}{EI}\int_0^{\pi/2} M_1 \frac{\partial M_1}{\partial H} R\,d\theta = 0 \tag{5.274}$$

Substitute Eqs. 5.272 and 5.273 into Eq. 5.274.

$$\delta_{CX} = \frac{1}{EI}\int_0^{\pi/2} [HR(1-\cos\theta)+M_c - VR\sin\theta](R)(1-\cos\theta)R\,d\theta = 0$$

Integrating the above equation and collecting terms

$$\frac{R^3}{EI}\left(0.3561H + 0.5707\frac{M_c}{R} - 0.500\,V\right) = 0 \tag{5.275}$$

This is the first equation that contains the two unknowns, H and M_c.

The second equation can be determined from the relative slope at point C, which can be written as follows:

$$\theta = \frac{1}{EI}\int_0^{\pi/2} M_1 \frac{\partial M_1}{\partial M_c} R\,d\theta = 0 \tag{5.276}$$

From Eq. 5.272

$$\frac{\partial M_1}{\partial M_c} = 1 \tag{5.277}$$

Substituting Eqs. 5.272 and 5.277 into Eq. 5.276

$$\theta = \frac{1}{EI}\int_0^{\pi/2} [HR(1-\cos\theta) + M_c - VR\sin\theta](1)R\,d\theta = 0$$

Integrating the above equation and collecting terms

$$\frac{R^2}{EI}\left(0.5707\,H + 1.5707\frac{M_c}{R} - V\right) = 0 \tag{5.278}$$

Only the items in the brackets in Eqs. 5.275 and 5.278 can be zero so these two equations can be used to solve for H and M_c. These two equations are repeated below.

$$0.3561\,H + 0.5707\frac{M_c}{R} = 0.500\,V$$

$$0.5707\,H + 1.5707\frac{M_c}{R} = V$$

Solving these two equations simultaneously results in the following:

$$H = 0.458\,P \tag{5.279}$$

$$M_c = 0.152\,PR \tag{5.280}$$

The vertical deflection due to force V can now be determined from the following equation.

$$\delta = \frac{1}{EI}\int_0^{\pi/2} M_1 \frac{\partial M_1}{\partial V} R\,d\theta \tag{5.281}$$

From Eq. 5.272:

$$\frac{\partial M_1}{\partial V} = -R \sin\theta \tag{5.282}$$

Substituting Eqs. 5.272 and 5.282 into Eq. 5.281

$$\delta = \frac{1}{EI}\int_0^{\pi/2} [HR(1-\cos\theta) + M_c - VR\sin\theta]\,(-R\sin\theta)R\,d\theta$$

Integrating the above equation and collecting terms

$$\delta = \frac{R^3}{EI}\left(-\frac{H}{2} - \frac{M_c}{R} + \frac{\pi}{4}V\right) \tag{5.283}$$

Substituting Eqs. 5.279 and 5.280 into Eq. 5.283

$$\delta = 0.0115\frac{PR^3}{EI} = \frac{PR^3}{87EI} \tag{5.284}$$

5.25. Strain Energy—Circular Arc with Free Ends

If the circular arc has ends that are free to move laterally but are restrained from rotating, it is desirable to find the value of one of the moments before the deflection due to a vertical load can be determined. Since the system is symmetrical, only one half of it may be used in the analysis (Fig. 5.76).

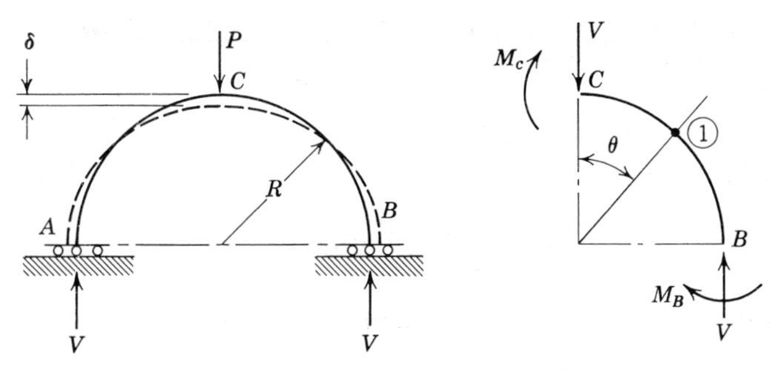

FIGURE 5.76. Circular arc with free ends and a vertical load.

The unknown moment M_c can be determined by considering the relative slope at point C, which will be zero. Using Castigliano's theorem, this can be written as follows:

$$\theta_c = \frac{1}{EI}\int_0^{\pi/2} M_1 \frac{\partial M_1}{\partial M_c} R\, d\theta = 0 \tag{5.285}$$

The bending moment at point 1 will be

$$M_1 = M_c - VR \sin\theta \tag{5.286}$$

$$\frac{\partial M_1}{\partial M_c} = 1 \tag{5.287}$$

Substituting Eqs. 5.286 and 5.287 into Eq. 5.285

$$\theta_c = \frac{1}{EI}\int_0^{\pi/2} (M_c - VR\sin\theta)\ (1) R\, d\theta = 0$$

Integrating the above equation

$$\frac{R}{EI}\left(\frac{\pi}{2} M_c - VR\right) = 0 \tag{5.288}$$

Only the terms in the brackets can be zero so

$$M_c = \frac{2}{\pi} VR$$

Since

$$V = \frac{P}{2} \tag{5.289}$$

$$M_c = \frac{RP}{\pi} = 0.318PR \tag{5.290}$$

The deflection at point C can now be determined as follows:

$$\delta = \frac{1}{EI}\int_0^{\pi/2} M_1 \frac{\partial M_1}{\partial V} R\, d\theta \qquad (5.291)$$

From Eq. 5.286

$$\frac{\partial M_1}{\partial V} = -R \sin\theta \qquad (5.292)$$

Substituting Eqs. 5.286 and 5.292 into Eq. 5.291

$$\delta = \frac{1}{EI}\int_0^{\pi/2} (M_c - VR\sin\theta)(-R\sin\theta)R\, d\theta$$

Integrating the above equation and collecting terms

$$\delta = \frac{R^2}{EI}\left(-M_c + \frac{\pi}{4}VR\right)$$

Substituting Eqs. 5.289 and 5.290 into the above expression

$$\delta = \frac{0.0744PR^3}{EI} = \frac{PR^3}{13.44EI} \qquad (5.293)$$

The moment at point B can be determined by taking moments about point C or B in Fig. 5.76.

$$M_B + M_C = VR \qquad (5.294)$$

Substituting Eqs. 5.289 and 5.290 into the above expression

$$M_B = (0.500 - 0.318)PR = 0.182PR \qquad (5.295)$$

The horizontal deflection at point B, due to the vertical load P, can be determined by using a fictitious load H at point B in the direction of the deflection (Fig. 5.77). This fictitious load will be set equal to zero later, since it really does not exist.

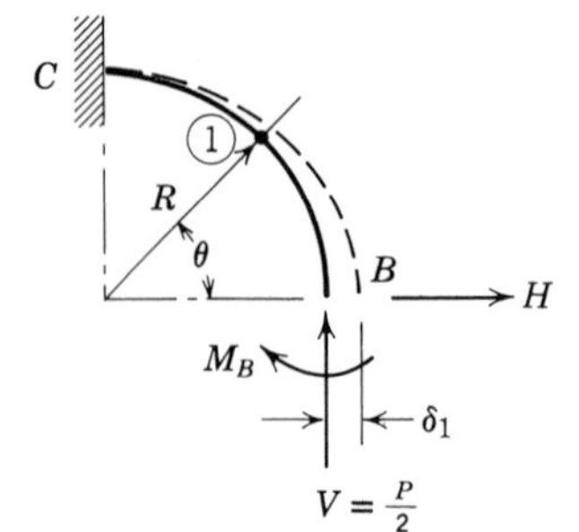

FIGURE 5.77. Fictitious horizontal load H used to find deflection.

The deflection at point B, due to the force H will be as follows:

$$\delta_1 = \frac{1}{EI}\int_0^{\pi/2} M_1 \frac{\partial M_1}{\partial H} R\, d\theta \tag{5.296}$$

The bending moment at point 1 will be

$$M_1 = VR(1-\cos\theta) - M_B + HR\sin\theta \tag{5.297}$$

$$\frac{\partial M_1}{\partial H} = R\sin\theta \tag{5.298}$$

Substituting Eqs. 5.297 and 5.298 into Eq. 5.296

$$\delta_1 = \frac{1}{EI}\int_0^{\pi/2} [VR(1-\cos\theta) - M_B + HR\sin\theta](R\sin\theta)R\, d\theta$$

$$\delta_1 = \frac{R^2}{EI}\int_0^{\pi/2} [VR(\sin\theta - \sin\theta\cos\theta) - M_B\sin\theta + HR\sin^2\theta]\, d\theta$$

Since H is fictitious, let it go to zero, then integrate the above equation and collect terms.

$$\delta_1 = \frac{R^2}{EI}\left(\frac{VR}{2} - M_B\right)$$

Substituting Eqs. 5.289 and 5.295 into the equation above

$$\delta_1 = 0.0683\frac{PR^3}{EI} \tag{5.299}$$

5.26. Circular Arc with Lateral Load and Hinged Ends

Castigliano's theorem can be used to analyze a circular arc with hinged ends and a lateral load (Fig. 5.78).

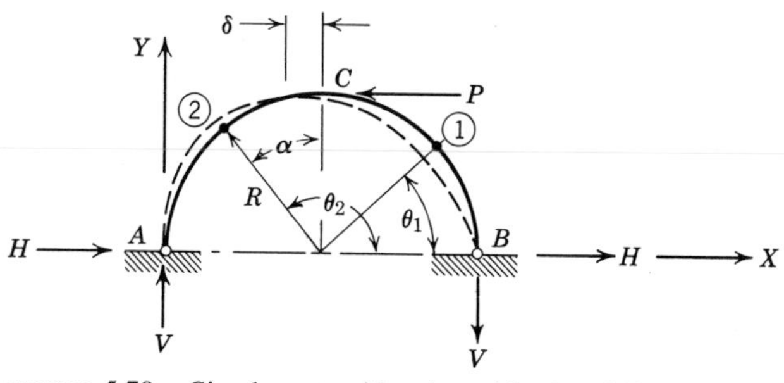

FIGURE 5.78. Circular arc with a lateral load and hinged ends.

Since the slope at point C will change when the arc deforms under the load P, bending moments on both sides of the load at points 1 and 2 will be determined.

Determine the horizontal reaction by considering the sum of all the forces in the X direction, the vertical reaction by considering the forces in the Y direction, and moments about point A. This results in the following:

$$H = \frac{P}{2} \tag{5.300}$$

$$V = \frac{P}{2} \tag{5.301}$$

The bending moment at point 1 is for $0 \leqslant \theta_1 \leqslant \pi/2$

$$M_1 = HR \sin \theta_1 - VR(1 - \cos \theta_1) \tag{5.302}$$

Substituting Eqs. 5.300 and 5.301 into Eq. 5.302

$$M_1 = \tfrac{1}{2} PR(\sin \theta_1 - 1 + \cos \theta_1) \tag{5.303}$$

The bending moment at point 2 is for $\pi/2 \leqslant \theta_2 \leqslant \pi$

$$M_2 = PR(1 - \cos \alpha) + HR \cos \alpha - VR(1 + \sin \alpha)$$

Since $\theta_2 = 90° + \alpha$ then

$$\cos \theta_2 = -\sin \alpha \qquad \text{and} \qquad \sin \theta_2 = \cos \alpha$$

so

$$M_2 = PR(1 - \sin \theta_2) + HR \sin \theta_2 - VR(1 - \cos \theta_2) \tag{5.304}$$

Notice that Eq. 5.304 is exactly the same as Eq. 5.302 except for the expression with P and the terms with θ_1 or θ_2.

Substituting Eqs. 5.300 and 5.301 into Eq. 5.304

$$M_2 = \tfrac{1}{2} Pr(1 - \sin \theta_2 + \cos \theta_2) \tag{5.305}$$

From Eq. 5.303

$$\frac{\partial M_1}{\partial P} = \frac{R}{2}(\sin \theta_1 - 1 + \cos \theta_1) \tag{5.306}$$

From Eq. 5.305

$$\frac{\partial M_2}{\partial P} = \frac{R}{2}(1 - \sin \theta_2 + \cos \theta_2) \tag{5.307}$$

The lateral deflection produced by the force P can be determined from the expression

$$\delta = \frac{1}{EI} \int_0^{\pi/2} M_1 \frac{\partial M_1}{\partial P} R\, d\theta_1 + \frac{1}{EI} \int_{\pi/2}^{\pi} M_2 \frac{\partial M_2}{\partial P} R\, d\theta_2 \tag{5.308}$$

Substituting Eqs. 5.303, 5.305, 5.306, and 5.307 into the above

$$\delta = \frac{1}{EI}\int_0^{\pi/2}\left[\frac{PR}{2}(\sin\theta_1 - 1 + \cos\theta_1)\left(\frac{R}{2}\right)(\sin\theta_1 - 1 + \cos\theta_1)\right] R\,d\theta_1$$

$$+\frac{1}{EI}\int_{\pi/2}^{\pi}\left[\frac{PR}{2}(1 - \sin\theta_2 + \cos\theta_2)\left(\frac{R}{2}\right)(1 - \sin\theta_2 + \cos\theta_2)\right] R\,d\theta_2$$

Both halves of the above equation end up with the same value when they are integrated.

$$\delta = \frac{PR^3}{4EI}(\pi - 3) + \frac{PR^3}{4EI}(\pi - 3)$$

$$\delta = 0.0708\frac{PR^3}{EI} = \frac{PR^3}{14.12EI} \tag{5.309}$$

5.27. Edge-Guide Springs for Printed-Circuit Boards

Circular arc sections are very often found in guides that support the edges of printed-circuit boards. These edge-guides provide accurate circuit-board alignment to ensure proper engagement of plug-in electrical connectors. These guides also add rigidity to the edges of the circuit board, which helps to increase the stiffness during vibration. In addition, these guides may also be required to conduct heat away from the circuit board, thus they are often called thermal clips. A typical board edge-guide installation is shown in Fig. 5.79. An enlarged view of the edge-guide might appear as shown in Fig. 5.80.

It is important to know the spring rate of the edge-guide since the natural frequency of the circuit board will be influenced by the stiffness of the guide. The dynamic loads and stresses developed in the guide will also be influenced by its stiffness.

A good approximation of the spring rate, for small deflections, can be obtained by analyzing only one half of the guide (Fig. 5.81).

Assuming a concentrated load at the center of the inside edge on the guide, the deflection can be determined by using Castigliano's theorem. The strain energy must be considered for the straight portions and the curved portions of the guide, as shown in Fig. 5.82.

Considering only bending deflections, Eqs. 5.200 and 5.203 can be used as follows:

$$\delta = \frac{1}{EI}\left(\int_A^B M_1\frac{\partial M_1}{\partial P}\,dX_1 + \int_0^{\pi} M_2\frac{\partial M_2}{\partial P}R\,d\theta + \int_D^C M_3\frac{\partial M_3}{\partial P}\,dX_3\right.$$

$$\left.+\int_D^E M_4\frac{\partial M_4}{\partial P}\,dX_4\right) \tag{5.310}$$

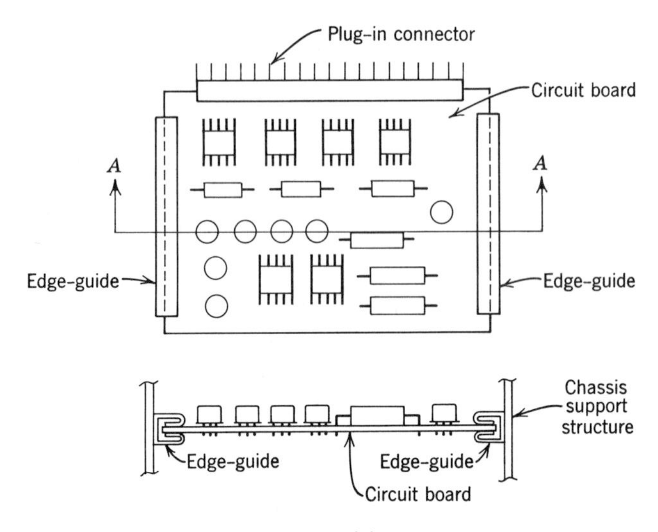

FIGURE 5.79. Printed circuit board with guides that support opposite edges.

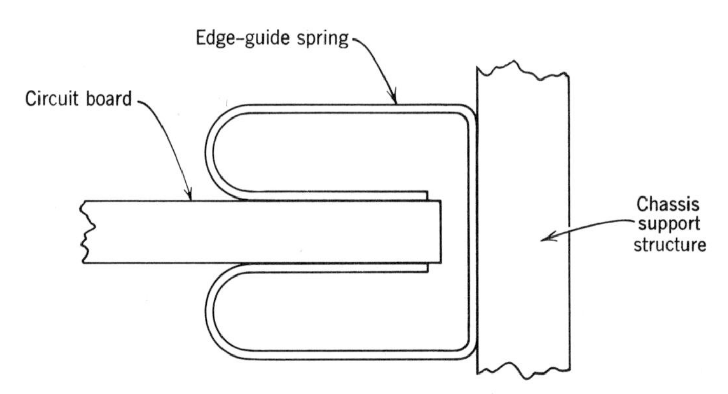

FIGURE 5.80. Enlarged view of a board edge guide.

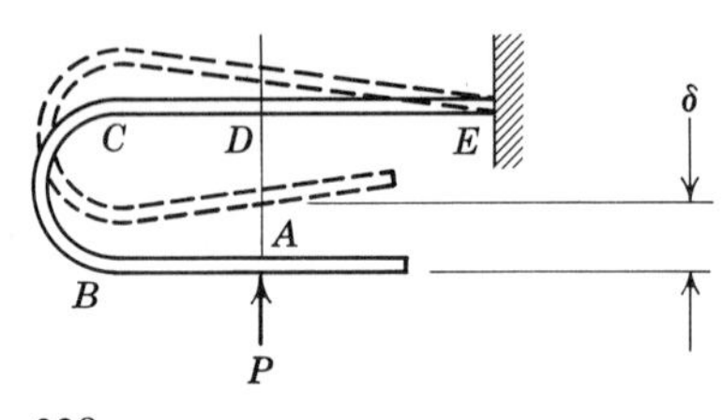

FIGURE 5.81. Deflection mode of a board edge guide.

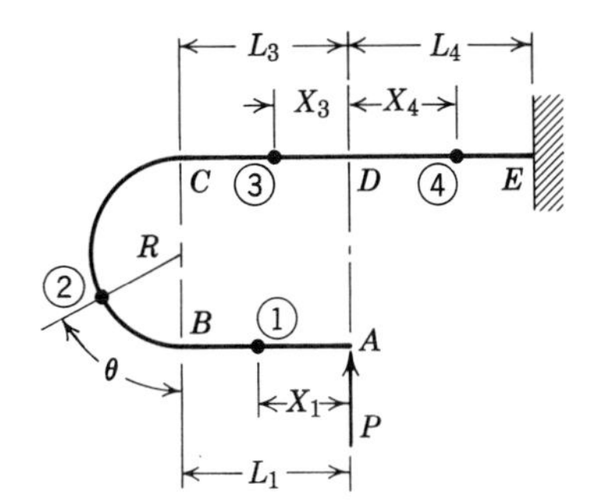

FIGURE 5.82. Model of half of the board edge guide.

The bending moment at point 1 is

$$M_1 = PX_1$$

$$\frac{\partial M_1}{\partial P} = X_1 \tag{5.311}$$

The bending moment at point 2 is

$$M_2 = P(L_1 + R \sin \theta) \tag{5.312}$$

$$\frac{\partial M_2}{\partial P} = L_1 + R \sin \theta \tag{5.313}$$

The bending moment at point 3 is

$$M_3 = PX_3 \tag{5.314}$$

$$\frac{\partial M_3}{\partial P} = X_3 \tag{5.315}$$

The bending moment at point 4 is

$$M_4 = -PX_4 \tag{5.316}$$

$$\frac{\partial M_4}{\partial P} = -X_4 \tag{5.317}$$

Substituting Eqs. 5.311–5.317 into Eq. 5.310

$$\delta = \frac{1}{EI}\left(\int_0^{L_1} (PX)(X)\, dX + \int_0^{\pi} P(L_1 + R \sin \theta)(L_1 + R \sin \theta) R\, d\theta \right.$$

$$\left. + \int_0^{L_3} (PX)(X)\, dX + \int_0^{L_4} (-PX)(-X)\, dX \right)$$

Integrating the above equation and collecting terms

$$\delta = \frac{P}{EI}\left(\frac{L_1^3}{3} + L_1^2 R\pi + 4L_1R^2 + \frac{\pi}{2}R^3 + \frac{L_3^3}{3} + \frac{L_4^3}{3}\right) \tag{5.318}$$

Beryllium copper is normally used for springs of this type because of its high strength and good fatigue life. A typical board edge-guide for a 0.062-in.-thick printed-circuit board will probably have dimensions similar to those shown in Fig. 5.83.

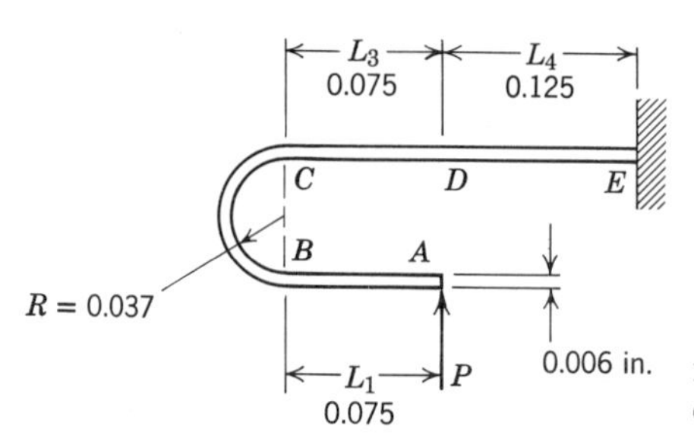

FIGURE 5.83. Geometry of half of the board edge guide.

Edge-guide lengths can vary from 1 to about 10 in. For convenience, consider a unit length of 1 in.

$$I = \frac{bh^3}{12} = \frac{(1.0)(0.006)^3}{12} = 0.018 \times 10^{-6}\text{ in.}^4$$

$$E = 18 \times 10^6\text{ lb/in.}^2\text{ modulus of elasticity}$$

$$L_1 = 0.075, \quad L_3 = 0.075, \quad L_4 = 0.125\text{ in.}$$

If a unit load of 1 lb is used, the deflection and spring rate can be determined from Eq. 5.318.

$$\delta = \frac{1.0}{(18 \times 10^6)(0.018 \times 10^{-6})}\left[\frac{(0.075)^3}{3} + (0.075)^2(0.037)\pi + 4(0.075)(0.037)^2 + \frac{\pi}{2}(0.037)^3 + \frac{(0.075)^3}{3} + \frac{(0.125)^3}{3}\right]$$

$$\delta = 6.41 \times 10^{-3}\text{ in. deflection} \tag{5.319}$$

If the board edge-guide is 3.0 in. long, the spring rate can be determined as follows:

$$K = \frac{3.0}{\delta} = \frac{3.0}{6.41 \times 10^{-3}} = 468\text{ lb/in.} \tag{5.320}$$

This is the spring rate of only one half of the board edge-guide, as shown in Figs. 5.81–5.83. The full edge-guide appears as shown in Fig. 5.80. If there is no preload in this edge-guide, a small displacement of the printed-circuit board edge will compress one half of the edge-guide slightly so only one half of the guide will be in contact with the circuit board (Fig. 5.84).

When there is no preload in the edge-guide it acts like a linear spring and the load/deflection curve will be a straight line (Fig. 5.85).

If the edge-guide is preloaded, a small displacement of the circuit-board edge will compress one half of the guide and allow the opposite half to expand. Both halves of the edge-guide will then remain in contact with the circuit board, as long as the preload is not exceeded.

During the period that both halves of the edge-guide are in contact with the circuit board, the effective spring rate of the guide will be two times the spring rate of half a guide. When the displacement exceeds the preload, then only one half of the guide is in contact with one side of the circuit board and the spring rate is reduced to that of half a guide (Fig. 5.86).

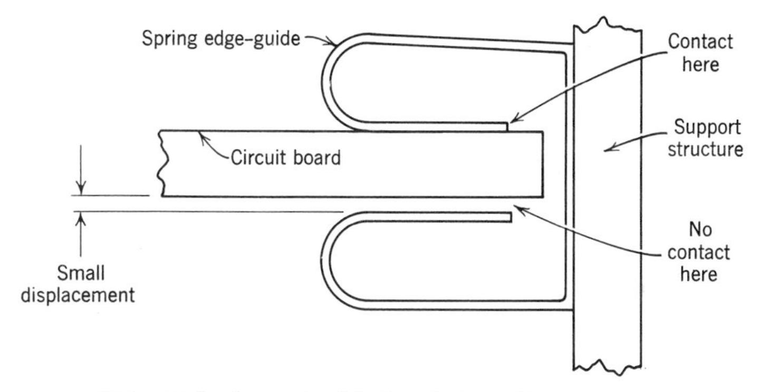

FIGURE 5.84. Deflection mode of the board edge guide.

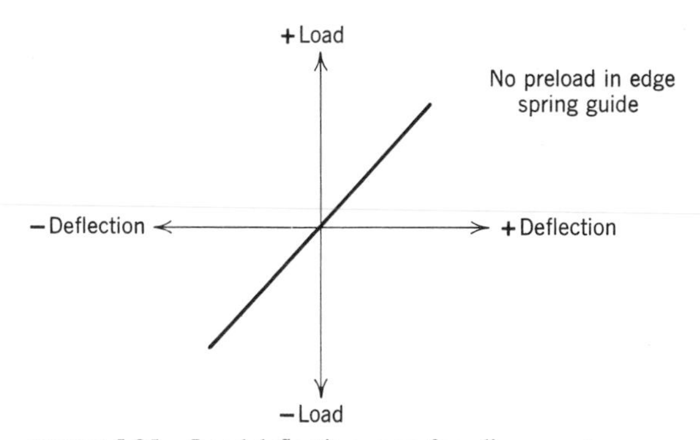

FIGURE 5.85. Load deflection curve for a linear system.

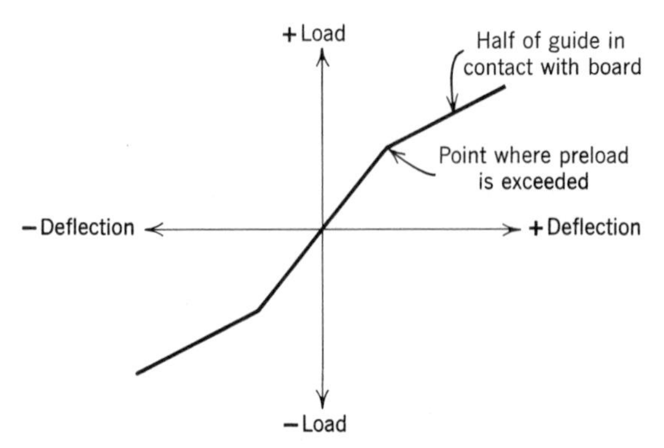

FIGURE 5.86. Load deflection curve for a non-linear system.

The maximum bending stress in the spring card edge-guide will occur at point E in Fig. 5.83. The stress will be determined by the dynamic load, the geometry, and possible stress concentration factors, depending upon the way in which the edge-guide is fabricated and fastened to the support structure.

The board edge-guides will act as springs along the edges of the circuit board, which will influence the resonant frequency of the board. The stiffness of the circuit board and the stiffness of the guides should be adjusted so that their resonances are separated by at least one octave. A rigid guide will reduce the dynamic deflections and stresses, but the high friction forces may make it very difficult to replace a circuit board. A compliant guide may make it simple to replace a board, but the large dynamic deflections may fracture the pin contacts in electrical connectors that are not floated.

Consider a printed-circuit board installation as shown in Fig. 5.79, with edge-guides that have a spring rate of 468 lb/in. (as shown by Eq. 5.320). If the edge-guides have no preload, and if a floating plug-in connector is used in the electronic box, then the natural frequency of a rigid circuit board on the spring edge-guides can be determined by assuming the combination is similar to a single-degree-of-freedom system. The natural frequency will then be as follows:

$$f_n = \frac{1}{2\pi}\left(\frac{Kg}{W}\right)^{1/2} \qquad \text{(see Eq. 2.7)}$$

where $K = 2(468) = 936$ lb/in. (see Eq. 5.320)
$W = 0.50$ lb circuit-board weight
$g = 386$ in./sec^2 acceleration of gravity

Substituting into the frequency equation

$$f_n = \frac{1}{2\pi}\left[\frac{(936)(386)}{0.50}\right]^{1/2} = 135 \text{ Hz}$$

If the natural frequency of the circuit board is twice this value for the Octave Rule, then the circuit-board resonance will be about 270 Hz, which is high for this type of installation and may require ribs.

Dynamic stresses in the edge-guide can be determined from the dynamic loading on the circuit board. If a floating plug-in chassis connector is used, all of the dynamic load will pass through the edge-guides. Assuming a sinusoidal vibration input acceleration of 8G's peak, the transmissibility of the spring edge guides at resonance may be approximated by the relation

$$Q = \tfrac{1}{2}(f_n)^{1/2} = \tfrac{1}{2}(135)^{1/2} = 5.8$$

The dynamic load in one spring guide can then be determined by

$$P_d = \frac{W G_{in} Q}{2}$$

where $W = 0.50$ lb circuit-board weight
$G_{in} = 8.0$ peak input acceleration G force
$Q = 5.8$ transmissibility of edge guides

Substituting into the dynamic load equation

$$P_d = \frac{(0.50)(8.0)(5.8)}{2} = 11.6 \text{ lb}$$

The single-amplitude displacement of the spring edge-guide can be determined from the spring rate of the guide.

$$\delta = \frac{P_d}{K} = \frac{11.6}{468} = 0.0248\text{-in. single amplitude}$$

The single-amplitude displacement of the spring edge-guide can also be obtained from the frequency and the G force with the use of Eq. 2.42.

$$\delta = \frac{9.8 G_{in} Q}{f_n^2}$$

where $G_{in} = 8.0$ peak input acceleration
$Q = 5.8$ transmissibility of edge guides
$f_n = 135$ Hz resonant frequency of board on guides

Substituting into the above equation

$$\delta = \frac{(9.8)(8.0)(5.8)}{(135)^2} = 0.0249 \text{ in.}$$

The dynamic bending stresses at point E in Fig. 5.83 can be determined with the use of the standard bending stress equation, which can be obtained from a structural handbook as follows (neglecting stress concentration factors).

$$S_E = \frac{M_E c}{I}$$

where $M_E = P_d L_4 = 11.6(0.125) = 1.45$ lb in. bending moment

$$c = \frac{h}{2} = \frac{0.006}{2} = 0.003 \text{ in.}$$

$$I = \frac{bh^3}{12} = \frac{(3.0)(0.006)^3}{12} = 54.0 \times 10^{-9} \text{ in.}^4$$

Substituting into the bending stress equation

$$S_E = \frac{(1.45)(0.003)}{54.0 \times 10^{-9}} = 80{,}500 \text{ lb/in.}^2 \tag{5.321}$$

Circuit-board spring-guides of this type may fail during vibration, due to the high stress levels and the large number of fatigue cycles that can be accumulated during resonant dwell conditions required by qualification tests. For example, consider a resonant dwell condition of 1 hr at a frequency of 135 Hz. It is possible to estimate if a fatigue failure will occur by considering Miner's cumulative fatigue theory [23]. This theory assumes that every structural member has a certain fatigue life and every stress reversal uses up part of this life. This is often written in terms of a fatigue cycle ratio R_n which theoretically equals 1.0 at failure. This fatigue cycle ratio can be expressed in the following terms:

$$R_n = \frac{n}{N} = 1.0 \text{ for failure} \tag{5.322}$$

where n = actual number of accumulated stress reversal cycles
N = number of fatigue cycles required for a failure

The number of fatigue cycles accumulated in a 1-hr resonant dwell at 135 Hz is

$$n = 135 \text{ cycles/sec} \times 3600 \text{ sec/hr} = 4.86 \times 10^5 \text{ cycles} \tag{5.323}$$

The number of fatigue cycles required to produce a fatigue failure in the beryllium copper spring edge-guide at a stress level of 80,500 psi can be determined from the S–N curves shown in Chapter 10, Fig. 10.20c if heat-treated half-hard beryllium copper is used.

$$N = 8.0 \times 10^5 \text{ cycles for a failure} \tag{5.324}$$

Substituting Eqs. 5.323 and 5.324 into Eq. 5.322

$$R_n = \frac{4.86 \times 10^5}{8.0 \times 10^5} = 0.61$$

Since this value is less than 1.0, the edge guide will probably not fail unless the test is longer than one hour, or unless another beryllium copper alloy is used. For a slightly more conservative analysis, a fatigue cycle ratio of 0.70 is often used instead of 1.0.

Another type of spring guide that is often used to align and support the edges of a printed circuit board is the wavy spring strip (Fig. 5.87).

This type of spring is uaually used in a channel section that also supports the edge of the printed circuit board (Fig. 5.88).

Castigliano's theorem can be used to find the deflection and the spring rate of the wavy spring, by considering half of one wave (Fig. 5.89).

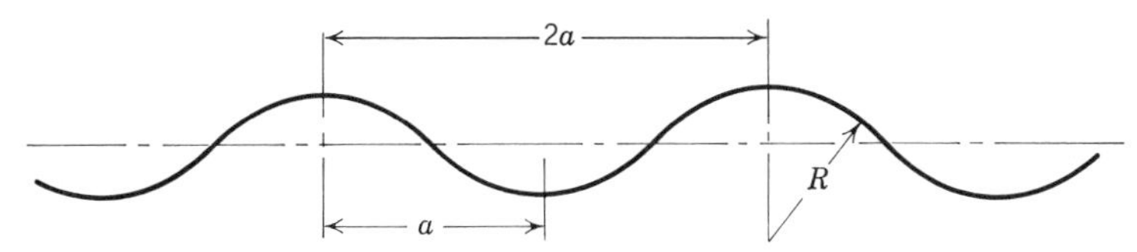

FIGURE 5.87. A wavy spring from a board edge guide.

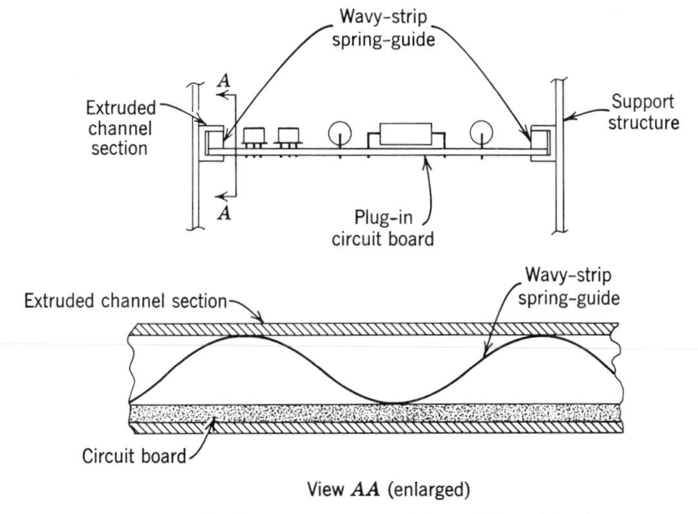

FIGURE 5.88. A circuit board supported by guides with wavy springs.

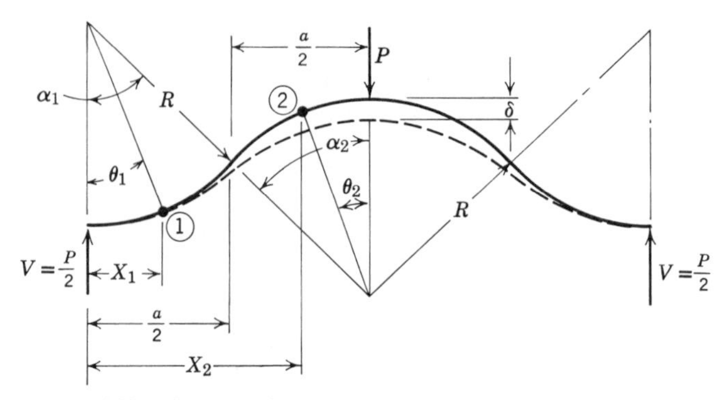

FIGURE 5.89. One wave in a wavy spring.

The bending moment at point 1 is

$$M_1 = VR \sin \theta_1$$
$$\frac{\partial M_1}{\partial V} = R \sin \theta_1 \tag{5.325}$$

The bending moment at point 2 is

$$M_2 = V\left(\frac{a}{2}+\frac{a}{2}-R \sin \theta_2\right) = V(a-R \sin \theta_2) \tag{5.326}$$

$$\frac{\partial M_2}{\partial V} = a-R \sin \theta_2 \tag{5.327}$$

The deflection due to force P acting on one full wave will be the same as the deflection of force V acting on half of one wave. Notice the point of inflection where the radius changes from a concave section to a convex section. The deflection due to force V acting on half a wave will then be as follows:

$$\delta = \frac{1}{EI}\int_0^{\alpha_1} M_1 \frac{\partial M_1}{\partial V} R\, d\theta_1 + \frac{1}{EI}\int_0^{\alpha_2} M_2 \frac{\partial M_2}{\partial V} R\, d\theta_2 \tag{5.328}$$

Substituting Eqs. 5.325–5.327 into Eq. 5.328

$$\delta = \frac{1}{EI}\int_0^{\alpha_1} (VR \sin \theta_1)(R \sin \theta_1) R\, d\theta_1$$
$$+\frac{1}{EI}\int_0^{\alpha_2} V(a-R \sin \theta_2)(a-R \sin \theta_2) R\, d\theta_2$$

Integrate the above equation and notice that $\alpha_1 = \alpha_2$ and $V = P/2$ because of symmetry.

$$\delta = \frac{PR}{EI}\left\{R^2\left(\frac{\alpha}{2} - \frac{\sin 2\alpha}{4}\right) + a\left[\frac{a\alpha}{2} + R(\cos\alpha - 1)\right]\right\} \tag{5.329}$$

The spring rate of a beryllium copper spring, with 3 waves, can be determined for the geometry shown in Fig. 5.90.

where $E = 18 \times 10^6$ lb/in.^{2} beryllium copper

$$I = \frac{bh^3}{12} = \frac{(0.165)(0.008)^3}{12} = 7.04 \times 10^{-9}\text{ in.}^4$$

$$\alpha = 33.5^\circ = 0.585\text{ rad}$$

$$a = 0.53\text{ in.}$$

Substituting into Eq. 5.329

$$\delta = \frac{0.478}{EI}\left\{(0.478)^2\left(\frac{0.585}{2} - \frac{0.921}{4}\right) + 0.53\left[\frac{(0.53)(0.585)}{2} + 0.478(0.835 - 1.0)\right]\right\}$$

$$\delta = \frac{0.0264P}{EI}\text{ for one spring wave} \tag{5.330}$$

The spring rate for one spring wave is

$$K = \frac{P}{\delta} = \frac{EI}{0.0264}$$

This deflection equation can be checked very quickly to see if it is approximately correct, by considering half of one spring wave as a cantilevered beam. The deflection is then

$$\delta_c = \frac{Va^3}{3EI} = \frac{P(0.53)^3}{6EI} = \frac{0.0248P}{EI} \tag{5.331}$$

In this case, because the spring is so flat, a cantilevered beam happens to be a very good approximation, as shown by comparing the results with Eq. 5.330.

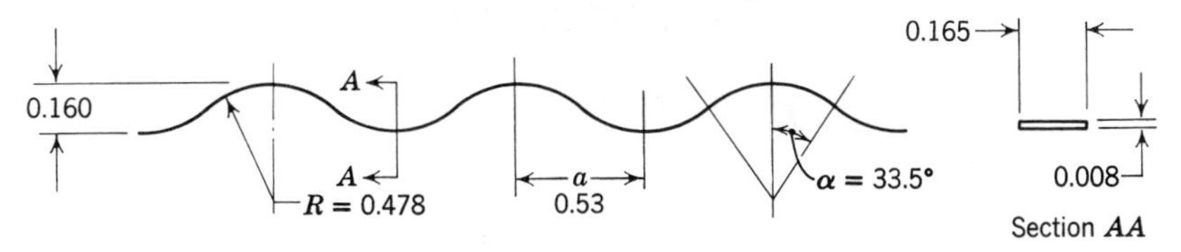

FIGURE 5.90. Dimensions for a typical wavy spring.

The spring rate for three waves can be determined from Eq. 5.330.

$$K = (3)\frac{EI}{0.0264} = \frac{3(18 \times 10^6)(7.04 \times 10^{-9})}{0.0264} = 14.4 \text{ lb/in.} \quad (5.332)$$

5.28. Gyro Flex-Leads

Flex-leads are thin, precision spring members that are often used on rate gyros. Several leads, usually about three or four, are generally used to perform the following functions:

1. supply electrical power to the spinning gyro rotor;
2. measure the angular rate or angular velocity by balancing out the gyroscopic torque developed from the input angular rates;
3. support one or both ends of the gimbal system that houses the spinning rotor.

Some rate gyros use a torsion bar in place of the flex-leads to measure angular rates. In both cases the spring device limits the angular displacement about the gimbal axis, which is perpendicular to the spin axis of the rotor. A pick-off device is usually arranged to measure the angular displacement of the gimbal axis, or output axis, so that it is proportional to

FIGURE 5.91. An airborne electronic package for a simplified inertial-guidance system (courtesy Norden division of United Aircraft).

the angular rate. When the angular input rate is zero, the gyro flex-leads will return the gimbal to its neutral position.

The accuracy of the rate gyro depends, to a great extent, upon the accuracy and precision of the flex-leads, so it is important to know the characteristics of these flex-leads when they act as springs.

A cross-section through the rate gyro with four flex-leads may look something like Fig. 5.92, and a single flex-lead shown in more detail may appear as shown in Fig. 5.93.

The gyro output axis or gimbal axis will generally be supported by ball or jewel bearings so that only rotation of the axis is permitted, with no translation of the axis. As the precession torque generated by an input angular rate, or angular velocity, forces the gimbal to rotate about the output axis, the gyro flex-lead is forced to move from position B to B′ along the radius r and through the angle α as shown in Fig. 5.93. Since the end of the flex-lead is rigidly fastened to the gimbal at point B, the rotation

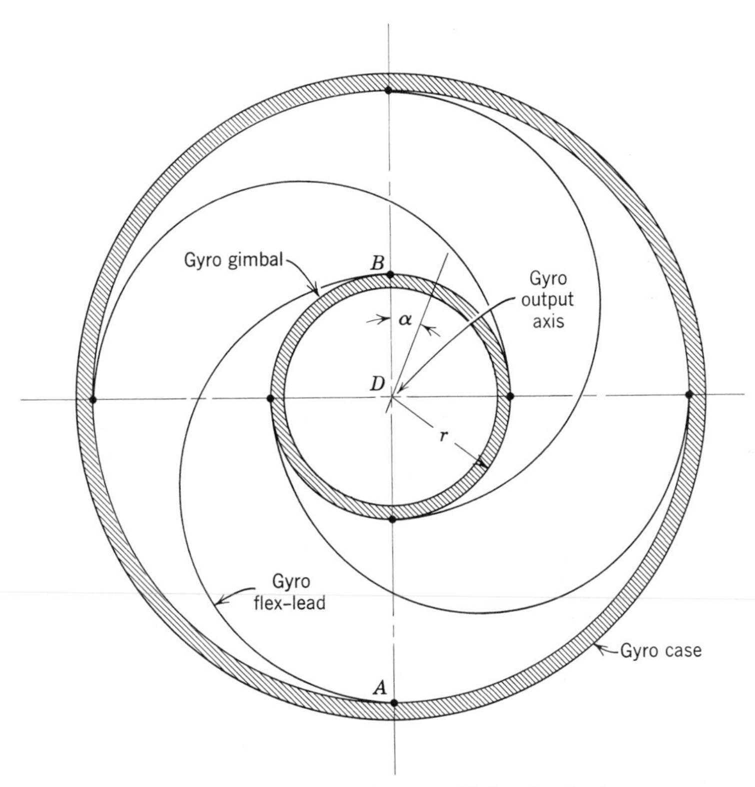

FIGURE 5.92. Typical cross section of a gyro with four flex-leads.

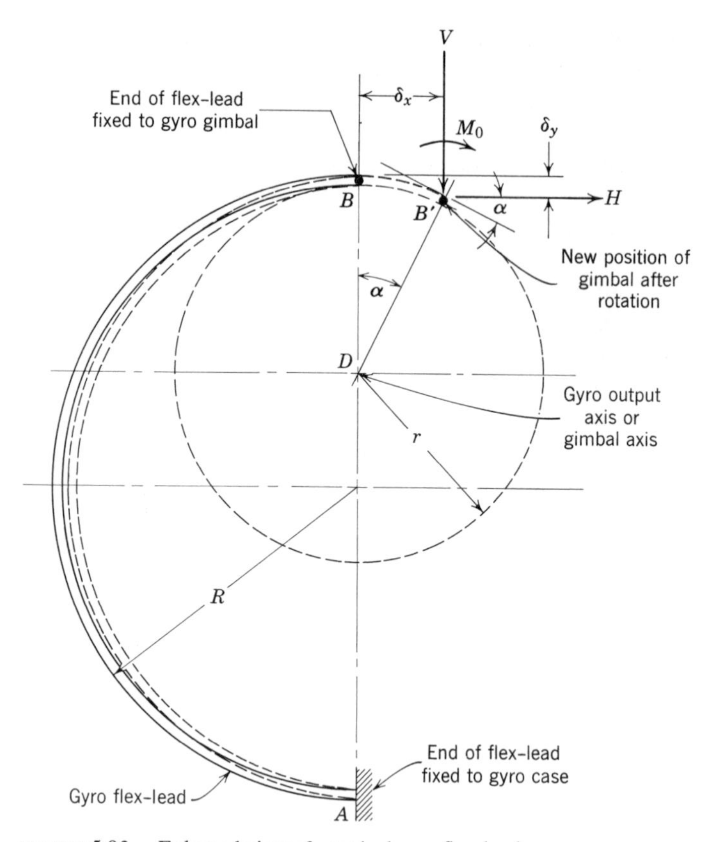

FIGURE 5.93. Enlarged view of a typical gyro flex-lead.

through the angle α forces point B to rotate through the same angle α. In addition to the rotation, point B will translate through distances δ_x and δ_y.

The angular spring rate of the gyro flex-lead can be determined with the use of Castigliano's theorem. The flex-leads can be approximated as a curved beam with a horizontal force H, a vertical force V and a moment M_0 at point B. The flex-lead is assumed to be fixed at point A (Fig. 5.94).

The horizontal deflection at point B, due to force H, can be determined by considering only bending deflections. This results in an equation with three unknowns, so three equations must be developed to solve for them. Starting with Eqs. 5.200 and 5.203, the horizontal deflection is

$$\delta_X = \frac{1}{EI} \int_0^{\pi} M_1 \frac{\partial M_1}{\partial H} R \, d\theta \tag{5.333}$$

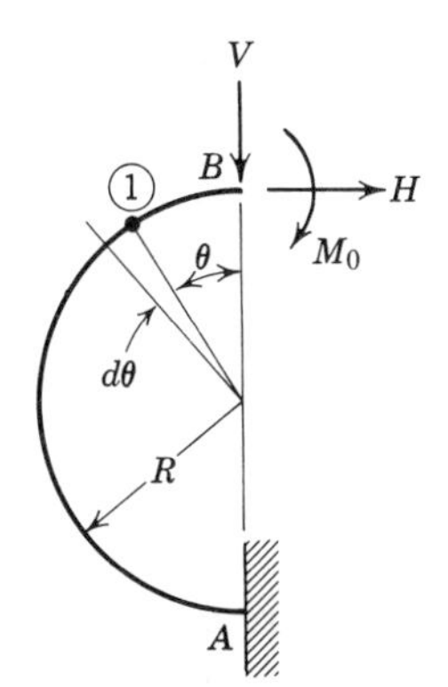

FIGURE 5.94. Model of a gyro flex-lead.

The bending moment at point 1 is

$$M_1 = VR \sin\theta + HR(1-\cos\theta) + M_0 \tag{5.334}$$

$$\frac{\partial M_1}{\partial H} = R(1-\cos\theta)$$

Substituting into Eq. 5.333

$$\delta_X = \frac{1}{EI}\int_0^\pi [VR\sin\theta + HR(1-\cos\theta)+M_0][R(1-\cos\theta)]\,R\,d\theta$$

$$\delta_X = \frac{R^3}{EI}\int_0^\pi \left[V(\sin\theta - \sin\theta\cos\theta) + H(1-2\cos\theta+\cos^2\theta) + \frac{M_0}{R}(1-\cos\theta)\right] d\theta$$

Integrating the above equation and collecting terms

$$\delta_X = \frac{R^3}{EI}\left(2V + \frac{3}{2}H\pi + \pi\frac{M_0}{R}\right) \tag{5.335}$$

This is the first of three equations required for a solution. The second equation can be determined for the vertical deflection at point B due to the force V as follows:

$$\delta_Y = \frac{1}{EI}\int_0^\pi M_1 \frac{\partial M_1}{\partial V} R\,d\theta \tag{5.336}$$

From Eq. 5.334

$$\frac{\partial M_1}{\partial V} = R\sin\theta \tag{5.337}$$

Substituting Eqs. 5.334 and 5.337 into Eq. 5.336

$$\delta_Y = \frac{1}{EI} \int_0^{\pi} [VR \sin\theta + HR(1-\cos\theta) + M_0]\ (R\sin\theta)\ R\ d\theta$$

Integrating the above equation and collecting terms

$$\delta_Y = \frac{R^3}{EI}\left(\frac{\pi}{2}V + 2H + 2\frac{M_0}{R}\right) \tag{5.338}$$

This is the second equation required for a solution. The third equation can be determined from the angle of rotation at point B due to the moment M_0 as follows:

$$\alpha = \frac{1}{EI}\int_0^{\pi} M_1 \frac{\partial M_1}{\partial M_0} R\ d\theta \tag{5.339}$$

From Eq. 5.334

$$\frac{\partial M_1}{\partial M_0} = 1 \tag{5.340}$$

Substituting Eqs. 5.334 and 5.340 into Eq. 5.339

$$\alpha = \frac{1}{EI}\int_0^{\pi} [VR\sin\theta + HR(1-\cos\theta) + M_0](1)R\ d\theta$$

Integrating the above equation and collecting terms

$$\alpha = \frac{R^2}{EI}\left(2V + H\pi + \pi\frac{M_0}{R}\right) \tag{5.341}$$

An examination of Fig. 5.93 shows that deflections δ_X and δ_Y can be expressed as

$$\delta_X = r\sin\alpha$$

$$\delta_Y = r(1-\cos\alpha)$$

If the angle α is small, the sine of the angle can be replaced by the angle without too much error.

$$\delta_X \cong r\alpha \tag{5.342}$$

$$\delta_Y = r(1-\cos\alpha) = \frac{r\sin^2\alpha}{1+\cos\alpha} \cong \frac{r\alpha^2}{2} \tag{5.343}$$

Substitute Eqs. 5.342 and 5.343 into Eqs. 5.335 and 5.338. This results in three equations with three unknowns, V, H, and M_0.

$$\frac{2RV}{\pi}+\frac{3}{2}RH+M_0=\frac{r\alpha EI}{R^2\pi} \qquad \text{(see Eq. 5.335)}$$

$$\frac{\pi RV}{2}+2RH+2M_0=\frac{r\alpha^2 EI}{2R^2} \qquad \text{(see Eq. 5.338)}$$

$$\frac{2RV}{\pi}+RH+M_0=\frac{\alpha EI}{R\pi} \qquad \text{(see Eq. 5.341)}$$

Let

$$a=\frac{r\alpha EI}{R^2\pi}$$

$$b=\frac{r\alpha^2 EI}{2R^2}$$

$$c=\frac{\alpha EI}{R\pi}$$

Then the above equations can be expressed in the form of a matrix.

$$\begin{bmatrix} \frac{2R}{\pi} & 1.5R & 1 \\ \frac{\pi}{2}R & 2R & 2 \\ \frac{2R}{\pi} & R & 1 \end{bmatrix} \begin{bmatrix} V \\ H \\ M_0 \end{bmatrix} = \begin{bmatrix} a \\ b \\ c \end{bmatrix} \tag{5.344}$$

With a little algebra it can be shown that the matrix is symmetrical about the diagonal. The three unknowns in the matrix can be determined by using determinants to solve the three simultaneous equations as follows:

$$\frac{V}{\alpha}=\frac{\begin{bmatrix} a & 1.5R & 1 \\ b & 2R & 2 \\ c & R & 1 \end{bmatrix}}{\begin{bmatrix} \frac{2R}{\pi} & 1.5R & 1 \\ \frac{\pi R}{2} & 2R & 2 \\ \frac{2R}{\pi} & R & 1 \end{bmatrix}}=\frac{EI(4R-\pi R\alpha)}{R^3(8-\pi^2)} \tag{5.345}$$

$$\frac{H}{\alpha}=\frac{\begin{bmatrix}\frac{2R}{\pi} & a & 1\\ \frac{\pi R}{2} & b & 2\\ \frac{2R}{\pi} & c & 1\end{bmatrix}}{\begin{bmatrix}\frac{2R}{\pi} & 1.5R & 1\\ \frac{\pi R}{2} & 2R & 2\\ \frac{2R}{\pi} & R & 1\end{bmatrix}}=\frac{2EI}{\pi R^2}\left(\frac{r}{R}-1\right) \tag{5.346}$$

$$\frac{M_0}{\alpha}=\frac{\begin{bmatrix}\frac{2R}{\pi} & 1.5R & a\\ \frac{\pi R}{2} & 2R & b\\ \frac{2R}{\pi} & R & c\end{bmatrix}}{\begin{bmatrix}\frac{2R}{\pi} & 1.5R & 1\\ \frac{\pi R}{2} & 2R & 2\\ \frac{2R}{\pi} & R & 1\end{bmatrix}}=\frac{EI}{\pi R}\left[-\frac{2r}{R}+\frac{2\pi r\alpha+16R-3\pi^2 R}{(8-\pi^2)R}\right]$$

For small angles, the angle α in the righthand side of the above equations is negligible so these equations can be simplified as follows:

$$\frac{V}{\alpha}=-\frac{2.15EI}{R^2} \tag{5.348}$$

$$\frac{H}{\alpha}=-\frac{0.637EI}{R^2}\left(1-\frac{r}{R}\right) \tag{5.349}$$

$$\frac{M_0}{\alpha}=\frac{0.318EI}{R}\left(7.30-\frac{2r}{R}\right) \tag{5.350}$$

Notice that V and H are negative in the above equations. This means that the forces V and H in Figs. 5.93 and 5.94 are shown in the wrong direction, but the direction of moment M_0 is correct.

The total moment per angular displacement α about the gyro output axis can be determined from Fig. 5.93 as follows:

$$\frac{M_T}{\alpha}=\frac{M_0}{\alpha}+\frac{H}{\alpha}r\cos\alpha+\frac{V}{\alpha}r\sin\alpha \tag{5.351}$$

For small angular displacements, the moment due to V is very small since $\sin \alpha$ approaches zero while $\cos \alpha$ approaches 1. Therefore, the force V can be neglected with a very small error.

Substituting Eqs. 5.349 and 5.350 into Eq. 5.351

$$\frac{M_T}{\alpha}=\frac{0.318EI}{R}\left(7.30-\frac{2r}{R}\right)-\frac{0.637EIr}{R^2}\left(1-\frac{r}{R}\right)$$

$$\frac{M_T}{\alpha}=\frac{0.318EI}{R}\left(7.30-\frac{4r}{R}+\frac{2r^2}{R^2}\right) \qquad (5.352)$$

This expression can be simplified with the use of the following:

$$N=R-r \qquad \text{or} \qquad r=R-N \qquad (5.353)$$

Substituting Eq. 5.353 into Eq. 5.352

$$\frac{M_T}{\alpha}=\frac{EI}{R}\left(1.68+\frac{0.636N^2}{R^2}\right) \qquad (5.354)$$

5.29. Sample Problem

A typical rate-gyro flex-lead is shown in Fig. 5.95. The angular spring rate for the flex-lead can be determined from Eq. 5.354 using the following information:

$E = 10.5 \times 10^6$ lb/in.^{2} aluminum flex-lead

$$I=\frac{bt^3}{12}=\frac{(0.030)(0.001)^3}{12}=0.25\times 10^{-11}\text{ in.}^4$$

$R = 0.80$ in. radius

$r = 0.53$ in. radius

$N = R - r = 0.27$ in.

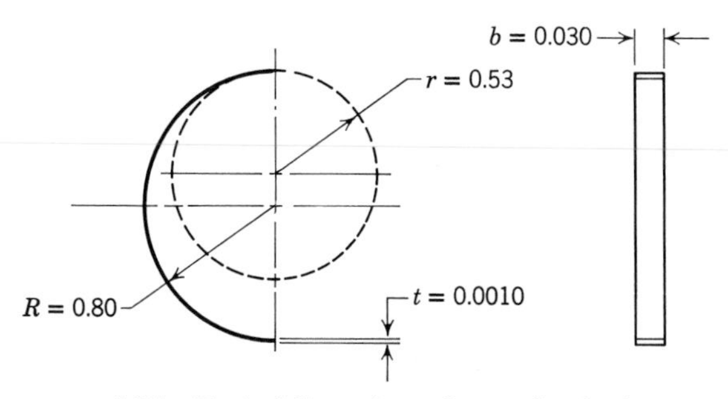

FIGURE 5.95. Typical dimensions of a gyro flex-lead.

Substituting into Eq. 5.354

$$\frac{M_T}{\alpha} = \frac{(10.5 \times 10^6)(0.25 \times 10^{-11})}{0.80} \left[1.68 + \frac{0.636(0.27)^2}{(0.80)^2} \right]$$

$$\frac{M_T}{\alpha} = 5.75 \times 10^{-5} \text{ in. lb/rad} \tag{5.355}$$

This angular spring rate can be expressed in metric units, which then becomes

$$\frac{M_T}{\alpha} = 0.065 \text{ dyne cm/millirad} \tag{5.356}$$

For four flex-leads in a gyro, the total angular spring rate is

$$K_\alpha = \frac{4M_T}{\alpha} = 4(0.065) = 0.260 \text{ dyne cm/millirad} \tag{5.357}$$

A very limited number of tests were run on one gyro to measure the angular spring rate by rotating the gyro gimbal through various angles and utilizing a torque feedback loop to measure the related torques. The test results in this single case were within 10% of the calculated value shown above.

5.30. Natural Frequencies of Bents and Arcs

The natural frequency of a rectangular bent or a circular arc can be approximated from the static deflection equation. For example, the natural frequency along the vertical axis of a uniform bent, with hinged ends, can be estimated by concentrating the entire weight at the center of the top member. This is then equivalent to a single-degree-of-freedom system where the elastic properties are in a weightless spring and all of the weight is concentrated in one mass (Fig. 5.96).

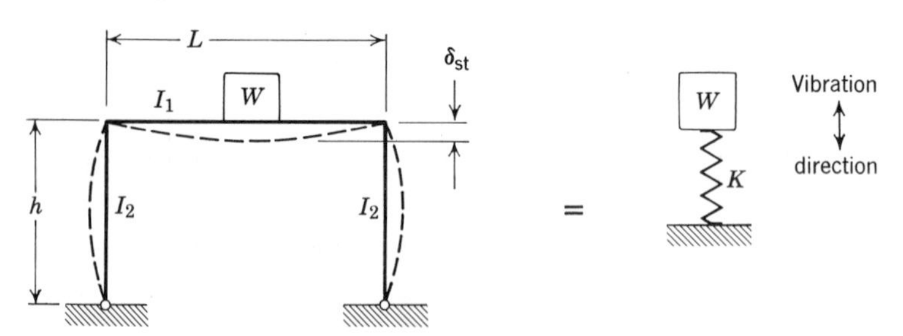

FIGURE 5.96. Bent as a single-degree-of-freedom system.

It was previously shown that the natural frequency of a single-degree-of-freedom system could be determined from the static-deflection equation as shown in Chapter 2, Eq. 2.10

$$f_n = \frac{1}{2\pi}\left(\frac{g}{\delta_{st}}\right)^{1/2} \tag{5.358}$$

where $g = 386$ in./sec^2 gravity

$$\delta_{st} = \frac{WL^3}{48EI_1}\left[1 - \frac{9}{4(2K+3)}\right] \qquad \text{(see Eq. 5.114)}$$

Substituting into Eq. 5.358 for the natural frequency

$$f_n = \frac{1}{2\pi}\left\{\frac{48EI_1 g}{WL^3\left[1 - \dfrac{9}{4(2K+3)}\right]}\right\}^{1/2} \tag{5.359}$$

A rectangular bent with fixed ends can be used to simulate a typical resistor or capacitor soldered to a printed-circuit board (Fig. 5.97). If the electronic component part is not attached to the circuit board with cement and if there is clearance between the board and the body, then it is possible to approximate the natural frequency of the component body from its static deflection on its electrical lead wires. Consider, for example, a RN-70 metal film resistor mounted as shown in Fig. 5.97 (see also Fig. 6.37).

The static deflection can be determined from Eq. 5.18 as

$$\delta_{st} = \frac{WL^3}{48EI_1}\left(1 - \frac{3}{2K+4}\right) \tag{5.360}$$

where $W = 0.0035$ lb for a RN-70 resistor

$L = 0.226$ in. length

$h = 0.174$ in. height

$K = \dfrac{hI_1}{LI_2} = 0.77$

$E = 15 \times 10^6$ lb/in.^{2} copper wire

$I_1 = I_2 = \dfrac{\pi d^4}{64} = \dfrac{\pi}{64}(0.031)^4 = 4.52 \times 10^{-8}$ in.^{4}

$$\delta_{st} = \frac{(3.5 \times 10^{-3})(2.26 \times 10^{-1})^3}{48(15 \times 10^6)(4.52 \times 10^{-8})}\left[1 - \frac{3}{2(0.77)+4}\right]$$

$$\delta_{st} = 0.570 \times 10^{-6} \text{ in.}$$

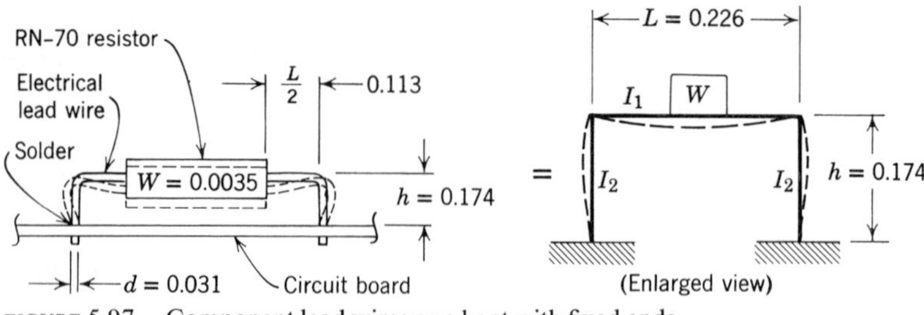

FIGURE 5.97. Component lead wires as a bent with fixed ends.

Substituting into Eq. 5.358 for the natural frequency

$$f_n = \frac{1}{2\pi}\left(\frac{3.86 \times 10^2}{0.570 \times 10^{-6}}\right)^{1/2} = 415 \text{ Hz} \tag{5.361}$$

In order to prevent a large build-up of forces during vibration, the Octave Rule should be followed (see Chapter 6). This rule suggests that the resonant frequency of each degree of freedom should be twice the resonant frequency of the structure to which it is attached. This prevents the possibility of two coincident resonances in two structures coupled to one another. Coincident resonances can multiply transmissibilities, which can lead to rapid fatigue failures.

If the Octave Rule is to be observed here, the printed-circuit board should have a maximum resonant frequency of about 207 Hz, which is half the resonant frequency of the resistor that is mounted on the circuit board.

In order to prevent resonances from developing in components mounted on printed-circuit boards, large resistors and large capacitors should be cemented or tied to the circuit board.

The natural frequency of a rectangular bent during vibration in the lateral direction can be determined from the static deflection shown by Eq. 5.74. If the total weight is concentrated at the top, the deflection will appear as in Fig. 5.98.

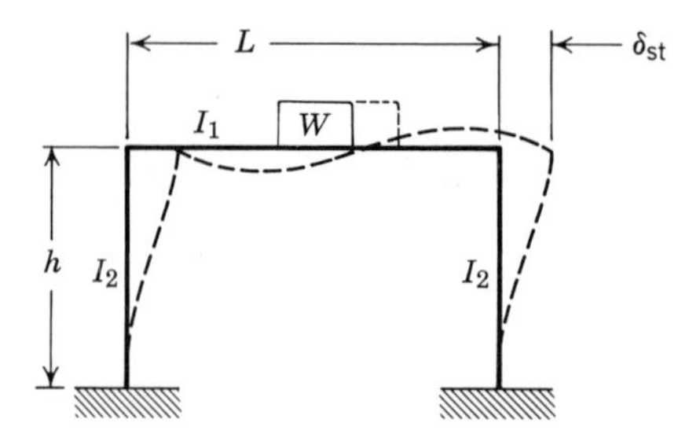

FIGURE 5.98. Lateral vibration of a bent with a concentrated load.

Then using Eq. 5.358, the natural frequency becomes

$$f_n = \frac{1}{2\pi}\left[\frac{24EI_2 g}{Wh^3\left(1+\dfrac{3}{6K+1}\right)}\right]^{1/2} \tag{5.362}$$

The natural frequency of a rectangular bent during vibration in the transverse direction can be used to approximate an electrical component part soldered to a printed-circuit board (Fig. 5.99).

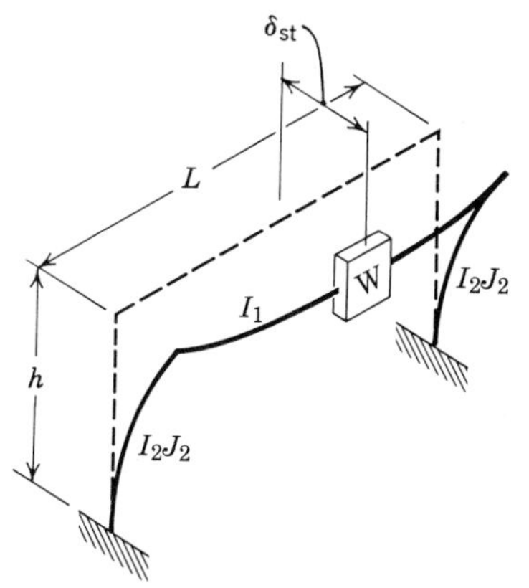

FIGURE 5.99. Transverse vibration of a bent with a concentrated load.

Equations 5.106 and 5.358 can be used to approximate the natural frequency as follows:

$$f_n = \frac{1}{2\pi}\left\{\frac{g}{\dfrac{W}{2}\left[\dfrac{L^3}{24EI_1}+\dfrac{h^3}{3EI_2}-\dfrac{L^4GJ_2}{32EI_1(2hEI_1+LGJ_2)}\right]}\right\}^{1/2} \tag{5.363}$$

The static-deflection equation can also be used to approximate the natural frequency of a circular arc. If the entire weight of the arc is concentrated at its center, then the arc can be considered as a weightless spring and the structure can be treated as a single-degree-of-freedom system. A circular arc with hinged ends is shown in Fig. 5.100.

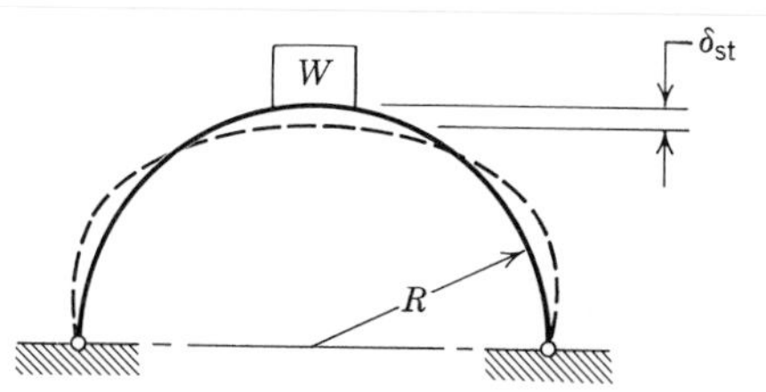

FIGURE 5.100. Vertical vibration of a circular arc with hinged ends.

Equations 5.270 and 5.358 can be used to approximate the natural frequency as follows:

$$f_n = \frac{1}{2\pi}\left(\frac{52.6EIg}{WR^3}\right)^{1/2} \tag{5.364}$$

For a circular arc with fixed ends and a concentrated load at the center, the structure will appear as in Fig. 5.101.

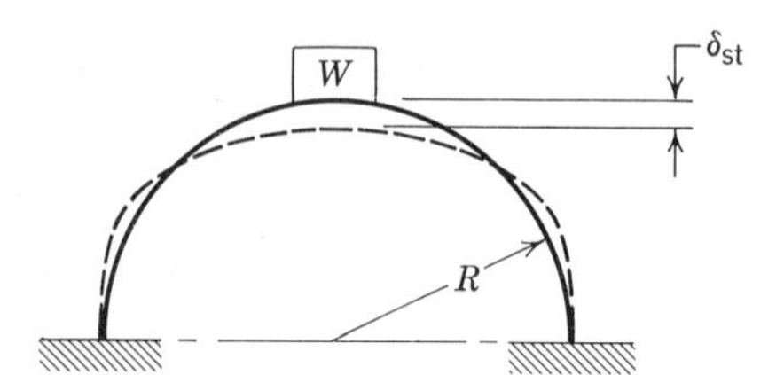

FIGURE 5.101. Vertical vibration of a circular arc with fixed ends.

Equations 5.284 and 5.358 can be used to determine the resonant frequency in the vertical direction as follows:

$$f_n = \frac{1}{2\pi}\left(\frac{87.0EIg}{WR^3}\right)^{1/2} \tag{5.365}$$

A circular arc with hinged ends vibrating in the lateral direction is shown in Fig. 5.102.

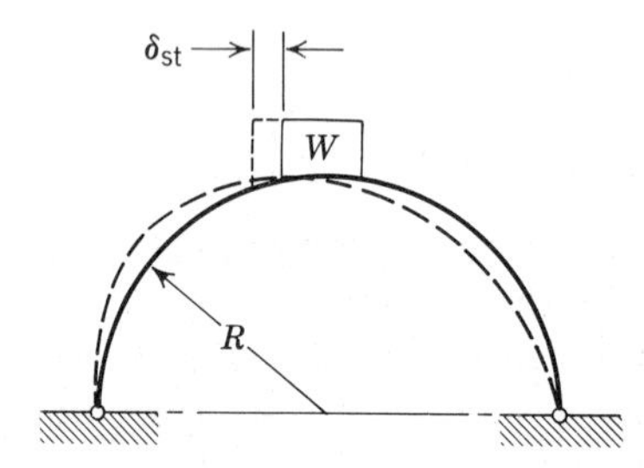

FIGURE 5.102. Lateral vibration of a circular arc with hinged ends.

The natural frequency can be approximated from Eqs. 5.309 and 5.358 as follows:

$$f_n = \frac{1}{2\pi}\left(\frac{14.12EIg}{WR^3}\right)^{1/2} \tag{5.366}$$

A circular arc with fixed ends and a concentrated load vibrating in the lateral direction is shown in Fig. 5.103.

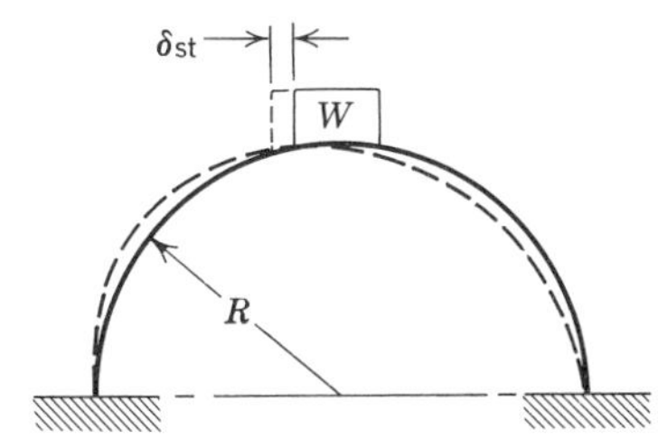

FIGURE 5.103. Lateral vibration of a circular arc with fixed ends.

The natural frequency can be approximated from Eq. 5.358 along with the static deflection as follows:

$$f_n = \frac{1}{2\pi}\left(\frac{54.0EIg}{WR^3}\right)^{1/2} \tag{5.367}$$

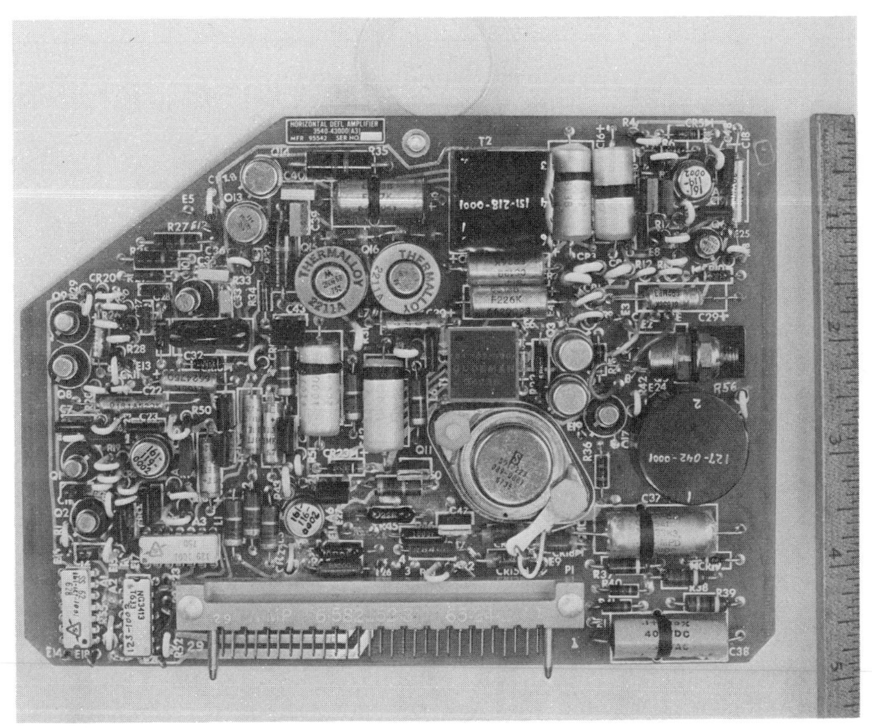

FIGURE 5.104. A plug-in printed-circuit board where lacing cord is used to tie large electronic component parts to the board (courtesy Norden division of United Aircraft).

CHAPTER 6

PRINTED-CIRCUIT BOARDS AND FLAT PLATES

6.1. Various Types of Printed-Circuit Boards

Electronic systems make extensive use of plug-in printed-circuit boards because they are very easy to service. Defective circuit boards can be removed and replaced quickly and easily without bothering with wires and a soldering iron. After a defective circuit board has been removed, trouble-shooting is easy and repairs can be made by a skilled technician.

Many different types of printed-circuit boards are manufactured by the electronics industry. Epoxy fiberglass is the most common material used, with laminated layers of copper on one or both sides of the board to form the electrical conductors. The overall printed-circuit-board thickness can vary from about 0.006 to 0.125 in. Board sizes can vary from about 2 to 16 in. Many different shapes can be found ranging from small squares to large circular plates and triangles, depending upon the shape of the electronic box used to support the circuit boards. Since electronic equipment is being packed into every available inch of space in most airplanes, missiles, and even television sets, the shape of the circuit board is often dictated by the geometry of the available space. The rectangular printed-circuit board is the most common shape used by the electronics industry, since this shape is easily adapted to the popular modular plug-in type of assembly, which utilizes an electrical connector along the bottom edge of the circuit board.

Printed-circuit boards with many high-power-dissipating components will run very hot unless the heat is removed. Therefore aluminum and copper, which have high thermal conductivities, are often bonded to the epoxy fiberglass circuit boards to act as heat sinks and to conduct away the heat.

Ribs are often added to printed-circuit boards that must operate in a severe vibration and shock environment. Ribs will increase the stiffness of the circuit board, which in turn will increase the resonant frequency. This will reduce the board deflections during resonant conditions, thus reducing the stresses developed in the electronic component parts mounted on the circuit board. Ribs can be fabricated of steel, aluminum, or epoxy fiberglass. If metal ribs are used, caution must be exercised to prevent short circuits across exposed electrical printed-circuit strips. Ribs can be bolted, riveted, soldered, cemented, welded, or cast integral with heat-sink plates.

Printed-circuit boards may be supported by the electronic box in many different ways depending upon such factors as the environment, weight, maintainability, accessibility, and cost. For example, in a vacuum environment where there is no air to provide convection heat transfer, heat will often be conducted from the circuit board to a heat exchanger. High-pressure thermal interfaces must then be provided, using materials that have a high thermal conductivity, in order to prevent excessive temperatures from developing in the electronic components mounted on the circuit boards.

Screws are ideal for providing high-pressure interfaces but they are generally time-consuming to install since they take many turns to insert and remove. Maintenance people prefer simply to unplug one module and plug in another. This reduces maintenance time substantially, since many electronic systems can easily contain over a hundred plug-in printed-circuit boards. Sometimes a fast-acting cam-and-wedge type of clamp is used to provide a high-pressure interface, if there is enough room available in the electronic box.

The manner in which the printed-circuit boards are supported in the electronic box can be an important factor in determining just how the boards will respond to vibration and shock. If the electronic box is fabricated from a light sheet metal and manufacturing tolerances are loose, a loose fit is desirable for the plug-in circuit boards. This will permit easy connector engagement, with a fixed connector position, when there is a substantial mismatch in the mating box connector. This type of installation is not desirable, however, for a system that will be subjected to severe vibration and shock. A loose circuit board will often develop high acceleration loads which will lead to high deflections and stresses in the electronic component parts mounted on the circuit boards.

The edges of the printed-circuit boards should be supported if they will be subjected to severe vibration or shock environments. Board edge-guides are available that will support a wide variety of printed-circuit boards. These guides are usually fabricated of beryllium copper, but they are also available in a wide range of metals and molded plastics.

A board edge-guide that grips the edge of a printed-circuit board firmly is very desirable. This firm grip can reduce deflections due to edge rotation and translation, which will increase the natural frequency of the circuit board. In addition, a firm grip will tend to dissipate more energy during vibration because of friction and relative motion between the edges of the board and the edge-guide. This, in turn, will reduce the transmissibility experienced by the printed-circuit board during resonance. A tight board edge-guide may require tight manufacturing tolerances, plus a floating chassis connector, to permit accurate connector alignment during the circuit board installation, to prevent connector damage.

The transmissibility developed by a printed-circuit board during resonance will depend upon many factors such as the board material, number and type of laminations in a multilayer board, natural frequency, type of mounting, type of electronic component parts mounted on the circuit board, acceleration *G* levels, type of conformal coating, type of connector, and the shape of the board. Slight modifications in the installation geometry of the circuit board can have a sharp affect on the transmissibilities and the mode shapes developed at resonance. For example, it was shown in Chapter 2, Eq. 2.76 that the transmissibility of a single-degree-of-freedom system will increase as the natural frequency increases. This can be demonstrated by changing the edge restraints of a printed-circuit board from a condition that is approximately simply supported to a condition that is approximately fixed. Vibration test data on one group of boards, which used Birtcher guides on two opposite sides of the board to simulate a simply supported condition, showed natural frequencies of about 260 Hz. The transmissibility was about 16 for a 5-*G*-peak sinusoidal-vibration input. This is about equal to the square root of the natural frequency. The opposite sides of the circuit boards were then clamped to simulate a fixed-edge condition. This increased the typical natural frequency to about 380 Hz and increased the typical transmissibility to about 25. This is equal to 1.28 times the square root of the typical natural frequency. The group of printed-circuit boards used in these tests had the dimensions and construction as shown in Fig. 6.1.

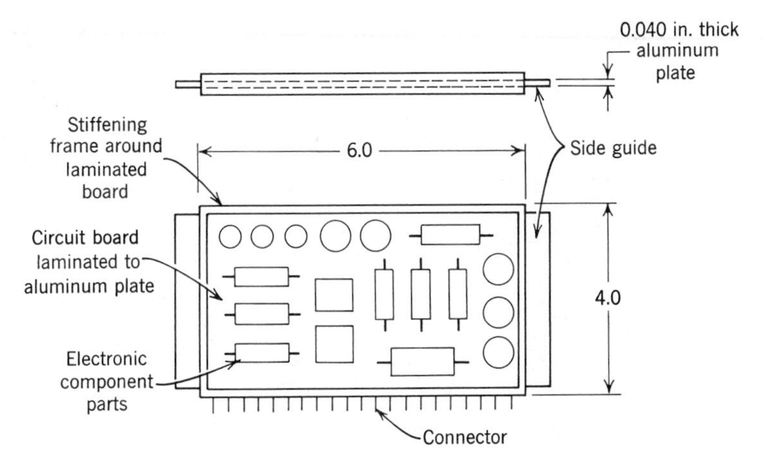

FIGURE 6.1 Laminated circuit board with a stiffening frame.

These tests were run in a rigid vibration-test fixture that included a mating connector for each circuit board. Accelerometers were mounted at several points on the vibration fixture and each circuit board to monitor the input and output acceleration *G* forces for the system (Fig. 6.2).

These same printed-circuit boards were then installed in an electronic box and the 5-*G*-peak sinusoidal-vibration tests were repeated. The circuit boards with Birtcher guides showed a typical natural frequency of about 225 Hz with a typical transmissibility of about 14. This is equal to 0.93 times the square root of the natural frequency. The circuit boards then had their opposite sides clamped to the chassis structure to simulate a fixed-edge condition. The natural frequency increased to about 340 Hz

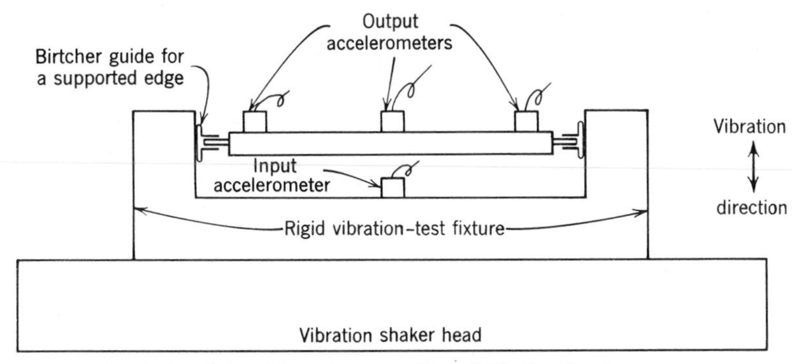

FIGURE 6.2. Printed circuit board mounted in a rigid vibration fixture.

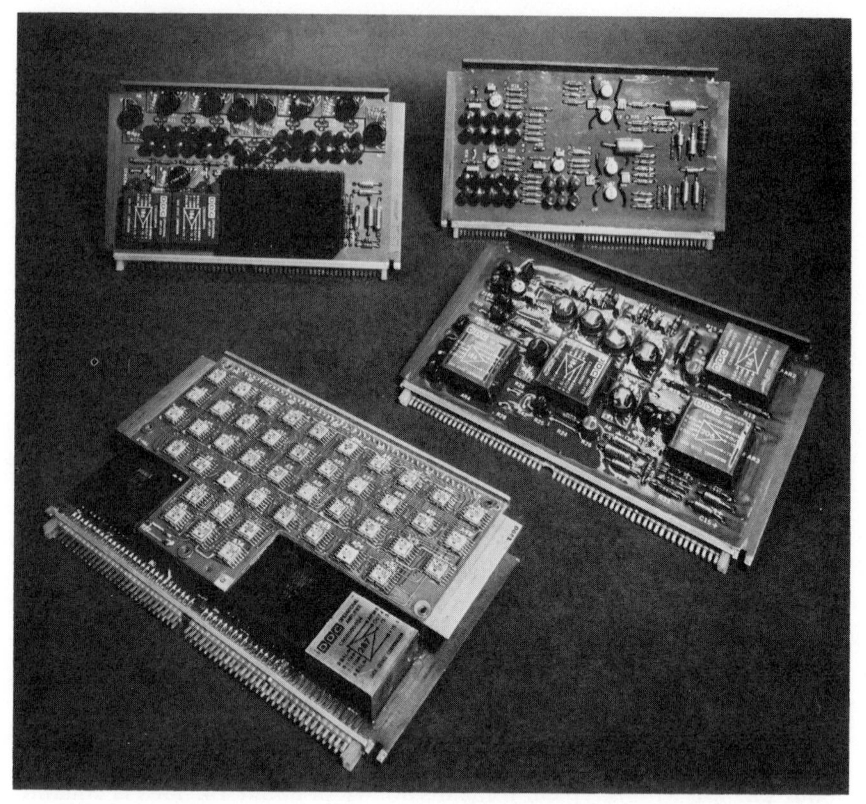

FIGURE 6.3. Plug-in assemblies using a thin printed-circuit board laminated to an aluminum extrusion to provide a rigid structure with good cooling (courtesy Kearfott division, The Singer Company.

and the transmissibility, from the chassis to the center of the circuit board increased to about 20. This is equal to 1.08 times the square root of the natural frequency. In all cases, the transmissibility of the circuit boards decreased slightly when they were taken out of the rigid test fixture and placed in a typical electronic box. Figure 6.3 shows several plug-in assemblies that use a thin printed-circuit board laminated to an aluminum extrusion.

6.2. Changes in Circuit-Board Edge Conditions

In order to calculate the natural frequency of a printed-circuit board, it is necessary to estimate the edge conditions of the board. Much insight can be obtained on edge conditions by using a strobe light to watch board motions during vibration. This shows very quickly whether the edges are translating or rotating. Sometimes edge conditions can change when the acceleration forces change. An edge that appears to be simply supported with a 2-*G*-peak input may appear to be free with a 5-*G*-peak input. The natural frequency with a 2-*G* input will then be different from the natural frequency with a 5-*G* input.

Sinusoidal vibration tests were run on a group of printed-circuit boards that had the opposite sides supported by a preloaded beryllium copper wavy spring which acted as a board edge-guide, in a channel section. The bottom edge of each board had a 100-pin connector that plugged into the sockets of a mating connector mounted in a rigid vibration fixture. The top edge of each board was restrained by three foam-rubber strips 1 in. by 0.125 in. thick and compressed about 60% when the board was installed. These rubber strips were used to restrain the unsupported top edge of the circuit board, to simulate the action of a cover using the same type of rubber strips. Strips of this type are convenient to use because they provide some support and damping without the high fabrication costs associated with close tolerances on machined parts. A sketch of this type of installation is shown in Fig. 6.4.

Vibration tests were run with a peak input of 2*G*s, 5*G*s, 10*G*s, and 15*G*s using a strobe light to observe the action of the circuit-board edges

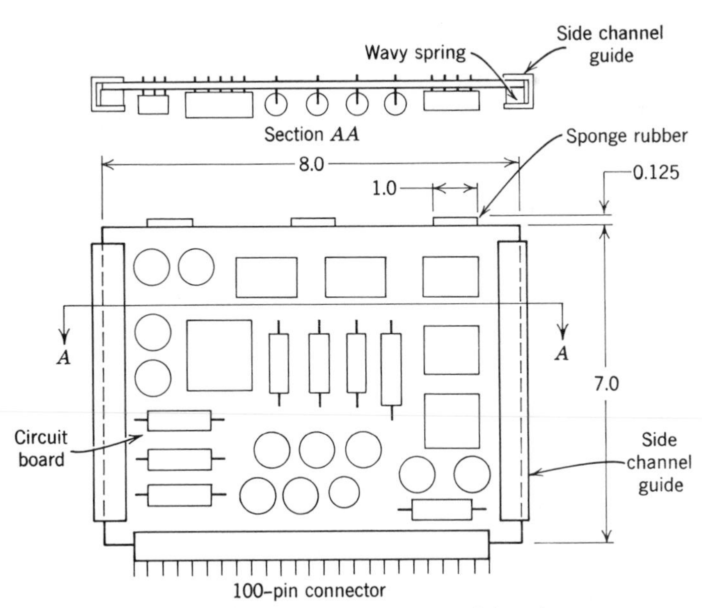

FIGURE 6.4. Typical installation of a printed circuit board.

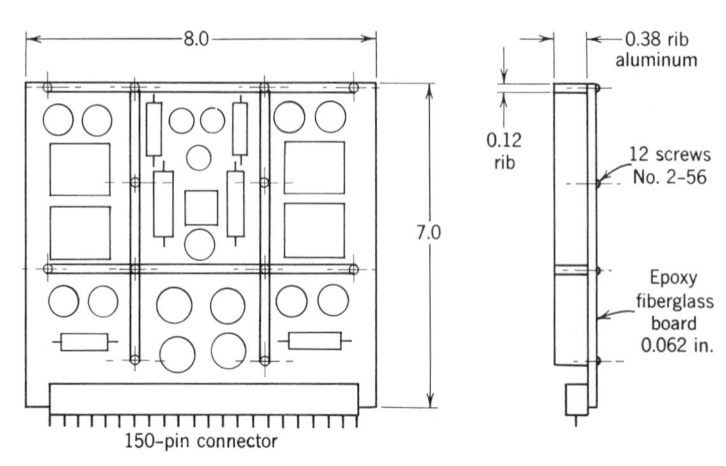

FIGURE 6.5. Printed circuit board with ribs fastened by screws.

during resonant conditions. The 2-*G* tests showed that all four edges of the circuit board appeared to be simply supported. The edges were observed to be rotating without translation. The 5-*G* tests showed that the sides of the circuit boards acted as though they were supported, but the top edge at the rubber pads and the bottom edge at the connector showed some translation in addition to the rotation. The 10-*G* tests still showed that the sides of the circuit boards acted as though they were simply supported, but the top edges at the rubber strips and the bottom edges at the connectors all experienced much more translation than rotation. The 15-*G* tests showed that some translation at the sides in addition to the rotation, while the top and bottom edges appeared to be free because there was very little rotation, only large translation amplitudes. Resonant dwell tests at a 10*G* peak resulted in many broken connector pins.

Stiffening ribs were added to some of the circuit boards to increase their resonant frequencies. Some of these ribs were screwed to the faces of the circuit boards (Fig. 6.5).

TABLE 6.1

Input *G* Level	Resonant Frequency (Hz)	Transmissibility (Q)
2	215 Hz	15.0
5	182 Hz	11.2
10	161 Hz	8.2

Vibration tests were run on these boards using the same type of foam-rubber strips at the top of each circuit board and the same type of wavy-spring channel-guides at the sides of each board. When the input G levels were changed, there were substantial changes in the natural frequencies and transmissibilities (Table 6.1).

6.3. Estimating the Transmissibility of a Printed-Circuit Board

Printed-circuit boards can have a wide variety of sizes and shapes along with many different mounting arrangements. In the early stages of a design, these circuit boards must be analyzed to make sure they will function properly in the required environment. One critical part of the analysis relates to the dynamic loads developed in these circuit boards at their fundamental resonant frequencies. These loads are closely associated with the transmissibilities developed by the circuit boards, but transmissibilities are very difficult to estimate without previous test data on similar circuit boards. Obviously, test data are the best sources for information on the transmissibility characteristics for various types of circuit boards. However there are times when test data are not available because of a new design or a geometry that has not been used before. Under these conditions it becomes necessary to estimate the response characteristics of the circuit board, at least until a model circuit board can be fabricated and tested.

There are a number of factors that should be considered when the transmissibility of a printed-circuit board must be estimated without the use of test data. These factors are all related to the damping characteristics of the board, which determine the amount of energy lost during the vibrating condition. When more energy is lost, or transformed to heat, there is less energy remaining and the transmissibility is lower, hence the dynamic loads and stresses are also lower.

The greatest energy losses are probably due to hysteresis and friction. Hysteresis losses are generally due to internal strains that are developed during bending deflections in the circuit board. Friction losses are generally due to relative motion between high-pressure interfaces such as mounting surfaces, stiffening ribs, and edge-guides. These energy losses are greatest when the deflections are greatest and smallest when the deflections are smallest. Since higher frequencies have smaller deflections, they will also have less damping. This means higher frequencies will usually have higher transmissibilities at resonant conditions.

Term such as "low" resonant frequency and "high" resonant frequency are only relative. In general, however, the term "low" resonant frequency applies to those circuit boards that have a resonant frequency below about 100 Hz. The term "high" resonant frequency applies to those circuit boards that have a resonant frequency above about 400 Hz. Most circuit boards appear to have their fundamental resonant frequencies between about 200 Hz and 300 Hz.

In Section 2.17, Eq. 2.76 it was shown that the transmissibility is related to the square root of the spring-rate, K. In Section 2.1, Eq. 2.7 it was shown that the natural frequency is also a function of the square root of the spring-rate, K. If there is no change in the damping characteristics, it appears that the transmissibility will be directly related to the natural frequency. Test data, however, show the transmissibility of a printed-circuit board can generally be related to the square root of the natural frequency of the board. For rectangular boards, the transmissibility will normally range from about 0.50–2.0 times the square root of the natural frequency, depending upon many factors. A small circuit board with small electronic component parts and no stiffening ribs, with a resonant frequency of about 400 Hz, can have a transmissibility as high as 2.0 times the square root of the natural frequency for a low input G force below a $2G$ peak. The transmissibility for this case will then be about 40. A large circuit board with large electronic component parts and several stiffening ribs, with a resonant frequency of about 100 Hz, can have a transmissibility as low at 0.50 times the square root of the natural frequency for a high input G force above a $10G$ peak. The transmissibility for this case will then be about 5.

Over the middle frequency ranges (200–300 Hz), the transmissibility will often be about equal to the square root of the natural frequency for a 5-G-peak sinusoidal-vibration input. Again, there are many other factors that will influence the transmissibility of a printed-circuit board. If there are no test data available on the particular type of circuit board being analyzed, then the approximations just outlined are suggested as a good starting point.

Some of the factors that should be considered when the transmissibility of a circuit board is being estimated are as follows:

1. The natural frequency of the circuit board: A high natural frequency means low displacements and low strains, so the transmissibilities are usually higher. Conversely, a low natural frequency means high displacements and high strains, so the transmissibilities are usually lower.

2. The input G force for sinusoidal vibration: A lower input G force means low displacements and low strains, so the transmissibilities are usually higher. A high input G force means high displacements and high strains, so the transmissibilities are usually lower.

3. Ribs: Riveted and bolted ribs will generally permit some relative motion to occur at the high-pressure interface between the ribs and the circuit board, which will dissipate energy and reduce the transmissibility at resonance. Welded, cast, and cemented ribs are usually stiffer than riveted and bolted ribs so that they raise the resonant frequency more, but they do not provide as much damping.

4. Circuit-board supports: Circuit-board edge-guides that grip the edges of the board firmly will provide a high-pressure interface that will dissipate energy and lower the transmissibility. If mounting bolts are used to fasten the boards, high-pressure interfaces in the bolt areas will tend to dissipate energy. More mounting bolts will dissipate more energy and increase the stiffness at the same time.

5. Circuit-board connectors: Circuit-board connectors, such as the edge type or the pin-and-socket type, will provide some damping that will tend to reduce the transmissibility. A longer connector will provide more support to the circuit-board edge while providing additional damping.

6. Type of electronic component part: Large electronic component parts that are in intimate contact with the circuit board will tend to dissipate more energy than small electronic component parts. This is because larger components cover a bigger span on the circuit board, resulting in larger relative deflections at the circuit-board interface over a larger area, so the damping is slightly greater.

7. Heat-sink strips: Heat-sink strips are often laminated to the circuit board to conduct the heat away from the electronic component parts. These heat-sink strips will act as laminations which will dissipate energy through relative interface shear deflections. At high frequencies, where the displacements are small, the damping is small so transmissibilities are not reduced very much.

8. Multiple-layer circuit boards; Many printed-circuit boards must use extra layers to provide all of the electrical interconnections required by the circuits. These extra layers will increase the damping slightly and thus tend to reduce the transmissibility. At high frequencies, where deflections are small, transmissibilities are not reduced very much.

9. Conformal coatings: Conformal coatings are generally used to protect electronic components from dust accumulation. Many manufacturers of electronic equipment also use conformal coatings to protect sensitive circuits from water vapor and condensation. A thin polyurethane or epoxy film about 0.005 in. thick is usually sprayed or dipped on the circuit board. Unless the adhesion is good and unless all of the sharp points are coated, water vapor can wick in between the coating and the circuit board and condense. Once this happens, the moisture will usually stay where it is unless it is baked out at high temperatures. Small amounts

FIGURE 6.6. A plug-in printed-circuit module using a laminated construction that consists of a thin circuit board, an EMI shield, a prepreg cement, and an aluminum heat-sink plate (courtesy Kearfott division, The Singer Company).

of moisture will not have much effect on the electrical operating characteristics of the circuits unless the circuits have a high impedance. Conformal coatings also act as a cement which provides good adhesion for electronic component parts to the circuit board. During vibration, relative motion between the components and the circuit board will produce strains in the conformal coating and will tend to provide some damping.

6.4. Natural Frequency Using a Trigonometric Series

Most printed-circuit boards can be approximated as flat rectangular plates with different edge conditions and different loading conditions. General plate equations can then be used to determine the strain energy and the kinetic energy of the vibrating plate, which leads to the natural-frequency equation. One very convenient method for analyzing plates is the Rayleigh method [28]. A deflection curve is assumed in such a manner that it satisfies the geometric boundary conditions, which are the deflection and the slope for a particular plate. Once these boundary conditions are satisfied, the assumed deflection curve is used to obtain the strain energy and the kinetic energy of the particular plate. If there is no energy dissipated, the strain energy will be equal to the kinetic energy and the approximate natural frequency can be determined.

The Rayleigh method results in a natural frequency that is slightly higher than the true natural frequency for a given set of conditions, unless the exact deflection curve is used. This then results in the exact natural frequency equation.

Consider a flat, rectangular plate, with four simply supported edges and a uniformly distributed load, being vibrated in a direction perpendicular to the plane of the plate (Fig. 6.7).

The deflection curve for the simply supported plate can be represented by a double trigonometric series.

$$Z = \sum_{\substack{m \\ 1,3,5}}^{\infty} \sum_{\substack{n \\ 1,3,5}}^{\infty} A_{mn} \sin \frac{m\pi X}{a} \sin \frac{n\pi Y}{b}$$

Extensive vibration test data on printed-circuit boards show that most of the damage occurs at the fundamental resonant mode where the displacements and the stresses are the greatest. The above equation can then be simplified to the following expression:

$$Z = Z_0 \sin \frac{\pi X}{a} \sin \frac{\pi Y}{b} \tag{6.1}$$

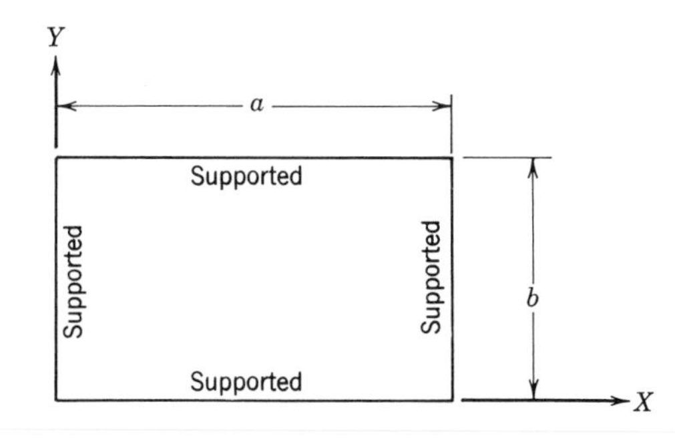

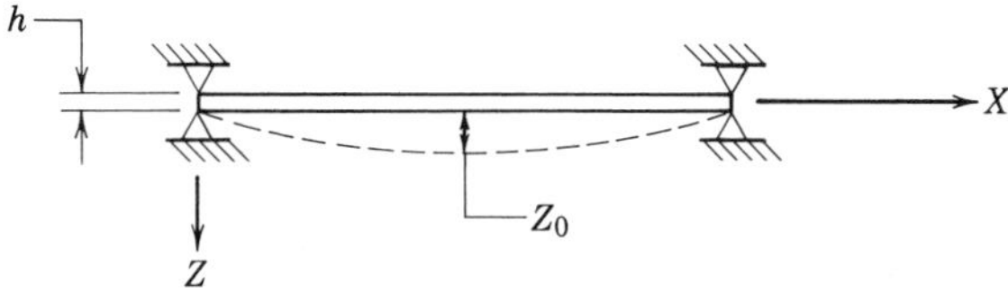

FIGURE 6.7. A flat rectangular plate supported on four sides.

The assumed deflection curve must be checked to make sure it meets the requirements for the geometric deflection boundary conditions. This means the edges of the plate must have a zero deflection and the center of the plate must have the maximum deflection. Then from Eq. 6.1

at
$$X=0 \quad \text{and} \quad y=0, \quad Z=0$$
$$X=0 \quad \text{and} \quad Y=b, \quad Z=0$$
$$X=a \quad \text{and} \quad Y=0, \quad Z=0$$
$$X=a \quad \text{and} \quad Y=b, \quad Z=0$$
$$X=\frac{a}{2} \quad \text{and} \quad Y=\frac{b}{2}, \quad Z=Z_0$$

The deflection boundary conditions are satisfied: there are no deflections at the edges and the maximum deflection is at the center.

Next check the geometric boundary conditions for the slope of the plate at different points. There must be a finite slope at all four edges but the center of the plate must have a zero slope. Considering first the slope along the X axis, the required equation can be determined from the partial derivative of Z with respect to X. Then from Eq. 6.1

$$\theta_X=\frac{\partial Z}{\partial X}=Z_0\frac{\pi}{a}\cos\frac{\pi X}{a}\sin\frac{\pi Y}{b}$$

at
$$X=0 \quad \text{and} \quad Y=\frac{b}{2}, \quad \theta_X=Z_0\frac{\pi}{a}$$
$$X=\frac{a}{2} \quad \text{and} \quad Y=\frac{b}{2}, \quad \theta_X=0$$
$$X=a \quad \text{and} \quad Y=\frac{b}{2}, \quad \theta_X=-Z_0\frac{\pi}{a}$$

The slope boundary conditions are satisfied at the edges and at the center of the plate along the X axis. The same check can be made for the slope along the Y axis, which will show the boundary conditions are satisfied.

The total strain energy V of the vibrating plate can be represented in the following form [4]:

$$V=\frac{D}{2}\int_0^a\int_0^b\left[\left(\frac{\partial^2 Z}{\partial X^2}\right)^2+\left(\frac{\partial^2 Z}{\partial Y^2}\right)^2+2\mu\left(\frac{\partial^2 Z}{\partial X^2}\right)\left(\frac{\partial^2 Z}{\partial Y^2}\right)\right.$$
$$\left.+2(1-\mu)\left(\frac{\partial^2 Z}{\partial X\partial Y}\right)^2\right]dX\,dY \tag{6.2}$$

where $D = \dfrac{Eh^3}{12(1-\mu^2)}$ plate stiffness factor

E = modulus of elasticity, lb/in.^{2}

h = plate thickness, in.

μ = Poisson's ratio, dimensionless

The total kinetic energy T of the vibrating plate can be represented in the following form [4]:

$$T = \frac{\rho\Omega^2}{2}\int_0^a\int_0^b Z^2\,dX\,dY \tag{6.3}$$

where $\rho = \dfrac{W}{abg} = \dfrac{\nu h}{g}$ mass per unit area

W = total weight of plate, lb

ν = material density, lb/in.^{3}

a = length of plate, in.

b = width of plate, in.

h = plate thickness, in.

g = acceleration of gravity, 386 in/sec^2

Ω = circular frequency, rad/sec

Performing the operations on Eq. 6.1 required by Eq. 6.2

$$\frac{\partial Z}{\partial X} = Z_0\frac{\pi}{a}\cos\frac{\pi X}{a}\sin\frac{\pi Y}{b} \tag{6.4}$$

$$\frac{\partial^2 Z}{\partial X^2} = -Z_0\frac{\pi^2}{a^2}\sin\frac{\pi X}{a}\sin\frac{\pi Y}{b} \tag{6.5}$$

$$\left(\frac{\partial^2 Z}{\partial X^2}\right)^2 = Z_0{}^2\frac{\pi^4}{a^4}\sin^2\frac{\pi X}{a}\sin^2\frac{\pi Y}{b} \tag{6.6}$$

$$\frac{\partial Z}{\partial Y} = Z_0\frac{\pi}{b}\sin\frac{\pi X}{a}\cos\frac{\pi Y}{b} \tag{6.7}$$

$$\frac{\partial^2 Z}{\partial Y^2} = -Z_0\frac{\pi^2}{b^2}\sin\frac{\pi X}{a}\sin\frac{\pi Y}{b} \tag{6.8}$$

$$\left(\frac{\partial^2 Z}{\partial Y^2}\right)^2 = Z_0{}^2\frac{\pi^4}{b^4}\sin^2\frac{\pi X}{a}\sin^2\frac{\pi X}{b} \tag{6.9}$$

From Eqs. 6.5 and 6.8

$$\left(\frac{\partial^2 Z}{\partial X^2}\right)\left(\frac{\partial^2 Z}{\partial Y^2}\right) = Z_0^2\frac{\pi^4}{a^2b^2}\sin^2\frac{\pi X}{a}\sin^2\frac{\pi Y}{b} \tag{6.10}$$

From Eq. 6.4 or 6.7

$$\frac{\partial^2 Z}{\partial X \partial Y} = Z_0 \frac{\pi^2}{ab} \cos \frac{\pi X}{a} \cos \frac{\pi Y}{b}$$

$$\left(\frac{\partial^2 Z}{\partial X \partial Y}\right)^2 = Z_0^2 \frac{\pi^4}{a^2 b^2} \cos^2 \frac{\pi X}{a} \cos^2 \frac{\pi Y}{b} \tag{6.11}$$

Since these equations must be integrated, the following relations are used

$$\left.\begin{aligned} \int_0^a \sin^2 \frac{\pi X}{a}\, dX &= \frac{a}{2} \\ \int_0^b \sin^2 \frac{\pi Y}{b}\, dY &= \frac{b}{2} \end{aligned}\right\} \tag{6.12}$$

Following Eq. 6.2 and integrating Eq. 6.6 using Eq. 6.12

$$\int_0^a \int_0^b \left(\frac{\partial^2 Z}{\partial X^2}\right)^2 dX\, dY = Z_0^2 \frac{\pi^4}{a^4} \left(\frac{ab}{4}\right) \tag{6.13}$$

Following Eq. 6.2 and integrating Eq. 6.9 using Eq. 6.12

$$\int_0^a \int_0^b \left(\frac{\partial^2 Z}{\partial Y^2}\right)^2 dX\, dY = Z_0^2 \frac{\pi^4}{b^4} \left(\frac{ab}{4}\right) \tag{6.14}$$

Following Eq. 6.2 and integrating Eq. 6.10 using Eq. 6.12

$$\int_0^a \int_0^b \left(\frac{\partial^2 Z}{\partial X^2}\right)\left(\frac{\partial^2 Z}{\partial Y^2}\right) dX\, dY = Z_0^2 \frac{\pi^4}{a^2 b^2} \left(\frac{ab}{4}\right) \tag{6.15}$$

Following Eq. 6.2 and integrating Eq. 6.11 using Eq. 6.12

$$\int_0^a \int_0^b \left(\frac{\partial^2 Z}{\partial X \partial Y}\right)^2 dX\, dY = Z_0^2 \frac{\pi^4}{a^2 b^2} \left(\frac{ab}{4}\right) \tag{6.16}$$

Substituting Eqs. 6.13–6.16 into Eq. 6.2 for the strain energy of the vibrating plate

$$V = \frac{D}{2} \left(\frac{ab Z_0^2 \pi^4}{4}\right) \left[\frac{1}{a^4} + \frac{1}{b^4} + \frac{2\mu}{a^2 b^2} + \frac{2(1-\mu)}{a^2 b^2}\right]$$

$$V = \frac{\pi^4 D Z_0^2 ab}{8} \left(\frac{1}{a^4} + \frac{2}{a^2 b^2} + \frac{1}{b^4}\right) \tag{6.17}$$

The kinetic energy of the vibrating plate can be determined from deflection Eq. 6.1 along with kinetic energy Eq. 6.3.

$$Z^2 = Z_0^2 \sin^2 \frac{\pi X}{a} \sin^2 \frac{\pi Y}{b} \tag{6.18}$$

Substituting Eq. 6.12 into Eq. 6.18 for the integration

$$\int_0^a \int_0^b Z^2 \, dX \, dY = Z_0^2 \frac{ab}{4} \tag{6.19}$$

Substituting Eq. 6.19 into Eq. 6.3 for the kinetic energy of the vibrating plate

$$T = \frac{\rho \Omega^2}{2} \left(Z_0^2 \frac{ab}{4} \right) \tag{6.20}$$

Since the strain energy of the vibrating plate must equal the kinetic energy at resonance, if there is no energy dissipated, then Eq. 6.17 must be equal to Eq. 6.20.

$$\frac{\pi^4 D Z_0^2 ab}{8} \left(\frac{1}{a^2} + \frac{1}{b^2} \right)^2 = \frac{\rho \Omega^2 Z_0^2 ab}{8}$$

$$\Omega^2 = \frac{\pi^4 D}{\rho} \left(\frac{1}{a^2} + \frac{1}{b^2} \right)^2$$

Solving for the natural frequency of the rectangular plate

$$f_n = \frac{\Omega}{2\pi} = \frac{\pi}{2} \sqrt{\frac{D}{\rho}} \left(\frac{1}{a^2} + \frac{1}{b^2} \right) \tag{6.21}$$

In this particular case the natural-frequency equation shown above is exact using the Rayleigh method because the deflection curve used for the vibrating plate is exact for the boundary conditions.

6.5. Natural Frequency Using a Polynomial Series

The natural-frequency equation for a uniform, flat, rectangular plate simply supported on four edges can also be derived with the use of a polynomial series. Consider the coordinate system for the rectangular plate when the X and Y axes are taken at the center of the plate (Fig. 6.8).

The deflection curve for the simply supported plate can be represented by a double polynomial series as follows [30]:

$$Z = h \sum_{m=0}^{\infty} \sum_{n=0}^{\infty} \left(1 - \frac{X^2}{c^2}\right)\left(1 - \frac{Y^2}{d^2}\right)\left(\frac{X}{c}\right)^m \left(\frac{Y}{d}\right)^n A_{mn} \cos \Omega t$$

Since most of the damage on a printed-circuit board will occur at the

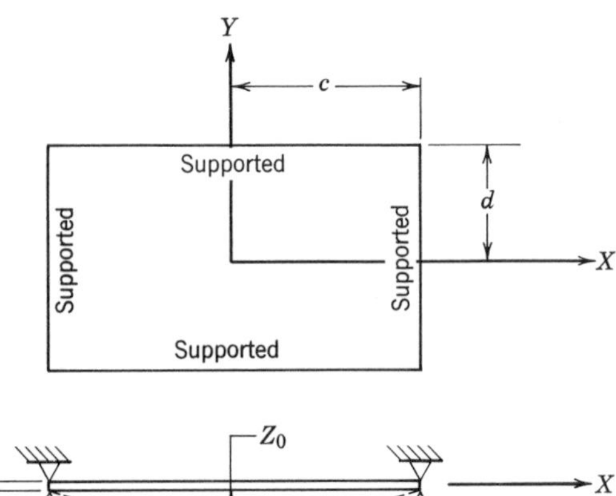

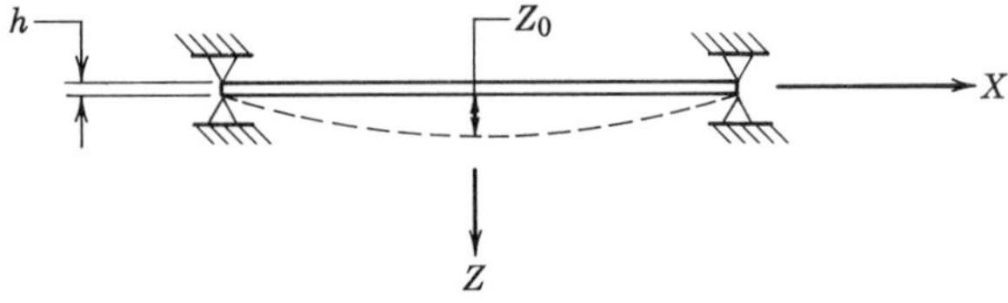

FIGURE 6.8. A simply supported plate with the axes at the center.

fundamental resonant mode, where the displacements and the stresses are the greatest, the above equation can be simplified to the following:

$$Z = Z_0\left(1-\frac{X^2}{c^2}\right)\left(1-\frac{Y^2}{d^2}\right) \tag{6.22}$$

The assumed deflection curve must be checked to make sure it meets the requirements for the geometric deflection boundary conditions. This means the edges of the plate must have a zero deflection and the center of the plate must have the maximum deflection. Then from Eq. 6.22

at

$$\begin{aligned} X&=0 \quad \text{and} \quad Y=0, \quad Z=Z_0 \\ X&=0 \quad \text{and} \quad Y=d, \quad Z=0 \\ X&=c \quad \text{and} \quad Y=d, \quad Z=0 \\ X&=c \quad \text{and} \quad Y=0, \quad Z=0 \end{aligned}$$

The deflection boundary conditions are satisfied since there are no deflections at the edges and the maximum deflection is at the center.

Next check the geometric boundary conditions for the slope of the plate at different points. There must be a finite slope at all four edges but the center of the plate must have a zero slope. Considering first the slope along the X axis, the required equation can be determined from the partial derivative of Z with respect to X. Then from Eq. 6.22

$$\theta_X = \frac{\partial Z}{\partial X} = -Z_0\frac{2X}{c^2}\left(1-\frac{Y^2}{d^2}\right)$$

at

$$X = 0 \quad \text{and} \quad Y = 0, \quad \theta_X = 0$$

$$X = c \quad \text{and} \quad Y = 0, \quad \theta_X = -Z_0 \left(\frac{2}{c}\right)$$

$$X = -c \quad \text{and} \quad Y = 0, \quad \theta_X = Z_0 \left(\frac{2}{c}\right)$$

$$X = 0 \quad \text{and} \quad Y = d, \quad \theta_X = 0$$

The slope boundary conditions are satisfied at the edges and at the center of the plate along the X axis. The same check can be made for the slope along the Y axis.

The strain energy equation (Eq. 6.2) and the kinetic energy equation (Eq. 6.3) can also be used with the polynomial equation (Eq. 6.22). Performing the operations on Eq. 6.22 as required by Eq. 6.2 for the strain energy

$$\frac{\partial Z}{\partial X} = -Z_0 \frac{2X}{c^2}\left(1 - \frac{Y^2}{d^2}\right) \tag{6.23}$$

$$\frac{\partial^2 Z}{\partial X^2} = -Z_0 \frac{2}{c^2}\left(1 - \frac{Y^2}{d^2}\right) \tag{6.24}$$

$$\left(\frac{\partial^2 Z}{\partial X^2}\right)^2 = Z_0^2 \frac{4}{c^4}\left(1 - \frac{2Y^2}{d^2} + \frac{Y^4}{d^4}\right) \tag{6.25}$$

$$\frac{\partial Z}{\partial Y} = -Z_0 \frac{2Y}{d^2}\left(1 - \frac{X^2}{c^2}\right) \tag{6.26}$$

$$\frac{\partial^2 Z}{\partial Y^2} = -Z_0 \frac{2}{d^2}\left(1 - \frac{X^2}{c^2}\right) \tag{6.27}$$

$$\left(\frac{\partial^2 Z}{\partial Y^2}\right)^2 = Z_0^2 \frac{4}{d^4}\left(1 - \frac{2X^2}{c^2} + \frac{X^4}{c^4}\right) \tag{6.28}$$

From Eqs. 6.24 and 6.27

$$\left(\frac{\partial^2 Z}{\partial X^2}\right)\left(\frac{\partial^2 Z}{\partial Y^2}\right) = Z_0^2 \frac{4}{c^2 d^2}\left(1 - \frac{X^2}{c^2} - \frac{Y^2}{d^2} + \frac{X^2 Y^2}{c^2 d^2}\right) \tag{6.29}$$

From Eq. 6.23 or 6.26

$$\frac{\partial^2 Z}{\partial X \partial Y} = Z_0 \frac{2X}{c^2}\frac{2Y}{d^2}$$

$$\left(\frac{\partial^2 Z}{\partial X \partial Y}\right)^2 = Z_0^2 \left(\frac{16 X^2 Y^2}{c^4 d^4}\right) \tag{6.30}$$

When the integration is performed according to Eq. 6.2 for the strain energy, observe that the X and Y reference axes for this geometry are not at the edges of the plate; they are at the center of the plate. Therefore, to consider the full plate

$$\int_{-c}^{+c}\int_{-d}^{+d} f(Z)\,dX\,dY = 4\int_{0}^{c}\int_{0}^{d} f(Z)\,dX\,dY \tag{6.31}$$

From Eqs. 6.2, 6.25, and 6.31

$$4\int_{0}^{c}\int_{0}^{d}\left(\frac{\partial^2 Z}{\partial X^2}\right)^2 dX\,dY = \frac{16Z_0^2cd}{c^4}\left(1-\frac{2}{3}+\frac{1}{5}\right) \tag{6.32}$$

From Eqs. 6.2, 6.28, and 6.31

$$4\int_{0}^{c}\int_{0}^{d}\left(\frac{\partial^2 Z}{\partial Y^2}\right)^2 dX\,dY = \frac{16Z_0^2cd}{d^4}\left(1-\frac{2}{3}+\frac{1}{5}\right) \tag{6.33}$$

From Eqs. 6.2, 6.29, and 6.31

$$4\int_{0}^{c}\int_{0}^{d}\left(\frac{\partial^2 Z}{\partial X^2}\right)\left(\frac{\partial^2 Z}{\partial Y^2}\right) dX\,dY = \frac{16Z_0^2cd}{c^2d^2}\left(1-\frac{1}{3}-\frac{1}{3}+\frac{1}{9}\right) \tag{6.34}$$

From Eqs. 6.2, 6.30, and 6.31

$$4\int_{0}^{c}\int_{0}^{d}\left(\frac{\partial^2 Z}{\partial X\partial Y}\right)^2 dX\,dY = \frac{64Z_0^2cd}{c^2d^2}\left(\frac{1}{3}\right)\left(\frac{1}{3}\right) \tag{6.35}$$

The strain energy of the plate can be obtained by substituting Eqs. 6.32–6.35 into Eq. 6.2.

$$V = \frac{16DZ_0^2cd}{2}\left[\frac{8}{15c^4}+\frac{8}{15d^4}+\frac{2\mu(4)}{9c^2d^2}+\frac{2(1-\mu)(4)}{9c^2d^2}\right]$$

$$V = 64DZ_0^2cd\left[\frac{1}{15c^4}+\frac{1}{9c^2d^2}+\frac{1}{15d^4}\right] \tag{6.36}$$

The kinetic energy of the plate can be determined from Eqs. 6.3 and 6.22 as follows:

$$Z^2 = Z_0^2\left(1-\frac{2X^2}{c^2}-\frac{2Y^2}{d^2}+\frac{4X^2Y^2}{c^2d^2}-\frac{2X^4Y^2}{c^4d^2}-\frac{2X^2Y^4}{c^2d^4}+\frac{X^4}{c^4}+\frac{Y^4}{d^4}+\frac{X^4Y^4}{c^4d^4}\right) \tag{6.37}$$

Since the X and Y reference axes are through the center of the plate and integration must take place over the entire plate area, Eq. 6.31 must be used with the kinetic energy Eq. 6.3. Then from Eq. 6.37

$$4\int_{0}^{c}\int_{0}^{d} Z^2\,dX\,dY = 4Z_0^2\,cd\left(1-\frac{2}{3}-\frac{2}{3}+\frac{4}{9}-\frac{2}{15}-\frac{2}{15}+\frac{1}{5}+\frac{1}{5}+\frac{1}{25}\right) \tag{6.38}$$

Substituting Eq. 6.38 into Eq. 6.3 for the kinetic energy of the vibrating plate

$$T = 0.566Z_0^2cd\rho\Omega^2 \tag{6.39}$$

Since the strain energy must equal the kinetic energy at resonance, Eq. 6.36 must be equal to Eq. 6.39.

$$64DZ_0^2cd\left[\frac{1}{15c^4}+\frac{1}{9c^2d^2}+\frac{1}{15d^4}\right]=0.566Z_0^2cd\rho\Omega^2$$

$$\Omega^2=\frac{64D}{0.566\rho}\left(\frac{1}{15c^4}+\frac{1}{9c^2d^2}+\frac{1}{15d^4}\right)$$

Solving for the natural frequency of the plate

$$f_n=\frac{\Omega}{2\pi}=\frac{5.31}{\pi}\left[\frac{D}{\rho}\left(\frac{1}{15c^4}+\frac{1}{9c^2d^2}+\frac{1}{15d^4}\right)\right]^{1/2} \tag{6.40}$$

6.6. Sample Problem

Consider a rectangular printed-circuit board that will be subjected to a 2-G-peak sinusoidal-vibration test. With a low input G level, the four edges of the circuit board will act as though they are simply supported. The board has a uniformly distributed weight of 1 lb, with dimensions as shown in Fig. 6.9.

Using the natural frequency equation derived with the use of a trigonometric expression

$$f_n=\frac{\pi}{2}\left(\frac{D}{\rho}\right)^{1/2}\left(\frac{1}{a^2}+\frac{1}{b^2}\right) \quad \text{(see Eq. 6.21)}$$

where $a=8.0$ in. board length

$b=7.0$ in. board width

$W=1.0$ lb. board weight

$g=386$ in/sec^2 gravity

$h=0.062$ in. board thickness

$E=2.0\times10^6$ lb/in.^{2} epoxy fiberglass

$\mu=0.12$, Poisson's ratio, dimensionless

$$D=\frac{Eh^3}{12(1-\mu^2)}=\frac{2\times10^6(0.062)^3}{12[1-(0.12)^2]}=40.1 \text{ lb in.}$$

$$\rho=\frac{\text{mass}}{\text{area}}=\frac{W}{gab}=\frac{1.0}{(386)(8.0)(7.0)}=0.463\times10^{-4}\text{ lb sec}^2/\text{in.}^3$$

Substituting into Eq. 6.21

$$f_n=\frac{\pi}{2}\left(\frac{40.1}{0.463\times10^{-4}}\right)^{1/2}\left(\frac{1}{64}+\frac{1}{49}\right)$$

$$f_n=52.6 \text{ Hz} \tag{6.41}$$

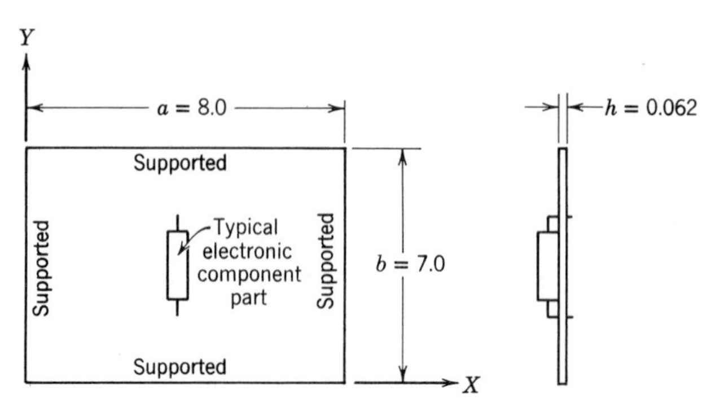

FIGURE 6.9. Dimensions of a simply supported rectangular plate.

The natural frequency of the circuit board can also be determined from the equation derived with the use of a polynomial expression.

$$f_n = \frac{5.31}{\pi}\left[\frac{D}{\rho}\left(\frac{1}{15c^4} + \frac{1}{9c^2d^2} + \frac{1}{15d^4}\right)\right]^{1/2} \qquad \text{(see Eq. 6.40)}$$

where $D = 40.1$ lb in.

$$\rho = 0.463 \times 10^{-4} \text{ lb sec}^2/\text{in.}^3$$

$$c = \frac{a}{2} = \frac{8.0}{2} = 4.0 \text{ in.}$$

$$d = \frac{b}{2} = \frac{7.0}{2} = 3.5 \text{ in.}$$

Substituting into Eq. 6.40

$$f_n = \frac{5.31}{\pi}\left\{\frac{40.1}{0.463 \times 10^{-4}}\left[\frac{1}{15(4.0)^4} + \frac{1}{9(4.0)^2(3.5)^2} + \frac{1}{15(3.5)^4}\right]\right\}^{1/2}$$

$$f_n = 56.1 \text{ Hz} \tag{6.42}$$

6.7. Natural-Frequency Equations Derived Using the Rayleigh Method

The Rayleigh method is very convenient for determining approximate resonant frequencies of rectangular plates with various edge conditions. All that is required is to assume a deflection curve that satisfies the geometric boundary conditions of deflection and slope for the particular plate.

Consider the case of a rectangular plate that is fixed on all four edges. The deflection will be zero at all four edges and maximum at the center. The slope will also be zero at the four edges and zero at the center. If a

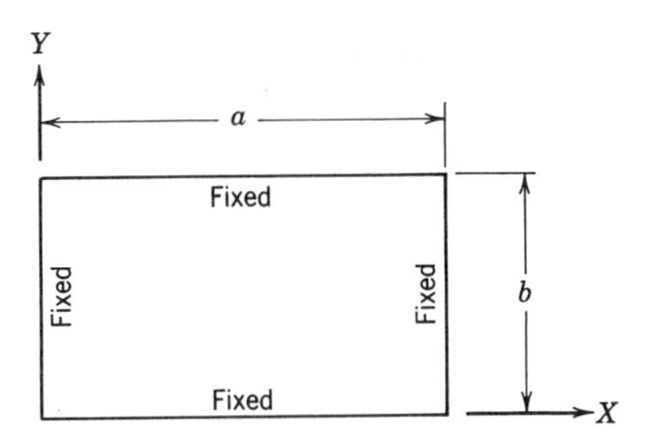

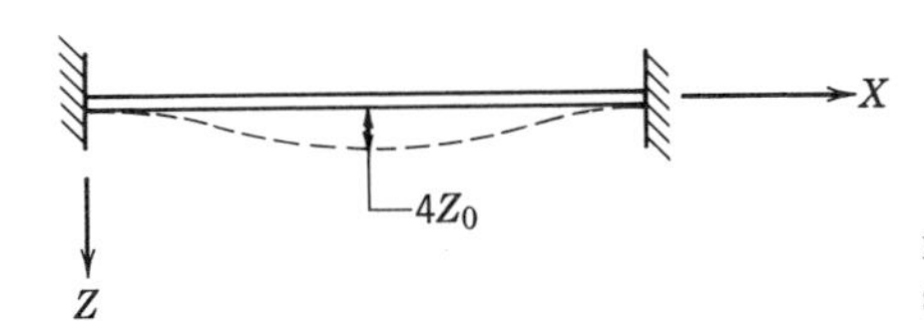

FIGURE 6.10. A rectangular plate with four sides fixed.

trigonometric expression is used to describe this plate, the coordinate axes can be placed at the edges of the plate as shown in Fig. 6.10.

One trigonometric expression that will satisfy these boundary conditions is

$$Z = Z_0\left(1-\cos\frac{2\pi X}{a}\right)\left(1-\cos\frac{2\pi Y}{b}\right)$$

This deflection curve can be examined to make sure it meets the geometric deflection requirements.

At
$$X=0 \text{ and } Y=0, \quad Z=0$$
$$X=0 \text{ and } Y=b, \quad Z=0$$
$$X=a \text{ and } Y=b, \quad Z=0$$
$$X=\frac{a}{2} \text{ and } Y=\frac{b}{2}, \quad Z=4Z_0$$

Now examine this curve to make sure it meets the geometric slope requirements. Consider first the slope along the X axis, which can be determined by taking the partial derivative of Z with respect to X.

$$\theta_X = \frac{\partial Z}{\partial X} = \frac{2\pi}{a}\sin\frac{2\pi X}{a}\left(1-\cos\frac{2\pi Y}{b}\right)$$

at
$$X=0 \text{ and } Y=\frac{b}{2}, \quad \theta_X=0$$
$$X=a \text{ and } Y=\frac{b}{2}, \quad \theta_X=0$$
$$X=\frac{a}{2} \text{ and } Y=\frac{b}{2}, \quad \theta_X=0$$

The same check can be made for the slope along the Y axis, which shows the curve will satisfy the geometric slope requirements. Then using Eqs. 6.2 and 6.3 for the strain energy and the kinetic energy, the fundamental natural frequency becomes

$$f_n = \frac{\pi}{1.5}\left[\frac{D}{\rho}\left(\frac{3}{a^4}+\frac{2}{a^2b^2}+\frac{3}{b^4}\right)\right]^{1/2}$$

A polynomial expression can also be used to describe a rectangular plate fixed on all four sides. The coordinate axes for this plate can be shifted to the center (Fig. 6.11).

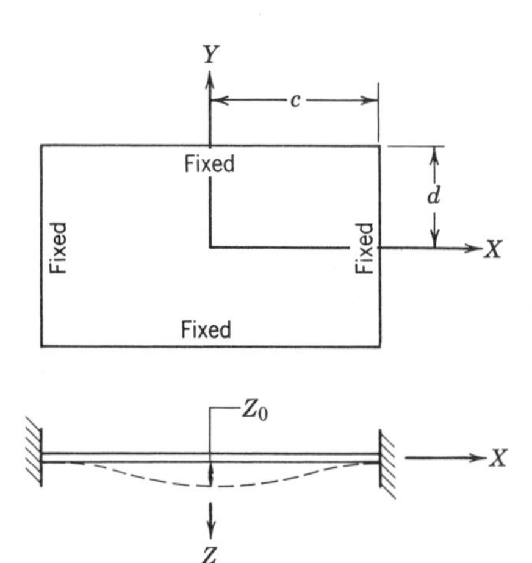

FIGURE 6.11. Axes at the center of a plate with four fixed edges.

One polynomial expression that will satisfy these boundary conditions is

$$Z = Z_0\left(1-\frac{X^2}{c^2}\right)^2\left(1-\frac{Y^2}{d^2}\right)^2$$

This polymonial will result in the following natural frequency

$$f_n = \frac{1.96}{\pi}\left[\frac{D}{\rho}\left(\frac{2.01}{c^4}+\frac{1}{c^2d^2}+\frac{2.01}{d^4}\right)\right]^{1/2}$$

The natural frequency of a rectangular plate with three supported edges and one free edge can be derived using a trigonometric function or a polynomial function. A sketch of the coordinate axes for each plate, along with the deflection expression and the resulting natural-frequency equations for a trigonometric function and for a polynomial function, are shown in Fig. 6.12.

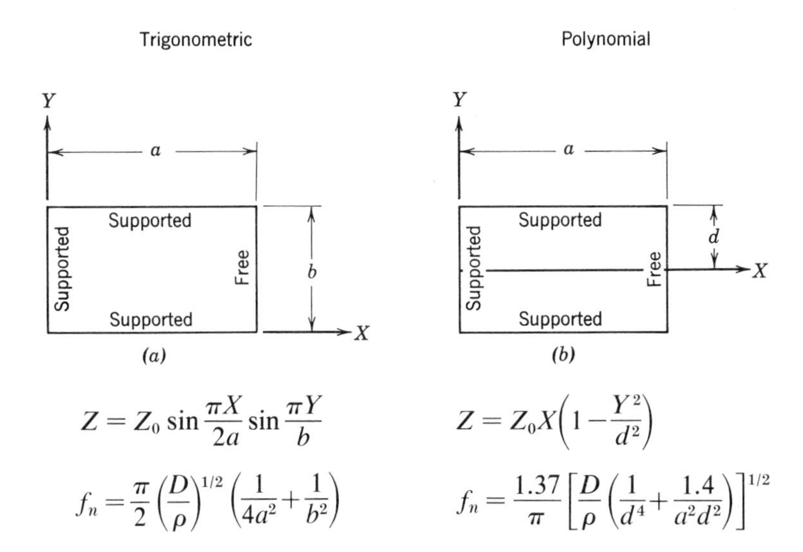

FIGURE 6.12. Three sides supported and one side free: (*a*) trigonometric, (*b*) polynomial.

A comparison can be made of the deflection equation and the resulting natural-frequency equation using a trigonometric function or a polynomial function for plates with combinations of free edges, supported edges, and fixed edges, as shown in Figs. 6.13 and 6.14.

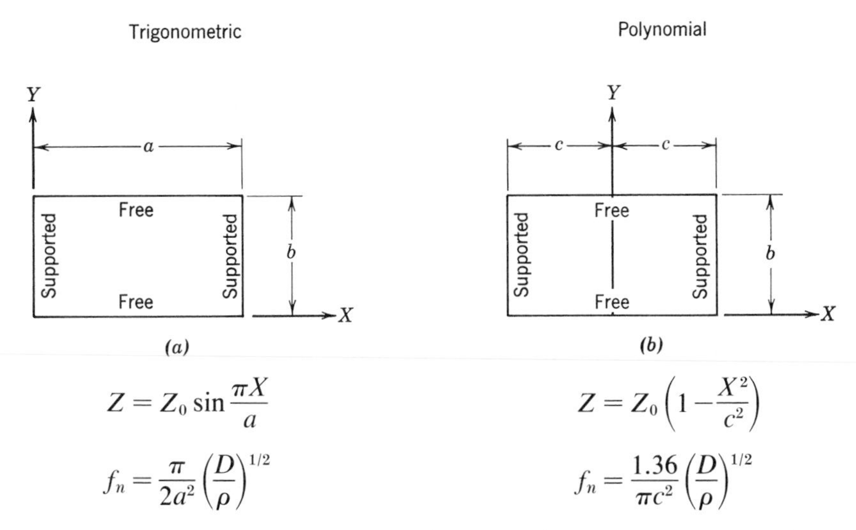

FIGURE 6.13. Two opposite edges free and two opposite edges supported: (*a*) trigonometric, (*b*) polynomial.

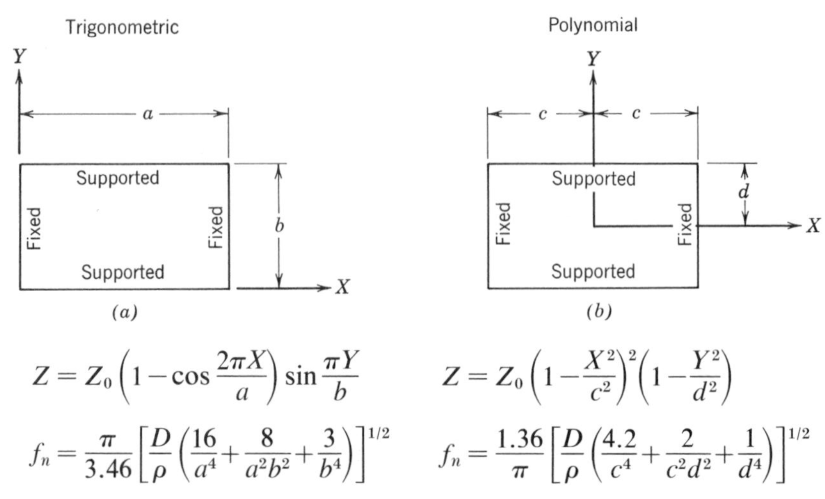

FIGURE 6.14. Two opposite edges fixed and two opposite edges supported: (*a*) trigonometric (*b*) polynomial.

Trigonometric functions can also be combined with polynomial functions in the same deflection equation to describe the deflection of a uniform rectangular plate. For example, considering a plate that is simply supported on two opposite edges and fixed on two opposite edges, the coordinate axes and the deflection equation that meets the geometric boundary conditions will be as shown in Fig. 6.15.

$$Z = Z_0\left(1-\cos\frac{2\pi X}{a}\right)\left(1-\frac{Y^2}{d^2}\right)$$

There has been an extensive amount of work on the derivation of natural-frequency equations for uniform, flat, rectangular plates with various edge conditions. Little [30], Warburton [31], and Laura [33] made extensive use of trigonometric and polynomial series to determine the various resonant modes of different plates.

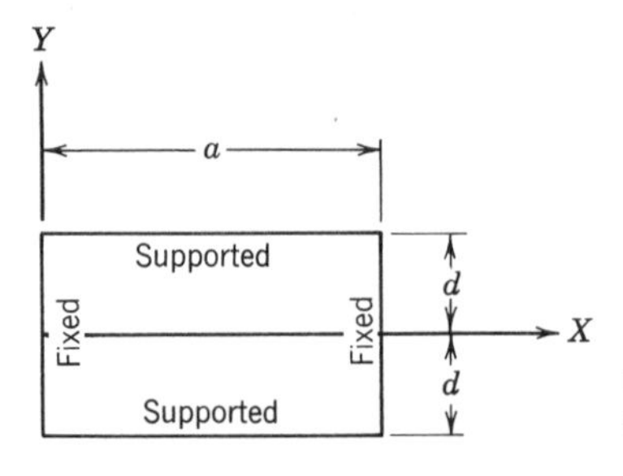

FIGURE 6.15. Combining a trigonometric function tion and a polynomial function.

If only the fundamental resonant mode of a uniform, flat, rectangular plate is desired, then the simplified types of trigonometric and polynomial expressions, just outlined, can be combined to determine the fundamental-resonant-frequency equations for many different types of plates. Many of these equations for plates with various edge conditions are shown in Figs. 6.16–6.18.

Free edge   Supported edge   Fixed edge

Equation	Plate
$f_n = \frac{\pi}{8}\left(\frac{D}{\rho}\right)^{1/2}\left(\frac{1}{a^2}+\frac{1}{b^2}\right)$	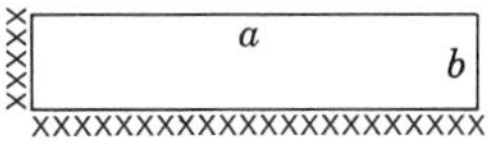
$f_n = \frac{\pi}{2}\left(\frac{D}{\rho}\right)^{1/2}\left(\frac{1}{4a^2}+\frac{1}{b^2}\right)$	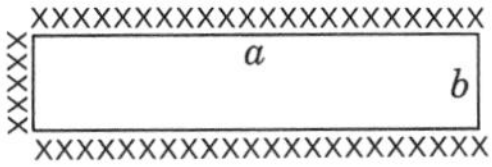
$f_n = \frac{\pi}{2}\left(\frac{D}{\rho}\right)^{1/2}\left(\frac{1}{a^2}+\frac{1}{b^2}\right)$	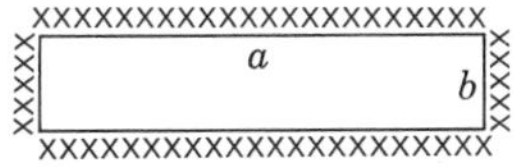
$f_n = \frac{\pi}{5.42}\left[\frac{D}{\rho}\left(\frac{1}{a^4}+\frac{3.2}{a^2b^2}+\frac{1}{b^4}\right)\right]^{1/2}$	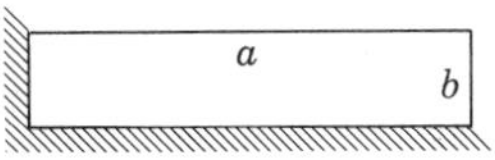
$f_n = \frac{\pi}{3}\left[\frac{D}{\rho}\left(\frac{0.75}{a^4}+\frac{2}{a^2b^2}+\frac{12}{b^4}\right)\right]^{1/2}$	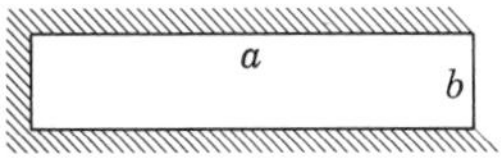
$f_n = \frac{\pi}{1.5}\left[\frac{D}{\rho}\left(\frac{3}{a^4}+\frac{2}{a^2b^2}+\frac{3}{b^4}\right)\right]^{1/2}$	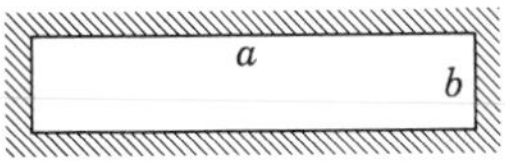
$f_n = \frac{\pi}{3.46}\left[\frac{D}{\rho}\left(\frac{16}{a^4}+\frac{8}{a^2b^2}+\frac{3}{b^4}\right)\right]^{1/2}$	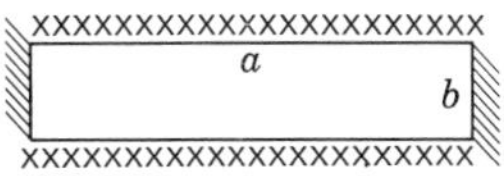

FIGURE 6.16. Natural frequency equations for uniform plates.

Free edge — Supported edge (XXXXXXXXXXXXXXXXX) — Fixed edge

Equation	Plate
$f_n = \frac{\pi}{2}\left[\frac{D}{\rho}\left(\frac{2.08}{a^2b^2}\right)\right]^{1/2}$	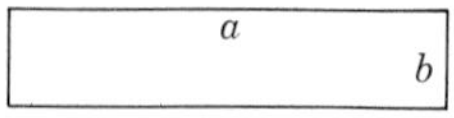 a, b
$f_n = \frac{0.56}{a^2}\left(\frac{D}{\rho}\right)^{1/2}$	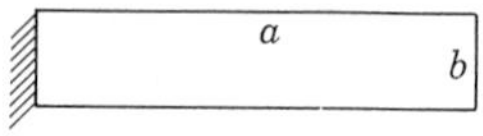 a, b
$f_n = \frac{3.55}{a^2}\left(\frac{D}{\rho}\right)^{1/2}$	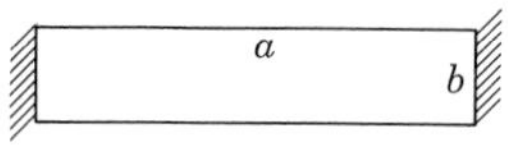 a, b
$f_n = \frac{0.78\pi}{a^2}\left(\frac{D}{\rho}\right)^{1/2}$	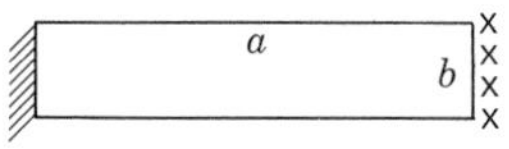 a, b
$f_n = \frac{\pi}{2a^2}\left(\frac{D}{\rho}\right)^{1/2}$	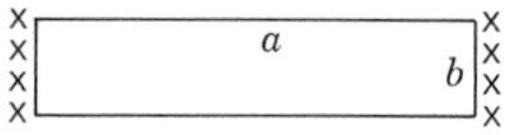 a, b
$f_n = \frac{\pi}{1.74}\left[\frac{D}{\rho}\left(\frac{4}{a^4}+\frac{1}{2a^2b^2}+\frac{1}{64b^4}\right)\right]^{1/2}$	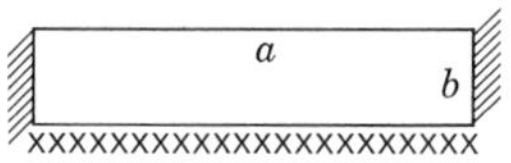 a, b
$f_n = \frac{\pi}{2}\left[\frac{D}{\rho}\left(\frac{0.127}{a^4}+\frac{0.20}{a^2b^2}\right)\right]^{1/2}$	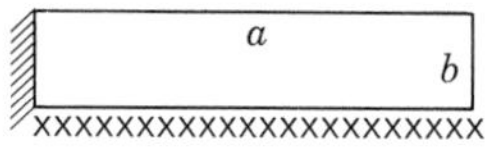 a, b

FIGURE 6.17. Natural frequency equations for uniform plates.

Free edge — Supported edge (xxxxxxxx) — Fixed edge (hatched)

Equation	Plate
$f_n = \frac{\pi}{2}\left[\frac{D}{\rho}\left(\frac{1}{a^4}+\frac{0.608}{a^2b^2}+\frac{0.126}{b^4}\right)\right]^{1/2}$	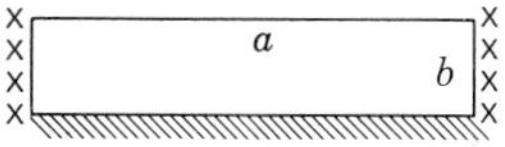
$f_n = \frac{\pi}{2}\left[\frac{D}{\rho}\left(\frac{2.45}{a^4}+\frac{2.90}{a^2b^2}+\frac{5.13}{b^4}\right)\right]^{1/2}$	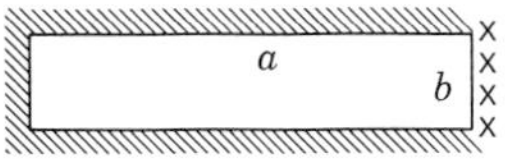
$f_n = \frac{\pi}{2}\left[\frac{D}{\rho}\left(\frac{0.127}{a^4}+\frac{0.707}{a^2b^2}+\frac{2.44}{b^4}\right)\right]^{1/2}$	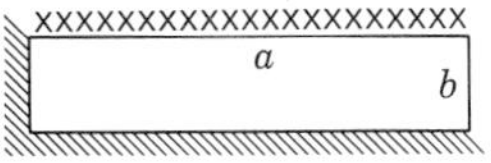
$f_n = \frac{\pi}{2}\left[\frac{D}{\rho}\left(\frac{2.45}{a^4}+\frac{2.68}{a^2b^2}+\frac{2.45}{b^4}\right)\right]^{1/2}$	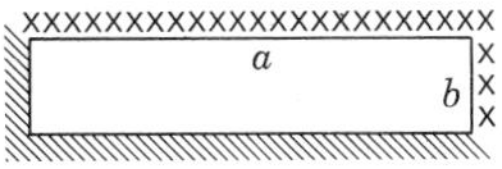
$f_n = \frac{\pi}{2}\left[\frac{D}{\rho}\left(\frac{2.45}{a^4}+\frac{2.32}{a^2b^2}+\frac{1}{b^4}\right)\right]^{1/2}$	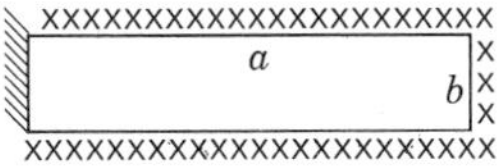
$f_n = \frac{4.50}{\pi a^2}\left(\frac{D}{\rho}\right)^{1/2}$	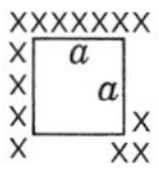
$f_n = \frac{1.13}{a^2}\left(\frac{D}{\rho}\right)^{1/2}$	

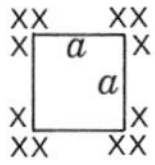

FIGURE 6.18. Natural frequency equations for uniform plates.

6.8. Electrical Lead Wire Stresses

Electronic component parts such as resistors, capacitors, diodes, and flatpack integrated circuits are often mounted on printed-circuit boards by means of their electrical lead wires. In Chapter 5, Section 5.1, it was pointed out that printed-circuit-board resonances may lead to failures in the electrical lead wires, or their solder joints, due to large deflections and bending stresses that are developed as the circuit board bends back and forth. Although it is often possible to have component resonances where the electrical lead wires act as springs and the component body acts as a bouncing mass, this condition does not have to occur in order to develop fatigue failures in the electrical lead wires. When this condition does occur, it is a simple matter to tie the component body with lacing cord.

Printed-circuit-board resonances can lead to high bending stresses in the lead wires on electronic component parts. Every time the circuit board bends, it forces the electrical lead wires to bend since the wires are normally soldered to the circuit board. The greater the circuit-board deflection, the greater the bending stresses in the lead wires. If the component body is cemented to the circuit board, the relative motion between the board and the component will be substantially reduced.

Stresses in the component lead wires can be determined by examining the geometry of the electronic component part, the natural frequency of the circuit board, and the acceleration G forces. For example, considering the geometry, components such as resistors, capacitors, diodes, and flatpacks have relatively small electrical lead wires compared to a large component body. Almost all of the relative deflection between the circuit board and the component will, therefore, occur in the electrical lead wires. If the body of the component is ignored and only the electrical-lead-wire deflections are examined, the resulting errors will be very small.

The relative slope of the electrical lead wires, where they are soldered to the circuit board, will not change as the board vibrates up and down. The wires will always remain perpendicular to the board. Also, each electrical lead wire will always remain perpendicular to the component body where the lead joins the body because the body is so much stiffer than the lead. Therefore, as the circuit board bends back and forth during vibration, the electrical lead wires will bend as shown in Fig. 5.2. If only the electrical lead wires are considered, they will approximate the shape of the bent shown in Fig. 5.3. This particular condition was analyzed in Chapter 5 where equations were derived to show bending moments and stresses at various points in the electrical lead wires. These equations can be used to determine the bending stresses and fatigue characteristics of the lead wires on many different types of components.

Consider a typical tantalytic capacitor, type CS 12-B ("B" case size) mounted at the center of the printed-circuit board (Fig. 6.9). If the capacitor is not cemented, tied, or otherwise fastened to the circuit board, then as the circuit board reaches its peak amplitude during a resonant condition, the electrical lead wires will bend approximately as shown in Fig. 6.19.

The smaller dimension of the rectangular circuit board will produce the maximum electrical-lead-wire stress, due to the more rapid rate of change in the curvature of the board along the shorter span for the given deflection.

The relative displacement of the electrical lead wires, with respect to the circuit board, will be determined by the maximum displacement of the circuit board at its resonance. Assuming the circuit board acts like a single-degree-of-freedom system at its fundamental resonant mode, and assuming a 2-G-peak sinusoidal-vibration input, the maximum single-amplitude displacement at the center of the circuit board can be determined if the natural frequency and the transmissibility are known.

The natural frequency for this particular circuit board was calculated in the sample problem shown in Section 6.6. Two different methods were used to solve for the natural frequency, as shown by Eqs. 6.41 and 6.42. Using an approximate average of the two resonances results in a value of about 55 Hz; this value will be used here.

The transmissibility for this circuit board at its resonance can be approximated by using the methods outlined in Section 6.3. Since the input G force is quite low, only 2 Gs peak, and since there are no ribs or large electronic component parts mounted on the circuit board, the transmissibility will probably be about equal to the square root of the natural frequency.

$$Q = (f_n)^{1/2} = (55)^{1/2} = 7.4 \qquad \text{(use a value of 8.0)}$$

The maximum single-amplitude displacement at the center of the board can be determined from the following equation:

$$\delta = \frac{9.8 G_{\text{in}} Q}{f_n^2} \qquad \text{(see Eq. 5.43)}$$

where $G_{\text{in}} = 2.0\ G$ peak input acceleration
$Q = 8.0$ approximate transmissibility
$f_n = 55$ Hz resonant frequency of board

Substituting into the above equation

$$\delta = \frac{(9.8)(2.0)(8.0)}{(55)^2} = 0.0519 \text{ in.} \tag{6.43}$$

The relative deflection between the capacitor lead wire and the circuit board can be determined for the simply supported circuit board by approximating its dynamic deflection curve with a half sine wave.

$$\delta_c = \delta\left(1 - \sin\frac{\pi Y}{b}\right) \tag{6.44}$$

where $\delta = 0.0519$ in. (see Eq. 6.43)
$Y = 2.94$ in.
$b = 7.0$ in. board width

$$\delta_c = 0.0519\left[1 - \sin\frac{\pi(2.94)}{7.0}\right]$$

$$\delta_c = 0.00161 \text{ in.} \tag{6.45}$$

This represents the total deflection due to the combined loading on the electrical lead wire, which is acting like a bent, as shown in Fig. 5.3*a*.

The slope of the printed-circuit board, at the junction with the soldered lead wire, can be determined from the first derivative of the deflection equation, for the board shown in Fig. 6.19.

$$Z = \delta \sin\frac{\pi Y}{b}$$

$$\theta_c = \frac{dZ}{dY} = \delta\frac{\pi}{b}\cos\frac{\pi Y}{b} \tag{6.46}$$

where $\delta = 0.0519$ in. (see Eq. 6.43)
$b = 7.0$ in. board width
$Y = 2.94$ in. dimension to wire

$$\theta_c = (0.0519)\frac{\pi}{7.0}\cos\frac{\pi(2.94)}{7.0}$$

$$\theta_c = 0.00578 \text{ rad} \tag{6.47}$$

The effective height of the component lead wire will depend upon the type of printed-circuit board being used and the clearance between the component body and the circuit board. If plated-through holes are used, the effective height of the wire will be measured from the top surface of the circuit board to the first bend in the lead wire. The top surface of the board is used because during the dip-soldering operation, solder will wick up the plated-through hole and support the electrical lead wire at the top surface. If printed-circuit runs are used only on the bottom surface of the circuit board, then the height of the electrical lead wire will be measured from the solder joint on the bottom surface, since there will be no support at the top surface (Fig. 6.20).

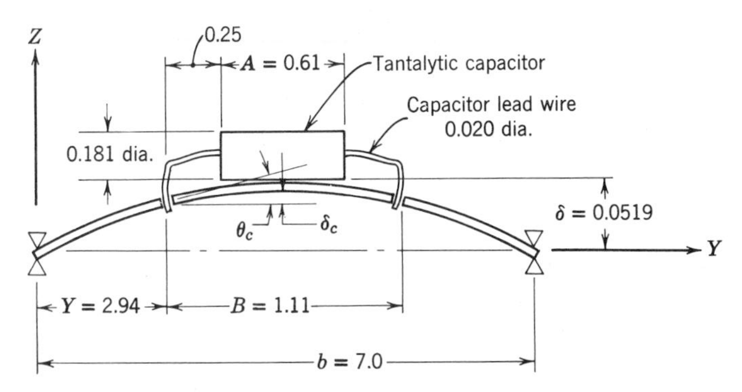

FIGURE 6.19. A tantalytic capacitor mounted on a circuit board.

The mechanical strength of a joint that is soldered at the top and the bottom of the board is far superior to the joint that is soldered on only one side of the board. Vibration fatigue failures are more likely to occur in the electrical lead wires and not in the solder when plated-through holes are used and the leads are soldered on both sides of the board. When printed circuits are used on one side of the board only, bending moments in the lead wires can produce rapid fatigue failures in the solder because of shear tear-out fatigue. Since solder has a lower fatigue strength than copper, the copper wire may not fail where a solder joint will fail. See Chapter 5, Section 5.13 for more details.

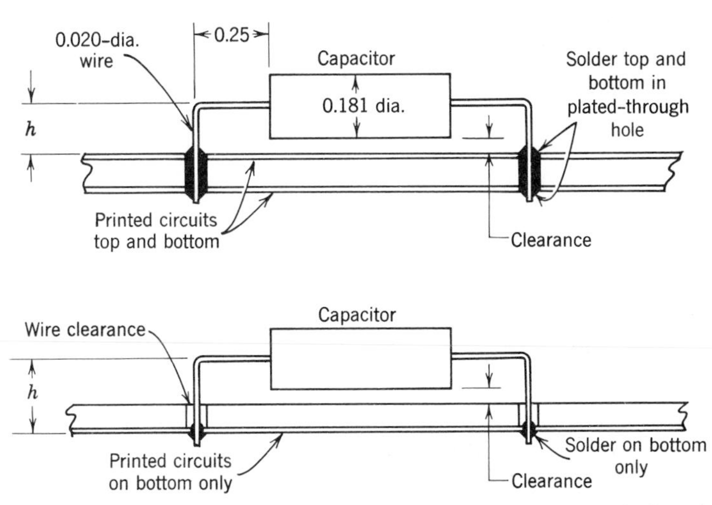

FIGURE 6.20. Effective lead height for components mounted on single and double sided circuit boards.

Thermal stresses should also be considered. Thermal stresses may be higher in the solder joints of boards with plated-through holes and printed circuits on both sides than in boards with printed circuits on only one side. The different thermal expansion coefficients of the electrical lead wire, the copper in the plated-through hole, and the solder may result in high stresses over a wide temperature range. These stresses may also be higher in a thicker board using plated-through holes because of the greater change in length for a given temperature change. Thermal stresses will not be considered in this book, but the reader should be aware of the problems that can develop due to thermal stresses that may be blamed on vibration stresses.

If the body of the component is ignored and only the electrical lead wires are considered, they will assume the shape of a rectangular bent. As the circuit board bends during vibration, the wire bent will deform in a complex bending mode that is described in Chapter 5, Section 5.1 and shown in Fig. 5.3.

If there is no clearance between the circuit board and the component body, and if plated-through holes are used, the physical dimensions for the electrical lead wires in the shape of a bent will appear as shown in Fig. 6.21.

The equivalent dynamic vertical load P can be determined from the deflection equations that are derived in Chapter 5, Sections 5.2–5.4. The combined deflection is shown by Eq. 5.35 and is repeated here for convenience.

$$\delta_c = \frac{PL^3}{48EI_1}\left(1 - \frac{3}{2K+4}\right) - \frac{\theta_c L}{4}\left[2 - \left(\frac{3+2K}{2+K}\right)\right]$$

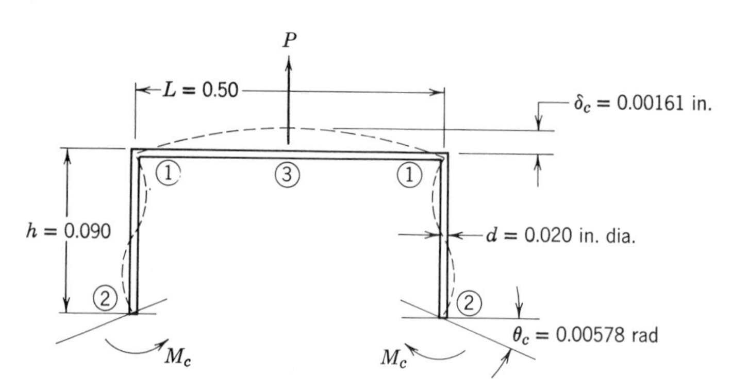

FIGURE 6.21. Dimensions for electrical lead wires in plated through holes.

where $h = 0.090$ in., for a plated-through hole

$L = 0.50$ in., length of top member

$d = 0.020$ in., wire diameter

$$I_1 = I_2 = \frac{\pi d^4}{64} = \frac{\pi}{64}(0.020)^4 = 0.785 \times 10^{-8} \text{ in.}^4$$

$$K = \frac{h}{L} = \frac{0.090}{0.50} = 0.180, \text{ dimensionless}$$

$E = 30 \times 10^6$ lb/in.^{2}, nickel lead wire

$\theta_c = 0.00578$ rad (see Eq. 6.47)

$\delta_c = 0.00161$ in. (see Eq. 6.45)

Substituting into Eq. 5.35:

$$0.00161 = \frac{P(0.50)^3}{48(30 \times 10^6)(0.785 \times 10^{-8})}\left(1 - \frac{3}{4.36}\right)$$

$$-\frac{(0.00578)(0.50)}{4}\left(2 - \frac{3.36}{2.18}\right)$$

$$1.61 \times 10^{-3} = 3.45 \times 10^{-3}P - 0.332 \times 10^{-3}$$

$$P = 0.563 \text{ lb} \tag{6.48}$$

The bending moment M_c developed in the electrical lead wires, due to bending of the circuit board, can be determined from Eq. 5.31.

$$M_c = \frac{2\theta_c E I_2}{h}\left(\frac{3+2K}{2+K}\right)$$

$$M_c = \frac{2(5.78 \times 10^{-3})(30 \times 10^6)(0.785 \times 10^{-8})}{0.090}\left(\frac{3.36}{2.18}\right)$$

$$M_c = 0.0465 \text{ lb in.} \tag{6.49}$$

The total bending moment at point 1, as shown in Fig. 6.21, can be determined from Eq. 5.36.

$$M_1 = \frac{PL}{4K+8} + \left(\frac{4EI_1\theta_c}{L} - KM_c\right)$$

$$M_1 = \frac{(0.563)(0.50)}{0.720+8} + \frac{4(1.36 \times 10^{-3})}{0.50} - (0.180)(0.0465)$$

$$M_1 = 0.0346 \text{ lb in.} \tag{6.50}$$

The total bending moment at point 2, as shown in Fig. 6.21, can be determined from Eq. 5.37.

$$M_2 = \frac{PL}{8K+16} + \frac{2EI_2\theta_c}{h}\left(\frac{3+2K}{2+K}\right)$$

$$M_2 = \frac{(0.563)(0.50)}{17.44} + \frac{2(1.36\times 10^{-3})}{0.090}\left(\frac{3.36}{2.18}\right)$$

$$M_2 = 0.0625 \text{ lb in.} \tag{6.51}$$

The total bending moment at point 3, as shown in Fig. 6.21, can be determined from Eq. 5.39.

$$M_3 = \frac{PL(K+1)}{4K+8} - \left(\frac{4EI_1\theta_c}{L} - KM_c\right)$$

$$M_3 = \frac{(0.563)(0.50)(1.18)}{8.72} - \left[\frac{4(1.36\times 10^{-3})}{0.50} - (0.180)(0.0465)\right]$$

$$M_3 = 0.0354 \text{ lb in.} \tag{6.52}$$

The total horizontal force H developed in the electrical lead wire, as shown in Figs. 5.5 and 5.11, can be determined from Eq. 5.38.

$$H_T = \frac{3PL}{2h(4K+8)} + \frac{M_c}{h}(1-K) + \frac{4EI_1\theta_c}{hL}$$

$$H_T = \frac{3(0.563)(0.50)}{2(0.090)(8.72)} + \frac{0.0465(0.82)}{0.090} + \frac{4(1.36\times 10^{-3})}{(0.090)(0.50)}$$

$$H_T = 1.081 \text{ lb} \tag{6.53}$$

The vertical load V in the wire, as shown in Fig. 5.5, can be determined from Eq. 5.15.

$$V = \frac{P}{2} = \frac{0.563}{2} = 0.281 \text{ lb} \tag{6.54}$$

When the circuit board is dip-soldered, a solder fillet will be formed around the lead wire and the printed-circuit pad on the circuit board. This solder fillet will add support to the lead wire due to the effective increase in the wire diameter as the solder wicks up the lead.

Solder fillets can vary in size, depending upon the method used. Dip soldering and wave soldering will produce a relatively uniform joint, with very little solder wicking up the lead above the circuit board. Hand soldering, on the other hand, will often result in less uniform solder joints with larger masses of solder and more wicking up the lead wire above the circuit board. See Chapter 5, Section 5.13 for more on solder-joint stresses.

Although it is rather difficult to generalize on the size and shape of an average solder joint, some typical measured dimensions of one group of solder joints formed by dip soldering are shown in Fig. 6.22.

The maximum bending moment in the capacitor lead wire is at point 2, shown by Eq. 6.51 as 0.0625 lb in. This is the point where the electrical lead wire is soldered to the printed-circuit board. An examination of this solder joint, as shown in Fig. 6.22, shows there is a large solder fillet radius at this point which will tend to reduce the bending stresses at the joint.

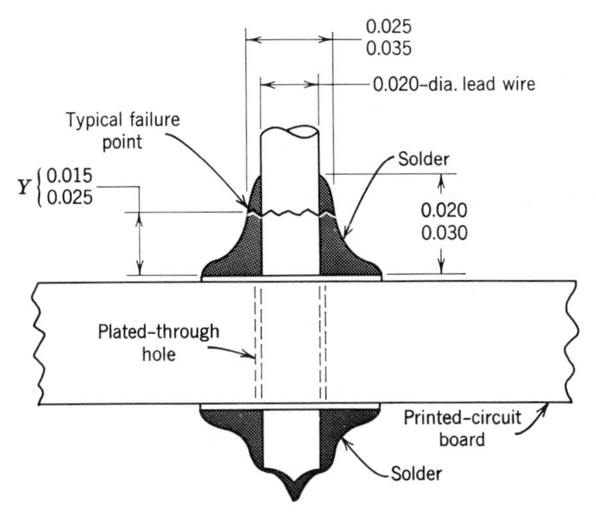

FIGURE 6.22. Typical measured dimensions on one group of solder joints.

If there is no solder fillet at the solder joint and the capacitor is not cemented to the circuit board, then the bending stress in the electrical lead wire at point 2 can be determined from the standard bending stress equation.

$$S_2 = \frac{M_2 c}{I_2}$$

where $M_2 = 0.0625$ lb in. (see Eq. 6.51)

$$c = \frac{d}{2} = \frac{0.020}{2} = 0.010 \text{ in. wire radius}$$

$$I_1 = I_2 = 0.785 \times 10^{-8} \text{ in.}^4 \text{ wire moment of inertia}$$

$$S_2 = \frac{(0.0625)(0.010)}{0.785 \times 10^{-8}} = 79{,}700 \text{ lb/in.}^2 \tag{6.55}$$

Extensive vibration tests have shown that if there is a solder fillet at the joint and it happens to have the geometry shown in Fig. 6.22, some failures may occur at a distance Y, which is about two thirds of the height of the solder joint above the circuit board. The bending moment in the wire at this point can be determined from Figs. 5.5 and 5.11 by taking moments about point Y in both figures as follows:

$$M_Y = M_A - H_P Y + M_c - H_M Y$$

In Chapter 5, Section 5.4, it was shown that

$$M_2 = M_A + M_c \qquad \text{and} \qquad H_T = H_P + H_M$$

Substituting these values into the equation above

$$M_Y = M_2 - H_T Y \tag{6.56}$$

If the two-thirds point on the solder joint is taken as 0.015 in. then the moment at point Y can be determined by substituting Eqs. 6.51 and 6.53 into Eq. 6.56.

$$M_Y = 0.0625 - (1.81)(0.015) = 0.0463 \text{ lb in.} \tag{6.57}$$

At point Y on the electrical lead wire, the outside diameter is increased by the solder wicking up above the fillet. If this diameter is about 0.025 in., as shown in Fig. 6.22, then the bending stress in the solder at point Y is

$$S_{bs} = \frac{M_Y c_s}{I_s}$$

where $M_Y = 0.0463$ lb in. bending moment at point Y

$$c_s = \frac{0.025}{2} = 0.0125 \text{ in. outer radius of solder fillet}$$

$$I_s = \frac{\pi d^4}{64} = \frac{\pi}{64}(0.025)^4 = 1.92 \times 10^{-8} \text{ in.}^4$$

$$S_{bs} = \frac{(0.0463)(0.0125)}{1.92 \times 10^{-8}} = 30{,}200 \text{ lb/in.}^2 \tag{6.58}$$

Although this is a very high stress for solder, which would normally fail rapidly at this level, it is not the stress in the nickel lead wire. The stress in the lead wire itself can be approximated from the diameter of the wire, since the bending moment stays the same.

$$S_{bw} = \frac{M_Y c_w}{I_2}$$

where $M_Y = 0.0463$ lb in. (see Eq. 6.57)

$$c_w = \frac{d}{2} = \frac{0.020}{2} = 0.010 \text{ in. radius of lead wire}$$

$$I_1 = I_2 = 0.785 \times 10^{-8} \text{ in.}^4 \text{ wire alone, without solder}$$

$$S_{bw} = \frac{(0.0463)(0.010)}{0.785 \times 10^{-8}} = 59{,}000 \text{ lb/in.}^2 \quad (6.59)$$

This stress level is not very high for nickel wire, so a rapid failure will not occur because the nickel wire increases the strength of the section at point Y. The effects of fatigue are considered in more detail in the next section.

If the capacitor body is not cemented or tied to the circuit board, the bending stress at point 1, as shown in Fig. 6.21, can be determined from the bending moment at point 1 as follows:

$$S_1 = \frac{M_1 c_w}{I_1}$$

where $M_1 = 0.0346$ lb in. (see Eq. 6.50)

$$c_w = \frac{d}{2} = \frac{0.020}{2} = 0.010 \text{ in. wire radius}$$

$$I_1 = I_2 = 0.785 \times 10^{-8} \text{ in.}^4 \text{ wire moment of inertia}$$

$$S_1 = \frac{(0.0346)(0.010)}{0.785 \times 10^{-8}} = 44{,}100 \text{ lb/in.}^2 \quad (6.60)$$

This is a very sensitive area because it is at the bend in the lead wire. If the bend is sharp, a stress-concentration factor should be used. Also, if a sharp tool is used to form the bend, and the wire is scratched or if tool marks are carelessly left in the wire surface, stress cencentrations can result which will substantially shorten the fatigue life. Otherwise, the bend radius is normally great enough so stress concentrations are small.

A stress-concentration factor must be used to determine the maximum bending stress at point 3, as shown in Fig. 6.21. When the body of the capacitor is not fastened to the circuit board the bending stress is

$$S_{b3} = K_t \frac{M_3 c_w}{I_1}$$

where $M_3 = 0.0354$ lb in. (see Eq. 6.52)

$$c_w = \frac{0.020}{2} = 0.010 \text{ in. wire radius}$$

$$I_1 = I_2 = 0.785 \times 10^{-8} \text{ in.}^4$$

K_t = theoreitical stress concentration factor where the electrical lead wire joins the component body as shown in Fig. 6.23 (from Peterson [18]).

$$\frac{r}{d}=\frac{0.005}{0.020}=0.25 \qquad \text{and} \qquad \frac{D}{d}=\frac{0.030}{0.020}=1.5$$

$$K_t = 1.34 \text{ (bending)}$$

$$S_{b3}=\frac{(1.34)(0.0354)(0.010)}{0.785\times10^{-8}}=60{,}500 \text{ lb/in.}^2 \tag{6.61}$$

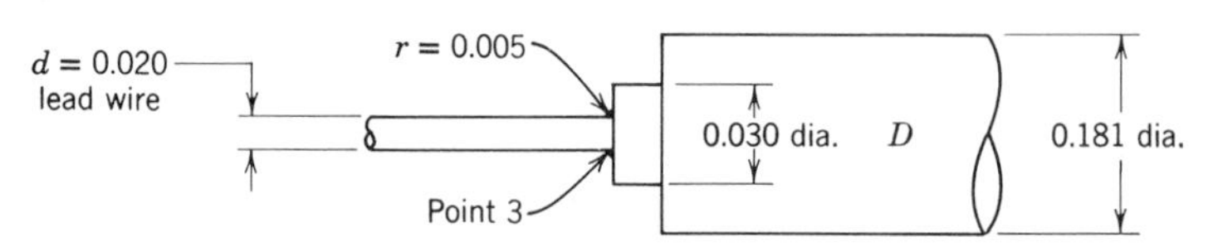

FIGURE 6.23. Enlarged view of electrical lead wire junction on the capacitor body.

In addition to the bending stresses in the electrical lead wires, tensile and shear stresses are also developed in the wires as the circuit board bends during vibration. The shear stresses in the wires are relatively small so they can be ignored. The tensile stresses, however, may not be as small and they act in the same direction as the bending stresses; they must be added directly to the bending stresses.

The maximum tensile stress is due to the horizontal force H_T which will occur in the wires at points 1 and 3 in Fig. 6.21. A stress concentration factor must also be included at point 3 for the tensile stress.

$$S_{t3}=K_t\frac{H_T}{A}$$

where $H_T = 1.081$ lb (see Eq. 6.53)

$$A=\frac{\pi}{4}d^2=\frac{\pi}{4}(0.020)^2=3.14\times10^{-4}\text{ in.}^2\text{ area}$$

K_t = theoretical stress concentration factor where the electrical lead wire joins the component body as shown in Fig. 6.23 (from Peterson [18]).

$$\frac{r}{d}=\frac{0.005}{0.020}=0.25 \qquad \text{and} \qquad \frac{D}{d}=\frac{0.030}{0.020}=1.5$$

$$K_t = 1.48 \text{ (tension)}$$

$$S_{t3}=\frac{(1.48)(1.081)}{3.14\times10^{-4}}=5100 \text{ lb/in.}^2 \tag{6.62}$$

The maximum axial stress at point 3 will be the sum of the bending and the tensile stresses.

$$S_{3\,\max} = S_{b3} + S_{t3} = 60{,}500 + 5100$$

$$S_{3\,\max} = 65{,}600 \text{ lb/in.}^2 \tag{6.63}$$

These electrical-lead-wire stresses are for the capacitor when it is mounted at the center of the circuit board. If the capacitor is not mounted at the center of the board, but off to one side, then the relative deflection between the circuit board and the lead wires will be reduced so that the dynamic bending and tensile stresses will also be reduced. The deflection of the circuit board and the relative deflection of the electrical lead wire can be determined from the general deflection equation of the circuit board when the edge boundary conditions are known.

If the capacitor shown in Fig. 6.19 is mounted near the edge of the circuit board and the body of the capacitor is not attached to the board in any way, a typical installation will appear as shown in Fig. 6.24.

The displacement of the circuit board, at the component body, can be determined from the plate deflection (equation 6.1) using simply supported sides as follows:

$$Z_1 = \delta \sin \frac{\pi X_1}{a} \sin \frac{\pi Y_1}{b}$$

where $\delta = 0.0519$ in. (see Eq. 6.43)
$X_1 = 1.5$ in. at component body
$a = 8.0$ in. board length
$b = 7.0$ in. board width
$Y_1 = 3.5$ in. center of component body

$$Z_1 = 0.0519 \sin \frac{\pi(1.5)}{8.0} \sin \frac{\pi(3.5)}{7.0}$$

$$Z_1 = 0.0288 \text{ in.} \tag{6.64}$$

An examination of Eqs. 6.43 through 6.62 shows that the deflections, loads, and stresses in the electrical lead wires are directly proportional to the deflections in the circuit board at the body of the component. Therefore the maximum axial (bending + tension) stress in the electrical lead wire at point 3, as shown in Fig. 6.23, can be determined with a direct ratio of the deflections using Eqs. 6.43, 6.63, and 6.64 as follows:

$$S'_{3\max} = 65{,}600 \frac{Z_1}{\delta}$$

$$S'_{3\max} = 65{,}600 \left(\frac{0.0288}{0.0519}\right) = 36{,}400 \text{ lb/in.}^2 \tag{6.65}$$

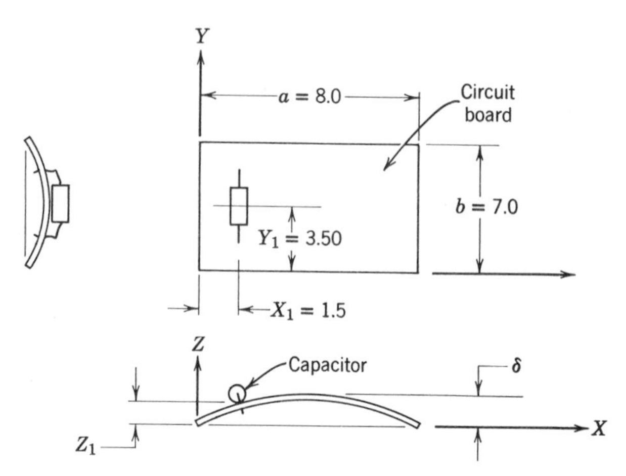

FIGURE 6.24. A component mounted near the edge of a circuit board.

6.9. Cumulative Fatigue Damage in Electrical Lead Wires

The fatigue life of the electrical lead wires can be approximated with the use of Miner's [23] theory on the accumulation of fatigue damage, explained in more detail in Chapter 10. This theory in its simple form states that when a structural member is subjected to an alternating load, every stress cycle uses up a small part of the useful life. When enough stress cycles have been accumulated, the effective fatigue life is used up and the member will fail. This can be expressed in terms of a fatigue cycle ratio R_n as follows:

$$R_n = \frac{n_1}{N_1} + \frac{n_2}{N_2} + \frac{n_3}{N_3} + \ldots = \sum \frac{n_i}{N_i} = 1.0 \tag{6.66}$$

In the above equation n_1, n_2, and n_3 are the actual number of completely reversed stress cycles accumulated at stress levels S_1, S_2, and S_3. The number of completely reversed stress cycles required to produce a failure at these same stress levels are shown by N_1, N_2, and N_3.

Although the number 1.0 is often mentioned for the fatigue cycle ratio, for a more conservative analysis a fatigue cycle ratio of about 0.70 may be used by stress analysts to determine whether a failure will occur.

The actual number of completely reversed stress cycles n accumulated will depend upon the resonant frequency and the amount of time spent at the resonant point. If a 30-min resonant dwell is required for a resonant frequency of 55 Hz, for example, the actual number of fatigue cycles accumulated at point 3 in Fig. 6.23 will be as follows:

$$n_1 = tf_n$$

where $t = 30 \text{ min} = 1800 \text{ sec}$

$f_n = 55$ Hz average of Eqs. 6.41 and 6.42

$$n_1 = 1800 \text{ sec } (55 \text{ cycles/sec}) = 99{,}000 \text{ fatigue cycles} \tag{6.67}$$

The number of completely reversed stress cycles N required to produce a failure in the electrical lead wires on the capacitor can be determined from the S–N fatigue curve for the wire material. For a 2-G-peak vibration input, Eq. 6.63 shows a maximum axial stress of 65,600 psi at point 3 in Fig. 6.23. This condition occurs when the capacitor is mounted by its electrical lead wires at the center of the circuit board. An examination of the S–N curve for type K wire, as shown in Fig. 10.10a, indicates an approximate fatigue life of

$$N_1 = 1.0 \times 10^4 \text{ fatigue cycles} \tag{6.68}$$

Substituting Eqs. 6.67 and 6.68 into Eq. 6.66 to determine the fatigue cycle ratio

$$R_n = \frac{9.9 \times 10^4}{1.0 \times 10^4} = 9.9$$

Since this value is much greater than 0.70, the capacitor lead wires will probably fail before the 30-min resonant-dwell period has been completed. If there are no nicks or scratches in the lead wires to create stress concentrations, the approximate fatigue life of the wires can be determined from Eq. 4.14 as follows:

$$t = \frac{N}{f_n}$$

where $N = 1.0 \times 10^4$ fatigue cycles to fail (see Eq. 6.68)

$f_n = 55$ Hz approximate resonant frequency

Substituting into the above equation

$$t = \frac{1.0 \times 10^4 \text{ fatigue cycles to fail}}{55 \text{ cycles/sec} \times 60 \text{ sec/min}} = 3.03 \text{ min to fail}$$

If the same tantalytic capacitor is mounted near the edge of the circuit board, as shown in Fig. 6.24, then the maximum axial stress in the electrical lead wire at point 3, as shown in Fig. 6.23, will be only 36,400 psi, as shown in Eq. 6.65. An examination of the S–N curve for type K wire, as shown in Fig. 10.10a, indicates the fatigue life is greater than 10 million cycles. Conservatively using this value then

$$N_1 = 10 \times 10^6 \text{ fatigue cycles to fail} \tag{6.69}$$

Substitute Eqs. 6.67 and 6.69 into Eq. 6.66. After 30 min

$$R_n = \frac{9.9 \times 10^4}{1000 \times 10^4} = 0.0099$$

Since this value is so much smaller than 0.70, the capacitor lead wires will not fail.

Dynamic stresses in the component lead wires can be reduced by reducing the relative motion between the circuit board and the component body or by providing a strain relief in the lead wires. A good strain relief is just a longer lead wire that will increase the length from the body of the component to the first bend in the wire. This length increase will reduce the bending stresses in the wire. It will also take up more space on the circuit board thereby reducing the number of components that can be mounted on the board.

If the natural frequency of the circuit board is increased, or if the body of the component is more firmly attached to the board, by tying or by cementing, then the relative motion between the component body and the circuit board will be reduced. Cementing will substantially increase the circuit-board stiffness in the immediate area of the component. This will reduce the relative board deflection which in turn reduce the relative lead wire deflections and stresses. A rigid cement, however, will make it more difficult to repair the circuit board if a component part must be removed.

Conformal coatings, which are normally used to protect a circuit board from dust and humidity, may also act as a cement to bond the components to the circuit board. When repairs must be made, some of these conformal coatings can be stripped with solvents. Some coatings can be broken down from the heat of an ordinary soldering iron. Epoxy coatings generally must be chipped away to replace a component part.

Lacing cord is often used to tie component parts to a circuit board. Although the lacing cord does help to reduce the relative motion between the circuit board and the component body, it is not really very effective if the relative motion is due to the circuit-board resonance. Tying the component body at both ends is slightly better than tying the body at its center. If the relative motion between the component body and the circuit board is due to a resonance in the component, where the body is bouncing on the lead wires which act as springs, then tying the body of the component to the circuit board with lacing cord will effectively eliminate the resonance.

Another form of vibration testing used very often for printed-circuit boards is the sweep test: the frequency is varied while the input G force is held constant. Consider the case where a sinusoidal sweep test must be run from 10 to 500 Hz and back again to 10 Hz several times, using a sweep rate of one octave a minute. This means that the frequency will be increased from 10 to 20 Hz in one minute, then from 20 to 40 Hz in one minute, up to 500 Hz. If a 2-G-peak sinusoidal input is used for the same circuit board with the same capacitor mounted at the center, then at a frequency of about 55 Hz the circuit-board peak resonance will develop and this will produce an alternating stress of 65,600 psi in the electrical lead wires, as shown by Eq. 6.63.

The approximate time it takes to sweep through the circuit-board resonance can be determined by considering the half-power points, as explained in Chapter 4 and shown in Fig. 4.6. Most of the damage will occur in the area of the fundamental resonant mode, which can be approximated by the bandwidth at the half-power points. The time it takes to sweep through the half-power points was shown by Eq. 4.16 as

$$t = \frac{\log_e \dfrac{1+\dfrac{1}{2Q}}{1-\dfrac{1}{2Q}}}{R \log_e 2}$$

where t = time in minutes
R = rate of sweep, 1 octave min
Q = 8.0 transmissibility as used in Eq. 6.43

Substituting into the above equation

$$t = \frac{\log_e \dfrac{1+\dfrac{1}{16}}{1-\dfrac{1}{16}}}{(1) \log_e 2} = 0.184 \text{ min} = 11.05 \text{ sec} \tag{6.70}$$

The actual number of fatigue cycles that will be accumulated in one single sweep from 10 to 500 Hz can be approximated by considering only the fundamental resonant mode and neglecting the higher harmonics. Then

$$n = f_n t = (55 \text{ cycles/sec})(11.05 \text{ sec})$$

$$n = 608 \text{ fatigue cycles accumulated in one sweep} \tag{6.71}$$

The approximate number of single sweeps through the resonant point required to produce a failure in the electrical lead wires will then be

$$\text{NS} = \frac{N}{n} = \frac{\text{cycles to fail}}{\text{cycles per sweep}} \tag{6.72}$$

Substituting Eq. 6.68 and 6.71 into Eq. 6.72

$$\mathrm{NS} = \frac{1 \times 10^4}{6 \times 10^2} = 16.7 \text{ single sweeps to fail} \tag{6.73}$$

This means that a fatigue failure can be expected in the capacitor lead wire at point 3 in Fig. 6.23, after about 17 sweeps through the circuit-board resonance, using a 2-G-peak sinusoidal vibration input, when the capacitor is mounted at the center of the circuit board.

6.10. Dynamic Stress in the Circuit Board

Dynamic bending stresses in a uniformly loaded printed-circuit board can be determined for the resonant condition by considering a sinusoidal load variation for the dynamic load over the board surface when the edges of the board are simply supported. This follows from the relative dynamic deflections which will be maximum at the center of the board and zero at the edges (Fig. 6.25).

The load intensity at any point can be expressed in a trigonometric form that satisfies the boundary conditions.

$$q = q_0 \sin\frac{\pi X}{a} \sin\frac{\pi Y}{b}$$

The differential equation for the deflection of the circuit board then becomes [29]

$$\frac{\partial^4 Z}{\partial X^4} + 2\frac{\partial^4 Z}{\partial X^2 \partial Y^2} + \frac{\partial^4 Z}{\partial Y^4} = \frac{q_0}{D} \sin\frac{\pi X}{a} \sin\frac{\pi Y}{b} \tag{6.74}$$

The deflection of the circuit board at any point can also be represented by a trigonometric expression similar to the one shown in Eq. 6.1.

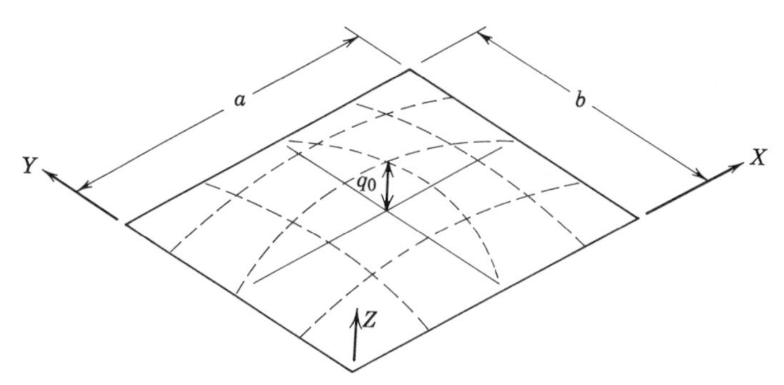

FIGURE 6.25. Dynamic deflection mode of a circuit board.

$$Z = A \sin \frac{\pi X}{a} \sin \frac{\pi Y}{b} \tag{6.75}$$

Performing the operations on Eq. 6.75 as required by Eq. 6.74

$$\frac{\partial^4 Z}{\partial X^4} = \frac{\pi^4}{a^4} A \sin \frac{\pi X}{a} \sin \frac{\pi Y}{b} \tag{6.76}$$

$$\frac{\partial^4 Z}{\partial Y^4} = \frac{\pi^4}{b^4} A \sin \frac{\pi X}{a} \sin \frac{\pi Y}{b} \tag{6.77}$$

$$\frac{\partial^4 Z}{\partial X^2 \partial Y^2} = \frac{\pi^4}{a^2 b^2} A \sin \frac{\pi X}{a} \sin \frac{\pi Y}{b} \tag{6.78}$$

The deflection form factor, A, can be determined by substituting Eqs. 6.76–6.78 into Eq. 6.74.

$$A = \frac{q_0}{\pi^4 D \left(\frac{1}{a^4} + \frac{2}{a^2 b^2} + \frac{1}{b^4} \right)} \tag{6.79}$$

Substitute Eq. 6.79 into Eq. 6.75 and the deflection at any point on the circuit board is

$$Z = \frac{q_0 \sin \frac{\pi X}{a} \sin \frac{\pi Y}{b}}{\pi^4 D \left(\frac{1}{a^4} + \frac{2}{a^2 b^2} + \frac{1}{b^4} \right)}$$

The maximum deflection, Z_0, will occur at the center of the board where $X = a/2$ and $Y = b/2$.

Substituting into the above equation

$$Z_0 = \frac{q_0}{\pi^4 D \left(\frac{1}{a^4} + \frac{2}{a^2 b^2} + \frac{1}{b^4} \right)} \tag{6.80}$$

Assuming that the circuit board acts like a single-degree-of-freedom system at its fundamental resonant mode, the maximum dynamic displacement can be approximated by Eq. 5.43 as follows:

$$Z_0 = \frac{9.8 G_{\text{out}}}{f_n^2}$$

Substitute this expression into Eq. 6.80 and solve for the maximum dynamic pressure intensity q_0 as follows:

$$q_0 = \frac{9.8 G_{\text{out}} \pi^4 D}{f_n^2} \left(\frac{1}{a^4} + \frac{2}{a^2 b^2} + \frac{1}{b^4} \right) \tag{6.81}$$

The plate stiffness factor, D, can be determined from the natural-frequency equation for a uniform, simply supported rectangular plate, using Eq. 6.21.

$$D = \frac{4 f_n^2 \rho}{\pi^2 \left(\dfrac{1}{a^4} + \dfrac{2}{a^2 b^2} + \dfrac{1}{b^4} \right)} \tag{6.82}$$

Substitute Eq. 6.82 into Eq. 6.81 and the maximum dynamic load intensity at the center of the plate simplifies to the following expression:

$$q_0 = \frac{W G_{\text{out}}}{ab} \tag{6.83}$$

The maximum dynamic bending moment in the rectangular circuit board will occur in the center. The bending moment must be greater for the section along the shorter plate dimension, b, which is M_Y because the bending is along the Y axis. The maximum bending moment then becomes [29].

$$M_Y = \frac{q_0 \left(\dfrac{\mu}{a^2} + \dfrac{1}{b^2} \right)}{\pi^2 \left(\dfrac{1}{a^2} + \dfrac{1}{b^2} \right)^2} \tag{6.84}$$

Considering the printed-circuit board shown in Section 6.8, the transmissibility at resonance was 8.0 for a 2-G-peak sinusoidal-vibration input. The maximum bending moment can then be determined as follows:

where $a = 8.0$ in. board length

$b = 7.0$ in. board width

$G_{\text{out}} = G_{\text{in}} Q = (2)(8.0) = 16$ at center

$W = 1.0$ lb board weight

$\mu = 0.12$ Poisson's ratio for G-10 epoxy fiberglass

$$q_0 = \frac{W G_{\text{out}}}{ab} = \frac{(1.0)(16)}{(8.0)(7.0)} = 0.286 \text{ lb/in.}^2$$

Substituting into Eq. 6.84 for the bending moment

$$M_Y = \frac{0.286 \left(\dfrac{0.12}{64} + \dfrac{1}{49} \right)}{\pi^2 \left(\dfrac{1}{64} + \dfrac{1}{49} \right)^2} = 0.496 \text{ in. lb/in.} \tag{6.85}$$

The dynamic bending stress at the center of the plate can then be determined from the standard plate equation. Since printed-circuit boards normally have holes in them for the component lead wires, a stress-concentration factor should be used.

$$S_b = \frac{6K_t M_Y}{h^2} \tag{6.86}$$

where $M_Y = 0.496$ in. lb/in. (see Eq. 6.85)
$h = 0.062$ in. circuit-board thickness
$K_t = 3.0$ theoretical stress-concentration factor for a small hole in the circuit board

Substituting into the above equation

$$S_b = \frac{6(3.0)(0.496)}{(0.062)^2} = 2320 \text{ lb/in.}^2 \tag{6.87}$$

An examination of the fatigue $S-N$ curve for the G-10 epoxy fiberglass shown in Chapter 10, Fig. 10.9*b*, shows that the circuit board will never fail with a stress so low.

6.11. Ribs on Printed-Circuit Boards

If the stiffness of a printed-circuit board is increased without a significant weight increase, the natural frequency will also increase and the deflection at the center of the board will decrease rapidly. A decrease in the circuit-board deflection means a decrease in the electrical-lead-wire stresses and the circuit-board stresses during vibration.

One simple method for increasing the stiffness of a circuit board is to add ribs. If the ribs are made of thin steel, copper, or brass, they can be soldered to the copper cladding on the circuit board. This forms a very stiff section because these materials have a high modulus of elasticity. These ribs can be undercut between supports to allow the printed circuits to run under the ribs without causing electrical shorts.

Consider the printed-circuit board shown in Fig. 6.9. If two steel ribs are soldered to the board in a symmetrical pattern along the length, the circuit board will appear as shown in Fig. 6.26.

If a 2-G-peak sinusoidal-vibration input is used to excite the circuit board, the four edges of the board will act as though they are simply supported. If a substantially higher acceleration G force is used, such as 10 Gs, the connector edge may act more like a free edge. Considering four simply supported edges, the natural frequency for the circuit board with ribs can be determined by making a slight modification in Eq. 6.21 to account for the different stiffness along the X and Y axes. The natural

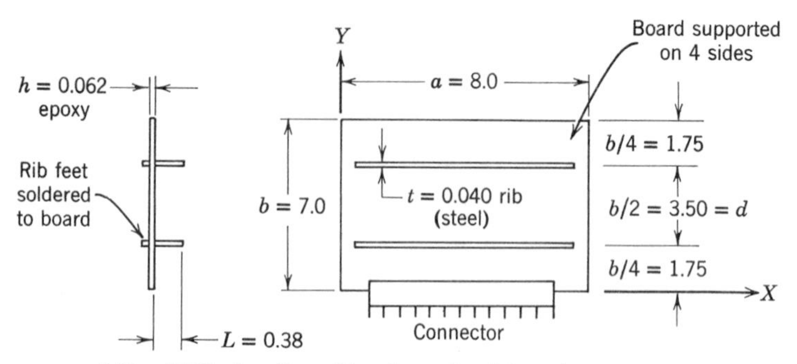

FIGURE 6.26. Stiffening ribs soldered to a circuit board.

frequency equation then takes the following form [29]:

$$f_n = \frac{\pi}{2}\left[\frac{1}{\rho}\left(\frac{D_X}{a^4} + \frac{4D_{XY}}{a^2b^2} + \frac{D_Y}{b^4}\right)\right]^{1/2} \tag{6.88}$$

The term D_X is the bending stiffness of the composite board along the X axis and D_Y is the bending stiffness along the Y axis. Since the circuit-board ribs are parallel to the X axis, they will increase the bending stiffness along the X axis but not along the Y axis. The torsional stiffness of the plate-and-rib combination is represented by D_{XY}.

If the ribs are in a symmetrical pattern on one side of the board, then the moment of inertia of one T section can be used to compute the bending stiffness along the X axis. A table (Table 6.2 and 6.3) is used to make it easier to compute the moment of inertia of the composite T section shown in Fig. 6.27.

TABLE 6.2

Item	Area	$E \times 10^6$	Z	$AE \times 10^6$	$AEZ \times 10^6$	$I_0 \times 10^{-3}$
1	0.2170	2.0	0.031	0.434	0.0134	$\frac{3.5(0.062)^3}{12} = 0.070$
2	0.0152	29.0	0.252	0.441	0.1110	$\frac{0.040(0.380)^3}{12} = 0.183$
				0.875	0.1244	

TABLE 6.3

Item	$EI_0 \times 10^3$	c	c^2	$AEc^2 \times 10^3$
1	0.140	0.111	0.0123	5.34
2	5.310	0.110	0.0121	5.34
	5.450			10.68

The centroid of the T section is at

$$\bar{Z} = \frac{\sum AEZ}{\sum AE} = \frac{0.1244 \times 10^6}{0.875 \times 10^6} = 0.142 \text{ in.}$$

The bending stiffness of the T section is

$$\sum EI = EI_0 + AEc^2 = 5.450 \times 10^3 + 10.68 \times 10^3$$

$$\sum EI = 16.13 \times 10^3 \text{ lb in.}^2 \tag{6.89}$$

The bending stiffness of the circuit board along the X axis, with the ribs, is

$$D_X = \frac{EI}{d} \tag{6.90}$$

where $EI = 16.13 \times 10^3$ lb in.^{2} (see Eq. 6.89)

$d = \frac{b}{2} = 3.5$ in. (see Fig. 6.26)

Substituting into the above equation

$$D_X = \frac{16.13 \times 10^3}{3.5} = 4610 \text{ lb in.} \tag{6.91}$$

The bending stiffness along the Y axis will be approximately the same as the epoxy board:

$$D_Y = \frac{Eh^3}{12(1-\mu^2)} \tag{6.92}$$

where $E = 2 \times 10^6$ lb/in.^{2} G-10 epoxy fiberglass

$h = 0.062$ in. circuit board thickness

$\mu = 0.12$ Poisson's ratio

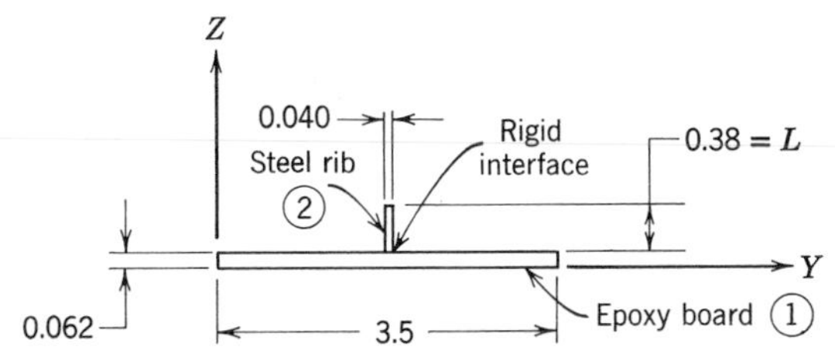

FIGURE 6.27. Dimensions for a rib section soldered to a circuit board.

Substituting into the above equation:

$$D_Y = \frac{(2 \times 10^6)(0.062)^3}{12[1-(0.12)^2]} = 40.2 \text{ lb in.} \tag{6.93}$$

The torsional stiffness can be determined by considering a unit board width along with one rib (Fig. 6.27). The subscripts e and r refer to the epoxy board and to the rib, respectively.

$$D_{XY} = G_e J_e + \frac{G_r J_r}{2d} \tag{6.94}$$

where $G_e = 0.90 \times 10^6$ lb/in.^{2} shear modulus, epoxy fiberglass

$J_e = \frac{1}{3} h^3$ unit torsional stiffness, epoxy fiberglass

$J_e = \frac{1}{3}(0.062)^3 = 79.3 \times 10^{-6}$ in.^{3}

$G_r = 12 \times 10^6$ lb/in.^{2} shear modulus, steel rib

$J_r = \frac{1}{3} Lt^3$ torsional stiffness, steel rib

$J_r = \frac{1}{3}(0.38)(0.040)^3 = 8.10 \times 10^{-6}$ in.^{4}

$d = \frac{b}{2} = 3.5$ in. rib spacing (see Fig. 6.26)

Substituting into the above equation

$$D_{XY} = (0.90 \times 10^6)(79.3 \times 10^{-6}) + \frac{(12 \times 10^6)(8.10 \times 10^{-6})}{2(3.5)}$$

$$D_{XY} = 85.3 \text{ lb in.} \tag{6.95}$$

With the addition of the two steel ribs, the circuit-board weight will increase to about 1.06 lb. The mass per unit area will then be:

$$\rho = \frac{\text{mass}}{\text{area}} = \frac{W}{gab} = \frac{1.06}{(386)(8.0)(7.0)}$$

$$\rho = 0.490 \times 10^{-4} \text{ lb sec}^2/\text{in.}^3 \tag{6.96}$$

The natural frequency of the circuit board with the two ribs can be determined by substituting Eqs. 6.91, 6.93, 6.95, and 6.96 into Eq. 6.88.

$$f_n = \frac{\pi}{2}\left\{\frac{1}{0.490 \times 10^{-4}}\left[\frac{4610}{(8.0)^4} + \frac{4(85.3)}{(8.0)^2(7.0)^2} + \frac{40.2}{(7.0)^4}\right]\right\}^{1/2}$$

$$f_n = 251 \text{ Hz} \tag{6.97}$$

It would appear from Eq. 6.97 that the natural frequency of the circuit board with two ribs will be about 251 Hz. However, since the center section of the circuit board, between the ribs, is only 0.062 in. thick, as shown in Fig. 6.26, it may be possible for this section to develop a resonance below 251 Hz. This can be checked by considering the center section of the circuit board, between the ribs, as a simply supported rectangular plate. Equation 6.21 can then be used to determine the natural frequency as follows:

$$f_n = \frac{\pi}{2}\left(\frac{D}{\rho}\right)^{1/2}\left(\frac{1}{a^2}+\frac{1}{b^2}\right) \qquad \text{(see Eq. 6.21)}$$

where $a = 8.0$ in. board length
$b = 3.5$ in. board width
$D = 40.2$ lb in.
$\rho = 0.463 \times 10^{-4}$ lb sec²/in.³

Substituting into the above equation

$$f_n = \frac{\pi}{2}\left(\frac{40.1}{0.463\times 10^{-4}}\right)^{1/2}\left[\frac{1}{(8.0)^2}+\frac{1}{(3.5)^2}\right]$$

$$f_n = 142 \text{ Hz} \tag{6.98}$$

Since the natural frequency of the unstiffened center section is only 142 Hz, a third rib will have to be added to raise the circuit-board fundamental resonant frequency to 251 Hz.

When the fundamental resonant frequency of the circuit board is increased to 251 Hz, the transmissibility of the circuit board will also increase. Considering the geometry and the mounting of the circuit board, if a low-input vibration-acceleration force with a 2-G peak is used, the transmissibility at resonance will probably be slightly greater than the square root of the natural frequency, as explained in Section 6.3. Vibration test data on one circuit board indicated that the transmissibility was related to the natural frequency as follows:

$$Q = 1.2(f_n)^{1/2}$$

$$Q = 1.2(251)^{1/2} = 19.0$$

The dynamic deflection at the center of the circuit board with the ribs can be determined from Eq. 5.43 by approximating the circuit board as a single-degree-of-freedom system during its fundamental resonant mode.

$$\delta_r = \frac{9.8 G_{\text{in}} Q}{f_n^2}$$

where $G_{\text{in}} = 2.0$ peak input acceleration
$Q = 19.0$ transmissibility at resonance
$f_n = 251$ Hz natural frequency of board with ribs

$$\delta_r = \frac{(9.8)(2.0)(19.0)}{(251)^2} = 0.00591 \text{ in.} \tag{6.99}$$

Comparing the circuit-board deflection with ribs, as shown by Eq. 6.99, to the deflection of the same circuit board without ribs, as shown by Eq. 6.43, indicates that the ribs will decrease the maximum dynamic deflection by a factor of about 8.77.

Electronic component parts mounted on the circuit board will experience lower stresses in their electrical lead wires when the circuit-board deflections are reduced. Equations 6.43–6.63 show these stresses are directly proportional to the circuit-board deflections at the fundamental resonant mode. This means that the deflections can be used in a direct ratio to determine the new stresses developed in the circuit-board components when the ribs are added.

The maximum axial stress in the capacitor lead wire at point 3, as shown in Fig. 6.23, can now be determined with the use of the two different board deflections as shown by Eqs. 6.43 and 6.99, along with the stress of 65,600 psi as shown by Eq. 6.63 for the board without ribs.

$$S_{3r} = 65{,}600\left(\frac{\delta_r}{\delta}\right)$$

$$S_{3r} = 65{,}600\left(\frac{0.00591}{0.0519}\right) = 7490\ \text{lb/in.}^2 \tag{6.100}$$

Adding ribs to the circuit board reduces the component-lead stresses to 7490 psi. An examination of the S–N fatigue curve for type K wire, as shown in Fig. 10.10a, shows that the capacitor lead wires will never fail.

6.12. Ribs Fastened to Circuit Boards with Screws

Extensive vibration tests on bolted assemblies show that bolted joints will experience a substantial amount of relative motion during major structural resonances. This relative motion may add damping to the system, which reduces the transmissibility at resonance. It also reduces the stiffness of the structure and this tends to reduce the natural frequency. The ability of two bolted interfaces to remain rigid is a function of the stiffness of the joint, the number of screws, the screw size, screw spacing, screw torque, interface conditions, the vibration force, and the vibration frequency. Figure 6.28 shows a printed-circuit board with stiffening ribs fastened with screws.

In order to correlate test data with a theoretical analysis, a bolted efficiency factor is very convenient. This efficiency factor will range from 0%, for a joint that has no physical connection, to 100%, for a joint that is welded. Evaluating the types of bolted joints normally found in electronic subassemblies such as brackets, covers, ribs, and stiffeners, which must

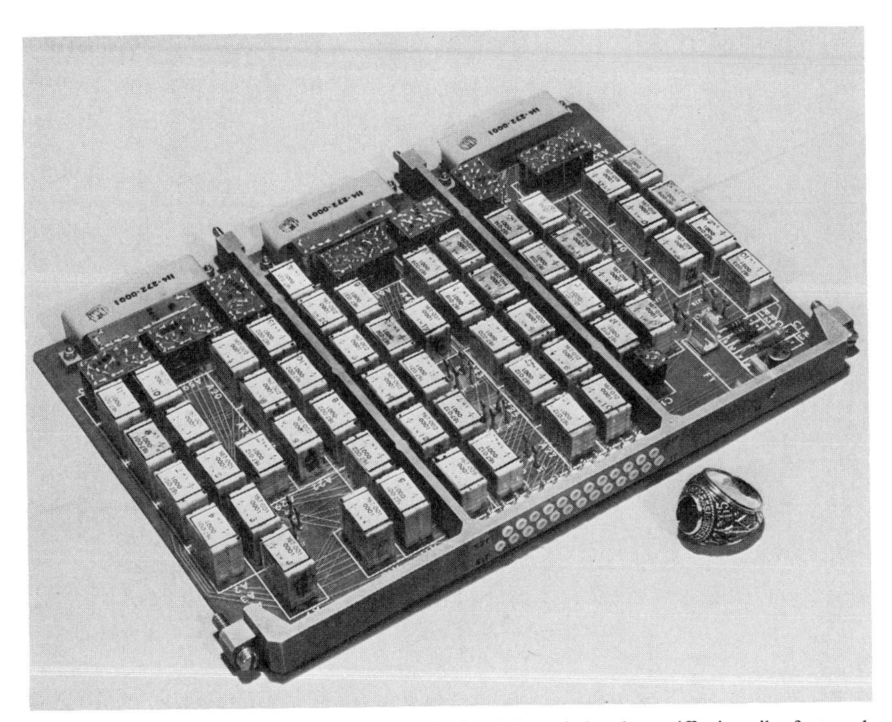

FIGURE 6.28. A multi-layer plug-in printed-circuit board that has stiffening ribs fastened to the board with screws (courtesy Norden division of United Aircraft).

transfer an alternating load, the typical efficiency of a bolted joint used for airborne electronic structures appears to be about 25%. This efficiency factor may be as high as 50% for small sheet-metal panels with large, closely spaced screws, and as low as 10% for large sheet-metal panels held down with quarter-turn quick-disconnect fasteners.

The natural frequency of a printed-circuit board, with stiffening ribs bolted to the board, can be determined by using a bolted efficiency factor. For example, if the ribs on the circuit board shown in Fig. 6.26 were bolted to the board, instead of soldered, the resulting T section through one rib will be similar to the section shown in Fig. 6.27, except for the efficiency of the bolted interface. The 25% efficiency factor for this bolted joint will be analyzed by reducing the effective thickness of the rib to 25% of its original thickness (Fig. 6.29).

The moment of inertia of the composite T-section, with the bolted rib, will be determined with the use of Tables 6.4 and 6.5. Notice that the equivalent thickness of the rib has been reduced from 0.040 to 0.010 in. to account for the 25% bolted efficiency factor.

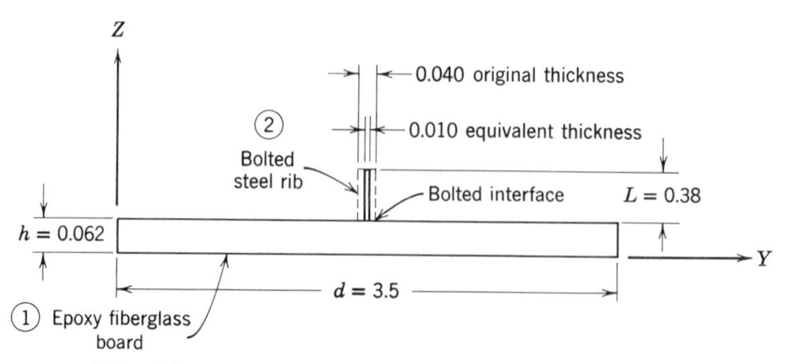

FIGURE 6.29. Dimensions for a rib section bolted to a circuit board.

TABLE 6.4

Item	Area	$E \times 10^6$	Z	$AE \times 10^6$	$AEZ \times 10^6$	$I_0 \times 10^{-3}$
1	0.2170	2.0	0.031	0.434	0.0134	$\dfrac{3.5(0.062)^3}{12} = 0.070$
2	0.0038	29.0	0.252	0.110	0.0277	$\dfrac{(0.010)(0.380)^3}{12} = 0.046$
				0.544	0.0411	

TABLE 6.5

Item	$EI_0 \times 10^3$	c	c^2	$AEc^2 \times 10^3$
1	0.140	0.0446	0.00199	0.864
2	1.335	0.1764	0.0311	3.420
	1.475			4.284

The centroid of the T-section is at

$$\bar{Z} = \frac{\sum AEZ}{\sum AE} = \frac{0.0411 \times 10^6}{0.544 \times 10^6} = 0.0756 \text{ in.}$$

The bending stiffness of the T-section is

$$\sum EI = EI_0 + AEc^2 = 1.475 \times 10^3 + 4.28 \times 10^3$$

$$\sum EI = 5.76 \times 10^3 \text{ lb in.}^2 \tag{6.101}$$

The bending stiffness of the circuit board along the X axis, with the bolted rib, is

$$D_X = \frac{EI}{d}$$

where $EI = 5.76 \times 10^3$ lb in.^{2} (see Eq. 6.101)

$$d = \frac{b}{2} = 3.5 \text{ in. (see Fig. 6.26)}$$

Substituting into the above equation

$$D_X = \frac{5.76 \times 10^3}{3.5} = 1645 \text{ lb in.} \tag{6.102}$$

The bending stiffness along the Y axis will be approximately the same as the epoxy board, which is shown by Eq. 6.93.

$$D_Y = 40.2 \text{ lb in.} \tag{6.103}$$

The torsional stiffness of the bolted T-section can be determined by considering a unit board width, along with one rib, as shown in Fig. 6.29. The subscripts e and r refer to the epoxy board and to the rib, respectively.

$$D_{XY} = G_e J_e + \frac{G_r J_r}{2d} \text{ (see Eq. 6.94)}$$

where $G_e = 0.90 \times 10^6$ lb/in.^{2} shear modulus epoxy fiberglass

$J_e = \frac{1}{3} h^3 = \frac{1}{3}(0.062)^3 = 79.3 \times 10^{-6}$ in.^{3}

$G_r = 12 \times 10^6$ lb/in.^{2} shear modulus, steel rib

$J_r = \frac{1}{3} L t^3 = \frac{1}{3}(0.38)(0.010)^3 = 0.126 \times 10^{-6}$ in.^{4}

$d = \frac{b}{2} = 3.5$ in. rib spacing (see Fig. 6.26)

Substituting into the above equation

$$D_{XY} = (0.90 \times 10^6)(79.3 \times 10^{-6}) + \frac{(12 \times 10^6)(0.126 \times 10^{-6})}{2(3.5)}$$

$$D_{XY} = 71.6 \text{ lb in.} \tag{6.104}$$

If the circuit board weight does not change, the mass per unit area will be the same as that shown in Eq. 6.96. The natural frequency of the circuit board with the bolted ribs can then be determined by substituting Eqs. 6.102–6.104 and 6.98 into Eq. 6.88.

$$f_n = \frac{\pi}{2} \left\{ \frac{1}{0.490 \times 10^{-4}} \left[\frac{1645}{(8.0)^4} + \frac{4(71.6)}{(8.0)^2(7.0)^2} + \frac{40.2}{(7.0)^4} \right] \right\}^{1/2}$$

$$f_n = 160 \text{ Hz} \tag{6.105}$$

Since the center section of the circuit board is not stiffened between the ribs, as shown in Fig. 6.26, it may be possible for this section to develop a resonance below 160 Hz. The natural frequency of this section was previously checked as a simply supported rectangular plate. These results are shown in Eq. 6.98, which shows a natural frequency of 142 Hz. If a natural frequency of 160 Hz is required, then a third rib will have to be bolted to the circuit board.

The deflection at the center of the circuit board, with the bolted ribs, can again be determined from Eq. 5.43 by approximating the board as a single-degree-of-freedom system at its resonance. The transmissibility developed with the bolted ribs should be slightly less than with the soldered ribs because of the additional damping at the rib interfaces and the bolt interfaces. Vibration test data on one group of circuit boards using a 2-G-peak sinusoidal-vibration input indicated the transmissibility was approximately related to the natural frequency as follows:

$$Q = (f_n)^{1/2}$$

$$Q = (160)^{1/2} = 12.7$$

Substituting into Eq. 5.43, the single amplitude displacement with the bolted ribs becomes

$$\delta_r = \frac{9.8 G_{\text{in}} Q}{f_n^2} = \frac{9.8(2.0)(12.7)}{(160)^2}$$

$$\delta_r = 0.00972 \text{ in.} \tag{6.106}$$

The stresses in the electrical lead wires for the capacitor mounted at the center of the circuit board can be determined using the same deflection-ration method as shown in Eq. 6.100.

6.13. Printed-Circuit Boards with Ribs in Two Directions

If two sets of ribs are fastened to one side of a circuit board in a symmetrical pattern, a T-section will have to be analyzed to determine the bending stiffness along the Y axis as well as the X axis (Fig. 6.30).

Again, be sure to check each plate section between the ribs to make sure each section has a natural frequency greater than the full-plate natural frequency including the ribs.

6.14. Proper Use of Ribs to Stiffen Plates and Circuit Boards

Ribs will not increase the stiffness of a plate or a circuit board if they are not properly used. In order for a rib to be effective, it must carry a load. Since all loads must eventually be transferred to the supports, the rib is most effective when it carries a load directly to the supports.

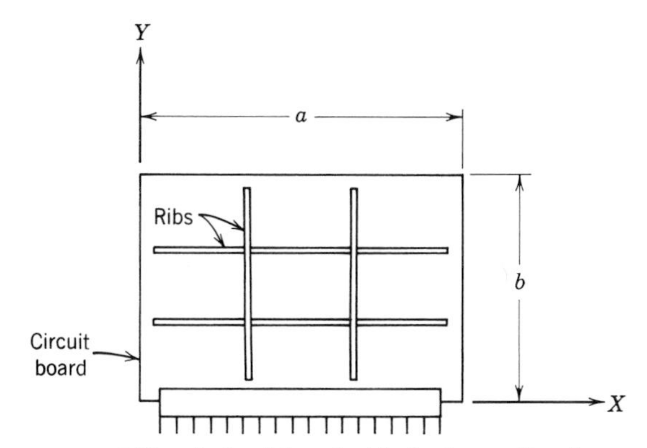

FIGURE 6.30. A circuit board with ribs in two directions.

For example, consider a flat square plate simply supported on two opposite edges and free on two opposite edges. If ribs are added to the plate so that they join the two free edges, the ribs will not carry the load directly to the supports and these ribs will not be very effective in stiffening the plate. If the ribs are added to the plate so that they join the two supported edges (Fig. 6.31), then these ribs will carry the load directly into the supports and they will be very effective in stiffening the plate.

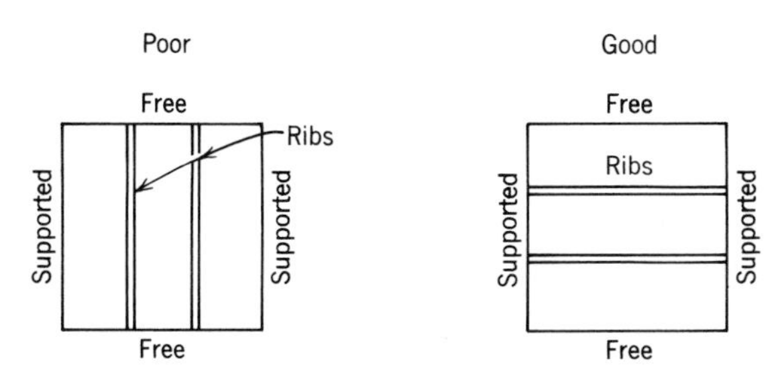

FIGURE 6.31. Adding ribs to effectively stiffen a flat plate.

If ribs cannot be carried directly into the supports, then a secondary member should be provided to carry the load to the supports to make the ribs more effective.

For example, consider a flat square plate simply supported on three sides and free on a fourth side. If ribs must be added to the plate so they end at the free edge, then an additional rib should be placed across the free edge to act as a secondary member which will carry the load to the supports (Fig. 6.32).

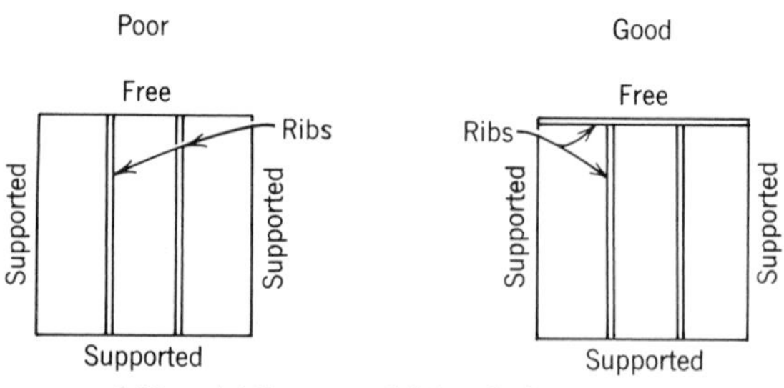

FIGURE 6.32. Adding an additional rib to stiffen a flat plate.

6.15. Estimating the Maximum Allowable Circuit-Board Deflections

The natural frequency of a printed-circuit board can be calculated if the size, weight, and boundary conditions at the edges of the circuit board can be defined. It is also possible to calculate the approximate stresses in the electrical lead wires of electronic component parts mounted on the printed-circuit board. These calculations, however, can involve a substantial amount of work when it is necessary to establish an optimum resonant frequency for a new circuit-board design. One relatively quick method, that works very well, is to base the maximum single-amplitude dynamic deflection at the center of the board on the length of the shorter side b, on a rectangular board. The displacement relation that can be used is as follows:

$$\delta = 0.003\, b \tag{6.107}$$

For example, the dynamic single-amplitude displacement at the center of a rectangular printed-circuit board 4 in. wide and 7 in. long, simply supported on four edges, should be limited to the following:

$$\delta_{\max} = 0.003(4) = 0.012 \text{ in. single amplitude}$$

Once the physical size of the circuit board has been determined and the natural frequency has been calculated for a new design, it is then possible to estimate the maximum allowable input G force the circuit board can withstand if the transmissibility is known. In Section 6.3 it was shown that the transmissibility can often be approximated as a function of the square root of the natural frequency. This relation can be expressed as follows:

$$Q = A(f_n)^{1/2} \tag{6.108}$$

In the above equation the A represents the resonance-amplitude factor which will vary from about 0.50 to about 2.0, depending upon the circuit-board mounting geometry and the input G force, as outlined in Section 6.3. The input G force can be expressed as a function of the transmissibility and the frequency as shown in Eq. 5.43.

$$G_{\text{in}} = \frac{f_n^2 \delta}{9.8Q} \tag{6.109}$$

Substituting Eq. 6.108 into Eq. 6.109

$$G_{\text{in}} = \frac{f_n^{1.5} \delta}{9.8A} \tag{6.110}$$

The above equation represents the maximum allowable input G force the circuit board can withstand.

Over the resonant-frequency range of 200–300 Hz, called the middle-frequency range in Section 6.3, the amplitude factor A is often equal to unity.

6.16. SAMPLE PROBLEM

Determine the maximum allowable peak-input G force the circuit board shown in Fig. 6.26 can withstand and still provide an infinite fatigue life for the electronic components mounted on the circuit board.

The maximum allowable single-amplitude displacement at the center of the circuit board will be determined from Eq. 6.107.

$$\delta = 0.003(7.0) = 0.021 \text{ in.} \tag{6.111}$$

$f_n = 251$ Hz (see Eq. 6.97)
$A = 1.0$ resonance amplitude factor approximated for the middle frequency range 200–300 Hz.

Substituting into Eq. 6.110, the maximum allowable peak input G force becomes

$$G_{\text{in}} = \frac{(251)^{1.5}(0.021)}{9.8(1.0)} = 8.56\, G \tag{6.112}$$

In the original analysis, a 2-G peak-input force was considered. This permitted the connector edge on the printed-circuit board to be treated as a simply supported edge. If an 8.56-G peak-input force is considered, the connector edge may no longer act as though it is simply supported and it may tend to act more like a free edge. This will tend to reduce the resonant frequency which, in turn, will increase the dynamic displacements and stresses. Also notice, if the input G force is increased, the transmissibility experienced at resonance will decrease. These factors should be considered when the dynamic deflections and stresses are calculated.

6.17. Curves for Determining the Maximum Allowable G Forces for Resistors

In the following section, G-force-versus-resonant-frequency curves for resistors are based on the use of plated-through holes for mounting the resistors. This technique permits the solder to wick up the electrical lead wires and support them at the top and bottom of the circuit board. If the solder joints are on only one side of the circuit board, bending moments in the electrical lead wires can cause shear tear-out failures in the solder

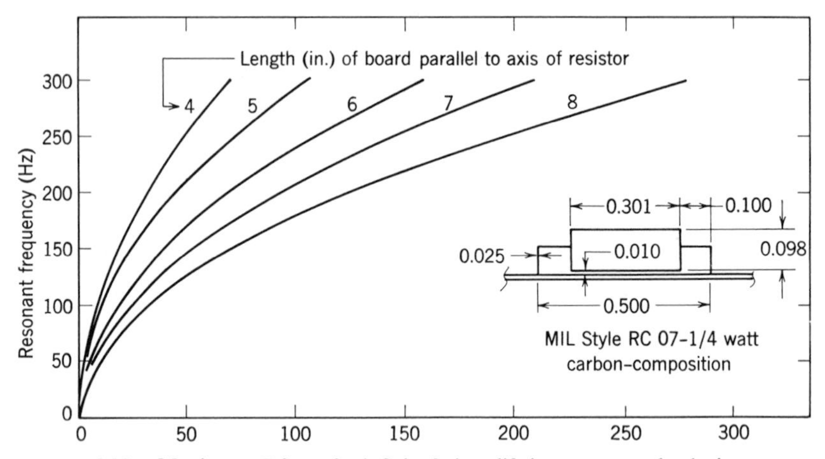

FIGURE 6.33. Maximum G force for infinite fatigue life in component lead wire.

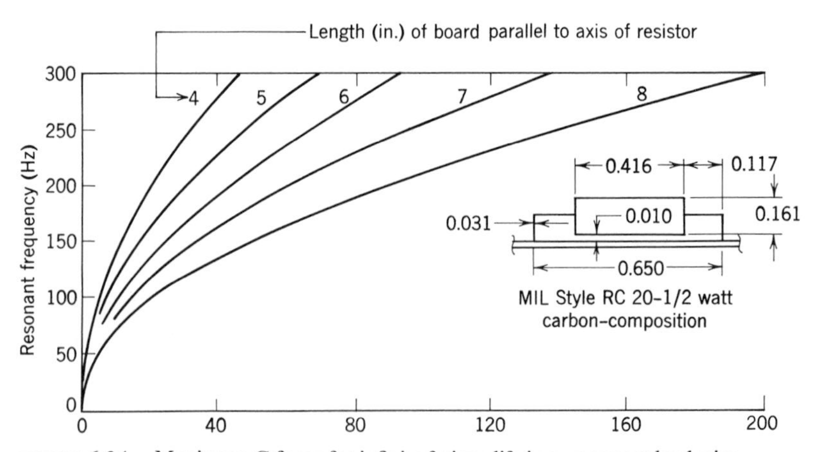

FIGURE 6.34. Maximum G force for infinite fatigue life in component lead wire.

joints at much lower G levels than those shown in the curves because the curves are based on stresses in the electrical lead wires [37].

Some companies use a minimum dimension of 0.10 in. from the body of the component to the first bend in the lead wire, in order to provide a reasonable strain relief for thermal and vibration stresses. A rectangular grid system spaced on 0.05-in. centers is also used to standardize the mounting pattern for electronic components. These two requirements, plus the length of the component body, determine the span between the electrical lead holes as shown in Figs. 6.33–6.38. The use of the curves can be demonstrated by considering the sample problem in Section 6.18.

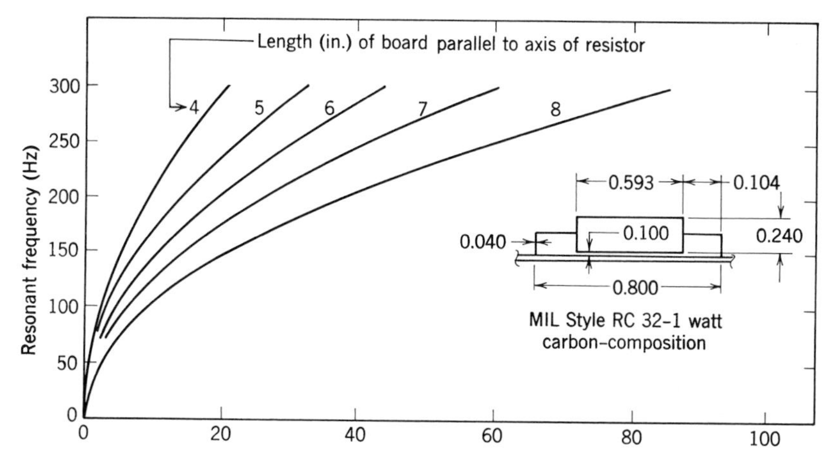

FIGURE 6.35. Maximum G force for infinite fatigue life in component lead wire.

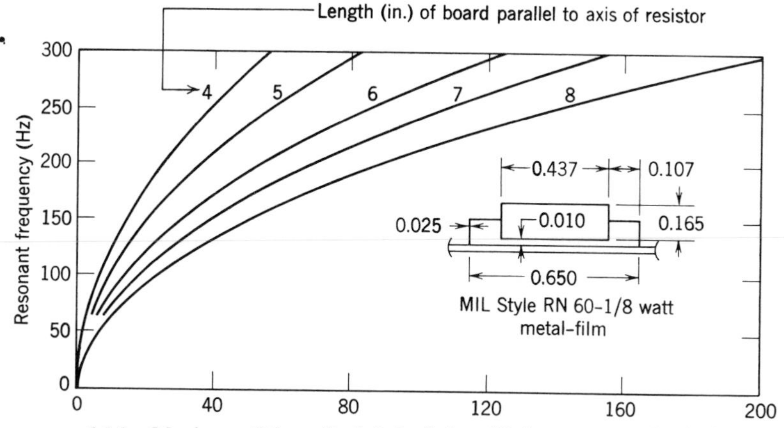

FIGURE 6.36. Maximum G force for infinite fatigue life in component lead wire.

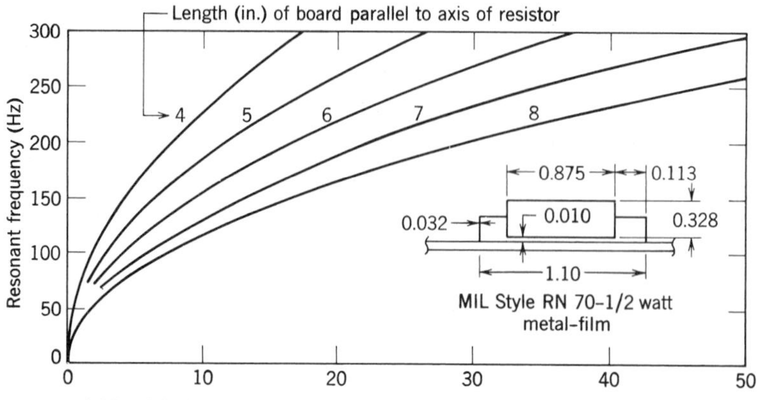

FIGURE 6.37. Maximum G force for infinite fatigue life in component lead wire.

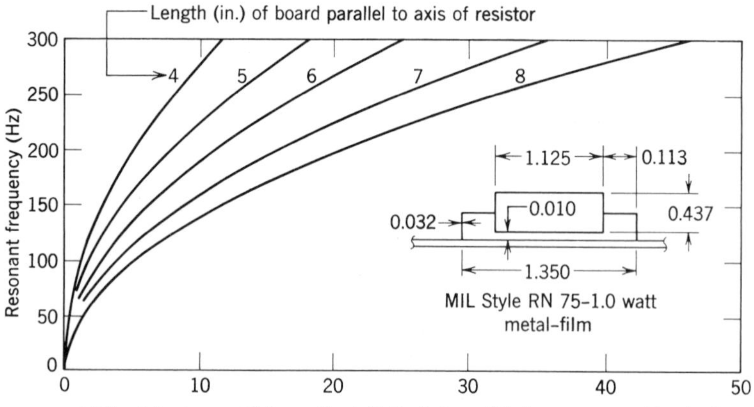

FIGURE 6.38. Maximum G force for infinite fatigue life in component lead wire.

6.18. Sample Problem Using the Resistor Curves

An electronic box must be subjected to a 5-G-peak sinusoidal-vibration test where it must dwell for long periods at major resonant points. The electronic box has many similar plug-in printed-circuit boards which have several $\frac{1}{2}$-watt carbon-composition resistors (MIL Style RC-20) mounted near the center of the board. The desired fatigue life is 10 million cycles, and the resistors are not cemented or tied to the circuit boards.

The problem is to determine whether the resistors on the circuit boards meet the design requirements. The printed-circuit board is shown in Fig. 6.39 and the details of the critical resistor mounted on the circuit board

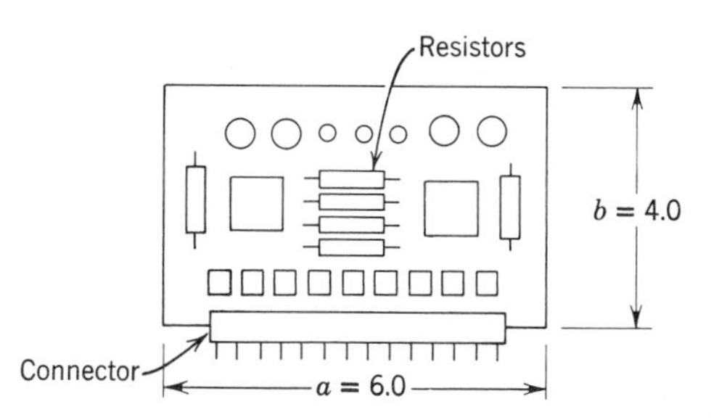

FIGURE 6.39. Dimensions for a circuit board supported on four sides.

are shown in Fig. 6.34, where the board is 0.062 in. thick, and weighs 0.25 lb.

The natural frequency of the circuit board can be determined from Eq. 6.21 as follows (for a board simply supported on four sides):

$$f_n = \frac{\pi}{2}\left(\frac{D}{\rho}\right)^{1/2}\left(\frac{1}{a^2}+\frac{1}{b^2}\right)$$

where $D = \dfrac{Eh^3}{12(1-\mu^2)} = \dfrac{2\times 10^6(0.062)^3}{12[1-(0.12)^2]} = 40.1 \text{ lb in.}$

$$\rho = \frac{W}{gab} = \frac{0.25}{(386)(6.0)(4.0)} = 0.27\times 10^{-4} \text{ lb sec}^2/\text{in.}^3$$

Substituting into the above equation

$$f_n = \frac{\pi}{2}\left(\frac{40.2}{0.27\times 10^{-4}}\right)^{1/2}\left(\frac{1}{36}+\frac{1}{16}\right) = 173 \text{ Hz}$$

The maximum G force at the center of the circuit board, where the resistors are mounted, depends on the transmissibility of the circuit board at resonance. If there are no test data available on similar boards, a good first order approximation for the transmissibility can be obtained by using the following relation:

$$Q = (f_n)^{1/2} = (173)^{1/2} = 13.1$$

The expected G force at the center of the circuit board is

$$G_{\text{out}} = G_{\text{in}} \times Q = 5(13.1) = 65.5 \text{ peak}$$

The maximum allowable G force for a 6-in.-long circuit board (with the axis of the resistor parallel to this edge) having a resonant frequency of 173 Hz is approximately $35G$ (Fig. 6.34). Since the expected G force is $65.5G$, the design is not satisfactory.

For an acceleration G force of $65.5G$ peak at the center of the circuit board, the same figure shows a resonant frequency of about 250 Hz. However, since the transmissibility increases with the resonant frequency, the resonant frequency of the circuit board must be increased even more, and therefore a resonant frequency of about 280 Hz is required. The expected G force at the center of the circuit board then becomes

$$G_{out} = 5(280)^{1/2} = 84G \text{ peak}$$

This will meet the requirements for a 6-in. circuit board as shown in Fig. 6.34.

In order to raise the resonant frequency of the circuit board from 173 to 280 Hz, several ribs could probably be added. These ribs could be metal or epoxy fiberglass, and they could be bolted, cemented, or, in the case of a copper or steel rib, soldered to the printed-circuit board.

Another possible solution is to use laminated circuit boards with high damping to reduce the transmissibility at resonance. These boards cost more and may be difficult to handle because plated-through holes cannot be used and dip soldering may create problems.

CHAPTER 7

ELECTRONIC BOXES

7.1. Preliminary Investigations

Electronic equipment can be packaged in many different configurations depending upon the space and shape factors available in missiles, airplanes, submarines, trains, autos, and ships. The rectangular type of box is generally the most common because it is usually much less expensive to fabricate, mounting is simple, and plug-in modules such as printed-circuit boards conform readily to this shape.

Before an extensive and expensive computerized analysis is made to determine the frequency response characteristics of the chassis, it is desirable to take a quick look at the different modes of vibration. Various bending and torsional modes can be examined to determine whether coupling may occur. Preliminary dynamic loads can also be determined to see whether panels will buckle, rivets shear, or welds crack.

Torsional vibration modes are often a source of trouble in an electronic chassis. These modes, which are generally ignored or overlooked, can result in low resonant frequencies. This in turn can lead to large deflections and stresses that can substantially reduce the fatigue life of the structure.

7.2. Different Types of Mounts

The type of mount can affect the response characteristics of a chassis. If quick installation and removal are desirable, the entire chassis can be inserted and removed by means of a single jacking screw located at the front of the chassis. Interface connectors can be placed at the rear of the chassis with floating mounts and alignment pins that will position the chassis and ensure accurate connector alignment (Fig. 7.1).

This type of installation is very easy to maintain, but very poor from a structural standpoint. The force required to insert the chassis and engage the rear connectors must be transferred from the rear connectors to the front jacking screw where the load is applied. This means that the entire jacking load must pass through the chassis structure, which must be made stiff enough to prevent buckling.

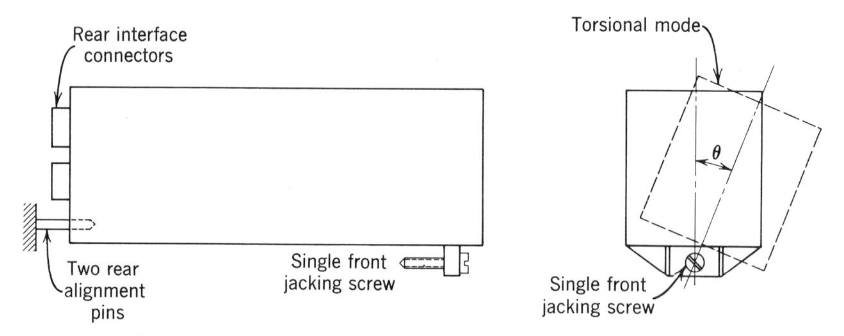

FIGURE 7.1. Chassis with a single jacking screw for inserting and removing the equipment.

The vibration response for this type of mount is poor because the entire chassis can rotate about the single jacking screw. Torsional modes will usually couple with bending modes, and this can result in very high transmissibilities. Tests on several different types of boxes, with similar mounts, have shown transmissibilities as high as 20 at the front outside corners, with a 2-G-peak sinusoidal-vibration input, when the resonant frequencies were above 150 Hz.

Another type of chassis mount that provides quick installation and removal is the ATR rack. This is a sheet-metal mounting rack supplied in several different lengths and widths in an attempt to standardize air transport rack equipment boxes. The back part of the rack has two spring-loaded alignment pins to position the box for connector engagement and to hold the box during vibration. The front of the rack usually has swivel screw mounts which swing up to hold the chassis after it has been installed (Fig. 7.2).

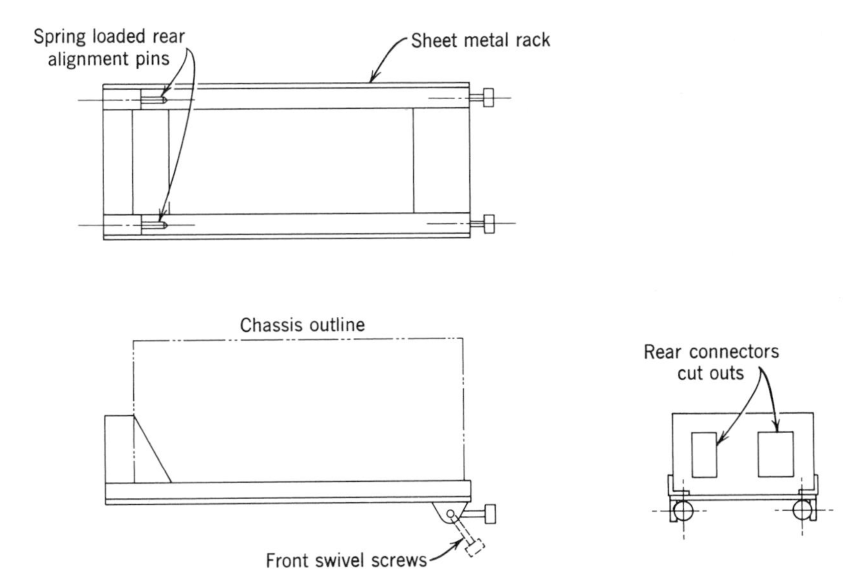

FIGURE 7.2. Air transport rack (ATR) for mounting electronic equipment.

The ATR mounting rack is usually used with vibration isolators. When the proper selection is made, the isolators can sharply reduce the dynamic loads in an electronic system. If a proper isolator selection is not made, the dynamic loads can be increased.

Quite often the ATR mounting rack will be used without isolators. This may lead to trouble unless the spring rate of the mounting rack is included in the response characteristics and the dynamic loading for the chassis. Since the ATR mounting rack is usually made of sheet aluminum, a high G-loading on a heavy chassis can fracture the sheet-metal rack or induce high acceleration G forces in the chassis itself. Any vibration level of about 2 G or greater should be examined very carefully when this type of mount is being considered.

A center-of-gravity (CG) mount should be used to reduce the possibility of coupling torsional modes with bending modes during vibration. When the CG lies on the mounting plane, torsional modes will still occur, but they will not couple with the bending modes during vibration along the X axis (see Fig. 7.3). This separation of resonances, or decoupling, will reduce the severity of the resonance thereby increasing the fatigue life of the structure.

A chassis with a high CG and a narrow cross-section should have its mounts examined very carefully. This type of geometry can lead to very high dynamic loads in the mounts and adjacent structure. This is because

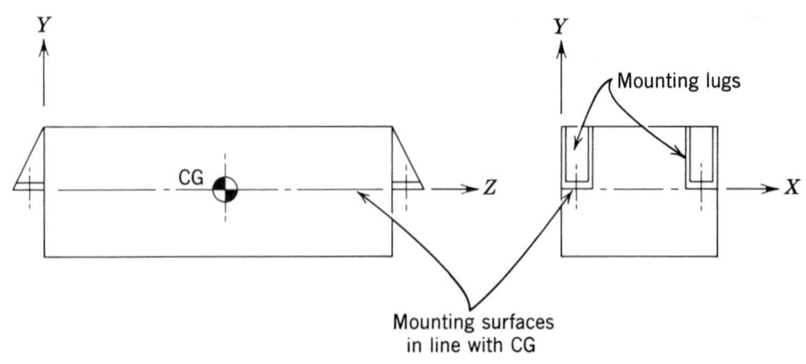

FIGURE 7.3. An electronic box with a CG mount.

of the severe overturning moments that can develop during vibration in the lateral direction (see Fig. 7.4).

There are many cases where the mounting geometry is dictated by the thermal environment. For example, in a space vehicle where vacuum conditions exist, many electronic systems are bolted directly to a cold plate to remove the heat. In order to transfer heat effectively in a vacuum, the interfaces must be flat and smooth. Also, a high interface pressure must be maintained. One good method for accomplishing this is to provide a stiff section with many bolts. The penalty is paid in the form of extra weight. Although the weight is required for thermal reasons, the dynamic characteristics of such a chassis will usually be quite good. Thick walls and a solid mount will provide the chassis with a high resonant frequency, which will result in small displacements and low structural stresses (Fig. 7.5).

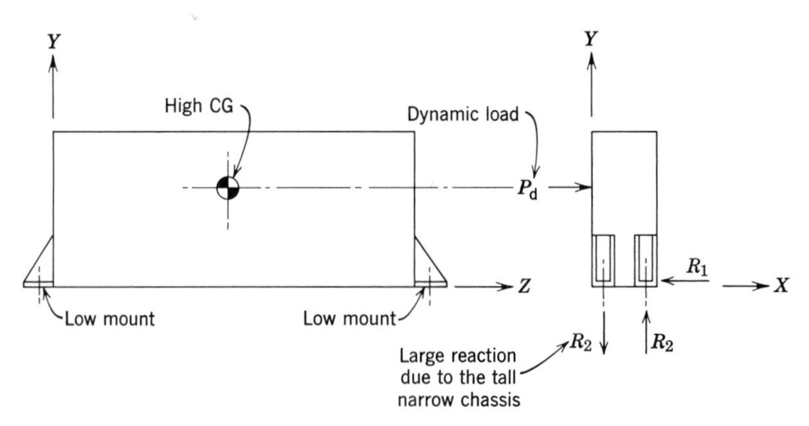

FIGURE 7.4. An electronic box with a low mount and a high CG.

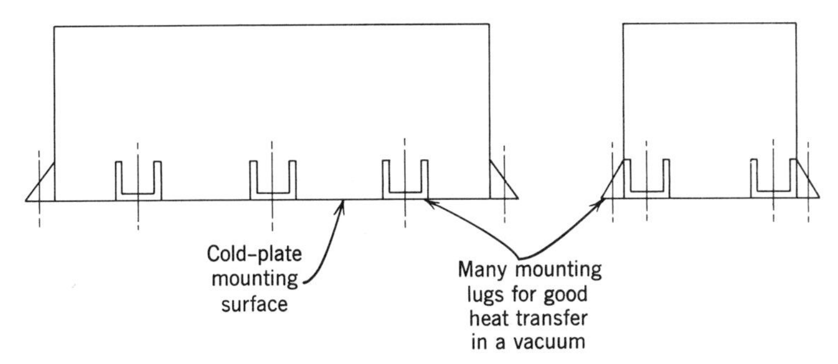

FIGURE 7.5. Many mounting lugs on an electronic box.

7.3. PRELIMINARY DYNAMIC ANALYSIS

A preliminary dynamic analysis of a complex chassis can very often be made by analyzing it as a simply supported beam. If the cross-section of the chassis is relatively constant and the chassis is supported at each end, the simple-beam analysis will often reveal many weak points in the initial design. Design changes can then be incorporated early in the program. This will reduce the amount of time spent in setting up comprehensive computer studies that would give the same results. After the preliminary design changes have been made, the computer program can then be used to provide a more accurate and detailed analysis of internal and external dynamic loads. An airborne electronic box with plug-in modules is shown in Fig. 7.6.

Consider an electronic box that uses a riveted sheet-aluminum chassis with top and bottom covers fastened by screws. The chassis will be fastened to the airframe structure by means of two mounting brackets at each end of the chassis (Fig. 7.7).

Since the chassis has a uniform cross-section with mounting brackets at each end, a preliminary analysis of the chassis can be made by approximating it as a simply supported beam. The high center of gravity indicates that the bending mode will couple with the torsional mode during vibration in the lateral direction along the X axis. This will tend to lower the fundamental resonant mode in this direction.

The moment of inertia of the cross-section along the X and Y axis will include the epoxy fiberglass interconnecting board (modulus of elasticity of 2×10^6 compared to 10×10^6 for aluminum). The calculations can be simplified by using an equivalent aluminum plate that has the same stiffness as the epoxy fiberglass plate. This is accomplished by adjusting the

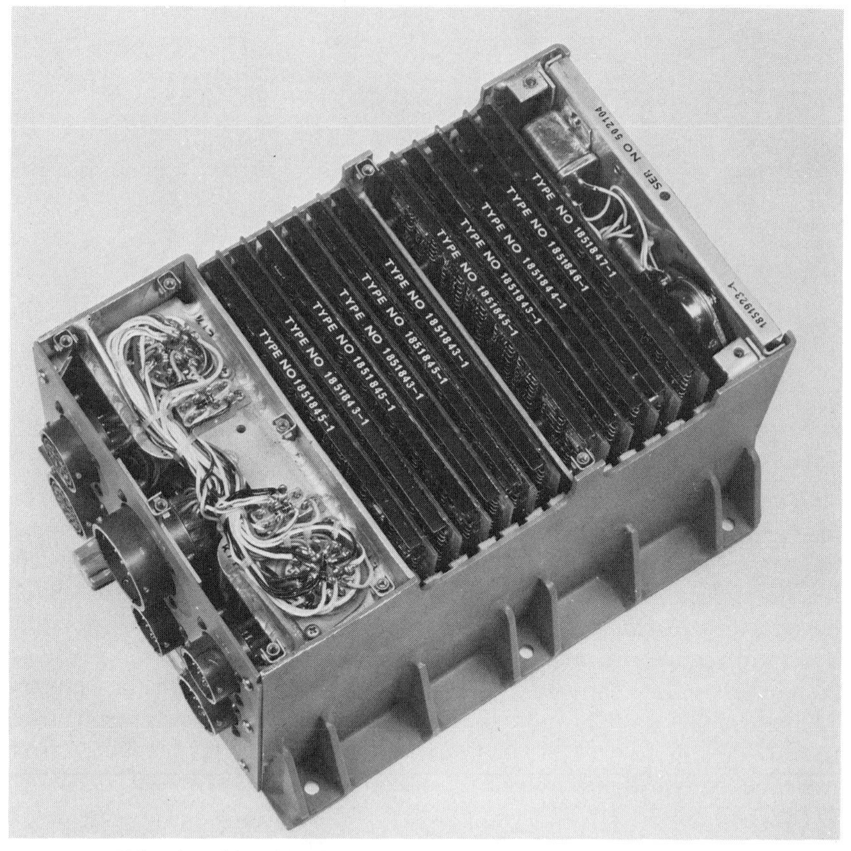

FIGURE 7.6. An airborne electronic box with many plug-in modules (courtesy The Bendix Corporation, Navigation and Control Division).

epoxy plate area and moment of inertia with the ratio of the modulus of elasticity of the two materials. The aluminum equivalent of the epoxy board area and stiffness is (see also Section 4.8)

$$A_{\text{alum}} = A_{\text{epox}} \times \frac{E_{\text{epoxy}}}{E_{\text{alum}}} = A_{\text{epox}} \times \frac{2}{10}$$

$$I_{\text{alum}} = I_{\text{epox}} \times \frac{E_{\text{epoxy}}}{E_{\text{alum}}} = I_{\text{epox}} \times \frac{2}{10} \qquad \left(\begin{array}{c}\text{for members with} \\ \text{the same height}\end{array}\right)$$

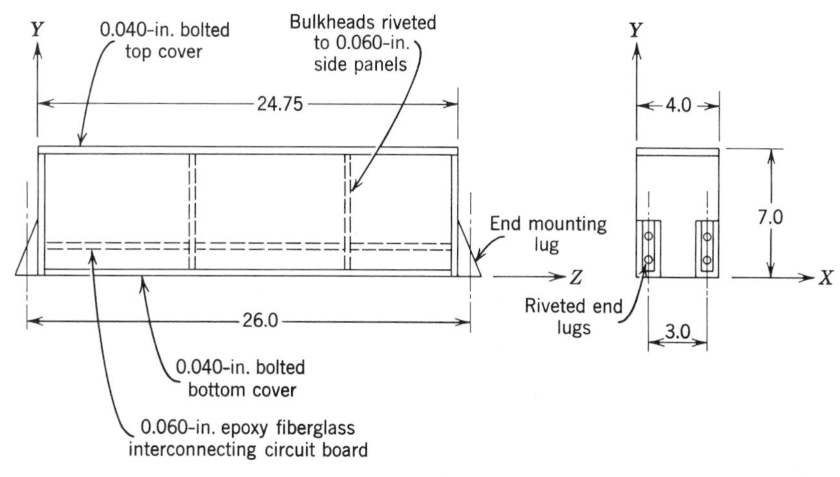

FIGURE 7.7. Dimensions of a tall narrow electronic box.

7.4. Bolted Covers

Bolted covers are used to provide access to the top and bottom sections of the chassis. The general tendency in the electronics industry is to provide convenient access to the electronic components. This usually means quarter-turn screws so that the covers can be removed and replaced quickly. Tests on this type of fastener indicate it is only about 10% efficient in its ability to hold two members together during vibration. This efficiency factor ranges from 0% for a joint that has no connection to 100% for a joint that is welded.

The epoxy fiberglass circuit-interconnecting board is also a bolted member. However, since this type of installation is usually of a more permanent type, regular screws, and more of them, are normally used. Therefore a bolted efficiency factor of 25% is used for this member.

The bolted efficiency factor of a heavy flanged cover, with many high-strength cap screws similar to the sketch shown in Fig. 7.30, is about 50%. This type of construction is often used for electronic boxes that are to be used in the hard vacuum of outer space. Since covers are not normally removed for maintenance in a space vehicle, many high-strength screws are often used to hold covers in place.

The cross-section of the electronic chassis is shown in Fig. 7.8.

The moment of inertia of the cross-section along the Y axis can be determined with the use of Table 7.1.

TABLE 7.1

Item	area (in.^{2})	Y	AY	$I_0 =$ in.^{4}	c	c^2	Ac^2
1	$6.92(0.060) = 0.415$	3.50	1.450	$\frac{0.06(6.92)^3}{12} = 1.65$	0.36	0.13	0.054
2	$6.92(0.060) = 0.415$	3.50	1.450	$\frac{0.06(6.92)^3}{12} = 1.65$	0.36	0.13	0.054
3[a]	$0.1(4.0)(0.040) = 0.016$	6.98	0.111	$\frac{0.1(4.0)(0.04)^3}{12} = 0$	3.84	14.8	0.237
4[a]	$0.1(4.0)(0.040) = 0.016$	0.02	0.000	$\frac{0.1(4.0)(0.04)^3}{12} = 0$	3.12	9.75	0.156
5	$0.75(0.060) = 0.045$	0.07	0.003	$\frac{0.75(0.06)^3}{12} = 0$	3.07	9.42	0.424
6	$0.75(0.060) = 0.045$	0.07	0.003	$\frac{0.75(0.06)^3}{12} = 0$	3.07	9.42	0.424
7[b]	$0.25(3.7)(0.060)(0.2) = 0.011$	1.00	0.011	$\frac{0.25(3.7)(0.06)^3(0.2)}{12} = 0$	2.14	4.59	0.050
	0.963		3.028	3.30			1.399

$$\bar{Y} = \frac{\sum AY}{\sum A} = \frac{3.028}{0.963} = 3.14 \text{ in.}$$

$$I_Y = I_0 + Ac^2 = 3.30 + 1.399 = 4.70 \text{ in.}^4 \tag{7.1}$$

[a] Includes a 10% bolted efficiency factor.
[b] Includes a 25% bolted efficiency factor and a correction factor for epoxy fiberglass.

The moment of inertia along the X axis can be determined with the use of table 7.2

TABLE 7.2

Item	Area (in.^{2})	X	AX	$I_0 =$ in.^{4}	c	c^2	Ac^2
1	0.415	0.030	0.012	$\frac{(6.92)(0.060)^3}{12} = 0.000$	1.97	3.89	1.61
2	0.415	3.970	1.647	$\frac{(6.92)(0.060)^3}{12} = 0.000$	1.97	3.89	1.61
3[a]	0.016	2.000	0.032	$\frac{(0.10)(0.04)(4.0)^3}{12} = 0.021$	0.00	0.00	0.00
4[a]	0.016	2.000	0.032	$\frac{(0.10)(0.04)(4.0)^3}{12} = 0.021$	0.00	0.00	0.00
5	0.045	0.435	0.019	$\frac{(0.06)(0.75)^3}{12} = 0.001$	1.57	2.46	0.11
6	0.045	3.565	0.160	$\frac{(0.06)(0.75)^3}{12} = 0.001$	1.56	2.44	0.11
7[b]	0.011	2.000	0.022	$\frac{0.25(0.060)(3.7)^3(0.2)}{12} = 0.001$	0.00	0.00	0.00
	0.963		1.924	0.045			3.44

$$\bar{X} = \frac{\sum AX}{\sum A} = \frac{1.924}{0.963} = 2.00 \text{ in.}$$

$$I_X = I_0 + Ac^2 = 0.045 + 3.44 = 3.48 \text{ in.}^4 \tag{7.2}$$

[a]Includes a 10% bolted efficiency factor.
[b]Includes a 25% bolted efficiency factor and a correction factor for epoxy-fiberglass.

For the preliminary analysis, assume the chassis is equivalent to a simply supported beam with a uniform load. The resonant frequency for vibration along the Y axis can then be determined from the standard beam equation.

$$f_n = \frac{\pi}{2}\left(\frac{EI_Y g}{WL^3}\right)^{1/2} \tag{7.3}$$

where $E = 10 \times 10^6$ lb/in.^{2} aluminum

$I_Y = 4.70$ in.^{4} (see Eq. 7.1)

$g = 386$ in./sec^2 gravity

$L = 26.0$ in. length

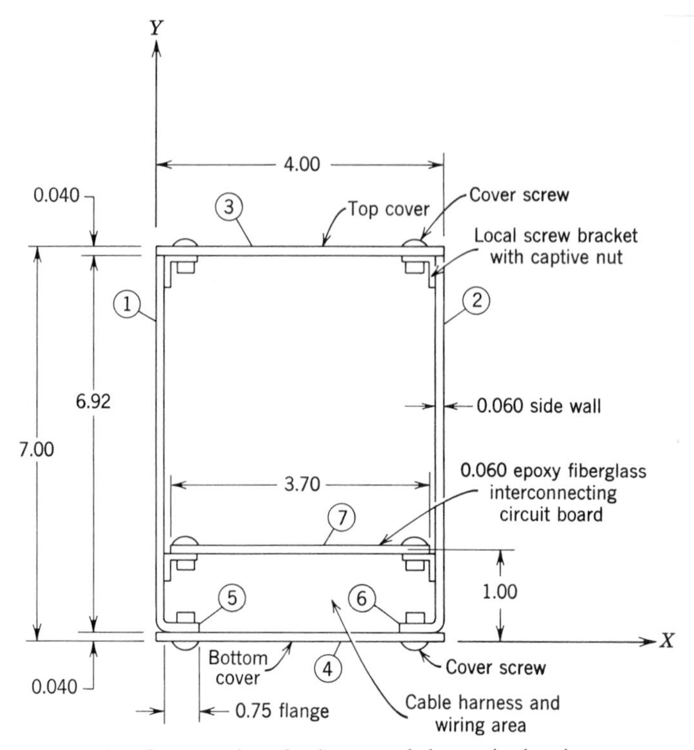

FIGURE 7.8. Cross section of a sheet metal electronic chassis.

W = weight approximated by box volume with an average density of 0.031 lb/in.^{3}

$$W = (4.0)(7.0)(24.75)(0.031) = 21.5 \text{ lb} \tag{7.4}$$

The natural frequency for vibration along the Y axis is

$$f_Y = \frac{\pi}{2}\left[\frac{(10 \times 10^6)(4.70)(386)}{(21.5)(26.0)^3}\right]^{1/2}$$

$$f_Y = 344 \text{ Hz} \tag{7.5}$$

The natural frequency for vibration along the X axis can be determined from the same standard beam equation, using the moment of inertia along the X axis.

$$I_x = 3.48 \text{ in.}^4 \text{ (see Eq. 7.2)}$$

Substituting into Eq. 7.3

$$f_X = \frac{\pi}{2}\left[\frac{(10\times 10^6)(3.48)(386)}{(21.5)(26.0)^3}\right]^{1/2}$$

$$f_X = 296 \text{ Hz} \tag{7.6}$$

7.5. Coupled Modes

The chassis mounts are located at the bottom edge of the box, at each end; this results in a high center of gravity. If an imaginary load is applied at the CG in a direction along the X axis, the chassis will tend to bend in the direction of the load. At the same time, the chassis will tend to rotate in the direction of the load as shown in Fig. 7.9. Since the chassis will tend to bend and rotate simultaneously under the action of a single load, it means that the bending mode will couple with the torsional mode during vibration along the X axis. This coupling will reduce the fundamental resonant mode of the chassis.

If a similar load is applied at the CG in the vertical direction, along the Y axis, the chassis will tend to move vertically with no rotation. This means that the vertical bending mode will probably not couple with the torsional mode.

In order to determine how the bending mode will couple with the torsional mode, the natural frequency of the chassis in the torsional mode will be computed and combined with the natural frequency of the chassis in the bending mode.

The natural frequency of the chassis in the torsional mode can be determined from the torsional frequency equation, by assuming the chassis is equivalent to a single-degree-of-freedom system. Then

$$f_n = \frac{1}{2\pi}\left(\frac{K_\theta}{I_m}\right)^{1/2} \quad \text{(see Eq. 2.27)} \tag{7.7}$$

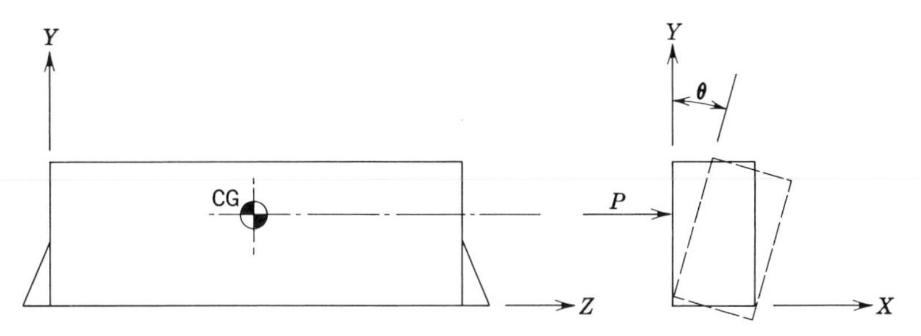

FIGURE 7.9. Testing how vibration modes may tend to couple.

The term K_θ is the torsional spring rate of the chassis. This can be determined by placing a unit torque of 1 lb in. at the CG of the chassis and calculating the angular deflection that results. If the CG of the chassis is at its center, one half of the chassis can be considered as shown in Fig. 7.10.

The angle through which the chassis will rotate can be determined as follows:

$$\theta = \frac{Ma}{GJ} \tag{7.8}$$

where $\frac{T}{2}$ = torsional moment on half a chassis = M

$a = L/2$ half of chassis length

G = Shear modulus

J = Torsional form factor

$$\theta = \frac{\left(\frac{T}{2}\right)\left(\frac{L}{2}\right)}{GJ} = \frac{TL}{4GJ} \tag{7.9}$$

The torsional spring rate of the chassis is defined as

$$K_\theta = \frac{T}{\theta} \tag{7.10}$$

Substituting Eq. 7.10 into Eq. 7.9

$$K_\theta = \frac{4GJ}{L} \tag{7.11}$$

where $G = 4 \times 10^6$ psi aluminum shear modulus

$L = 26.0$ in. chassis length

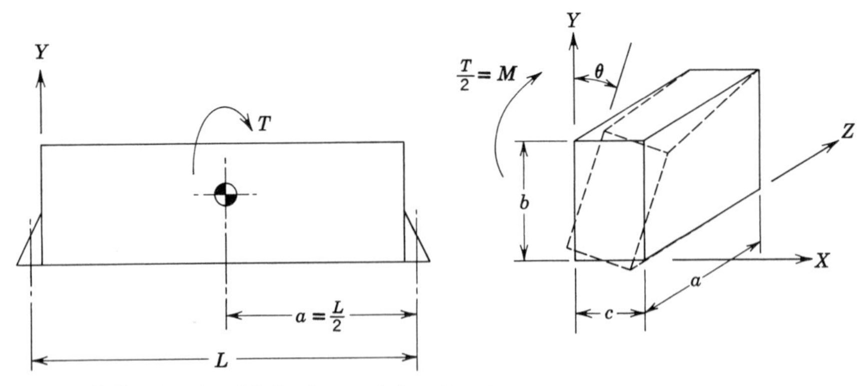

FIGURE 7.10. Torsional deflection mode in a chassis.

$$J = \frac{4A^2}{\oint \frac{ds}{t}} \text{ torsional form factor}$$

A good approximation for the bolted thin-wall rectangular section is (also see Roark [19])

$$J = \frac{I_X + I_Y}{2} = \frac{3.48 + 4.70}{2} = 4.09 \text{ in.}^4$$

Substituting into Eq. 7.11

$$K_\theta = \frac{(4)(4 \times 10^6)(4.09)}{26.0} = 2.52 \times 10^6 \text{ in. lb/rad} \tag{7.12}$$

The mass moment of inertia for the chassis will be taken with respect to its mounting point at the base, where the rotation will occur.

$$I_m = \frac{W}{12g}(4b^2 + c^2) \tag{7.13}$$

where $W = 21.5$ lb weight (see Eq. 7.4)
$g = 386$ in/sec² gravity
$b = 7.0$ in. chassis height
$c = 4.0$ in. chassis width

Substituting into Eq. 7.13

$$I_m = \frac{21.5}{12(386)}[4(7.0)^2 + (4.0)^2]$$

$$I_m = 0.983 \text{ lb in. sec}^2 \tag{7.14}$$

The rotational natural frequency can be determined by substituting Eqs. 7.12 and 7.14 into Eq. 7.7.

$$f_r = \frac{1}{2\pi}\sqrt{\frac{K_\theta}{I_m}} = \frac{1}{2\pi}\left(\frac{2.52 \times 10^6}{0.983}\right)^{1/2}$$

$$f_r = 255 \text{ Hz} \tag{7.15}$$

Dunkerley's equation (see Section 3.2) can be used to determine the approximate fundamental resonant frequency of the coupled mode when bending and torsion are combined. This method is conservative since it will result in a frequency that is slightly lower than the true frequency.

$$\frac{1}{f_c^2} = \frac{1}{f_X^2} + \frac{1}{f_r^2} \tag{7.16}$$

where $f_X = 296$ Hz bending along X axis (see Eq. 7.6)
$f_r = 255$ Hz rotation about base (see Eq. 7.15)

Substituting into Eq. 7.16

$$\frac{1}{f_c^2} = \frac{1}{(296)^2} + \frac{1}{(255)^2} = 0.267 \times 10^{-4}$$

$$f_c = \frac{1}{0.267 \times 10^{-4}} = 193 \text{ Hz} \tag{7.17}$$

Equation 7.17 shows that the fundamental resonant frequency of a chassis, with a low mount and a high CG, can be much lower than expected if coupling is ignored. A lower resonant frequency means greater displacements and stresses which will shorten the fatigue life.

7.6. Dynamic Loads in a Chassis

The dynamic loads developed in a chassis can be determined by considering the dynamic response characteristics of the chassis. If a sinusoidal vibration input is considered, the inertia loads developed at resonance will be maximum at the center of a uniformly loaded chassis and minimum at the supported ends. The transmissibility at the supported ends will be unity (one) if there is no relative motion between the chassis and its supports.

A good approximation for the dynamic load can be obtained with a sinusoidal load distribution acting along the chassis (Fig. 7.11).

The unit dynamic load distribution on the chassis can be represented by the expression

$$p = p_0 \sin \frac{\pi Z}{L} \tag{7.18}$$

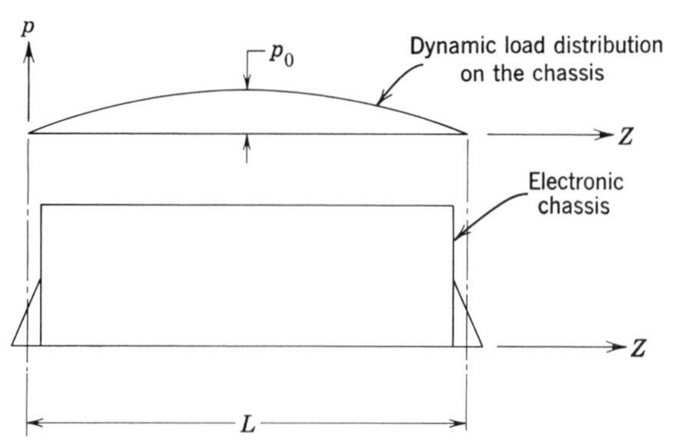

FIGURE 7.11. Vertical load distribution in a chassis.

The differential equation for this loading can be represented by the equation

$$\frac{d^4Y}{dZ^4} = \frac{p_0}{EI} \sin \frac{\pi Z}{L} \tag{7.19}$$

The boundary conditions for this equation will be satisfied by using the following expression where A is the deflection form factor.

$$Y = A \sin \frac{\pi Z}{L} \tag{7.20}$$

Taking the first four derivatives as indicated

$$\frac{dY}{dZ} = A \frac{\pi}{L} \cos \frac{\pi Z}{L}$$

$$\frac{d^2Y}{dZ^2} = -A \frac{\pi^2}{L^2} \sin \frac{\pi Z}{L} \tag{7.21}$$

$$\frac{d^3Y}{dZ^3} = -A \frac{\pi^3}{L^3} \cos \frac{\pi Z}{L}$$

$$\frac{d^4Y}{dZ^4} = A \frac{\pi^4}{L^4} \sin \frac{\pi Z}{L} \tag{7.22}$$

Substituting Eq. 7.22 into Eq. 7.19 and solving for A

$$A \frac{\pi^4}{L^4} \sin \frac{\pi Z}{L} = \frac{p_0}{EI} \sin \frac{\pi Z}{L}$$

$$A = \frac{p_0 L^4}{\pi^4 EI} \tag{7.23}$$

Substituting Eq. 7.23 into Eq. 7.20 gives the deflection in the chassis at any point due to the unit dynamic load.

$$Y = \frac{p_0 L^4}{\pi^4 EI} \sin \frac{\pi Z}{L}$$

The maximum deflection, due to the unit dynamic load, will occur at the center of the chassis where $Z = L/2$.

$$Y_{\max} = \frac{p_0 L^4}{\pi^4 EI} \tag{7.24}$$

The maximum deflection in the chassis can also be determined for the bending resonant mode by considering it to be equivalent to a single-degree-of-freedom system (Fig. 7.12).

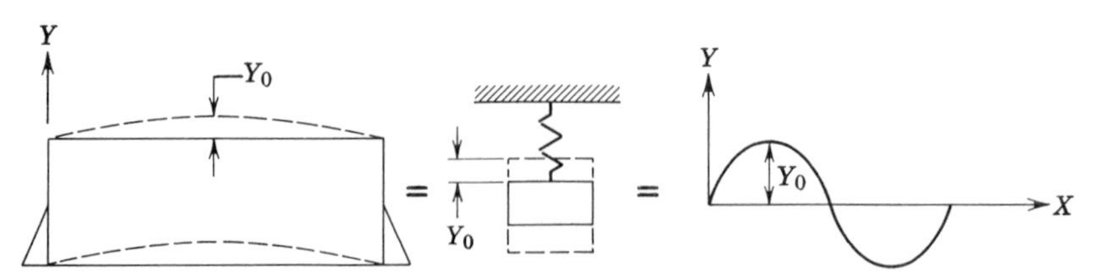

FIGURE 7.12. Chassis simulated as a single-degree-of-freedom system.

The maximum displacement can be determined from Eq. 2.42 as follows:

$$Y_0 = \frac{9.8G}{f^2} \tag{7.25}$$

Since Eqs. 7.24 and 7.25 both represent the maximum displacement of the chassis, they must be equal.

$$\frac{p_0 L^4}{\pi^4 EI} = \frac{9.8G}{f^2}$$

$$p_0 = \frac{9.8\pi^4 EIG}{L^4 f^2} \tag{7.26}$$

Equation 7.26 can be used to determine the maximum unit dynamic load acting on the chassis. The moment of inertia, I, can be determined from the natural-frequency equation for a uniform, simply supported beam, as shown by Eq. 7.3. The moment of inertia then becomes

$$I = \frac{4f^2 W L^3}{\pi^2 E g}$$

Substituting this expression into Eq. 7.26

$$p_0 = \frac{9.8\pi^4 EG}{L^4 f^2}\left(\frac{4f^2 W L^3}{\pi^2 E g}\right) = \frac{WG}{L} \tag{7.27}$$

for a 4-G-peak sinusoidal-vibration input to the electronic box

where $G_{in} = 4\ G$ peak input along the X axis
$Q = 8$ based on test data for this type of chassis and mount, with removable top and bottom covers
$G = G_{in} \times Q = 4(8) = 32\ G$ acceleration
$W = 21.5$ lb weight (see Eq. 7.4)
$L = 26.0$ in. length (see Fig. 7.6)

Substituting into Eq. 7.27, the maximum unit dynamic load becomes

$$p_0 = \frac{(21.5)(32)}{26.0} = 26.4 \text{ lb/in.} \tag{7.28}$$

The total dynamic load developed in the chassis during vibration along the X axis can be determined from the sinusoidal force distribution acting on the chassis as shown in Fig. 7.13. The area under the curve represents the total dynamic load acting on the chassis. This load must also pass through the chassis to the supports.

The total dynamic load is

$$P_d = \int_0^L p\, dZ \tag{7.29}$$

where

$$p = p_0 \sin \frac{\pi Z}{L} \qquad \text{(see Eq. 7.18)}$$

$$P_d = p_0 \int_0^L \sin \frac{\pi Z}{L}\, dZ = \frac{p_0 L}{\pi}\left(-\cos \frac{\pi Z}{L}\right)_0^L$$

$$P_d = \frac{2 p_0 L}{\pi} \tag{7.30}$$

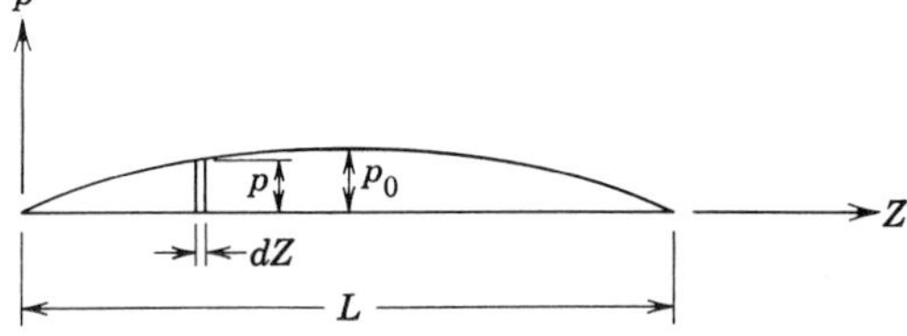

FIGURE 7.13. Dynamic load distribution in the chassis.

This conservatively assumes all of the internal masses are in resonance and in the same phase at the same time during vibration along the X axis where $p_0 = 26.4$ lb/in. (see Eq. 7.20), and $L = 26.0$ in. (see Fig. 7.6).

Substituting into Eq. 7.30, the total dynamic load acting at the CG of the chassis becomes:

$$P_d = \frac{2(26.4)(26.0)}{\pi} = 438 \text{ lb} \tag{7.31}$$

The bending moment for the chassis at any point can be determined from the following expression:

$$M = -EI \frac{d^2 Y}{dZ^2} \tag{7.32}$$

Substituting Eqs. 7.21 and 7.23 into Eq. 7.32

$$M = -EI\left(-\frac{\pi^2}{L^2}\left[\frac{p_0 L^4}{\pi^4 EI}\right]\sin\frac{\pi Z}{L}\right)$$

$$M = \frac{p_0 L^2}{\pi^2}\sin\frac{\pi Z}{L}$$

The maximum moment will occur at the center of the chassis where Z is $L/2$.

$$M_{\max} = \frac{p_0 L^2}{\pi^2} \tag{7.33}$$

where $p_0 = 26.4$ lb/in. dynamic load (see Eq. 7.28)
$L = 26.0$ in. length (see Fig. 7.6)

Substituting into Eq. 7.33 for the bending moment at the center of the chassis

$$M_{\max} = \frac{(26.4)(26.0)^2}{\pi^2} = 1820 \text{ lb in.} \tag{7.34}$$

7.7. Bending Stresses in the Chassis

The bending stresses in the side panels during vibration along the X axis can be determined from the standard bending stress equation

$$S_b = \frac{Mc}{I_X} \tag{7.35}$$

where $M = 1820$ lb in. (see Eq. 7.34)
$c = 2.0$ in. (see Fig. 7.8)
$I_X = 3.48$ in.^{4} (see Eq. 7.2)

Substituting into Eq. 7.35, the bending stress is

$$S_b = \frac{(1820)(2.0)}{3.48} = 1045 \text{ lb/in.}^2 \tag{7.36}$$

This is a relatively low stress level, so it would seem that the design is satisfactory. However the side panels must be checked to determine whether they can carry this stress load without buckling.

The critical buckling stress for the side walls of the chassis, under a compressive bending load, can be determined from the following equation [13]:

$$S_{\text{cr}} = KE\left(\frac{t}{b}\right)^2 \tag{7.37}$$

The K factor is a function of the edge restraints of the side panel subjected to the compressive bending load. A more detailed sketch of this panel is shown in Fig. 7.14.

The top edge of the side panel must be analyzed as a free edge, since there is no support. The bottom of the side panel can be considered supported if the bottom cover has only one screw in this area to keep this edge from rotating. The side panels are riveted to bulkheads which will prevent translation, but not rotation, so these edges must be considered as supported.

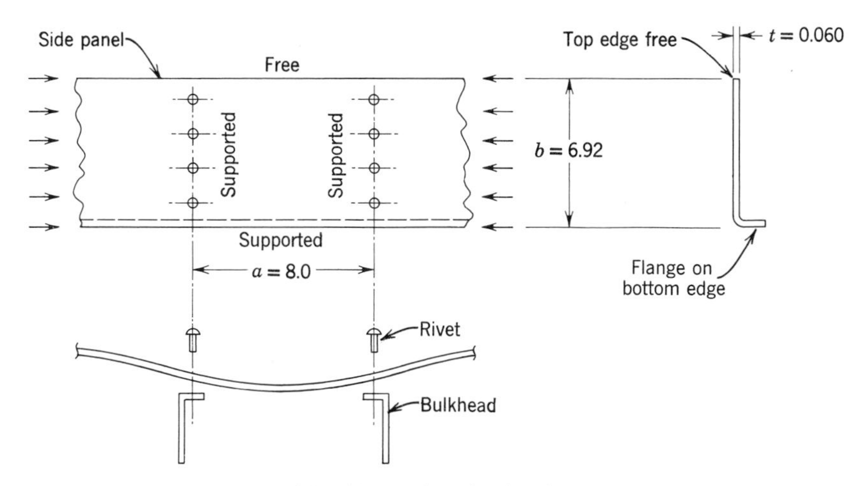

FIGURE 7.14. Dimensions of the side panel on the chassis.

The side panel will be analyzed as a flat plate, supported on opposite ends, with one side free and one side supported. The length-to-width ratio, or aspect ratio,

$$\frac{a}{b} = \frac{8.0}{6.92} = 1.15$$

From the curve shown in Fig. 7.15b

$$K = 1.2 \tag{7.38}$$

The following information is required:

$E = 10 \times 10^6$ psi modulus of elasticity, aluminum
$t = 0.060$ in. thickness
$b = 6.92$ in. width of panel

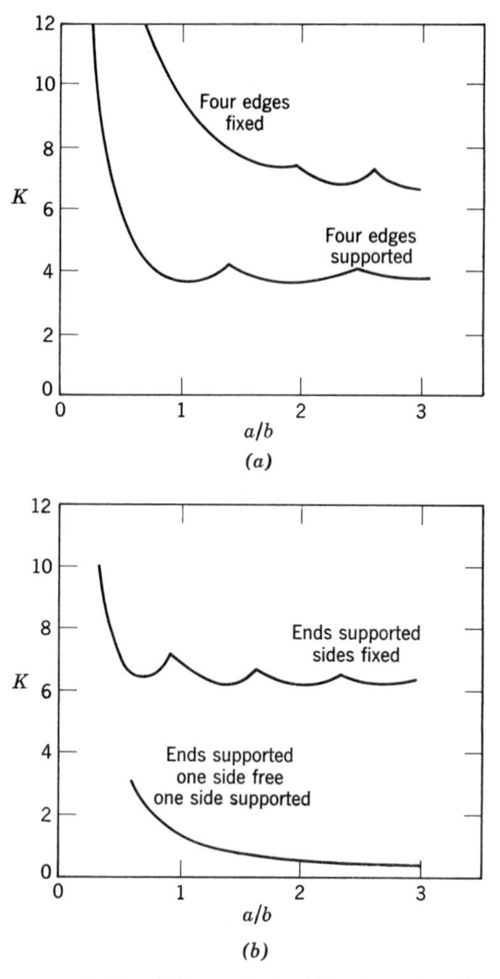

FIGURE 7.15. *K* factors for buckling due to bending.

Substituting into Eq. 7.37 for the critical buckling stress

$$S_{cr} = 1.2(10 \times 10^6)\left(\frac{0.060}{6.92}\right)^2$$

$$S_{cr} = 902 \text{ lb/in.}^2 \tag{7.39}$$

7.8. Buckling Stress Ratio for Bending

Stress ratios can be used to determine the margin of safety for buckling in the side panels, due to the compressive bending stress developed during vibration along the X axis [27].

The buckling stress ratio due to bending is defined as

$$R_b = \frac{S_b}{S_{cr}} \tag{7.40}$$

where

$S_{cr} = 902$ psi buckling stress (see Eq. 7.39)

$S_b = 1045$ psi panel bending stress (see Eq. 7.36)

Substituting into the above equation, this ratio is

$$R_b = \frac{1045}{902} = 1.16 \tag{7.41}$$

The margin of safety (MS) for buckling can be determined from the standard equation [27]

$$\text{MS} = \frac{1}{R_b} - 1 \tag{7.42}$$

$$\text{MS} = \frac{1}{1.16} - 1 = -0.138 \tag{7.43}$$

The negative margin of safety shows the side panel will buckle during vibration.

Obviously, if the side panels will buckle during vibration, they cannot carry the compressive load. This means the moment of inertia of the cross section, shown by Eq. 7.2 is no longer valid. The chassis, therefore, will have a coupled resonant frequency lower than the 193 Hz shown by Eq. 7.17. Also, the dynamic stresses will increase because the displacement shown by Eq. 7.25 will increase. Increasing the stresses can result in more rapid failures, since the fatigue life depends upon the stress level, as shown in Chapter 10.

A simple method for increasing the stiffness of the side panels, as well as the stiffness of the entire chassis, is to form a flange in the top edge of the panel (Fig. 7.16). If the top cover is then fastened at several spots

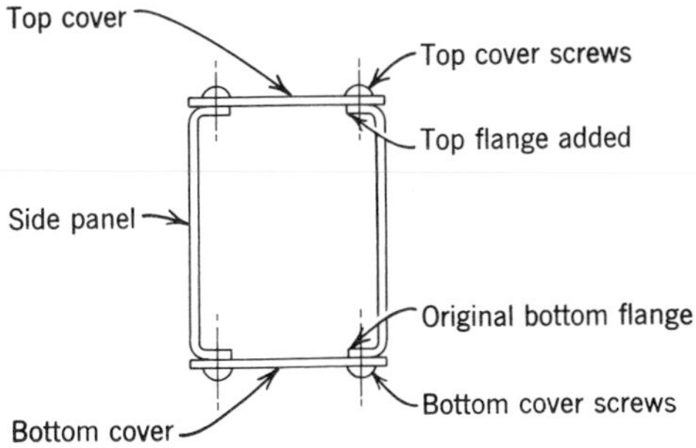

FIGURE 7.16. Cross section of the chassis with top flanges added.

along this flange, the side panel stiffness will be greatly increased. Enclosing a sheet metal chassis in this manner makes more effective use of the covers as shear panels. This will substantially increase the overall stiffness of the chassis with very little added weight. These top flanges will increase the stiffness of the side panels by making them equivalent to a flat plate that is supported on all four sides. The K factor for this geometry, as shown in Fig. 7.15, is

$$K = 3.7 \tag{7.44}$$

Substituting into Eq. 7.37 for the critical buckling stress

$$S_{cr} = 3.7(10 \times 10^6)\left(\frac{0.060}{6.92}\right)^2$$

$$S_{cr} = 2780 \text{ psi} \tag{7.45}$$

Adding the small flange to the top of the side panels, as shown in Fig. 7.16, will have very little effect on the moment of inertia of the cross-section of the chassis as is shown by Eqs. 7.1 and 7.2. Conservatively assuming that these values do not change with the addition of a small flange, there will also be no change in the dynamic load and the bending stress shown by Eq. 7.36. The buckling stress ratio due to bending, however, will change.

Substituting Eqs. 7.36 and 7.45 into Eq. 7.40, the buckling stress ratio due to bending, for the stiffer chassis, is

$$R_b = \frac{S_b}{S_{cr}} = \frac{1045}{2780} = 0.376 \tag{7.46}$$

Since the bending stress ratio for buckling is less than 1.0, it is obvious the side panels will no longer buckle because of an excessive compressive bending stress. There is, however, another stress acting on the side panels, in addition to the bending stress, which can also cause buckling. This is the torsional shear stress developed by the dynamic load during vibration along the X axis, as shown in Fig. 7.9. This torsional shear stress is due to the overturning moments developed by a high CG.

The buckling margin of safety for the side panels can be determined by combining the individual buckling stress ratios for bending and for torsional shear. The buckling stress ratio for bending was shown by Eq. 7.46. The buckling stress ratio for shear can be determined by analyzing the dynamic overturning moments and the resulting shear stresses that are developed in the side panels.

7.9. Torsional Stresses in the Chassis

During vibration along the X axis, overturning moments will force the chassis to rotate about its base in a torsional mode that will couple with the bending mode. This is due to a high CG which is the result of low mounting lugs at each end of the chassis. Torsional shear stresses will then develop as the chassis rotates about a longitudinal axis as shown in Fig. 7.17.

The cross-section of the electronic chassis, as shown in Figs. 7.8 and 7.16, is characteristic of the type of construction generally used in the electronics industry where thin sheet-metal boxes have covers that are fastened with a few screws. This does not form a rigid box structure. Since the top and bottom covers will slide as the box twists, these covers are obviously not 100% effective in eliminating relative motion. Therefore the box cannot be analyzed as a closed section. At the same time, the covers do provide some support, so the box cannot be analyzed as an open section.

Under these conditions, an efficiency factor can be used for the bolted covers, based on the ability of the bolted covers to resist relative sliding during vibration. This efficiency factor, as explained in the previous section on bolted covers, is a function of the number of bolts, bolt size, panel thickness, and so on. With this efficiency factor it is possible to approximate the shear flow and the shear stress in each panel by using the standard shear flow equations for a closed section. Equilibrium equations can then be written for the chassis cross-section using the shear center method, with the shear flow in each panel.

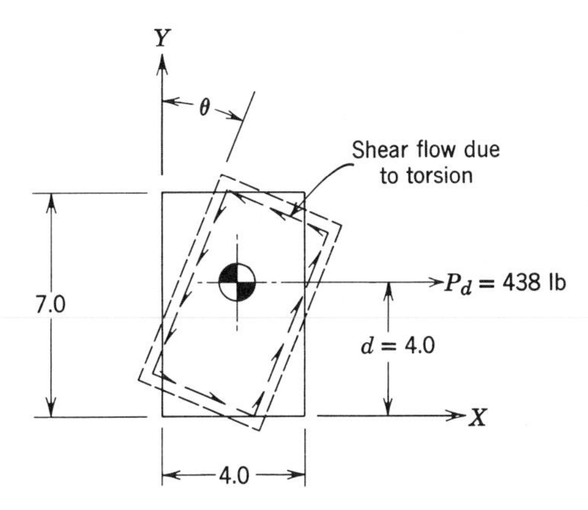

FIGURE 7.17. Shear flow in the chassis due to torsion.

Consider the cross-section of the chassis shown in Fig. 7.17. Due to symmetry, the shear center will lie at the center of the chassis.

The shear flow equation can be written in the form[13]

$$T = 2 \sum A_i q_i \tag{7.47}$$

where A = area of the triangles formed by the straight lines joining the shear center and the extremities of the panels

$A_1 = 2(\frac{1}{2})(7.0)(2.0) = 14.0$ in.^{2} (two sections)

$A_2 = 2(\frac{1}{2})(4.0)(3.5) = 14.0$ in.^{2} (two sections)

q = shear flow in each panel, lb/in.

P_d = 438.0 lb dynamic load (see Eq. 7.31)

d = 4.0 in. distance from mount to CG

$T = P_d \times d = 438(4.0) = 1752$ lbin. torque developed by a 4-G peak input along the X axis.

Substituting into Eq. 7.47

$$T = 2(A_1 q_1 + A_2 q_2)$$

$$1752 = 2(14) q_1 + 2(14) q_2$$

$$q_1 + q_2 = \frac{1752}{2(14)} = 62.5 \text{ lb/in.} \tag{7.48}$$

Notice that the epoxy fiberglass interconnecting circuit board was not included in the system (see Fig. 7.8). This is due to the intermittent spacing of the screw-mounting brackets for this epoxy board on the side walls of the chassis. The shear flow in the side walls will be affected only in the local area of the brackets. Ignoring the epoxy board will have very little effect on the shear flow in the system and it simplifies the problem substantially.

A second equation involving q_1 and q_2 can be established by considering the shear flow in each panel along with the bolted efficiency factor for each panel. Consider the cross-section of the chassis in Fig. 7.18. The shear flow will distribute itself in direct proportion to the shear-carrying ability of each panel. The panels that have a high shear stiffness will carry a higher proportion of the torsional load. Since the shear-carrying ability of each panel is related to its cross-sectional area and its bolted efficiency factor, the shear stiffness for each panel can be written as (see Fig. 7.8)

$$\left.\begin{aligned} K_1 &= A_1 e_1 \\ K_2 &= A_2 e_2 \end{aligned}\right\} \tag{7.49}$$

where $A_1 = (6.92)(0.060)(2) = 0.83$ in.^{2} area of 2 side panels

$A_2 = (4.0)(0.040)(2) = 0.32$ in.^{2} top and bottom panel area

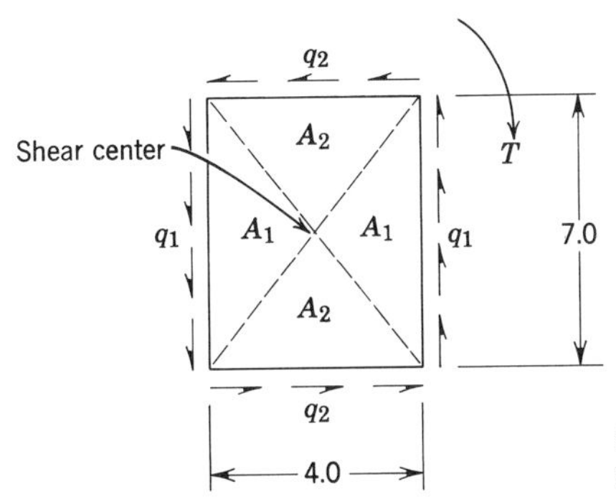

FIGURE 7.18. Shear flow pattern in the chassis cross section.

$e_1 = 100\%$ efficiency for the side panels
$e_2 = 10\%$ bolted efficiency for covers on top and bottom

Substituting into Eq. 7.49

$$K_1 = (0.83)(1.00) = 0.83$$

$$K_2 = (0.32)(0.10) = 0.032$$

The shear flow in each panel will be proportional to the shear stiffness.

$$\frac{q_1}{q_2} = \frac{K_1}{K_2} = \frac{0.83}{0.032} = 26.0$$

or

$$q_1 = 26.0q_2 \tag{7.50}$$

Substituting Eq. 7.50 into Eq. 7.48 and solving

$$26.0q_2 + q_2 = 62.5$$

$$\left.\begin{aligned} q_2 &= \frac{62.5}{27.0} = 2.32 \text{ lb/in.} \\ q_1 &= 26.0q_2 = 60.4 \text{ lb/in.} \end{aligned}\right\} \tag{7.51}$$

This shows that the side panels will carry a much greater shear load than the bolted top and bottom covers.

The shear stress in the side panel, induced by the torsional dynamic load, can be determined from the standard stress equation.

$$S_s = \frac{q_1}{t} \tag{7.52}$$

where $q_1 = 60.4$ lb/in. shear flow (see Eq. 7.51)
$t = 0.060$ in. panel thickness

Then

$$S_s = \frac{60.4}{0.060} = 1007 \text{ lb/in.}^2 \tag{7.53}$$

This is only an approximate value, since it is based on a cross-section with sharp corners and bolted members. However it still provides a means for evaluating the characteristics of the panels, even if approximate, to determine whether they may buckle due to the shear load.

7.10. Buckling Stress Ratio for Shear

The critical buckling stress, due to shear in the side panels, can be determined by considering the four edges of the side panel to be simply supported when there is a flange on the top and bottom edges of this panel. (See Figs. 7.14 and 7.16.)

With a flange on the top edge of the side panel, the shear flow will appear as shown in Figs. 7.17 and 7.19.

The critical buckling stress for the side walls, due to the action of the shear load, can be taken in the same form as the critical bulking stress for a compressive load. This equation becomes [13]

$$S_{\text{scr}} = K_s E \left(\frac{t}{b}\right)^2 \tag{7.54}$$

The K_s factor is a function of the edge restraints and the aspect ratio of the side panel subjected to shear. Figure 7.20 can be used to determine the K_s factor for flat sheet-metal panels simply supported on four sides.

$$\frac{a}{b} = \frac{8.0}{6.92} = 1.15$$

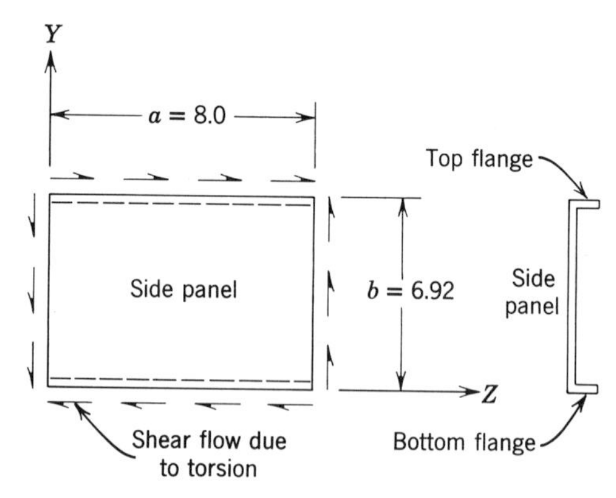

FIGURE 7.19. Shear flow pattern in the chassis side wall.

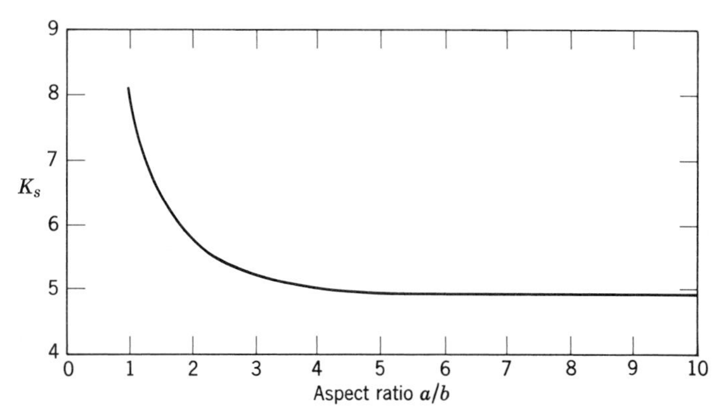

FIGURE 7.20. K_s factors for buckling due to shear.

The curve in Fig. 7.20 for this geometry is

$$K_s = 7.4 \tag{7.55}$$

also $E = 10 \times 10^6$ psi modulus of elasticity for aluminum
$t = 0.060$ in. thickness
$b = 6.92$ in. shortest dimension of panel

Substituting into Eq. 7.54 for the critical buckling stress

$$S_{scr} = 7.4(10 \times 10^6)\left(\frac{0.060}{6.92}\right)^2$$

$$S_{scr} = 5560 \text{ lb/in.}^2 \tag{7.56}$$

The buckling stress ratio for shear due to torsion is

$$R_s = \frac{S_s}{S_{scr}} \tag{7.57}$$

where $S_{scr} = 5{,}560$ psi (see Eq. 7.56)
$S_s = 1{,}007$ psi (see Eq. 7.53)

Substituting into Eq. (7.57)

$$R_s = \frac{1{,}007}{5{,}560} = 0.181 \tag{7.58}$$

7.11. Margin of Safety for Buckling

The margin of safety (MS) for buckling due to the combined action of the compressive bending and shear stresses for the stiffer chassis can be determined from the standard equation [27]

$$\text{MS} = \frac{1}{(R_b{}^2 + R_s{}^2)^{1/2}} - 1 \tag{7.59}$$

where $R_b = 0.376$ bending ratio (see Eq. 7.46)
$R_s = 0.181$ shear ratio (see Eq. 7.58)

Substituting into Eq. 7.59

$$\text{MS} = \frac{1}{[(0.376)^2 + (0.181)^2]^{1/2}} - 1$$

$$\text{MS} = +1.40 \tag{7.60}$$

The large positive margin of safety shows the side panels, with the flanges, will not buckle during vibration along the X axis with a 4-G peak input.

7.12. Dynamic Loads in the Chassis Mounting Lugs

Since the dynamic load must pass through the chassis to the end mounting lugs, the mounting lugs must be made strong enough to provide a good fatigue life. The dynamic loads can be determined from the acceleration G forces and the geometry of the supports, or lugs, as shown in Fig. 7.21.

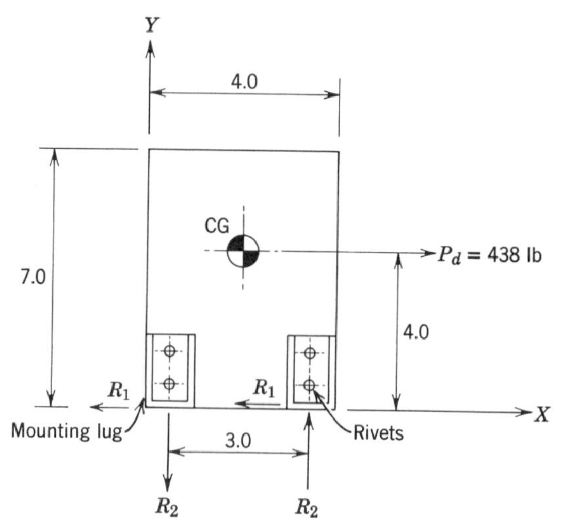

FIGURE 7.21. End view of chassis showing mounting lug locations.

Because of the high center of gravity (CG) and the narrow chassis, the most critical forces are developed during vibration along the X axis. The dynamic loads in the mounting lugs can be determined by considering the sum of all the forces acting in the X direction.

$$\sum F_X = 0$$

Then for four mounting lugs

$$R_1 = \frac{P_d}{4}$$

where $P_d = 438$ lb (see Eq. 7.31)

$$R_1 = \frac{438}{4} = 109.5 \text{ lb} \tag{7.61}$$

The vertical reactions R_2 can be determined by considering the moments about the base of the chassis. Since there are lugs at both ends of the chassis

$$\sum M = 0$$

$$R_2 = \frac{(438)(4.0)}{(2)(3.0)} = 292 \text{ lb} \tag{7.62}$$

Superposition is used to find the shear forces in the rivets that fasten the mounting lugs to the chassis. The horizontal force is considered first, then the vertical force, starting with a two-rivet installation, as shown in Fig. 7.22.

The horizontal force can be broken up into two components, direct shear and shear induced by an overturning moment.

The reaction due to direct shear is

$$R_3 = \frac{109.5}{2} = 54.7 \text{ lb} \tag{7.63}$$

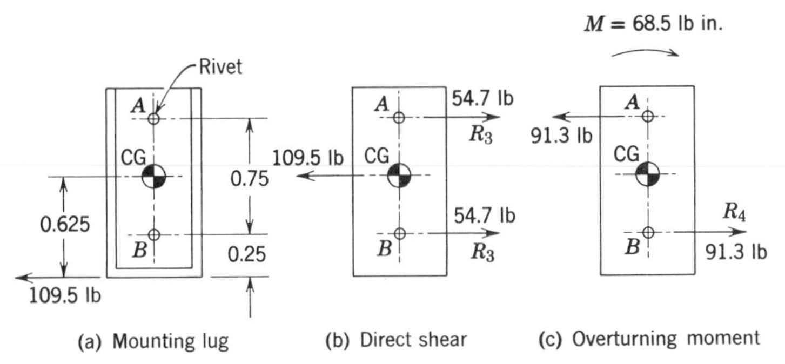

FIGURE 7.22. Dynamic loads in the chassis mounting lugs.

The reaction due to the overturning moment can be determined by taking moments about the center of gravity of the rivet pattern.

$$R_4 = \frac{(0.625)(109.5)}{0.75} = 91.3 \text{ lb} \tag{7.64}$$

The vertical force will divide equally between the two rivets in each mounting lug, as shown in Fig. 7.23.

$$R_5 = \frac{292}{2} = 146 \text{ lb} \tag{7.65}$$

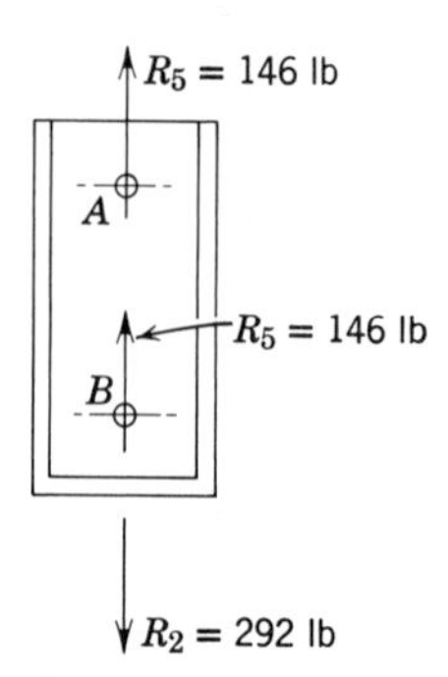

FIGURE 7.23. Vertical forces acting on the mounting lug rivets.

An examination of the dynamic loads induced in the mounting lug rivets, as shown in Figs. 7.22 and 7.23, shows the maximum loads will occur in rivet B. These loads are shown in Fig. 7.24.

The resultant shear load in rivet B is

$$R_r = [(146)^2 + (54.7 + 91.3)^2]^{1/2} = 206.5 \text{ lb} \tag{7.66}$$

Assume a 5/32-in. diameter rivet is used. The shear stress developed in rivet B is

$$S_s = \frac{R_r}{A} \tag{7.67}$$

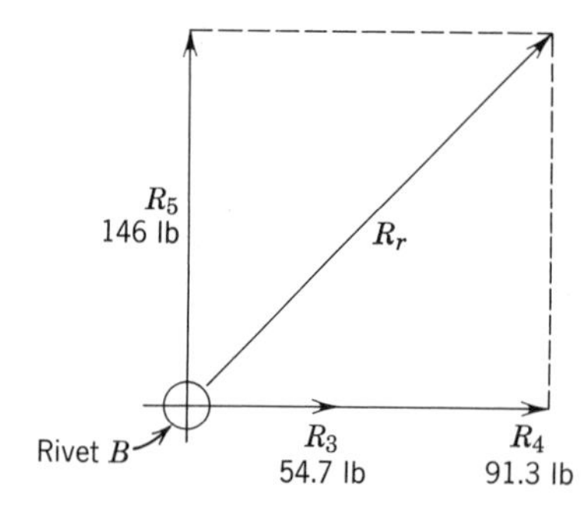

FIGURE 7.24. Dynamic loads acting on the mounting lug rivet.

where $R_r = 206.5$ lb (see Eq. 7.66)

$$A = \frac{\pi}{4}(0.156)^2 = 0.0191 \text{ in.}^2 \text{ rivet area}$$

Substituting into the equation above

$$S_s = \frac{206.5}{0.0191} = 10{,}810 \text{ lb/in.}^2 \quad (7.68)$$

Considering the rivet material to be a 2017–T31 aluminum alloy, the ultimate shear strength will be 34,000 psi. The fatigue strength at 500 million cycles will be about 11,300 psi, as shown by the S–N curve in Fig. 10.9*a*.

7.13. Margin of Safety for Fatigue

Since the entire dynamic load must pass through the mounting lugs on the chassis, it is desirable to provide a fatigue life of at least 5×10^8 cycles for the structure in this area. Therefore the margin of safety will be based

FIGURE 7.25. A long electronic box supported at both ends and containing a cathode-ray-tube display (courtesy Norden division of United Aircraft).

on the endurance limit of the material, which in this case is established at 5×10^8 cycles.

The margin of safety for the rivet in shear can be determined from the standard equation

$$\text{MS} = \frac{S_e}{S_s} - 1 \qquad (7.69)$$

where $S_e = 11{,}300$ psi endurance limit in shear for the rivet
$S_s = 10{,}810 \text{ lb/in.}^2$ (see Eq. 7.68)

Substituting into Eq. 7.69

$$\text{MS} = \frac{11{,}300}{10{,}810} - 1 = +0.045 \qquad (7.70)$$

The small positive margin of safety shows the $\frac{5}{32}$-in.-diameter rivet will be marginal. Since the mounting lugs are critical items in the primary load path, the attachment points should be made stronger. Consider a $\frac{3}{16}$-in.-diameter rivet of the same material. Then from Eq. 7.67

$$S_s = \frac{R_r}{A} \qquad (7.71)$$

where $R_r = 206.5$ lb (see Eq. 7.66)

$$A = \frac{\pi(0.187)^2}{4} = 0.0276 \text{ in.}^2 \qquad (7.72)$$

Substituting into the above equation

$$S_s = \frac{206.5}{0.0276} = 7490 \text{ lb/in.}^2 \qquad (7.73)$$

The margin of safety for this rivet, in shear, can be determined from Eq. 7.69 with the use of the 11,300-psi endurance limit.

$$\text{MS} = \frac{S_e}{S_s} - 1 = \frac{11{,}300}{7490} - 1 = +0.51 \qquad (7.74)$$

The large positive margin of safety shows that the $\frac{3}{16}$-in. diameter rivet will be satisfactory in shear fatigue.

Some small fabricating shops may have trouble bucking a large rivet, so the smaller rivet, in a three-rivet pattern, may be desirable. A sketch of a three-rivet pattern using the $\frac{5}{32}$-in.-diameter rivets is shown in Fig. 7.26. The method for determining the dynamic loads and stresses is the same as the methods previously outlined.

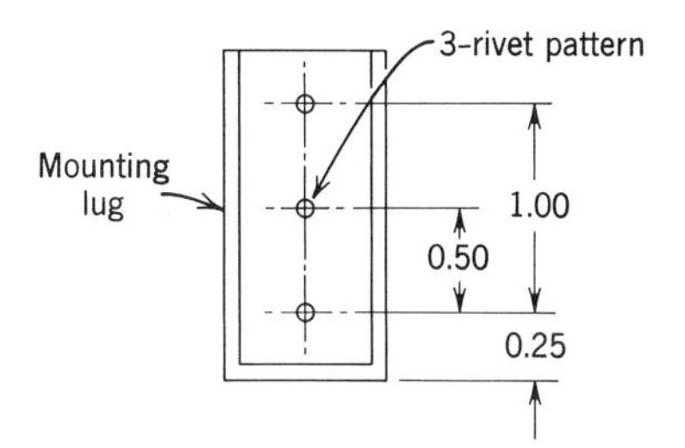

FIGURE 7.26. A mounting lug with a three rivet pattern.

7.14. RIVET-BEARING STRESSES

The bearing stresses developed in the end walls of the chassis must also be examined to determine whether the rivets will function properly. Considering the $\frac{3}{16}$-in.-diameter rivets, the bearing stress in the 0.060-in.-thick aluminum end walls can be determined for the two-rivet design from the following equation:

$$S_{\text{br}} = \frac{R_r}{A_{\text{br}}} \tag{7.75}$$

where $R_r = 206.5$ lb (see Eq. 7.66)
$A_{\text{br}} = (0.187)(0.060) = 0.0112$ in.^{2} bearing area

Substituting into the equation

$$S_{\text{br}} = \frac{206.5}{0.0112} = 18{,}450 \text{ lb/in.}^2 \tag{7.76}$$

The margin of safety can be determined with the use of Eq. 7.69 considering the end panels to be made of 6061-T6 aluminum alloy where $S_e = 15{,}000$ lb/in.^{2} endurance limit based on 5×10^8 cycles as shown in Fig. 10.14*b*,

$$S_{\text{br}} = 18{,}540 \text{ lb/in.}^2 \text{ (see Eq. 7.76).}$$

Substituting into Eq. 7.69

$$\text{MS} = \frac{S_e}{S_{\text{br}}} - 1 = \frac{15{,}000}{18{,}450} - 1 = -0.187 \tag{7.77}$$

The negative margin of safety indicates that the two-rivet design will be unsatisfactory for the bearing condition. Although the shear stresses were shown to be safe for the two-rivet design, the three-rivet pattern in Fig. 7.26 will be a much better choice.

7.15. Center-of-Gravity Mount

If a CG mount is used instead of a base mount, the lateral bending mode will not couple with the torsional mode during vibration along the X axis, as shown in Figs. 7.9 and 7.10. Instead, these resonant modes will be independent of each other (Fig. 7.27).

The natural frequency in the vertical direction, along the Y axis, will be the same as the frequency shown by Eq. 7.5: 344 Hz.

The natural frequency in the lateral direction, along the X axis, will be the same as the frequency shown by Eq. 7.6: 296 Hz.

The natural frequency of the chassis in the torsional mode can still be determined from Eq. 7.7, but the mass moment of inertia, as shown by Eq. 7.13 will change. Now the mass moment of inertia must be taken through the CG as follows:

$$I_m = \frac{W}{12g}(b^2 + c^2) \tag{7.78}$$

$$I_m = \frac{21.5}{12(386)}[(7.0)^2 + (4.0)^2] = 0.302 \text{ lb in sec}^2 \tag{7.79}$$

Substituting Eqs. 7.12 and 7.79 into Eq. 7.7 for the uncoupled rotational natural frequency

$$f_R = \frac{1}{2\pi}\left(\frac{K_\theta}{I_m}\right)^{1/2} = \frac{1}{2\pi}\left(\frac{2.52 \times 10^6}{0.302}\right)^{1/2}$$

$$f_R = 460 \text{ Hz} \tag{7.80}$$

With a CG mount, the bending resonant frequency and the rotational resonant frequency will be uncoupled. Therefore the dynamic stresses in the chassis must be examined for the bending resonant mode acting alone and for the rotational resonant mode acting alone.

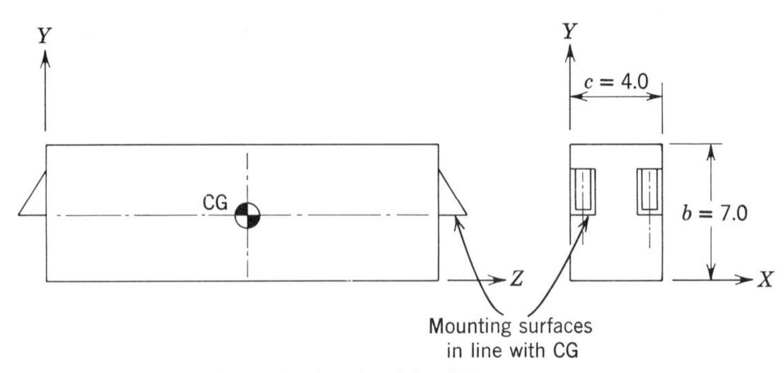

FIGURE 7.27. An electronic chassis with a CG mount.

Considering the bending resonant mode first, with a 4-G sinusoidal-vibration input and a CG mount, if the transmissibility at resonance is still 8, then Eq. 7.28 stays the same and the total dynamic load, as shown by Eq. 7.31 is still 438 lb. The bending stress of 1045 psi, as shown in Eq. 7.36, will still buckle the side panel when there is no flange at the top. This will result in a negative margin of safety, as shown in Eq. 7.43.

If a flange is added to the top edge of the side panel, as shown in Fig. 7.16, then the buckling stress ratio due to bending alone is 0.376 as shown by Eq. 7.46. The margin of safety for buckling, due to bending alone, can then be determined as follows:

$$\text{MS} = \frac{1}{R_b} - 1 \tag{7.81}$$

Substituting Eq. 7.46 into Eq. 7.81

$$\text{MS} = \frac{1}{0.376} - 1 = +1.66 \tag{7.82}$$

The large positive margin of safety shows that the side panels, with top and bottom flanges, will not buckle during vibration with a 4-G-peak acceleration input.

The dynamic loads in the chassis mounting lugs at each end of the chassis can be determined when the bending mode acts alone, by considering the forces in the CG mount as shown in Fig. 7.28.

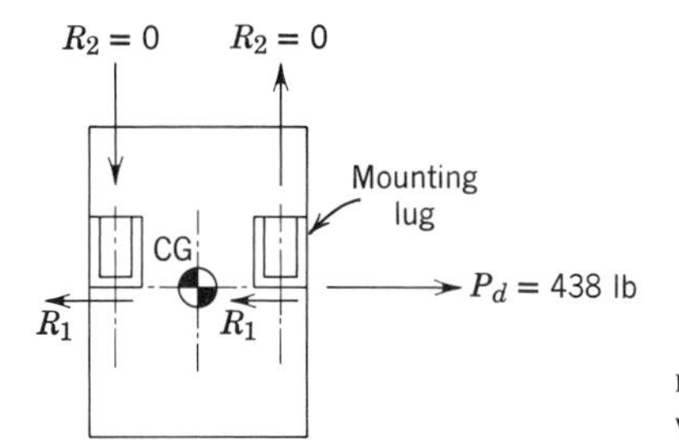

FIGURE 7.28. Dynamic loads acting on a chassis with a CG mount.

Considering the sum of the forces acting in the X direction

$$\sum F_X = 0$$

Then for four mounting lugs

$$R_1 = \frac{438}{4} = 109.5 \text{ lb} \tag{7.83}$$

There is no overturning moment because the center of gravity (CG) is in the mounting plane of the electronic box. Therefore

$$\sum M = 0 \qquad \text{and} \qquad R_2 = 0 \tag{7.84}$$

The shearing forces in the mounting-lug rivets are as shown in Fig. 7.22, due to the uncoupled bending mode of the chassis during vibration in the lateral direction. Again the greatest loads occur in rivet B. These rivet loads combine as shown in Fig. 7.29. The resultant shear load in rivet B is

$$R_r = R_3 + R_4 = 54.7 + 91.3 = 146 \text{ lbs} \tag{7.85}$$

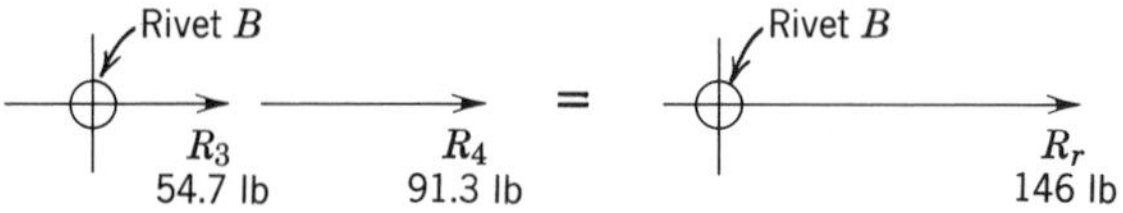

FIGURE 7.29. Dynamic loads acting on the mounting lug rivets.

Assuming a $\frac{5}{32}$-in.-diameter rivet, the rivet shear stress can be determined from Eq. 7.67.

$$S_s = \frac{R_r}{A} = \frac{146}{0.0191} = 7640 \text{ lb/in.}^2 \tag{7.86}$$

Assuming the same rivet material as before (2017-T31 aluminum), then the margin of safety for the rivet in shear can be determined from Eq. 7.69.

$$\text{MS} = \frac{S_e}{S_s} - 1 = \frac{11{,}300}{7640} - 1 = +0.48 \tag{7.87}$$

The positive margin of safety in shear shows the $\frac{5}{32}$-in.-diameter rivet will be satisfactory for the two-rivet pattern. The bearing margin of safety, however, will still be negative. The three-rivet pattern shown in Fig. 7.26 is therefore suggested for the CG mount.

Consider the uncoupled torsional resonant mode next. With the CG mount, the dynamic overturning moment will be determined by the location of the dynamic CG during vibration along the X axis. Since the dynamic CG will probably shift slightly as various elements of the chassis structure become excited during their resonances, a torsional resonant mode will occur. However, unless the dynamic balance is very poor, or unless a relatively large internal mass experiences a severe resonance, the torsional resonant mode of a CG-mounted chassis will be quite minor.

Stresses developed by this resonant mode will be low relative to the stresses developed by the uncoupled bending resonant mode. Therefore the structural design considerations will most likely be based on the bending mode characteristics of the chassis with a CG mount.

7.16. Sealed Electronic Boxes

Electronic boxes that must function in the vacuum conditions of outer space are usually heavier than electronic boxes that are designed to operate in the atmosphere. This is often due to the use of "O"-ring seals which may be required to maintain a low leak rate to hold a positive internal pressure or to keep out moisture. An O-ring seal requires a stiff flange to maintain a good seal, and a high internal pressure requires a stiff wall structure to reduce excessive deflections and stresses. In these cases the added stiffness means added weight.

Conductive heat transfer is often used to carry away the heat generated internally and a good heat-conduction path, having low thermal resistance, usually requires thicker walls. Heat exchangers may be used with circulating liquid coolants to remove the heat, so the weight of the complete electronic system and the volume may have to be increased.

Sometimes electronic boxes are purged with dry nitrogen and sealed with a positive internal pressure at sea-level conditions. In the vacuum of outer space, the pressure differential can easily reach a value of 20 psi above the initial starting value when internal heating is included. If large flat surfaces are used on the electronic box, heavy reinforcing ribs may have to be used to reduce deflections and stresses.

If the chassis is bolted directly to a liquid-cooled cold plate to remove the internal heat, then a flat, smooth, high-pressure interface must be provided between the chassis and the cold plate for efficient heat transfer. This usually means that several bolts will be used along the length of the chassis to ensure good contact with the cold plate.

A chassis that meets all of the above requirements will be heavier and more rugged than a chassis that is designed to operate in the atmosphere using natural convection cooling. This type of chassis, if it is welded, cast,

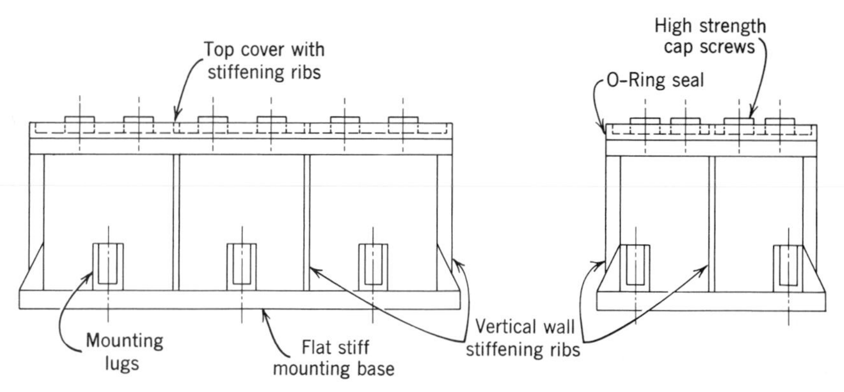

FIGURE 7.30. A typical chassis with rigid flanges for O ring seals and a flat stiff base for mounting on a spacecraft cold plate.

or dip brazed, will usually have a high resonant frequency because of the construction and mounting configuration. A sketch of a typical electronic box is shown in Fig. 7.30.

7.17. The Octave Rule

Acceleration G forces can build up very quickly when two coupled adjacent masses have similar resonant frequencies. Dynamic amplifications can lead to large deflections, high stresses, and rapid structural failures, so it is important to avoid coincident resonances in coupled masses. This can be accomplished by following the Octave Rule. This rule simply states that the resonant frequency of a substructure should be two (or more) times the resonant frequency of its support. For example, consider an electronic chassis with a natural frequency of 200 Hz which requires a transformer to be mounted to one of the walls. The natural frequency of the transformer on the wall should then be about 400 Hz. Here the transformer represents the substructure and the electronic chassis represents its support. Notice that the natural frequency should be doubled for each additional degree of freedom. The electronic chassis represents the first degree of freedom and the transformer represents the second degree of freedom. In order to double the natural frequency, the spring rate must be increased by a factor of four for each additional degree of freedom. This will avoid the possibility of coincident resonances that often lead to very high transmissibilities and stresses.

In order to make use of the Octave Rule, the electronic chassis assembly must be looked upon as being made up of many individual springs and masses. The chassis assembly may consist of printed-circuit boards for the small electronic component parts such as resistors, capacitors, transistors, and flatpack integrated circuits, and large stiff brackets to support more massive electronic components such as transformers, chokes, relays, motors, resolvers, disk memories, and ferrite-core memories. Figure 7.31 is a diagram of the Octave Rule, and shows how the natural frequency of each additional degree of freedom is doubled.

Although the structural configuration of many systems may be very complex, with odd-shaped sections, cut-outs, covers, recesses, inserts, protrusions, and notches, it is usually possible to approximate a complex structure with a series of simpler structures with known characteristics. Frequently the analogy may not be too accurate, but an approximation with the use of beams, plates, frames, and shells can often establish the groundwork for the preliminary design of an electronic system. Keeping in mind the basis for avoiding coupled resonant modes, the stiffness and, therefore, the section properties of many internal and external structural

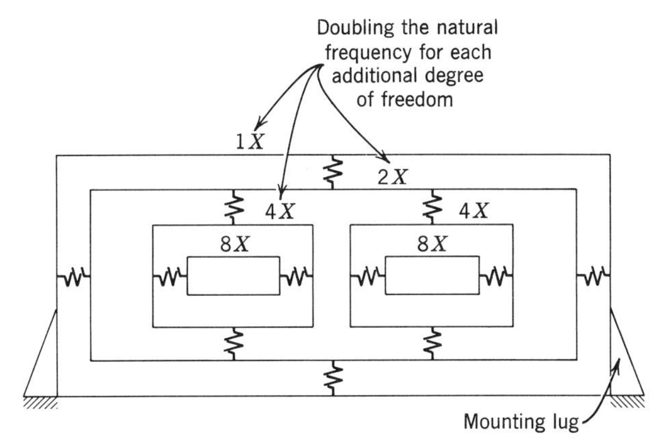

FIGURE 7.31. A model of the octave rule showing how the natural frequency is doubled for each additional degree of freedom.

members and the locations of sensitive electronic components can be established.

There are some cases where the Octave Rule works very well in reverse. This requires that the resonant frequency be reduced to one half for the additional degree of freedom. Since the purpose of the Octave Rule is to avoid coincident resonances, where amplifications can multiply very rapidly, any separation of individual resonances will help.

Consider, for example, a typical printed-circuit board plugged into an electronic chassis. If a relatively stiff chassis is used, the resonant frequency might be about 300 Hz. In order to follow the Octave Rule, the resonant frequency of the circuit board, which represents an additional degree of freedom with respect to the chassis, must be about 2 times 300, or 600 Hz. If the circuit board is quite large, perhaps 7 or 8 in. square, it will be quite difficult to obtain a resonance of 600 Hz without the addition of large and heavy ribs. However, if the Octave Rule is used in reverse, the circuit board can be designed to have a natural frequency of $\frac{1}{2}$ times 300, or 150 Hz. This situation is shown in Fig. 7.32, where the chassis has a natural frequency of 300 Hz and the plug-in printed-circuit board has a natural frequency of 150 Hz.

During a typical vibration test, when the forcing frequency is 150 Hz, the transmissibility of the chassis may be about 1.3. If the transmissibility of the plug-in circuit board is about 10 at this same frequency, then the board will see a transmissibility of 1.3 times 10 or 13. This means that the circuit board will receive an acceleration G force that is 13 times the input G force to the chassis. If the input to the chassis has a 4 G peak, then the circuit board will see a peak acceleration of 52 G at 150 Hz.

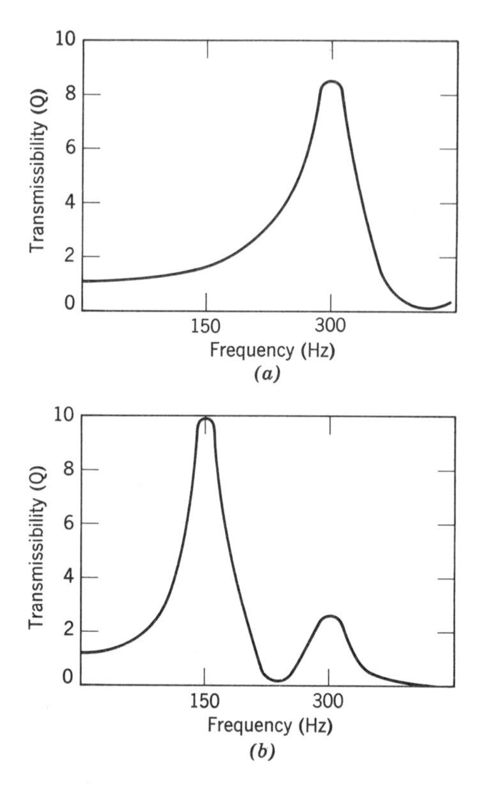

FIGURE 7.23. (a) Typical response of an electronic chassis; (*b*) typical response of a printed-circuit board.

When the forcing frequency is at 300 Hz, the major chassis resonance will become excited. The circuit board response at this frequency will not be very severe, since this forcing frequency is twice the fundamental resonant frequency of the circuit board. The possibility does exist, however, of exciting one of the higher harmonic resonances of the circuit board. But even if this does occur, the acceleration G forces and stresses will generally be much less severe than they will be for the fundamental resonant mode of the circuit board.

CHAPTER 8

MATHEMATICAL MODELS

8.1. Considerations in Establishing a Mathematical Model

Vibration analysis is normally associated with the extensive use of mathematics, thus the subject is normally considered to be very scientific in nature. And indeed it is! However there is one important area in vibration analysis that must be classified as more of an art than a science. This is the area of simulation and modeling. Here it is required to duplicate a three-dimensional electronic box with a mathematical model that will have the same vibration characteristics as the electronic box. This can often be accomplished with the use of a lumped-mass system using springs and dampers, where the inertial properties are in the masses, the elastic properties are in the springs, and the damping properties are in the dashpots.

Electronic boxes will, in many cases, have odd-shaped structural areas and compartments to support odd-shaped components such as cathode ray tubes, transformers, electric motors, linkages, gear systems, waveguides, magnetrons, rotating mirrors, and printed-circuit boards. The chassis itself often has an odd shape because it may have to fit into a corner next to a bulkhead or conform to the shape of a wing section.

Developing a mathematical model that will duplicate the exact vibration response characteristics of a complex electronic box is nearly impossible. However close approximations can be obtained, depending upon the number and arrangement of the masses, springs, and dampers. The accuracy of the system depends a great deal upon the ingenuity and imagination of the individual forming the mathematical model.

Analog and digital computers are the only practical tools that can effectively be used to determine the frequency-response characteristics of a complex system with many degrees of freedom, over a wide frequency band. Since the output of a computer is only as good as the input, it is necessary to develop a mathematical model that closely simulates the vibrational characteristics of the electronic box. A good model will require a comprehensive study of the basic box structure to determine the most probable load paths for the most massive elements in the system. This task can often be made simpler if it is remembered that the stiffer elements will carry a larger portion of the load. Also remember that the dynamic loads must eventually pass through the chassis mounting points where the chassis is fastened to its support structure.

It is probably better to start the preliminary mathematical model with more masses, springs, and dampers than are really necessary. Several attempts are normally required to establish a suitable model before a selection is finally made. Remember to examine the electronic box for vibration along each of the three mutually perpendicular axes: most dynamic environments will produce these loads and most qualification programs will require tests along these axes. Try to establish a separate model for each axis, then compare the models. Very often it is possible to make small changes in each of the three models to develop only one model which is representative of the electronic box for all three axes. This means only that the physical arrangement of the masses, springs, and dampers are the same. The values for the springs and dampers are likely to be different for each of the directions, but the values of the masses may stay the same.

Damping characteristics must be examined carefully for each axis because of the general lack of symmetry in electronic structures. For example, a bolted assembly dissipates a substantial amount of energy through friction along the axis that causes relative sliding at the bolted interface. Along axes where this relative sliding is small not as much energy is dissipated so damping values are much smaller.

8.2. Torsional Modes

One vibration mode, often overlooked even by experienced designers and engineers, which very often leads to problems is the torsional or rotational mode. This mode is usually not as obvious as the translational modes and requires an understanding of rotational inertia plus torsional stiffness characteristics of different types of structures. For example, I-beams and channels are very stiff in bending but they are very poor in torsion because they have open sections. Tubes and box beams, on the other hand, have closed sections so they are very good in torsion as well as bending.

Torsional modes often couple with bending modes and produce resonant frequencies that are much lower than expected. For example, a channel section, which is stiff in bending, is normally expected to have a high resonant frequency when it is vibrated along one of its stiff axes. However, since the channel section is poor in torsion, it may develop a low resonant frequency in its torsional mode during vibration along a stiff axis. If this low-frequency torsional mode is not foreseen, it can lead to unexpected large displacements and stresses which will shorten the ultimate fatigue life of the structure.

The same condition often occurs in an electronic chassis with a sheet-metal cover held in place with a few small screws. The chassis acts like an open channel section because the sheet-metal cover lacks rigidity. If the chassis is supported at the ends only, then during vibration in the vertical or lateral directions the natural frequency may be much lower than would be expected if the torsional modes are ignored.

Mathematical models for the torsional modes can also be made up of discrete masses, springs, and dampers. The inertial properties are again in the masses except that, instead of translational inertia, rotational inertia must be used. This is the same for the spring rates and the damping coefficients, where everything must be related to rotation instead of translation.

8.3. Breaking Up the Electronic Box into Lumped Elements

All mathematical models require the same decisions when being developed to simulate an electronic box. What parts of the structure should be designated as masses and what parts of the structure should be designated as springs?

There is no simple answer to this question, since it depends upon the geometry of each electronic box and the imagination of the individual. Large, rigid, heavy electronic components such as transformers, relays, magnetrons, motors, and chokes can be represented as masses without stretching the imagination. The chassis structure that supports the electronic equipment, however, must act as a mass and a spring at the same time. The mass of the chassis is due to the weight of the various structural elements used to support and enclose the electrical components and the components themselves. The masses will normally be lumped at several convenient discrete points along the length of the chassis, at intervals that are based on symmetry or the desire to examine the structural characteristics of the electronic chassis at a change in its cross-section, or perhaps to examine the dynamic loads in a bulkhead.

The spring elements in the chassis are often the same structural members that make up the masses, except that the springs are considered to have no mass. The springs are used to represent the stiffness of the structure and they are used to interconnect the masses.

The spring elements must be included in combinations of structures that can be analyzed. The most common structural elements are beams, plates, frames, rings, and shells; these have a wide variety of edge conditions that can vary from free to fixed and a variety of deflection modes that include bending, tension, shear, and torsion.

Beam elements are probably the easiest structural members to analyze and many good approximations can be made with plain beam members. Books such as the one by Roark [19] have many beam equations which can be used to determine spring rates.

A transformer bolted to a support bracket is shown in Fig. 8.1.

It does not take much imagination to consider the bracket as a beam that is simply supported at each end with the transformer acting as a concentrated load. The weight of the bracket might be included with the weight of the transformer as a total weight, W.

The spring rate for the bracket can be determined from the equation of a simply supported beam with a concentrated load.

$$K = \frac{48EI}{L^3} \tag{8.1}$$

The system can now be approximated as a single-degree-of-freedom one (Fig. 8.2).

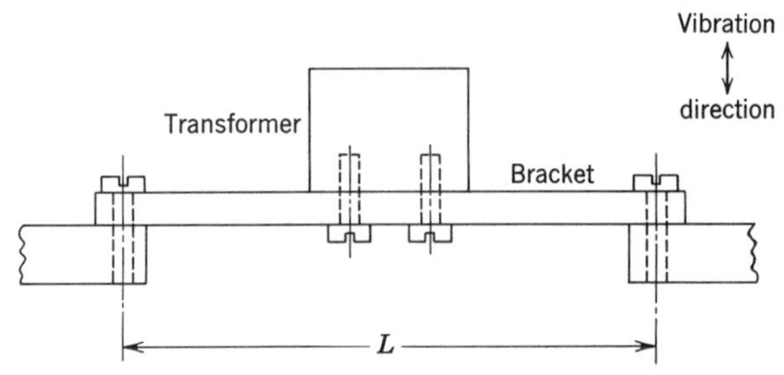

FIGURE 8.1. A transformer bolted to a support bracket.

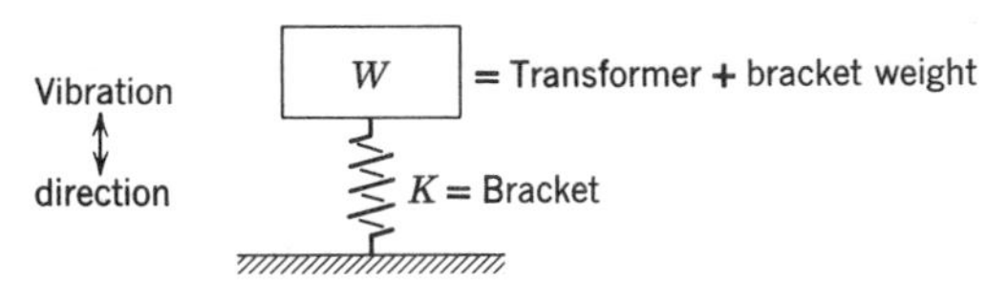

FIGURE 8.2. The transformer and bracket as a simple spring-mass system.

The same transformer might be mounted on the same bracket in a slightly different way, using a second bracket to actually support the transformer (Fig. 8.3). This structure can be approximated as a two-degrees-of-freedom system, where bracket *A* represents one degree of freedom with its own weight concentrated at the center of a massless beam (Fig. 8.4).

The transformer mounted on bracket *B* can represent the second degree of freedom. Bracket *B* can be approximated as a rectangular frame or bent with hinged ends and a concentrated load to represent the transformer (Fig. 8.5).

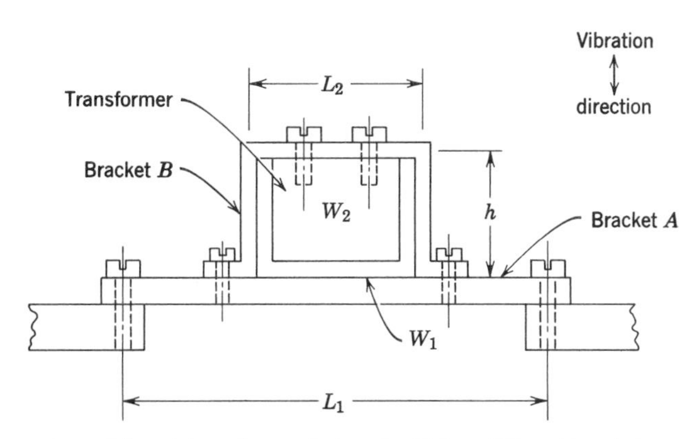

FIGURE 8.3. A transformer fastened with two different brackets.

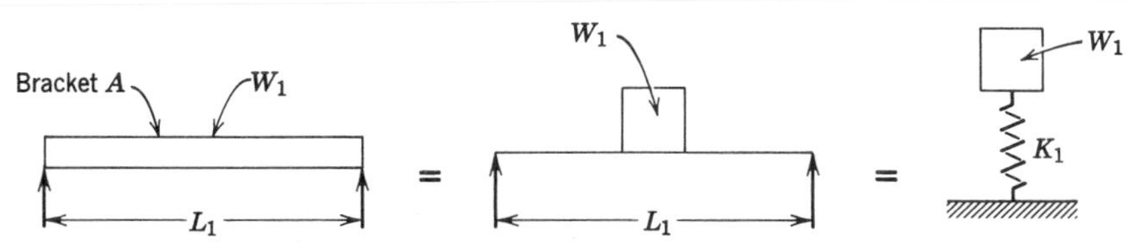

FIGURE 8.4. Bracket *A* simulated as a simple spring-mass system.

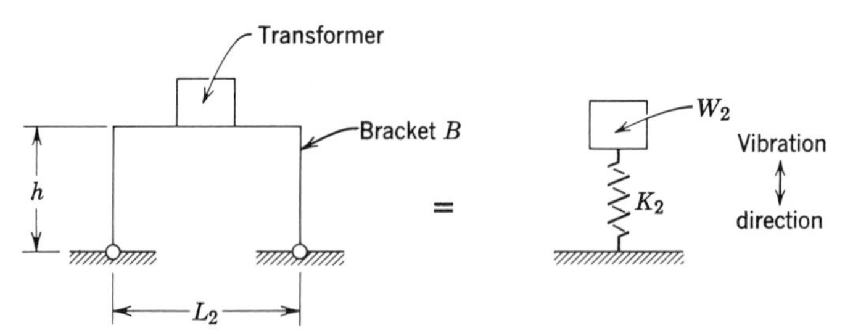

FIGURE 8.5. Bracket B simulated as a simple spring-mass system.

The weight of the transformer plus the weight of the bracket will be represented by the total weight, W_2. The spring rate of the bent can be obtained from Eq. 5.114 as shown in Chapter 5 and repeated below.

$$K_2 = \frac{48EI_1}{L_2^3\left[1 - \frac{9}{4(2K+3)}\right]} \tag{8.2}$$

Combining the two masses and two springs results in a structure that can now be approximated as the two-degrees-of-freedom system shown in Fig. 8.6.

The transformer and bracket assembly can be mounted in some type of electronic box; this adds another degree of freedom to the system. For example, consider the rack-and-panel mounting for an electronic box shown in Fig. 8.7. Here the transformer assembly is bolted across the bottom structural frame members of the electronic box.

During vibration in the vertical direction, it is necessary to determine how the deflections in the cantilevered chassis will occur, so that a logical

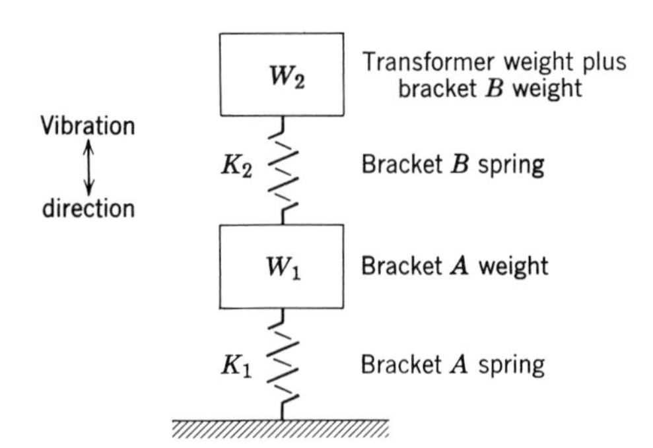

FIGURE 8.6. Brackets A and B combined into a two spring-mass system.

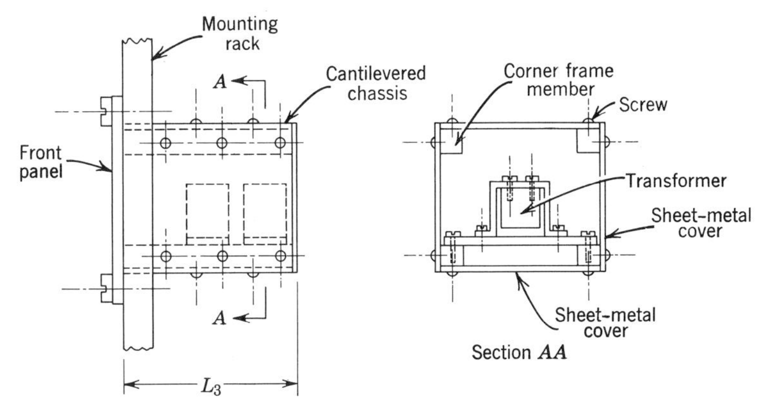

FIGURE 8.7. A typical rack and panel mounted electronic box.

spring system can be approximated and analyzed. An examination of the structure shows that the cantilevered frame members in the four corners of the chassis carry most of the loads to the front panel, which then transfers the loads to the mounting rack.

When the frame members deflect under the dynamic load, the sheet-metal covers also try to carry some of the load to the front panel. Since the sheet-metal covers are screwed to the chassis frame, there is a considerable amount of relative motion between the covers and the frame during the fundamental resonant mode. This tends to reduce the stiffness but it substantially increases the damping in this part of the system.

Two different deflection modes are examined for the cantilevered chassis structure, bending, and shear. Bending deflections are probably more appropriate here because the prime load-carrying member is the frame and the frame tends to bend during vibration. The sheet-metal covers stiffen the structure, but they are only about 25% efficient, because of the screws (as explained in Chapters 6 and 7).

The spring rate of the cantilevered chassis can be determined by approximating it as a cantilevered beam with an end mass. The spring rate is then

$$K_3 = \frac{3EI}{L_3{}^3} \tag{8.3}$$

The cantilevered chassis can be considered as another degree of freedom, hence the complete system now has three degrees of freedom, with three springs and three masses (Fig. 8.8). Dampers can be included to simulate the transmissibilities of the real structure.

In the construction of this mathematical model, notice that each degree of freedom consists of a specific weight, spring, and damper associated with a particular part of the electronic box. For the model shown in Fig. 8.8, this consists of the chassis as weight W_1; bracket A as W_2; and the transformer plus bracket B as W_3.

Using the same cantilevered chassis as shown in Fig. 8.7, if there are two identical sets of transformers and mounting brackets adjacent to one

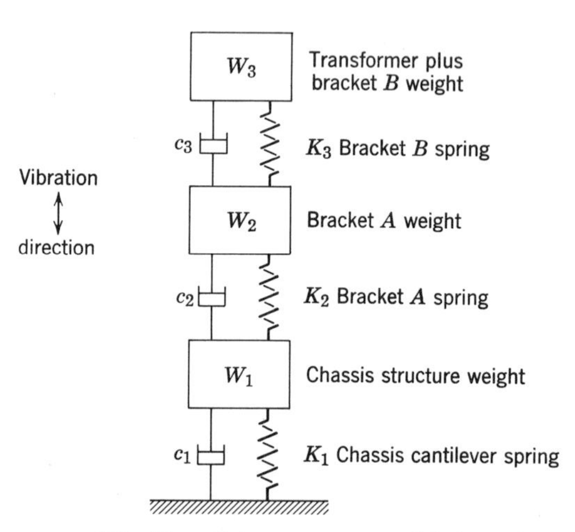

FIGURE 8.8. Transformer supports modeled as a three degree-of-freedom system.

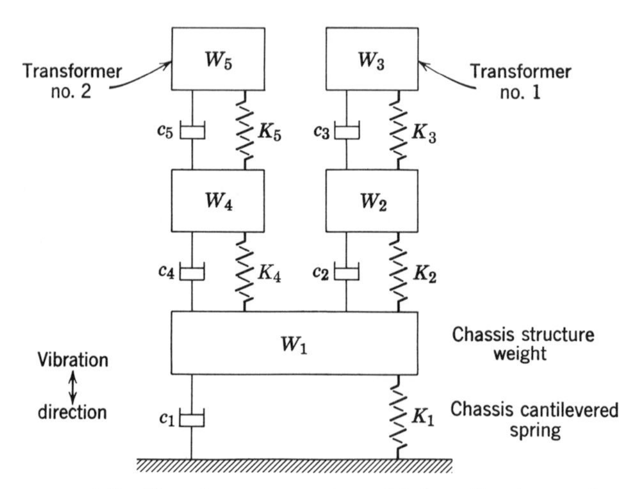

FIGURE 8.9. Transformer supports modeled as a five degree-of-freedom system.

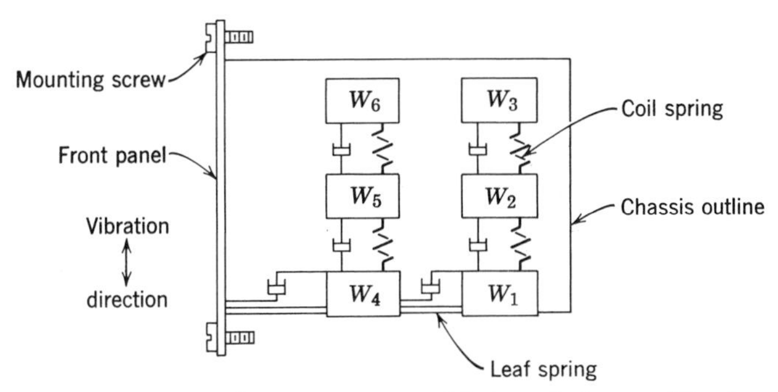

FIGURE 8.10. Transformer supports modeled as a six degree-of-freedom system.

FIGURE 8.11. A cantilevered electronic box with pin-fin heat exchangers on the top and bottom surfaces (courtesy Norden division of United Aircraft).

another, with both transformer sets fastened to the electronic chassis frame, then the mathematical model will appear as shown in Fig. 8.9.

The mathematical model for the cantilevered electronic box shown in Fig. 8.9 can be expanded even further by using two masses for the chassis structure. A leaf spring is used to interconnect the two masses and to

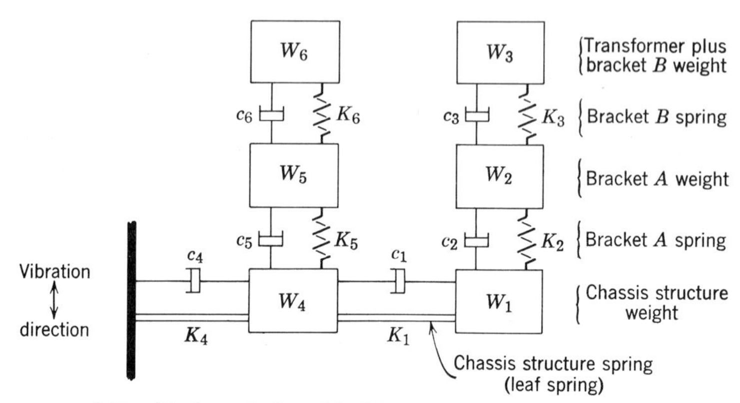

FIGURE 8.12. Mathematical model of the electronic chassis with brackets and transformers.

FIGURE 8.13. A large airborne electronic box divided into two sections, one for forced-convention cooling and one for natural-convection cooling. The natural-convection side has the perforated cover (courtesy Norden division of United aircraft.)

connect the two masses to the front panel. The leaf spring is intended to show a lateral transfer of the loads and deflections, while a coil spring is intended to show a vertical transfer of the loads and deflections.

An outline of the expanded mathematical model can be superimposed on the outline of the electronic box to show the relation of the lumped mass system to the box structure (Fig. 8.10). A cantilevered electronic box with pin-fin heat exchangers is shown in Fig. 8.11.

The mathematical model will finally be as shown in Fig. 8.12. For ease in drawing the chassis dampers, they are rotated 90° with no change in function.

This mathematical model considers only the translational vibration modes along each of the three mutually perpendicular axes. The mathematical model for the torsional modes can also be made up of discrete masses, springs, and dampers. The inertial properties are again in the masses, except that, instead of translational inertia, rotational inertia must be used. This is the same for the spring rates and damping coefficients, where everything must be related to rotation instead of translation.

8.4. Vibration Analysis of an Electronic Box Simulated by Seven Masses

An electronic box that contains a power-supply section, a printed-circuit-board section, and a ferrite-core-memory section is shown in Fig. 8.14.

In order to reduce the total number of masses in the system, the ferrite-core-memory section was clustered into two masses, the printed-circuit boards were clustered into a single mass, and the power supply racks were clustered into a single mass. The chassis structure was lumped into three masses, with one mass in each of the three sections of the chassis (Fig. 8.15).

8.5. Lumped Masses

The magnitude of the individual masses can be determined by calculating or weighing component parts and subassemblies. Consider the model of the chassis shown in Fig. 8.16. The total weight of the combined masses should equal the total weight of the complete system, including cables, connectors, and hardware.

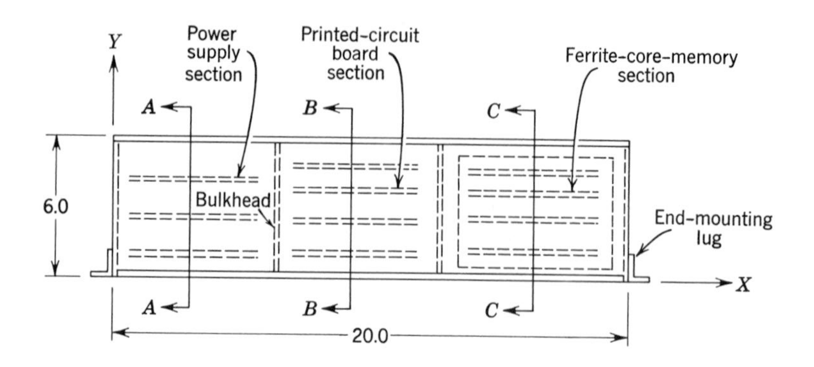

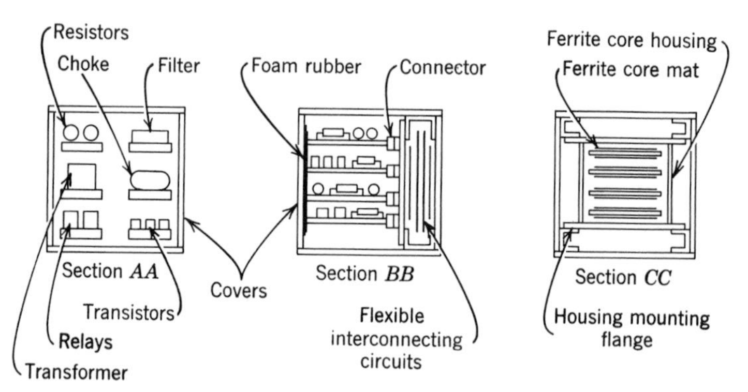

FIGURE 8.14. A typical electronic chassis with many different types of electronic components.

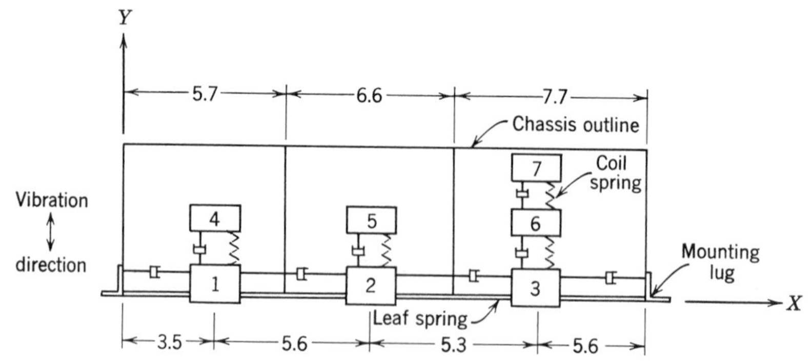

FIGURE 8.15. An electronic chassis broken up into discrete masses, springs and dampers.

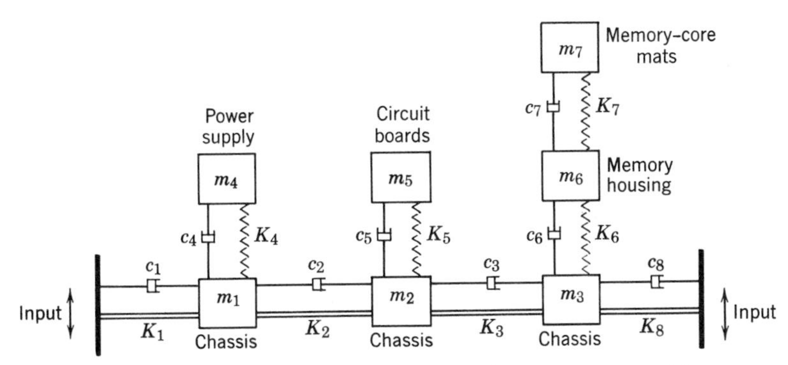

FIGURE 8.16. Mathematical model of the chassis with seven degrees-of-freedom.

The individual weights for this system are outlined as follows:

ITEM	DESCRIPTION	WEIGHT (lb)
W_1	Chassis structure	4.60
W_2	Chassis structure	2.80
W_3	Chassis structure	2.20
W_4	Power supply	2.70
W_5	Circuit boards	3.70
W_6	Memory core structure	4.80
W_7	Memory core mats	2.30
	Total weight	23.10 lb

The individual masses are determined from the weights as follows:

$$m = \frac{W}{g}$$

$$\left.\begin{aligned} m_1 &= 1.19 \times 10^{-2} \text{ lb sec}^2/\text{in.} \\ m_2 &= 0.725 \times 10^{-2} \\ m_3 &= 0.570 \times 10^{-2} \\ m_4 &= 0.700 \times 10^{-2} \\ m_5 &= 0.959 \times 10^{-2} \\ m_6 &= 1.24 \times 10^{-2} \\ m_7 &= 0.596 \times 10^{-2} \end{aligned}\right\} \tag{8.4}$$

8.6. CHASSIS SPRING RATES ALONG THE Y AXIS

The chassis springs are shown as K_1, K_2, K_3, and K_8. The spring rates can be determined with the use of the "deflection difference" method

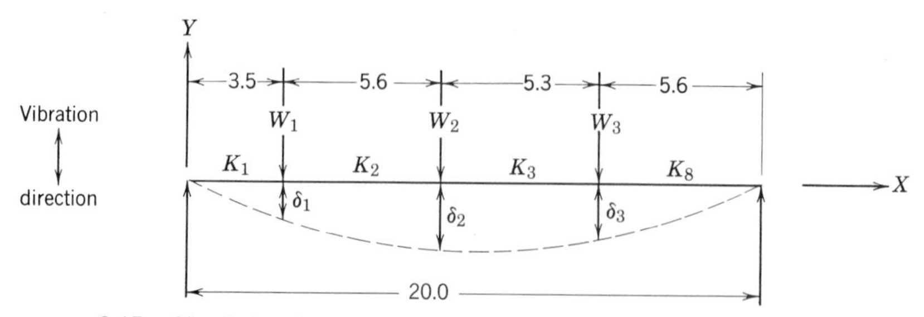

FIGURE 8.17. Simulating the chassis structure as a simply supported beam with three concentrated loads.

shown in Chapter 3, Eq. 3.105. The deflections at each chassis mass must be determined first. This can be done using the superposition techniques outlined in Chapter 3, Section 3.1. Deflections here are in the Y axis only (Fig. 8.17).

The moment of inertia of the chassis should be determined at several different locations and an average value computed using the method shown in Chapter 4, Section 4.6 and Chapter 7, Fig. 7.8.

The chassis spring rates can be determined as follows:

$$\left.\begin{aligned} K_1 &= \frac{W_1}{\delta_1} = 6.11 \times 10^5 \text{ lb/in.} \\ K_2 &= \frac{W_2}{\delta_2 - \delta_1} = 2.80 \times 10^5 \\ K_3 &= \frac{W_2}{\delta_2 - \delta_3} = 2.65 \times 10^5 \\ K_8 &= \frac{W_3}{\delta_3} = 1.41 \times 10^5 \end{aligned}\right\} \tag{8.5}$$

8.7. Power-Supply Spring Rate Along the Y Axis

Only one mass is used to represent the power supply, so only one spring is used to represent the combined stiffness of the structure. This may be difficult to do if there is a great variation in the different types of structures and the different types of components used in the power supply area. Several masses and several springs may then have to be used to represent this area.

Since there is sufficient uniformity in the various structural elements and components in this power supply, the natural frequency of one major section can be used for the entire power supply. Standard handbook equations are used where possible. For example, considering the structural members in the power supply as simply supported beams with uniform loads (Fig. 8.14), the natural frequency is that shown in Eq. 4.33 (see Fig. 8.18).

$$f_n = \frac{\pi}{2}\left(\frac{EIg}{WL^3}\right)^{1/2}$$

where $E = 10.5 \times 10^6$ lb/in.^{2} aluminum

$$I = \frac{bh^3}{12} = \frac{(3.0)(0.158)^3}{12} = 0.987 \times 10^{-3} \text{ in.}^4$$

Since the six elements in the power supply are all similar, the weight of each unit is

$$W = \frac{W_4}{6} = \frac{2.70}{6} = 0.45 \text{ lb}$$

$$L = 5.0 \text{ in.}$$

Substituting into the natural-frequency equation above

$$f_n = \frac{\pi}{2}\left[\frac{(10.5 \times 10^6)(0.987 \times 10^{-3})(386)}{0.45(5.0)^3}\right]^{1/2} = 420 \text{ Hz}$$

Approximating the beam element as a single-degree-of-freedom system means that a single spring and a single mass can be used. The spring rate can then be determined from the natural frequency, Eq. 2.7.

$$K = \frac{4\pi^2 f_n^2 W}{g} \tag{8.6}$$

Since one mass is to be used to represent the entire power-supply section, the weight W must represent the total power-supply weight, which is shown by W_4 as 2.70 lb. Then

$$K_4 = \frac{4(9.86)(420)^2(2.70)}{386} = 0.488 \times 10^5 \text{ lb/in.} \tag{8.7}$$

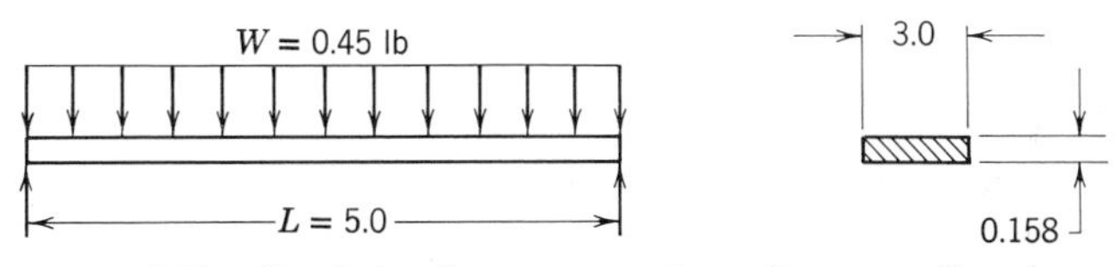

FIGURE 8.18. Simulating the power supply section as a uniform beam.

8.8. PRINTED-CIRCUIT-BOARD SPRING RATE ALONG THE *Y* AXIS

Only one mass is used to represent the printed-circuit-board section, so only one spring will be used to represent the stiffness. The circuit boards are all about the same size and weight, so one mass and one spring can be used to represent the group.

The spring rate for this group, during vibration in the vertical direction, can be determined by calculating the natural frequency of a typical single printed-circuit board. Using the natural-frequency equations shown in Chapter 6, a typical circuit board with four simply supported edges will have a natural frequency as follows:

$$f_n = \frac{\pi}{2}\left(\frac{D}{\rho}\right)^{1/2}\left(\frac{1}{a^2}+\frac{1}{b^2}\right)$$

Now considering the circuit board as a single-degree-of-freedom system, the spring rate can be determined from Eq. 8.6.

Since one mass is being used to represent all of the printed-circuit boards, the weight W must represent the total circuit-board weight; this is shown in Section 8.5 as W_5 with a weight of 3.70 lb. For this sample problem the spring rate for the circuit board group is

$$K_5 = 0.242 \times 10^5 \text{ lb/in.} \tag{8.8}$$

The lumping of several circuit boards into one mass is just for the convenience of reducing the total number of masses in this system. Using the resonant frequency of only one circuit board in determining the spring rate does not mean to suggest that all the circuit boards should be tuned to the same natural frequency. If the circuit boards all have the same natural frequency and are all moving in phase during resonance, they will develop a very large surge of kinetic energy which can easily lead to rapid fatigue failures.

8.9. MEMORY-CORE-HOUSING SPRING RATE ALONG THE *Y* AXIS

The memory core housing is in the form of a modular unit which is fastened to the chassis by eight mounting screws in eight mounting flanges (Fig. 8.19).

Considering the eight mounting flanges as eight similar cantilevered beams, the spring rate can be determined from the standard deflection equation as follows:

$$K_6 = \frac{3EI}{L^3}$$

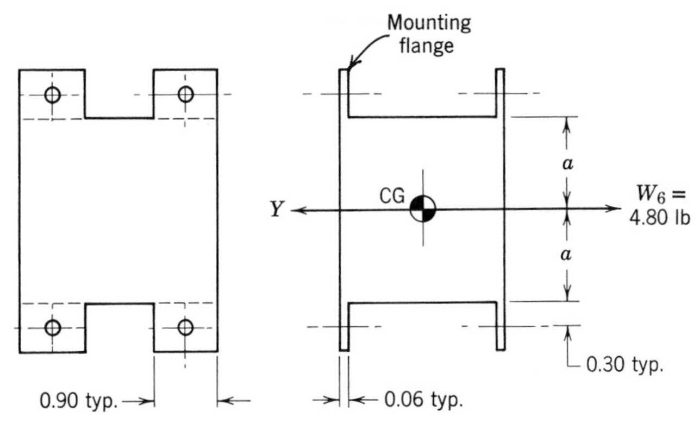

FIGURE 8.19. Geometry of the memory core housing.

where $E = 10.5 \times 10^6$ lb/in.^{2} aluminum modulus of elasticity

$$I = \frac{8bh^3}{12} = \frac{8(0.90)(0.06)^3}{12} = 1.29 \times 10^{-4} \text{ in.}^4$$

$L = 0.30$ in. length of flange

$$K_6 = \frac{3(10.5 \times 10^6)(1.29 \times 10^{-4})}{(0.30)^3} = 1.50 \times 10^5 \text{ lb/in.} \tag{8.9}$$

8.10. Memory-Core-Mat Spring Rate Along the *Y* Axis

Only one mass is used to represent the memory-core mats that support the ferrite cores. These mats are similar to rectangular flat plates with a uniformly distributed load.

Using the same method shown for the printed-circuit board area in Section 8.8, the spring rate for this sample problem is as follows:

$$K_7 = 0.482 \times 10^5 \text{ lb/in.} \tag{8.10}$$

8.11. Chassis Damping Coefficients Along the *Y* Axis

There is really no good way to calculate damping coefficients in a typical electronic structure because there are so many variables involved. Some of these variables include riveted assemblies with different interface pressures and different surface finishes, loose rivets, bolted assemblies where bolts are not torqued uniformly, loose screws, welded structures where the welding heat changes the heat-treatment and the temper of the metal, the effects of microscopic cracks, conformal coatings, tolerances in machined members, temperature changes that produce dimensional changes, warping, connector friction, cables rubbing, material changes, and stress levels.

Only test data and testing experience can really provide adequate information on damping. For the purpose of performing a vibration analysis, testing experience should be used to estimate the approximate transmissibility of different structural elements, based on the peak input G force and the resonant frequency. This must be done for each structural element in the system to ensure an accurate analysis.

Since each mass and spring is considered separately, it is convenient to consider the dampers separately. The method used here is to estimate the approximate transmissibility that each mass and spring has when acting as a single-degree-of-freedom system. This method is easily related to vibration tests, where accelerometers are mounted on the most massive elements in the system. The resonant peaks are sharply defined and convenient to use when the damping coefficients are calculated as shown in Chapter 2, Eq. 2.76.

Damping for the chassis structure shown in Fig. 8.14 is relatively high since the bolted side covers develop a substantial amount of relative motion during vibration in the vertical direction. There is also some relative motion in the cable and harness area, but most of this energy dissipation occurs at low frequencies.

A rigid vibration fixture should be used to support the electronic box to prevent fixture resonances from influencing the chassis resonances. Only the ends of the electronic box are fastened to the fixture, using the mounting lugs at each end of the box. This means that the bottom of the box is completely free to deflect without impacting against the vibration fixture. If the bottom surface of the electronic box impacts against the vibration fixture, the transmissibility will be sharply reduced.

The individual masses in the chassis structure are estimated to have a transmissibility of about 11.0 for the conditions outlined above. The damping coefficients can now be determined from Eqs. 2.76 and 8.5.

$$
\begin{aligned}
c_1 &= \frac{(K_1 m_1)^{1/2}}{Q_1} = \frac{[6.11 \times 10^5 (1.19 \times 10^{-2})]^{1/2}}{11.0} = 7.75 \text{ lb sec/in.} \\
c_2 &= \frac{(K_2 m_2)^{1/2}}{Q_2} = \frac{[2.80 \times 10^5 (0.725 \times 10^{-2})]^{1/2}}{11.0} = 4.10 \\
c_3 &= \frac{(K_3 m_2)^{1/2}}{Q_2} = \frac{[2.65 \times 10^5 (0.725 \times 10^{-2})]^{1/2}}{11.0} = 3.98 \\
c_8 &= \frac{(K_8 m_3)^{1/2}}{Q_3} = \frac{[1.41 \times 10^5 (0.570 \times 10^{-2})]^{1/2}}{11.0} = 2.58
\end{aligned}
\tag{8.11}
$$

8.12. Power-Supply Damping Coefficients

Damping for the power-supply components depends upon the amount of energy dissipated in this part of the system. This section normally contains many bolted and riveted interfaces due to the mounting of large components. Also, laminations with copper, which is a good heat conductor, may be used to control temperatures in critical areas. Laminations are also good energy dissipators and tend to reduce the transmissibilities at resonance.

For this particular model, the transmissibility for the power-supply section is estimated to be about 7.0 for vibration along the Y axis. The damping coefficient can be determined from Eq. 2.76 as follows:

$$c_4 = \frac{(K_4 m_4)^{1/2}}{Q_4} = \frac{[(0.488 \times 10^5)(0.700 \times 10^{-2})]^{1/2}}{7.0} = 2.64 \text{ lb sec/in.} \quad (8.12)$$

8.13. Circuit-Board Damping Coefficients

Damping for the plug-in circuit boards will depend upon the type of electrical connector, the type of board edge-guides, the circuit-board material, the type of electronic component parts, the number and type of stiffening ribs, the natural frequency, the acceleration G force, the conformal coating thickness, and any other factor that will dissipate energy.

The natural frequency of the typical circuit board used in this sample problem can be determined from the spring rate, shown by Eq. 8.8, and the mass, shown by Eq. 8.4, as follows:

$$f_n = \frac{1}{2\pi}\left(\frac{K_5}{m_5}\right)^{1/2} = \frac{1}{2\pi}\left(\frac{0.242 \times 10^5}{0.959 \times 10^{-2}}\right)^{1/2} = 253 \text{ Hz} \quad (8.13)$$

The transmissibility for many printed-circuit boards can often be approximated by using a value that is equal to the square root of the natural frequency. Considering the natural frequency shown above, the approximate transmissibility might be about 16. However, in this case, several stiffening ribs are riveted to the circuit board and several large potted modules, such as amplifiers and ladder networks, are mounted on these circuit boards. These items tend to increase the damping thereby decreasing the transmissibility. Because of the high damping characteristics associated with these features, the transmissibility is estimated to be about 7.0.

The damping coefficient for this lumped component can be determined from Eqs. 2.76, 8.8, and 8.4 as follows:

$$c_5 = \frac{(K_5 m_5)^{1/2}}{Q_5} = \frac{[(0.242 \times 10^5)(0.959 \times 10^{-2})]^{1/2}}{7.0} = 2.18 \text{ lb sec/in.} \quad (8.14)$$

8.14. Damping Coefficients for the Memory Module

Damping for the memory-core-housing structure, mass 6, is quite high due to the eight bolts that fasten the memory module to the chassis. The high-pressure interface provided by the mounting bolts develop a substantial amount of friction between the mounting flanges and the chassis due to the relative sliding motion that exists at the interface during the resonant condition.

A large number of electrical connections are required for the memory, which means that several connectors are used. Connectors dissipate a great deal of energy due to the large number of contacts and the high interface pressures required at the contacts to provide a good, low-resistance electrical path. Again, the relative sliding motion at resonance helps to reduce the transmissibility.

Considering the above factors, the transmissibility for the memory-core housing, during vibration in the vertical direction along the Y axis, is estimated to be about 6. The damping coefficient can then be determined from Eqs. 2.76, 8.9, and 8.4.

$$c_6 = \frac{(K_6 m_6)^{1/2}}{Q_6} = \frac{[(15.0\times 10^4)(1.24\times 10^{-2})]^{1/2}}{6} = 7.18 \text{ lb sec/in.} \qquad (8.15)$$

Damping for the memory-core mats, which are inside the memory core housing, is based on the damping characteristics of typical printed-circuit boards because both structures act like flat plates. The memory-core mats, however, do not plug in like printed-circuit boards. Instead, they are bolted to the memory-core-housing structure at several points around the perimeter of the rectangular mat plate.

A memory-core mat consists of hundreds of tiny ferrite cores that are threaded with very fine wires. Electrical connections are made to these fine wires on each mat and electrical connections are made between mats. During vibration, the thousands of tiny ferrite cores on all of the mats, and the thousands of wires, all dissipate energy and thus reduce the transmissibility at resonance.

The natural frequency of a typical mat in this system can be determined from Eqs. 8.10 and 8.4 as follows:

$$f_n = \frac{1}{2\pi}\left(\frac{K_7}{m_7}\right)^{1/2} = \frac{1}{2\pi}\left(\frac{0.482\times 10^5}{0.596\times 10^{-2}}\right)^{1/2} = 452 \text{ Hz} \qquad (8.16)$$

The transmissibility for many types of printed-circuit boards can often be approximated as being equal to the square root of the natural frequency. If this approximation is applied to the memory mats it results in a value of about 21. However an examination of a typical memory mat indicates there will be more energy dissipated in the mat than in a typical circuit board because of the large number of cores and wires. Therefore the transmissibility is estimated to be about 14.

The damping coefficient can now be determined from Eqs. 2.76, 8.10, and 8.4 as follows:

$$c_7 = \frac{(K_7 m_7)^{1/2}}{Q_7} = \frac{[(0.482 \times 10^5)(0.596 \times 10^{-2})]^{1/2}}{14} = 1.21 \text{ lb sec/in.} \quad (8.17)$$

8.15. Summary of Physical Constants for Vibration Along the Y Axis

A summary of the physical constants for the multiple spring, mass, and damper system, as shown in Fig. 8.16, is presented in Table 8.1. These constants represent the values that were determined for the electronic box shown in Fig. 8.14 and considering vibration in the vertical direction only (along the Y axis). The same mathematical model is used to represent the electronic box for vibration in the longitudinal direction, along the X axis; a whole new set of springs and dampers have to be computed. The same masses can be used here. The same mathematical model can then be used to represent the electronic box for vibration in the lateral direction, along the Z axis, where a third set of springs and dampers have to be computed. The same masses can also be used here.

TABLE 8.1. VIBRATION ALONG THE Y AXIS

Spring Rate (lb/in.)	Damping (lb sec/in.)	Mass (lb sec²/in.)
$K_1 = 6.11 \times 10^5$	$c_1 = 7.75$	$m_1 = 1.19 \times 10^{-2}$
$K_2 = 2.80 \times 10^5$	$c_2 = 4.10$	$m_2 = 0.725 \times 10^{-2}$
$K_3 = 2.65 \times 10^5$	$c_3 = 3.98$	$m_3 = 0.570 \times 10^{-2}$
$K_4 = 0.488 \times 10^5$	$c_4 = 2.64$	$m_4 = 0.700 \times 10^{-2}$
$K_5 = 0.242 \times 10^5$	$c_5 = 2.18$	$m_5 = 0.959 \times 10^{-2}$
$K_6 = 1.50 \times 10^5$	$c_6 = 7.18$	$m_6 = 1.24 \times 10^{-2}$
$K_7 = 0.482 \times 10^5$	$c_7 = 1.21$	$m_7 = 0.596 \times 10^{-2}$
$K_8 = 1.41 \times 10^5$	$c_8 = 2.58$	

8.16. Differential Equations of Motion for Vibration Along the Y Axis

The differential equations of motion for each mass must be determined for vibration along the Y axis. This can be accomplished by considering the inertia forces, the damper forces, and the spring forces acting along the Y axis, using the method as shown in Chapter 3, Section 3.5. The coordinate system for the mathematical model of the electronic box is shown in Fig. 8.20.

When the input vibration acceleration force is through both end supports of the electronic chassis at the same time and in the same phase, the equations of motion for each mass in the system are as follows:

For mass 1:

$$m_1\ddot{Y}_1 + c_1(\dot{Y}_1 - \dot{Y}_0) + c_2(\dot{Y}_1 - \dot{Y}_2) + c_4(\dot{Y}_1 - \dot{Y}_4) + K_1(Y_1 - Y_0) + K_2(Y_1 - Y_2) + K_4(Y_1 - Y_4) = 0$$

$$m_1\ddot{Y}_1 + \dot{Y}_1(c_1 + c_2 + c_4) - c_2\dot{Y}_2 - c_4\dot{Y}_4 + Y_1(K_1 + K_2 + K_4) - K_2Y_2 - K_4Y_4 = c_1\dot{Y}_0 + K_1Y_0 \tag{8.18}$$

For mass 2:

$$m_2\ddot{Y}_2 + c_2(\dot{Y}_2 - \dot{Y}_1) + c_3(\dot{Y}_2 - \dot{Y}_3) + c_5(\dot{Y}_2 - \dot{Y}_5) + K_2(Y_2 - Y_1) + K_3(Y_2 - Y_3) + K_5(Y_2 - Y_5) = 0$$

$$m_2\ddot{Y}_2 + \dot{Y}_2(c_2 + c_3 + c_5) - c_2\dot{Y}_1 - c_3\dot{Y}_3 - c_5\dot{Y}_5 + Y_2(K_2 + K_3 + K_5) - K_2Y_1 - K_3Y_3 - K_5Y_5 = 0 \tag{8.19}$$

For mass 3:

$$m_3\ddot{Y}_3 + c_3(\dot{Y}_3 - \dot{Y}_2) + c_6(\dot{Y}_3 - \dot{Y}_6) + c_8(\dot{Y}_3 - \dot{Y}_0) + K_3(Y_3 - Y_2) + K_6(Y_3 - Y_6) + K_8(Y_3 - Y_0) = 0$$

$$m_3\ddot{Y}_3 + \dot{Y}_3(c_3 + c_6 + c_8) - c_3\dot{Y}_2 - c_6\dot{Y}_6 + Y_3(K_3 + K_6 + K_8) - K_3Y_2 - K_6Y_6 = c_8\dot{Y}_0 + K_8Y_0 \tag{8.20}$$

For mass 4:

$$m_4\ddot{Y}_4 + c_4(\dot{Y}_4 - \dot{Y}_1) + K_4(Y_4 - Y_1) = 0$$

$$m_4\ddot{Y}_4 + c_4\dot{Y}_4 - c_4\dot{Y}_1 + K_4Y_4 - K_4Y_1 = 0 \tag{8.21}$$

For mass 5:

$$m_5\ddot{Y}_5 + c_5(\dot{Y}_5 - \dot{Y}_2) + K_5(Y_5 - Y_2) = 0$$

$$m_5\ddot{Y}_5 + c_5\dot{Y}_5 - c_5\dot{Y}_2 + K_5Y_5 - K_5Y_2 = 0 \tag{8.22}$$

For mass 6:

$$m_6\ddot{Y}_6 + c_6(\dot{Y}_6 - \dot{Y}_3) + c_7(\dot{Y}_6 - \dot{Y}_7) + K_6(Y_6 - Y_3) + K_7(Y_6 - Y_7) = 0$$

$$m_6\ddot{Y}_6 + \dot{Y}_6(c_6 + c_7) - c_6\dot{Y}_3 - c_7\dot{Y}_7 + Y_6(K_6 + K_7) - K_6Y_3 - K_7Y_7 = 0 \quad (8.23)$$

For mass 7:

$$m_7\ddot{Y}_7 + c_7(\dot{Y}_7 - \dot{Y}_6) + K_7(Y_7 - Y_6) = 0$$

$$m_7\ddot{Y}_7 + c_7\dot{Y}_7 - c_7\dot{Y}_6 + K_7Y_7 - K_7Y_6 = 0 \quad (8.24)$$

Equations 8.18–8.24 can be put into a convenient matrix form (Table 8.2).

TABLE 8.2. VIBRATION ALONG THE Y AXIS

Mass	Y_1	Y_2	Y_3	Y_4	Y_5	Y_6	Y_7		Y_0
1	m_1 $c_1+c_2+c_4$ $K_1+K_2+K_4$	$-c_2$ $-K_2$	0	$-c_4$ $-K_4$	0	0	0		c_1 K_1
2	$-c_2$ $-K_2$	m_2 $c_2+c_3+c_5$ $K_2+K_3+K_5$	$-c_3$ $-K_3$	0	$-c_5$ $-K_5$	0	0		0
3	0	$-c_3$ $-K_3$	m_3 $c_3+c_6+c_8$ $K_3+K_6+K_8$	0	0	$-c_6$ $-K_6$	0		c_8 K_8
4	$-c_4$ $-K_4$	0	0	m_4 c_4 K_4	0	0	0	=	0
5	0	$-c_5$ $-K_5$	0	0	m_5 c_5 K_5	0	0		0
6	0	0	$-c_6$ $-K_6$	0	0	m_6 c_6+c_7 K_6+K_7	$-c_7$ $-K_7$		0
7	0	0	0	0	0	$-c_7$ $-K_7$	m_7 c_7 K_7		0

Notice that this matrix is symmetrical about the diagonal that runs from the upper left corner to the lower right corner. This symmetry is typical for problems of this type.

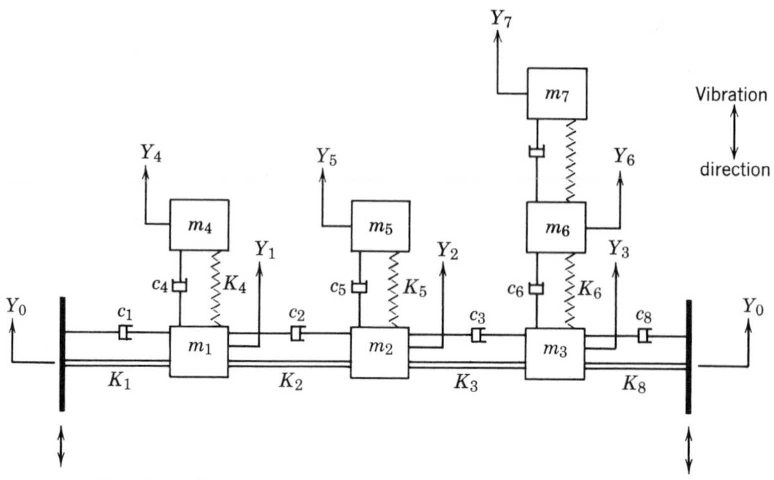

FIGURE 8.20. Coordinate system for the mathematical model of the electronic box.

8.17. LAGRANGE'S EQUATION

The differential equations of motion for the system shown in Fig. 8.20 can also be determined with the use of Lagrange's equation, which is shown by Eq. 3.128 and repeated here for convenience.

$$\frac{d}{dt}\left(\frac{\partial T}{\partial \dot{Y}_i}\right)-\frac{\partial T}{\partial Y_i}+\frac{\partial U}{\partial Y_i}+\frac{\partial D}{\partial \dot{Y}_i}=0 \tag{8.25}$$

The total kinetic energy T of the system is the sum of the kinetic energies of all the masses in the system.

$$T=\tfrac{1}{2}m_1\dot{Y}_1^2+\tfrac{1}{2}m_2\dot{Y}_2^2+\tfrac{1}{2}m_3\dot{Y}_3^2+\tfrac{1}{2}m_4\dot{Y}_4^2+\tfrac{1}{2}m_5\dot{Y}_5^2+\tfrac{1}{2}m_6\dot{Y}_6^2+\tfrac{1}{2}m_7\dot{Y}_7^2 \tag{8.26}$$

The total strain energy of the system is the sum of the strain energies in all of the springs.

$$\begin{aligned}U=&\tfrac{1}{2}K_1(Y_1-Y_0)^2+\tfrac{1}{2}K_2(Y_2-Y_1)^2+\tfrac{1}{2}K_3(Y_3-Y_2)^2+\tfrac{1}{2}K_4(Y_4-Y_1)^2\\&+\tfrac{1}{2}K_5(Y_5-Y_2)^2+\tfrac{1}{2}K_6(Y_6-Y_3)^2+\tfrac{1}{2}K_7(Y_7-Y_6)^2+\tfrac{1}{2}K_8(Y_0-Y_3)^2\end{aligned} \tag{8.27}$$

The total energy D dissipated in the dampers will be the sum of the dissipations in all of the dampers in the system.

$$\begin{aligned}D=&\tfrac{1}{2}c_1(\dot{Y}_1-\dot{Y}_0)^2+\tfrac{1}{2}c_2(\dot{Y}_2-\dot{Y}_1)^2+\tfrac{1}{2}c_3(\dot{Y}_3-\dot{Y}_2)^2+\tfrac{1}{2}c_4(\dot{Y}_4-\dot{Y}_1)^2\\&+\tfrac{1}{2}c_5(\dot{Y}_5-\dot{Y}_2)^2+\tfrac{1}{2}c_6(\dot{Y}_6-\dot{Y}_3)^2+\tfrac{1}{2}c_7(\dot{Y}_7-\dot{Y}_6)^2+\tfrac{1}{2}c_8(\dot{Y}_0-\dot{Y}_3)^2\end{aligned} \tag{8.28}$$

Following the operations indicated by Lagrange's equation, and starting with the kinetic energy equation (Eq. 8.26)

$$\frac{d}{dt}\left(\frac{\partial T}{\partial \dot{Y}_1}\right) = m_1\ddot{Y}_1 \quad \text{and} \quad \frac{\partial T}{\partial Y_1} = 0 \tag{8.29a}$$

$$\frac{d}{dt}\left(\frac{\partial T}{\partial \dot{Y}_2}\right) = m_2\ddot{Y}_2 \quad \text{and} \quad \frac{\partial T}{\partial Y_2} = 0 \tag{8.29b}$$

$$\frac{d}{dt}\left(\frac{\partial T}{\partial \dot{Y}_3}\right) = m_3\ddot{Y}_3 \quad \text{and} \quad \frac{\partial T}{\partial Y_3} = 0 \tag{8.29c}$$

$$\frac{d}{dt}\left(\frac{\partial T}{\partial \dot{Y}_4}\right) = m_4\ddot{Y}_4 \quad \text{and} \quad \frac{\partial T}{\partial Y_4} = 0 \tag{8.29d}$$

$$\frac{d}{dt}\left(\frac{\partial T}{\partial \dot{Y}_5}\right) = m_5\ddot{Y}_5 \quad \text{and} \quad \frac{\partial T}{\partial Y_5} = 0 \tag{8.29e}$$

$$\frac{d}{dt}\left(\frac{\partial T}{\partial \dot{Y}_6}\right) = m_6\ddot{Y}_6 \quad \text{and} \quad \frac{\partial T}{\partial Y_6} = 0 \tag{8.29f}$$

$$\frac{d}{dt}\left(\frac{\partial T}{\partial \dot{Y}_7}\right) = m_7\ddot{Y}_7 \quad \text{and} \quad \frac{\partial T}{\partial Y_7} = 0 \tag{8.29g}$$

Following the operations indicated by Lagrange's equation (8.25), and utilizing the strain energy equation (8.27)

$$\frac{\partial U}{\partial Y_1} = K_1(Y_1 - Y_0) - K_2(Y_2 - Y_1) - K_4(Y_4 - Y_1) \tag{8.30a}$$

$$\frac{\partial U}{\partial Y_2} = K_2(Y_2 - Y_1) - K_3(Y_3 - Y_2) - K_5(Y_5 - Y_2) \tag{8.30b}$$

$$\frac{\partial U}{\partial Y_3} = K_3(Y_3 - Y_2) - K_6(Y_6 - Y_3) - K_8(Y_0 - Y_3) \tag{8.30c}$$

$$\frac{\partial U}{\partial Y_4} = K_4(Y_4 - Y_1) \tag{8.30d}$$

$$\frac{\partial U}{\partial Y_5} = K_5(Y_5 - Y_2) \tag{8.30e}$$

$$\frac{\partial U}{\partial Y_6} = K_6(Y_6 - Y_3) - K_7(Y_7 - Y_6) \tag{8.30f}$$

$$\frac{\partial U}{\partial Y_7} = K_7(Y_7 - Y_6) \tag{8.30g}$$

Following the operations indicated by Lagrange's equation (8.25), and utilizing the dissipation energy equation (8.28)

$$\frac{\partial D}{\partial \dot{Y}_1} = c_1(\dot{Y}_1 - \dot{Y}_0) - c_2(\dot{Y}_2 - \dot{Y}_1) - c_4(\dot{Y}_4 - \dot{Y}_1) \tag{8.31a}$$

$$\frac{\partial D}{\partial \dot{Y}_2} = c_2(\dot{Y}_2 - \dot{Y}_1) - c_3(\dot{Y}_3 - \dot{Y}_2) - c_5(\dot{Y}_5 - \dot{Y}_2) \tag{8.31b}$$

$$\frac{\partial D}{\partial \dot{Y}_3} = c_3(\dot{Y}_3 - \dot{Y}_2) - c_6(\dot{Y}_6 - \dot{Y}_3) - c_8(\dot{Y}_0 - \dot{Y}_3) \tag{8.31c}$$

$$\frac{\partial D}{\partial \dot{Y}_4} = c_4(\dot{Y}_4 - \dot{Y}_1) \tag{8.31d}$$

$$\frac{\partial D}{\partial \dot{Y}_5} = c_5(\dot{Y}_5 - \dot{Y}_2) \tag{8.31e}$$

$$\frac{\partial D}{\partial \dot{Y}_6} = c_6(\dot{Y}_6 - \dot{Y}_3) - c_7(\dot{Y}_7 - \dot{Y}_6) \tag{8.31f}$$

$$\frac{\partial D}{\partial \dot{Y}_7} = c_7(\dot{Y}_7 - \dot{Y}_6) \tag{8.31g}$$

Substituting Eqs. 8.29–8.31 into Eq. 8.25 for the conditions where $i = 1$

$$m_1\ddot{Y}_1 + c_1(\dot{Y}_1 - \dot{Y}_0) - c_2(\dot{Y}_2 - \dot{Y}_1) - c_4(\dot{Y}_4 - \dot{Y}_1) + K_1(Y_1 - Y_0) - K_2(Y_2 - Y_1) - K_4(Y_4 - Y_1) = 0$$

$$m_1\ddot{Y}_1 + \dot{Y}_1(c_1 + c_2 + c_4) - c_2\dot{Y}_2 - c_4\dot{Y}_4 + Y_1(K_1 + K_2 + K_4) - K_2Y_2 - K_4Y_4 = c_1\dot{Y}_0 + K_1Y_0 \tag{8.32}$$

This is the same as Eq. 8.18.

Substituting Eqs. 8.29–8.31 into Eq. 8.25 for the conditions where $i = 2$

$$m_2\ddot{Y}_2 + c_2(\dot{Y}_2 - \dot{Y}_1) - c_3(\dot{Y}_3 - \dot{Y}_2) - c_5(\dot{Y}_5 - \dot{Y}_2) + K_2(Y_2 - Y_1) - K_3(Y_3 - Y_2) - K_5(Y_5 - Y_2) = 0$$

$$m_2\ddot{Y}_2 + \dot{Y}_2(c_2 + c_3 + c_5) - c_2\dot{Y}_1 - c_3\dot{Y}_3 - c_5\dot{Y}_5 + Y_2(K_2 + K_3 + K_5) - K_2Y_1 - K_3Y_3 - K_5Y_5 = 0 \tag{8.33}$$

This is the same as Eq. 8.19.

Substituting Eqs. 8.29–8.31 into Eq. 8.25 for the conditions where $i=3$

$$m_3\ddot{Y}_3+c_3(\dot{Y}_3-\dot{Y}_2)-c_6(\dot{Y}_6-\dot{Y}_3)-c_8(\dot{Y}_0-\dot{Y}_3)+K_3(Y_3-Y_2)$$
$$-K_6(Y_6-Y_3)-K_8(Y_0-Y_3)=0$$
$$m_3\ddot{Y}_3+\dot{Y}_3(c_3+c_6+c_8)-c_3\dot{Y}_2-c_6\dot{Y}_6+Y_3(K_3+K_6+K_8)-K_3Y_2$$
$$-K_6Y_6=c_8\dot{Y}_0+K_8Y_0 \tag{8.34}$$

This is the same as Eq. 8.20.

Substituting Eqs. 8.29–8.31 into Eq. 8.25 for the conditions where $i=4$

$$m_4\ddot{Y}_4+c_4(\dot{Y}_4-\dot{Y}_1)+K_4(Y_4-Y_1)=0$$
$$m_4\ddot{Y}_4+c_4\dot{Y}_4-c_4\dot{Y}_1+K_4Y_4-K_4Y_1=0 \tag{8.35}$$

This is the same as Eq. 8.21.

Substituting Eqs. 8.29–8.31 into Eq. 8.25 for the conditions where $i=5$

$$m_5\ddot{Y}_5+c_5(\dot{Y}_5-\dot{Y}_2)+K_5(Y_5-Y_2)=0$$
$$m_5\ddot{Y}_5+c_5\dot{Y}_5-c_5\dot{Y}_2+K_5Y_5-K_5Y_2=0 \tag{8.36}$$

This is the same as Eq. 8.22.

Substituting Eqs. 8.29–8.31 into Eq. 8.25 for the conditions where $i=6$

$$m_6\ddot{Y}_6+c_6(\dot{Y}_6-\dot{Y}_3)-c_7(\dot{Y}_7-\dot{Y}_6)+K_6(Y_6-Y_3)-K_7(Y_7-Y_6)=0$$
$$m_6\ddot{Y}_6+\dot{Y}_6(c_6+c_7)-c_6\dot{Y}_3-c_7\dot{Y}_7+Y_6(K_6+K_7)-K_6Y_3-K_7Y_7=0 \tag{8.37}$$

This is the same as Eq. 8.23.

Substituting Eqs. 8.29–8.31 into Eq. 8.25 for the conditions where $i=7$

$$m_7\ddot{Y}_7+c_7(\dot{Y}_7-\dot{Y}_6)+K_7(Y_7-Y_6)=0$$
$$m_7\ddot{Y}_7+c_7\dot{Y}_7-c_7\dot{Y}_6+K_7K_7-K_7Y_6=0 \tag{8.38}$$

This is the same as Eq. 8.24.

If the physical constants from Table 8.1 are substituted into the matrix shown in Table 8.2, the results will be a set of seven simultaneous differential equations that represent the equations of motion for the mathematical model of the electronic box shown in Figs. 8.14 and 8.20 during vibration in the vertical direction along the Y axis. The completed matrix is shown in Table 8.3.

TABLE 8.3. VIBRATION ALONG THE Y AXIS

	Column 1	Column 2	Column 3	Column 4	Column 5	Column 6	Column 7		Column 8
Row 1	1.19×10^{-2} 14.49 9.40×10^5	−4.10 -2.80×10^5	0	−2.64 -0.488×10^5	0	0	0		7.75 6.11×10^5
Row 2	−4.10 -2.80×10^5	0.725×10^{-2} 10.26 5.69×10^5	−3.98 -2.65×10^5	0	−2.18 -0.242×10^5	0	0		0
Row 3	0	−3.98 -2.65×10^5	0.570×10^{-2} 13.74 5.56×10^5	0	0	−7.18 -1.50×10^5	0		2.58 1.41×10^5
Row 4	−2.64 -0.488×10^5	0	0	0.700×10^{-2} 2.64 0.488×10^5	0	0	0	=	0
Row 5	0	−2.18 -0.242×10^5	0	0	0.959×10^{-2} 2.18 0.242×10^5	0	0		0
Row 6	0	0	−7.18 -1.50×10^5	0	0	1.24×10^{-2} 8.39 1.98×10^5	−1.21 -0.482×10^5		0
Row 7	0	0	0	0	0	−1.21 -0.482×10^5	0.596×10^{-2} 1.21 0.482×10^5		0

There are three items in the diagonal boxes that run from the upper left to the lower right. The top entry in each of these boxes is the mass coefficient from the second order differential, $\ddot{Y}$. The middle entry in each of these boxes is the damping coefficient from the first order differential, $\dot{Y}$. The bottom entry is the spring rate.

In the remaining boxes that have two items, the top entry is the damping coefficient from the first order differential, $\dot{Y}$. The bottom entry is the spring rate.

8.18. Setting Up the Digital Computer to Determine the Frequency Response

The matrix shown in Table 8.3 can be solved very conveniently with the use of a high-speed digital computer. If the solution is obtained at many frequency points, the result will be a frequency response of every mass in the system. A plot of these points for each mass will then result in a transmissibility curve for each mass. This family of curves will be very similar to the family of curves that will be generated by running a sinusoidal vibration test on the electronic box, where seven accelerometers are mounted on the seven major structural masses as shown in Figs. 8.15 and 8.16.

IBM has a computer program called LISA, (Linear Systems Analysis). This program is very convenient to use for solving steady-state and transient problems associated with vibration and shock. This program is available to IBM customers using the IBM 360 and 370 series computers. It is simple to set up and provides plots of the transmissibilities and phase angles for each mass in the system. A sample problem using the LISA program is shown in Section 3.9 for additional reference.

The following is a listing of the data cards that will be required to solve for the frequency response of each mass from 0 to 2000 Hz in 20-Hz increments and to obtain a plot of the transmissibility and phase angle for each mass. Each line in the listing represents one data card using the data shown in Table 8.3. The letter S stands for symmetry in the matrix and its use reduces the amount of data that must be entered.

The LISA program can handle up to 50 masses when the proper functions are used. The following sample program will handle only the 7 masses shown with the use of the polynomial function "POLY," but the computer operation is very fast.

The LISA program runs the calculations only in groups of four, so the first group of four masses are represented by V1, V2, V3, and V4. In order to run the additional three masses V5, V6, and V7, a second entry of the same data must be made as shown.

```
TITLE, (up to 66 spaces can be used for any title)
READ, MATRIX
1,1,2,1.19E-2,14.49,9.40E5
1,2,1,S,-4.10,-2.80E5
1,4,1,S,-2.64,-.488E5
1,8,1,7.75,6.11E5
2,2,2,.725E-2,10.26,5.69E5
2,3,1,S,-3.98,-2.65E5
2,5,1,S,-2.18,-.242E5
3,3,2,.570E-2,13.74,5.56E5
3,6,1,S,-7.18,-1.50E5
3,8,1,2.58,1.41E5
4,4,2,.700E-2,2.64,.488E5
5,5,2,.959E-2,2.18,.242E5
6,6,2,1.24E-2,8.39,1.98E5
6,7,1,S,-1.21,-.482E5
7,7,2,.596E-2,1.21,.482E5
DEFINE, V1 = V(1)
DEFINE, V2 = V(2)
DEFINE, V3 = V(3)
DEFINE, V4 = V(4)
COMPUTE, POLY, V1, V2, V3, V4
DATA, FREQUENCY = 0, 2000, 20, CPS, LIN
COMPUTE, BODE, V1, V2, V3, V4
LABEL, (up to 66 spaces can be used for any label)
PPLOT
DEFINE, V5 = V(5)
DEFINE, V6 = V(6)
DEFINE, V7 = V(7)
COMPUTE, POLY, V5, V6, V7
DATA, FREQUENCY = 0,2000,20,CPS,LIN
COMPUTE, BODE, V5, V6, V7
LABEL, (up to 66 spaces for the label)
PPLOT
EXIT
```

The three numbers in the first three columns of the data input, such as: 1,1,2, or 1,2,1, represent, respectively, the row of the matrix, the column of the matrix, and the highest order of the differential in the data (as shown in Table 8.3). For example, $\ddot{Y}$ is a second-order differential which appears only in the diagonal of the matrix and is shown at the top position of its

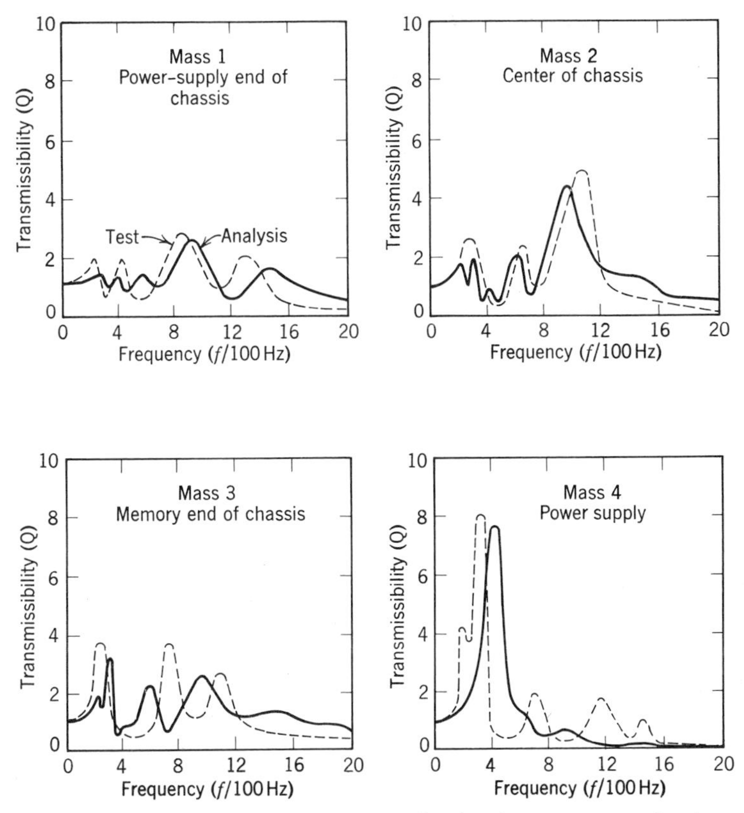

FIGURE 8.21. Frequency response curves showing the computer results along with the test results for masses 1–4.

box in the matrix, so a 2 is entered. A $\dot{Y}$ is a first-order differential, so a 1 is entered in the third position following the row and the column.

A total computer running time of 1.54 min was required for the solution of 100 discrete frequency points for each of the seven masses in the mathematical model, using a 360-50 IBM computer.

A plot of transmissibility versus frequency for each mass, as determined by the computer, is shown in Figs. 8.21 and 8.22 with a solid line. A series of vibration tests were also run on the same electronic box, as shown in Fig. 8.14, and these results are also shown in Figs. 8.21 and 8.22 with a broken line. The test data were obtained from accelerometers placed on structural and mass sections that closely simulated the computer mathematical model as shown in Fig. 8.16, using a 2-G-peak and a 5-G-peak sinusoidal-vibration input to a very rigid vibration fixture.

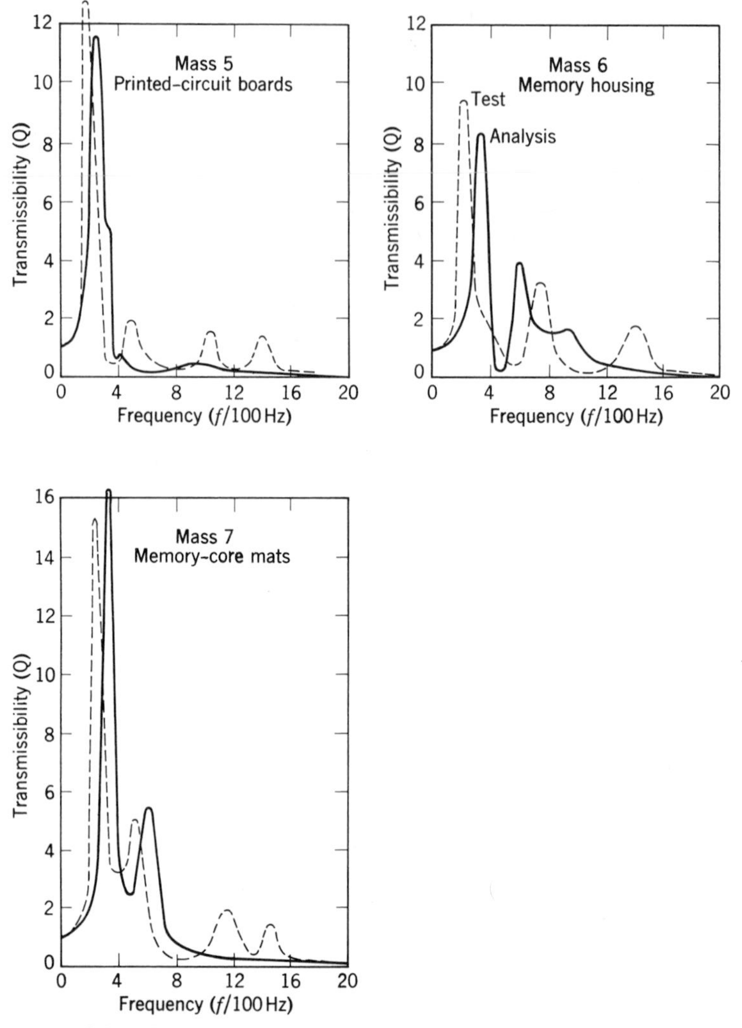

FIGURE 8.22. Frequency response curves showing the computer results along with the test results for masses 5–7.

8.19. Dynamic Forces in the Springs

In order to determine the dynamic stresses that will be developed in various sections of the electronic box during vibration along the Y axis, it is necessary to determine the dynamic forces in the springs. The springs represent the stiffness of various structural elements in the electronic box which were approximated as combinations of beams, plates, and frames.

Consider the mounting lugs at each end of the chassis as shown in Fig. 8.14. The dynamic loads must pass through these end lugs because they are used to support the electronic box. The dynamic loads must be those loads that are developed in springs K_1 and K_8 at each end of the box, as shown in Figs. 8.15 and 8.16, because the load path runs through these springs.

Equations for determining the dynamic loads in the springs of lumped-mass systems are derived in Chapter 3. The equation for the load in the spring that is attached to the support is shown by Eq. 3.157 as

$$P_1 = \frac{9.8 G_{in} K_1}{f^2} (Q_1^2 - 2Q_1 \cos \phi_1 + 1)^{1/2} \quad (8.39)$$

Considering a 4.0-G-peak sinusoidal-vibration input to the electronic box, the spring forces can be determined at a frequency of 300 Hz. Figure 8.21 shows that the maximum mounting-lug forces will probably occur near this frequency.

Using the results of the computer program, the dynamic load can be determined in spring K_1 as follows (see Figs. 8.21 and 8.22):

where $Q_1 = 1.43$ from the computer results for mass 1
$\phi_1 = 16.5°$ from the computer results for mass 1
$f = 300$ Hz (see Fig. 8.21)
$G_{in} = 4.0$-G-peak input acceleration
$K_1 = 6.11 \times 10^5$ lb/in. (see Eq. 8.5)

Substituting into Eq. 8.39 for the force in spring K_1

$$P_1 = \frac{9.8(4.0)(6.11 \times 10^5)}{(300)^2} [(1.43)^2 - 2(1.43) \cos 16.5^0 + 1]^{1/2}$$

$$P_1 = 146 \text{ lb} \quad (8.40)$$

The dynamic load in spring K_8 can be determined from the results of the computer program as follows (see Figs. 8.21 and 8.22):

$Q_3 = 3.15$ from the computer results for mass 3
$\phi_3 = 34.6^0$ from the computer results for mass 3
$f = 300$ Hz (see Fig. 8.21)
$K_8 = 1.41 \times 10^5$ lb/in. (see Eq. 8.5)
$G_{in} = 4.0$ peak

Equation 3.157 can be modified to determine the dynamic load in spring K_8 as follows:

$$P_8 = \frac{9.8 G_{in} K_8}{f^2} (Q_3^2 - 2Q_3 \cos \phi_3 + 1)^{1/2} \quad (8.41)$$

Substituting into the above equation for the force in spring K_8

$$P_8 = \frac{9.8(4.0)(1.41 \times 10^5)}{(300)^2} [(3.15)^2 - 2(3.15) \cos 34.6^0 + 1]^{1/2}$$

$$P_8 = 147 \text{ lb} \tag{8.42}$$

The dynamic loads in the power-supply section of the electronic box must be determined by an equation that considers a spring between two moving masses. This equation is also derived in Chapter 3 (Eq. 3.163). For spring K_4 between masses 4 and 1, this equation is

$$P_4 = \frac{9.8 G_{in} K_4}{f^2} [Q_4{}^2 - 2Q_4 Q_1 (\cos \phi_4 \cos \phi_1 + \sin \phi_4 \sin \phi_1) + Q_1{}^2]^{1/2} \tag{8.43}$$

Using the results of the computer program, it appears that the maximum dynamic loads in the power-supply section will occur at about 400 Hz (Fig. 8.21). The data in the computer run are based on 20-Hz increments, so the peak transmissibilities may not show up. Another computer run should be made with 2-Hz increments in the resonant area to determine the peak transmissibility values for each mass.

However using the data from the computer run at a frequency of 400 Hz gives a good approximation of the dynamic loads, since the structure is near its peak resonance. Then

$f = 400$ Hz (see Fig. 8.21)
$Q_4 = 7.6$ computer results for mass 4
$\phi_4 = 74.2^0$ computer results for mass 4
$Q_1 = 1.2$
$\phi_1 = 26.6^0$
$G_{in} = 4.0$ peak
$K_4 = 0.488 \times 10^5$ lb/in. (see Eq. 8.7)

Substituting into Eq. 8.43 for the force in spring K_4

$$P_4 = \frac{9.8(4.0)(0.488 \times 10^5)}{(400)^2}$$
$$\{(7.6)^2 - 2(7.6)(1.2)[(0.272)(0.894) + (0.962)(0.448)] + (1.2)^2\}^{1/2}$$

$$P_4 = 82.5 \text{ lb} \tag{8.44}$$

The dynamic loads in the memory-core-housing spring K_6, shown in Fig. 8.19, can be determined from Eq. 3.163 as follows:

$$P_6 = \frac{9.8 G_{in} K_6}{f^2} [Q_6{}^2 - 2Q_6 Q_3 (\cos \phi_6 \cos \phi_3 + \sin \phi_6 \sin \phi_3) + Q_3{}^2]^{1/2} \tag{8.45}$$

An examination of the curves in Fig. 8.22 shows the maximum dynamic loads will probably occur at about 320 Hz. Using the results from the computer run

$$
\begin{aligned}
f &= 320 \text{ Hz} \\
Q_6 &= 8.3 \text{ from computer results for mass 6} \\
\phi_6 &= 122.5^0 \text{ from computer results for mass 6} \\
Q_3 &= 3.0 \\
\phi_3 &= 110^0 \\
G_{in} &= 4.0 \text{ peak} \\
K_6 &= 1.50 \times 10^5 \text{ lb/in. (see Eq. 8.9)}
\end{aligned}
$$

Substituting into the above equation for the force in spring K_6

$$
P_6 = \frac{9.8(4.0)(1.50 \times 10^5)}{(320)^2} \{(8.3)^2 - 2(8.3)(3.0)[(-0.537)(-0.342) + (0.843)(0.940)] + (3.0)^2\}^{1/2}
$$

$$
P_6 = 312 \text{ lb} \tag{8.46}
$$

FIGURE 8.23. A "head-up display" electronic box that displays information for the pilot of an airplane (courtesy Norden division of United Aircraft).

8.20. Dynamic Stresses in the Springs

The dynamic load in spring K_1 was shown to be 146 lb in Eq. 8.40. This load must pass through the mounting lugs shown in Fig. 8.15. An end view of the electronic box showing the details of the mounting lugs is shown in Fig. 8.24.

There are two mounting lugs at each end of the chassis, so each lug will take half of the total load acting at the end, when the CG of the chassis is centered between the lugs.

$$R_1 = \frac{P_1}{2} = \frac{146}{2} = 73.0 \text{ lb} \tag{8.47}$$

Since the mounting lugs are riveted to the chassis, the dynamic loads must pass through the rivets, placing them in shear. The rivet shear stress for a 0.182-in.-diameter rivet is then

$$S_1 = \frac{R_1}{A} \tag{8.48}$$

where $A = 2(\pi/4)(0.182)^2 = 0.052 \text{ in.}^2$ shear area of 2 rivets. Substitute Eq. 8.47 and the above shear area into Eq. 8.48

$$S_1 = \frac{73.0}{0.052} = 1400 \text{ lb/in.}^2 \tag{8.49}$$

An examination of the fatigue curves for aluminum rivets, as shown in Fig. 10.9*a*, shows the rivets will probably have an infinite fatigue life during vibration in the vertical direction.

The dynamic stresses in the power-supply section of the electronic box

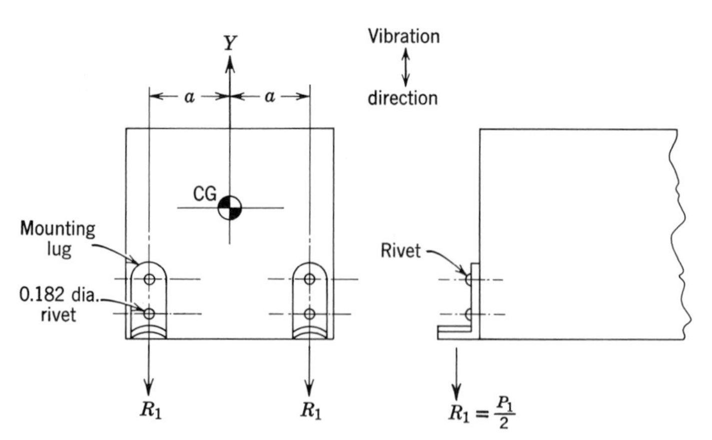

FIGURE 8.24. Mounting lugs at the end of the chassis.

can be determined from the dynamic load of 82.5 lb in spring K_4 as shown by Eq. 8.44. In Section 8.7, where the power-supply structure was analyzed, it was pointed out that six elements of the power-supply section were lumped together in order to simplify the dynamic analysis. The dynamic load of 82.5 lb, acting on the power-supply section, really acts on the 6 elements lumped together. Because each element has the geometry shown in Fig. 8.18, the dynamic stress can be computed for all six elements combined.

Since the vibration analysis was based on the use of concentrated masses, where single-degree-of-freedom systems were used to approximate uniformly distributed loads, the stress analysis is also based on the use of concentrated masses. The dynamic loading and the bending moment then appear as shown in Fig. 8.25.

The bending stress in spring K_4 can be determined from the standard bending stress equation as follows:

$$S_4 = \frac{Mc}{I}$$

where $M = 41.25\ (2.5) = 103$ lb in. maximum bending moment

$$c = \frac{0.158}{2} = 0.079 \text{ in.}$$

$$I = (6)\frac{bh^3}{12} = \frac{(6)(3.0)(0.158)^3}{12} = 5.92 \times 10^{-3} \text{in.}^4$$

$$S_4 = \frac{(103)(0.079)}{0.592 \times 10^{-3}} = 13{,}700 \text{ lb/in.}^2 \tag{8.50}$$

An examination of the fatigue curves for aluminum alloys, as shown in Figs. 10.12–10.14, shows that the power-supply section has an infinite fatigue life for vibration in the vertical direction.

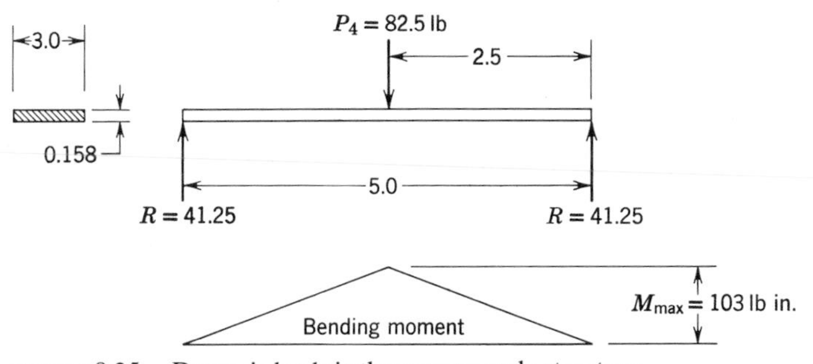

FIGURE 8.25. Dynamic loads in the power supply structure.

The dynamic stresses in the memory-core housing, spring K_6, can be determined from the dynamic load of 312 lb developed in the spring (Eq. 8.46). The geometry of the mounting flanges is shown in Fig. 8.26.

Since the structure is symmetrical, the dynamic load on one mounting flange is

$$R_6 = \frac{312}{8} = 39.0 \text{ lb}$$

The dynamic bending stress at the base of the flange can be determined by considering each flange as a cantilevered beam. If the mating structure supporting the flanges is very stiff and large screws are used to hold the mounting flanges, then the mounting flanges will not act as cantilevered beams, and the bending moments at the base of the flange will be reduced.

Approximating the mounting flange as a cantilevered beam and ignoring the effects of stress concentrations, the bending stress is

$$S_6 = \frac{Mc}{I}$$

where $M = 39\,(0.30) = 11.7$ lb in.

$$c = \frac{0.06}{2} = 0.030 \text{ in.}$$

$$I = \frac{bh^3}{12} = \frac{(0.90)\,(0.06)^3}{12} = 1.62 \times 10^{-5} \text{ in.}^4$$

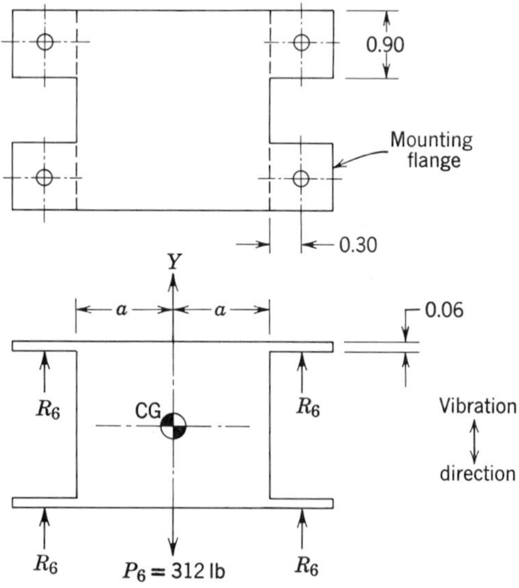

FIGURE 8.26. Dynamic loads in the memory core housing structure.

Substituting into the above equation

$$S_6 = \frac{(11.7)(0.03)}{1.62 \times 10^{-5}} = 21{,}600 \text{ lb/in.}^2 \tag{8.51}$$

The fatigue life for this mounting flange is examined in the following section.

8.21. Predicting the Fatigue Life

Once the dynamic stresses have been determined, the effective fatigue life of the different structural elements can be approximated by considering the type of vibration environment that will be imposed. For military equipment, specifications such as MIL-E-5400, MIL-T-5422, MIL-STD-810, MIL-STD-740, MIL-E-5272, and MIL-STD-167 list the types of tests and describe the way in which the tests are to be performed. The tests usually require 30-min dwells at each of the four major resonant points, in each of the three mutually perpendicular axes; this adds up to a total of 6 hr that must be spent dwelling at the most severe resonant points. Many millions of fatigue cycles, at high stress levels, can be accumulated rapidly under these conditions.

In order to determine whether a structural member can withstand the number of fatigue cycles expected during a specific test, "Miner's" cumulative fatigue ratio can be used. (See Chapter 10 for more details.) This ratio compares the number of fatigue cycles, n, accumulated at different stress levels, to the number of fatigue cycles, N, required to produce a failure at each of these stress levels. This is usually written as follows:

$$R_n = \frac{n_1}{N_1} + \frac{n_2}{N_2} + \frac{n_3}{N_3} + \cdots = \sum \frac{n_i}{N_i} = 1.0 \tag{8.52}$$

Failure is supposed to occur when the fatigue-cycle ratio R_n is equal to 1.0, at this point all of the fatigue life is "used up." Freudenthal[16] and Crandall[1] indicate that a value of 1.0 may be too high in many cases and suggest substantially lower values.

Vibration test experience with electronic equipment used in airplanes indicates that many systems developed failures when the fatigue-cycle ratio R_n was between about 0.70 and 1.3. If a conservative design ap-approach is desirable, then the maximum fatigue-cycle ratio should be about

$$R_n = 0.70 \tag{8.53}$$

The use of the fatigue-cycle ratio can be demonstrated by considering a qualification test program for the electronic box shown in Fig. 8.14. This program will consist of a sinusoidal-vibration test series in three parts, as follows:

Part 1: An input of 4.0 G peak with a 30-min resonant dwell.

Part 2: An input of 6.0 G peak with a 1.0-min resonant dwell.

Part 3: Four sweeps from 5 Hz to 500 Hz (with a 4.0-G-peak input) and then back down to 5 Hz, sweeping at a rate of 1.0 octave/min (5 to 10 Hz in 1 min, 10 to 20 Hz in 1 min, etc.).

An examination of the dynamic stresses determined in Section 8.20 for a 4.0-G-peak input shows that Eq. 8.51 has the highest stress with a value of 21,600 psi in the memory-core-housing mounting flange. This stress is developed at a frequency of 320 Hz, as shown in Eq. 8.46.

Considering part 1 of the qualification test series, the number of fatigue cycles, n, that are accumulated by a 30-min resonant swell at 320 Hz are as follows:

$$n_1 = 30 \text{ min } (60 \text{ sec/min}) (320 \text{ cycles/sec}) = 5.76 \times 10^5 \text{ actual fatigue cycles} \tag{8.54}$$

The approximate number of fatigue cycles, N, required to produce a fatigue failure in the mounting flange on the memory-core housing can be determined from the S–N curve for 6061-T4 aluminum alloy as shown in Chapter 10, Fig. 10.14*b*, for a bending stress of 21,600 psi, as follows:

$$N_1 = 6.0 \times 10^5 \text{ cycles} \tag{8.55}$$

Considering part 2 of the qualification test series, with an input of 6.0 G peak, the bending stress in the memory-core-housing mounting flange can be determined from a direct ratio of the acceleration G forces, using Eq. 8.51.

$$S_6 = \frac{6.0}{4.0} (21{,}600) = 32{,}400 \text{ lb/in.}^2$$

The actual number of fatigue cycles accumulated by a 1-min resonant dwell at 320 Hz are

$$n_2 = 1.0 \text{ min } (60 \text{ sec/min})(320 \text{ cycles/sec}) = 1.92 \times 10^4 \text{ actual fatigue cycles} \tag{8.56}$$

The approximate number of cycles required to produce a fatigue failure with a bending stress of 32,400 psi can be determined from Fig. 10.14*b*.

$$N_2 = 5.0 \times 10^3 \text{ cycles} \tag{8.57}$$

Considering part 3 of the qualification test series, the time it takes to sweep through the resonance at 320 Hz can be approximated by considering the bandwidth at the half-power points as shown in Chapter 4, Fig.

4.8, using a sweep rate of 1 octave/min. The time it takes to sweep through the resonance is shown by Eq. 4.16 as

$$t = \frac{\log_e \dfrac{1+\dfrac{1}{2Q}}{1-\dfrac{1}{2Q}}}{R \log_e 2}$$

where $Q = 8.3$ (see Q_6 shown in Eq. 8.46)
$R = 1.0$ sweep rate in octaves/min

Substituting into the above equation

$$t = \frac{\log_e \dfrac{1+\dfrac{1}{16.6}}{1-\dfrac{1}{16.6}}}{(1.0) \log_e 2} = 0.176 \text{ min} = 10.6 \text{ sec}$$

The actual number of fatigue cycles accumulated by four sweeps from 5 Hz to 500 Hz and back to 5 Hz, considering only the resonant band between the half-power points, is

$$n_3 = 4(2)\,(10.6 \text{ sec})\,(320 \text{ cycles/sec}) = 2.71 \times 10^4 \text{ fatigue cycles} \tag{8.58}$$

With a 4.0-G peak input, the approximate number of fatigue cycles required to produce a fatigue failure is the same as that shown in Eq. 8.55.

$$N_3 = 6.0 \times 10^5 \text{ cycles} \tag{8.59}$$

The fatigue cycle ratio R_n for the memory-core-housing mounting flange can be determined by substituting Eqs. 8.54–8.59 into Eq. 8.52.

$$R_n = \frac{5.76 \times 10^5}{6.0 \times 10^5} + \frac{1.92 \times 10^4}{5.0 \times 10^3} + \frac{2.71 \times 10^4}{6.0 \times 10^5}$$

$$R_n = 0.961 + 3.84 + 0.045 = 4.846 \tag{8.60}$$

Since the fatigue cycle ratio is substantially greater than the value of 0.70, as shown by Eq. 8.53, the mounting flange is unsatisfactory.

If the mounting-flange thickness can be increased, or ribs can be added, or a stronger material can be used, then it may be possible to reduce the fatigue-cycle ratio to a value below 0.70 and thus provide a satisfactory fatigue life.

The analysis shown in the preceding sections considered vibration only in the vertical direction, along the Y axis. In order to complete the analysis of this particular electronic box, a vibration study must also be made along the longitudinal axis and along the lateral axis. During vibration along the lateral axis, the torsional mode couples with the lateral bending mode because the mounting lugs at the ends of the chassis are well below the center of gravity (CG) of the box (Fig. 8.14). This coupling tends to reduce the resonant frequency during vibration along this axis. See Chapter 7 for more details on coupled modes.

8.22. Expanding the Mathematical Model

The electronic box shown in Fig. 8.14 can be modeled by considering each of the four memory-core mats as concentrated masses and each of the four printed-circuit boards as concentrated masses. The power-supply section can be lumped into six individual masses if all of the six sections are quite different, or into three concentrated masses if there is some symmetry. If the chassis structure is still represented by three concentrated masses, then the mathematical model of the electronic box appears as shown in Figs. 8.27 and 8.28.

Adding additional springs, masses, and dampers permits the dynamic loads to be determined at many additional points in the system. This results in a more accurate picture of the stress distribution throughout the electronic chassis. The additional springs require a substantial amount of additional work to calculate the required spring rates and the additional masses add to the number of simultaneous differential equations that must be solved by the computer.

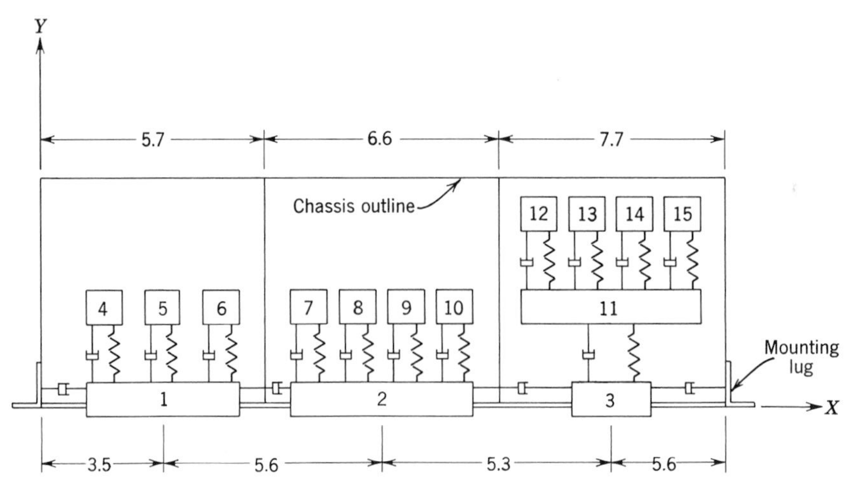

FIGURE 8.27. Simulating the electronic chassis as a 15 mass system.

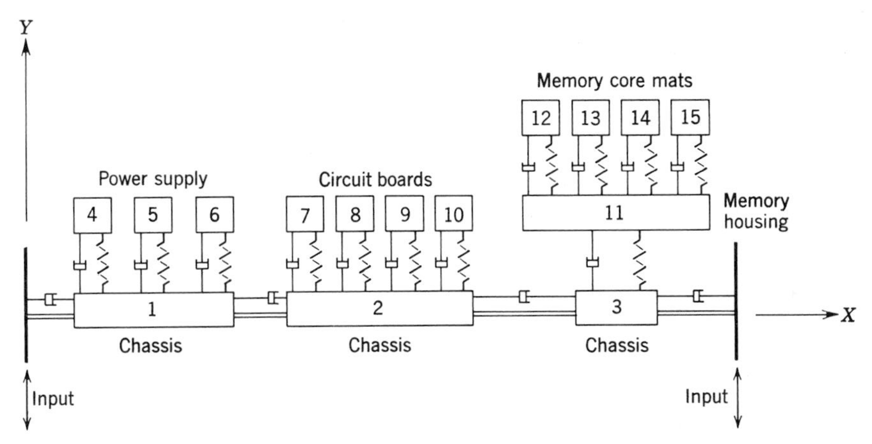

FIGURE 8.28. Mathematical model of the electronic chassis with 15-degrees-of-freedom.

8.23. Mathematical Model for an Infrared Scanning Radiometer

An electro-mechanical system that requires the use of a slowly rotating mirror is shown in Fig. 8.29 to demonstrate how a system of this type can be analyzed by breaking it up into discrete masses, springs, and dampers. This device contains a polished beryllium mirror which picks up infrared signals that are converted to record temperatures.

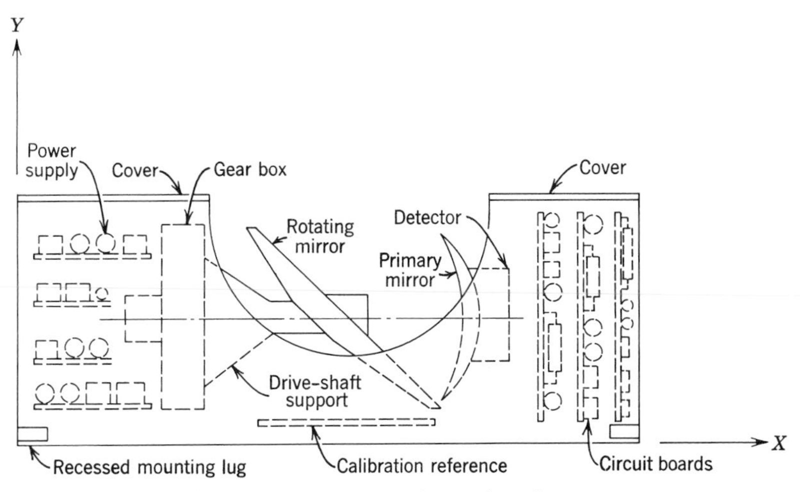

FIGURE 8.29. An electronic box with a slowly rotating mirror.

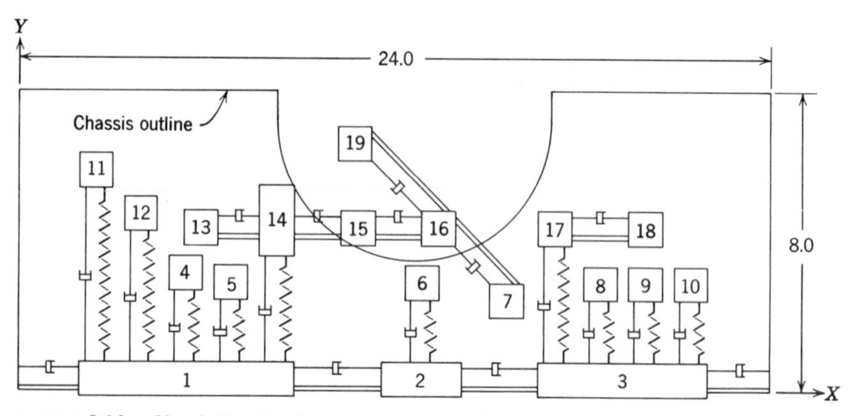

FIGURE 8.30. Simulating the electronic box as a 19 mass system.

Because mathematical modeling is more of an art than a science, different people will tend to model a complex system in different ways. The mathematical model presented in Figs. 8.30 and 8.31 was used to analyze this particular structure, using the methods outlined in this chapter. Vibration tests were then run on a mock-up of the structure, in each of the three mutually perpendicular axes, to determine the frequency response characteristics of the system. The results showed good agreement for the major resonances in the system, indicating that this particular mathematical model is representative of the actual system. Figure 8.32 shows an electronic box with a mirror that is used to detect energy in the infrared range.

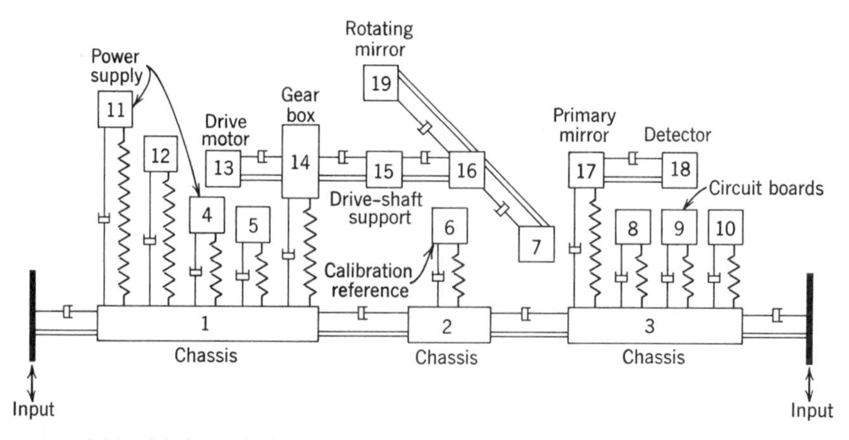

FIGURE 8.31. Mathematical model of the electronic box with 19-degrees-of-freedom.

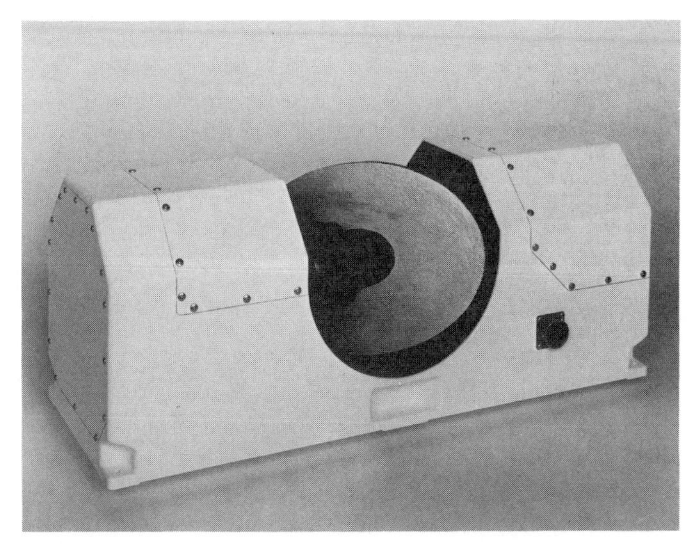

FIGURE 8.32. An electronic box with a rotating beryllium mirror that is used to detect energy in the infrared range (courtesy Barnes Engineering Company).

CHAPTER 9

VIBRATION FIXTURES AND VIBRATION TESTING

9.1. Vibration-Simulation Equipment

Vibration frequencies for high-speed vehicles, such as missiles and airplanes, generally extend to 1000 Hz and often up to 5000 Hz. Most specifications for these vehicles require tests that range from about 5 to 2000 Hz. In order to produce harmonic motion over such a broad frequency range, electrodynamic shakers are generally used. These shakers, or vibration machines, are very much like a loudspeaker which has a moving coil. Instead of connecting to a speaker, the moving coil is connected to an armature which becomes the shaker head that simulates the desired harmonic motion.

The armature has a driving coil which is cylindrical in shape and the shaker head is usually an extension of this shape. The shaker head itself must be a very rigid member in order to control the displacement amplitudes of the head at high frequencies. Electrodynamic shakers that are capable of producing frequencies up to 2000 Hz normally have their shaker head resonances above 3000 Hz.

In order to provide a pure translational motion for the shaker head,

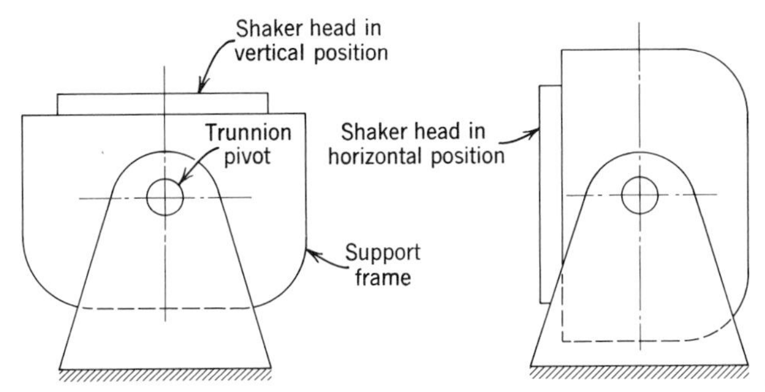

FIGURE 9.1. A typical vibration machine.

various types of flexures or springs, metal and rubber, are often used between the shaker head and the support frame. Even with these devices, many rotational modes may often develop in the shaker head during severe resonant conditions.

Vibration machines are generally rated in terms of the peak force in pounds, based on sinusoidal wave motion. These machines are available with various force ratings that range from 25 to 25,000 lb. The choice of the vibration machine would depend upon the maximum weight of the system to be tested and the maximum acceleration force required by the test.

Most vibration tests on electronic units require acceleration inputs along each of the three mutually perpendicular axes of the system. In order to provide this type of adaptability, the vibration machine usually has the shaker head mounted on a trunnion that permits the head to be rotated and locked in different positions (Fig. 9.1).

9.2. MOUNTING THE VIBRATION MACHINE

Vibration machines are usually mounted in two different ways, depending upon their function. If a mobile test system is required and the shaker must be moved to different locations frequently, the vibration machine should be mounted on vibration isolators. Large masses cannot be vibrated effectively with large acceleration G forces using this type of installation. In order to keep the shaker from dissipating a great deal of energy shaking itself, the vibration machine should be mounted on a large concrete block at least ten times heavier than the machine. This large mass should be isolated from the building structure because low-frequency resonances may damage the building structure.

9.3. Vibration Test Fixtures

The shaker head on a vibration machine usually has some form of a hole pattern that permits the installation of machine screws. These holes can often be used to mount small electronic components for vibration testing. Large electronic boxes require some sort of a mechanical adapter that will permit the shaker head to transfer vibrational motion to the electronic box. This adapter, or vibration test fixture as it is commonly called, should really be an extension of the armature in the form of a very rigid structure that can transfer the required force at the required frequency.

An optimum fixture would have the lowest natural frequency about 50% higher than the highest required forcing frequency in order to avoid fixture resonances during the test. For example, if the vibration test requirements have a range of 5–500 Hz, the vibration test fixture should have its fundamental resonant mode at about 750 Hz when the effective mass of the test specimen is included in the loading on the test fixture. Since most electronic boxes are not too heavy, a vibration-fixture resonance of 750 Hz is not too difficult to obtain. However if the vibration test must go to 2000 Hz, the desirable fixture resonance would be about 3000 Hz. This may be very difficult to obtain unless the test specimen happens to be quite light, probably less than about 20 lb.

If the test specimen is relatively heavy, perhaps greater than 50 lb, a test fixture with a resonance of 3000 Hz may be so massive that the force required to vibrate the test specimen may exceed the force rating on the available vibration machine. Under these conditions, a compromise has to be made. Either reduce the required force input to the test system or reduce the weight of the fixture and try to live with the resulting resonances that may develop in the fixture.

Very often severe fixture resonances can be reduced by introducing a highly damped fixture structure. This may be in the form of laminated structures where energy is dissipated at several interfaces. Even laminated wooden fixtures have been used successfully. Highly damped castings such as zirconium magnesium are often used for structures which require high damping and stiffness with a relatively light weight.

If there is any doubt as to why it is desirable to keep the natural frequency of a fixture at least 50% higher than the highest forcing frequency, remember that a resonance can magnify acceleration forces. If an improperly designed vibration fixture is used to support a sensitive electronic component during a 5-G sinusoidal-vibration test, it is possible for this component to receive 100Gs if the fixture has a transmissibility of 20 at its resonance. If the fixture were not properly monitored with accelerometers, a casual observer could conclude that this component failed at 5G.

Before any vibration fixtures are designed, it is necessary to understand the basic fundamentals of vibration. This requires a familiarity with the natural-frequency formulas for simple systems such as beams, plates, and multiple spring–mass systems. Without this knowledge even the best designer will be groping in the dark. The end result may be an inadequate fixture that must be redesigned or modified.

Rigorous mathematical solutions are not necessary to solve for the natural frequencies of various types of structures. If the natural frequency of a particular structure cannot be found in a reference book, it may be possible to derive the necessary equation using approximate methods such as the Rayleigh method, or work and strain-energy methods.

9.4. Basic Fixture Design Considerations

Sharp changes in the cross-section of any fixture should be avoided. Sharp changes in the cross-section result in a reduction of the effective spring rate without a proportional reduction in the mass. This will result in a lower natural frequency for the fixture.

Fixture designs should be kept simple, since this keeps costs down and permits a more accurate analysis to be made with standard handbook equations.

Always consider the stiffness-to-weight ratio of the fixture to make maximum use of the fixture mass. Understand the characteristics of the basic frequency equations to know what factors affect the resonant frequency. For example, the natural frequency of a uniform cantilevered rod during vibration along its longitudinal axis (Fig. 9.2) is

$$f_n = \frac{1}{4}\left(\frac{AEg}{WL}\right)^{1/2} \tag{9.1}$$

where A = cross-section area, in.^{2}
E = modulus of elasticity, lb/in.^{2}
g = gravity
W = weight, lb
L = rod length, in.

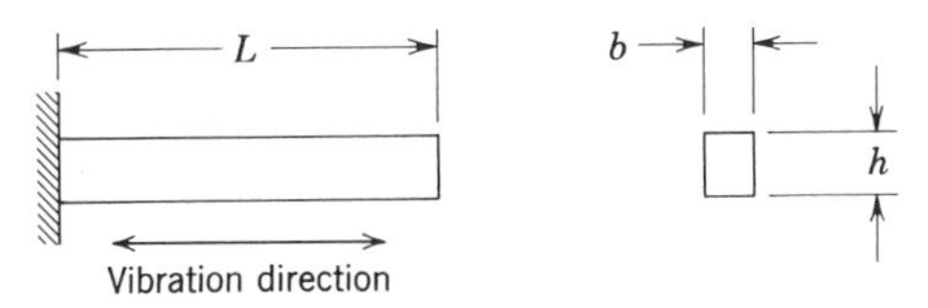

FIGURE 9.2. Longitudinal vibration of a cantilevered rod.

An examination of Eq. 9.1 shows that if the cross-sectional area A is doubled, the weight W will also be doubled so there is no increase in the resonant frequency.

If the rod material is magnesium, aluminum, or steel, there will be no difference in the natural frequency. This is because the ratio of the modulus of elasticity to the density of these three materials is just about the same (see Chapter 4, Section 4.4).

If the cantilevered rod in Fig. 9.2 is vibrated in the vertical direction, its natural frequency is

$$f_n = \frac{1.76}{\pi}\left(\frac{EIg}{WL^3}\right)^{1/2} \tag{9.2}$$

where I = moment of inertia of the cross-section = $bh^3/12$

An examination of Eq. 9.2 shows that if the beam depth is doubled the weight is doubled, but the moment of inertia of the cross-section increases eight times so the natural frequency is doubled.

Fixtures should be kept as small as possible to keep masses low and spring rates high. It is better to make fixtures out of castings, solid plates, or welded assemblies instead of bolted assemblies. At high frequencies, bolted assemblies tend to slide and separate so that the calculated stiffness does not really exist. If a large increase in stiffness is not required, bolted assemblies provide a substantial amount of damping during resonant conditions that effectively reduces transmissibilities. If bolted assemblies are also cemented with epoxy cements, a very rigid fixture can be fabricated.

Vibration test data on bolted assemblies indicate that the typical efficiency of a bolted joint is about 25%. This will vary, of course, depending upon the relative stiffness of the structure as well as the size, spacing, and number of bolts. With many bolts this factor may go as high as 50%, but very few designers seem to use enough properly spaced, large bolts to reach an efficiency factor of 50%.

A convenient method for determining the moment of inertia for a bolted assembly is to reduce the effective width of the bolted member to 25% of

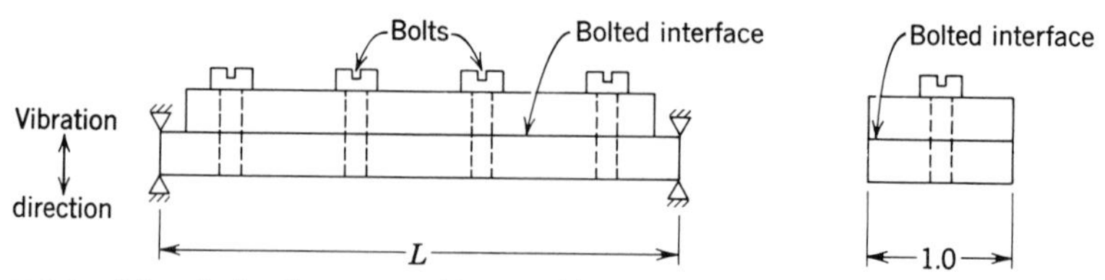

FIGURE 9.3. A simply supported beam with two bolted sections.

its original width. A simply supported beam made up of two bolted sections 1 in. wide is shown in Fig. 9.3.

The effective moment of inertia of the bolted assembly can be approximated by reducing the width of the top member to 0.25 in. when the bolted efficiency factor is 25% (Fig. 9.4).

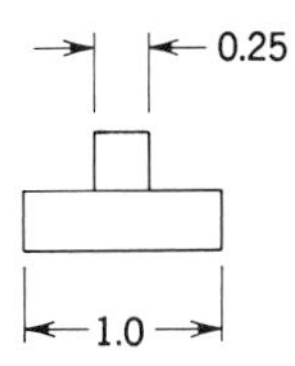

FIGURE 9.4. The equivalent bolted beam considering a 25% bolted efficiency factor.

9.5. EFFECTIVE SPRING RATES FOR BOLTS

Bolts are generally threaded into tapped holes to a depth of about two diameters so there appear to be many threads retaining the bolt. However a close examination of the engaged threads reveals that, for a normal class 2 fit, only a few threads are actually holding the bolt. If these threads happen to be near the tip of the bolt, the effective length of the bolt could extend well into the tapped hole. When bolts are loaded in an axial direction and the spring rate of the bolt must be determined, it is suggested that the effective length of the bolt should be considered as extending at least one diameter into the tapped hole.

All bolts should be installed with a predetermined torque value, which depends upon the bolt material and the function. The torque should be checked periodically on the bolts that can influence the vibration characteristics of the system because bolts that are inserted and removed often tend to loosen more easily during vibration.

If a bolt is assumed to be similar to a rod subjected to an axial load (see Fig. 9.5), the spring rate is

$$K = \frac{AE}{L} \tag{9.3}$$

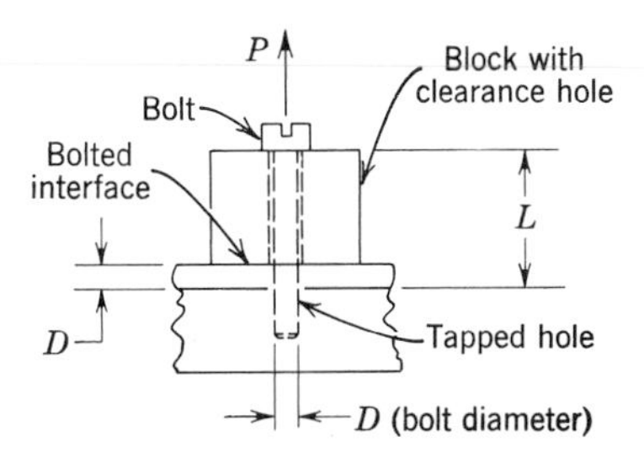

FIGURE 9.5. Effective length of a bolt in a tapped hole.

where A = area of cross-section
E = modulus of elasticity
L = length (extending one diameter into tapped hole)

Equation 9.3 shows the spring rate is not affected by the bolt torque. The torque determines the preload on the bolt; this preload must be exceeded (for metal-to-metal interfaces) by the external dynamic load before the bolted interfaces will separate.

Although Eq. 9.3 shows that the preload torque does not have an effect on the spring rate, an examination of the load-deflection curves, with and without a preload as shown in Fig. 9.6, shows the preload tends to increase the effective spring rate. (The preload will produce a nonlinear system which is beyond the scope of this book.)

The spring rate $\Delta P/\Delta X$ is the same for both systems shown in Fig. 9.6. However the apparent spring rate, which is the slope along line AB, is higher for the preloaded spring. The natural frequency is slightly higher if a resonance occurs with a preloaded spring than with no preload. The transmissibility with a preloaded spring is much lower than the transmissibility of a spring without a preload. This is due to the additional energy dissipated by the slapping action of the interfaces on the preloaded bolts.

There are times when it is desirable to increase the spring rate of a bolted system that has a fixed number of bolts and a fixed thread size. The spring rate of the bolts can often be increased by making the bolts out of beryllium which has a modulus of elasticity of 42 million. Sometimes shoulder bolts can be installed where the body of the bolt above the thread is larger than the thread diameter.

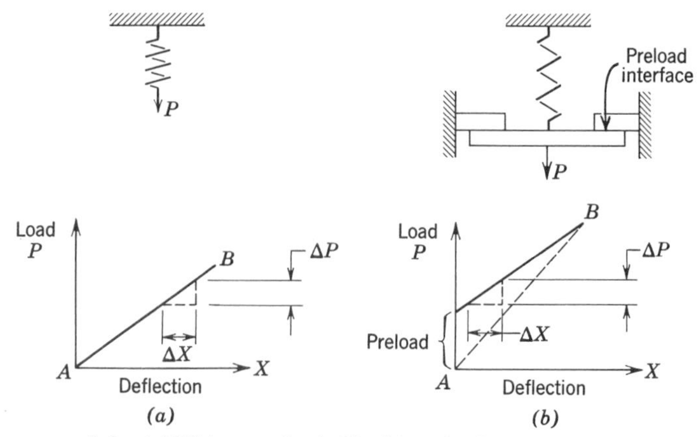

FIGURE 9.6. (*a*) Without preload; (*b*) with preload.

Long bolts are generally poor for vibration since they can "wind up" when they are installed. Under vibration these bolts can "unwind" so they may become loose very quickly.

9.6. Bolt Preload Torque

The relation between the applied torque and the resulting preload in a bolt can be determined from a simple approximate equation [9]

$$T = 0.2DP \tag{9.4}$$

where T = torque (in lb)
D = outside diameter of thread
P = axial load in induced bolt
0.2 = constant for most bolts and bolt materials

Only about 10% of the applied torque goes into tightening a bolt; about 90% is lost in friction. About 50% of the torque is lost in friction under the bolt head and about 40% of the torque is lost in friction in the bolt threads.

Tests on socket-head cap screws [10] show that a heavy lubricant has a noticeable effect on the torque–tension curve in the lower range, but has negligible effect under severe stress. These tests also showed that tension is an approximate straight-line function of the applied torque up to the yield point.

The ideal tightening torque should stress the bolt up to the elastic limit of the material. Because this condition is very difficult to obtain during mass production on an assembly line, the bolt torque should be limited to a value that will stress the bolt material to about 80% of the yield point. (See Chapter 1, Section 1.8 for recommended tightening torques.)

9.7. Rocking Modes and Overturning Moments

Severe rocking modes will often develop in the shaker head during vibration in the vertical direction. If the vibration fixture or the electronic box are either very tall or very broad (Figs. 9.7*a* and *b*), rocking modes are usually caused by an unbalanced condition or by a shift in the center of gravity during major resonant conditions.

Balancing a large fixture is very important, especially if the fixture is bolted directly to the shaker head. The balancing should be done with the actual electronic box or a dummy load attached to the fixture. Balancing weights if used, should be made a permanent part of the fixture or they may soon be lost, and forgotten until the shaker head is damaged. The importance of balancing every vibration fixture cannot be overempha-

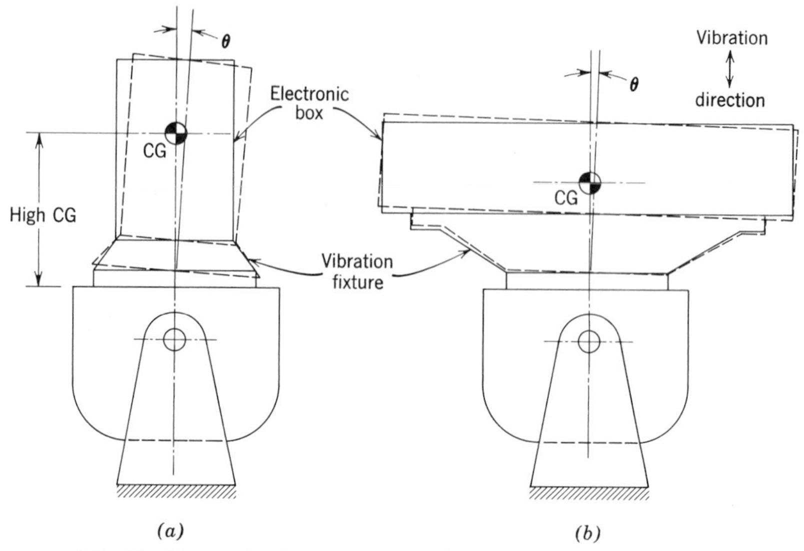

FIGURE 9.7. Rocking modes that may occur during vibration tests.

sized, yet it seems that many electronic vibration-testing laboratories never bother to balance their fixtures. This poor practice has led to many short circuits in armatures that have had their electrical insulation rubbed off by large overturning moments.

Rocking modes can also develop in relatively small fixtures that are unsymmetrical (Fig. 9.8). Although the combined center of gravity of the

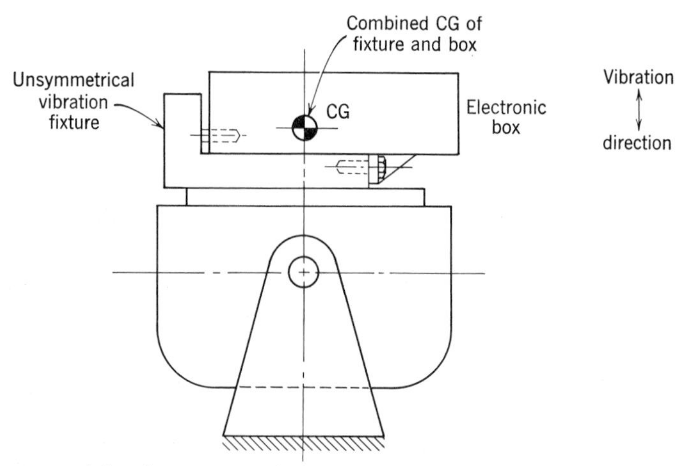

FIGURE 9.8. An unsymmetrical vibration fixture.

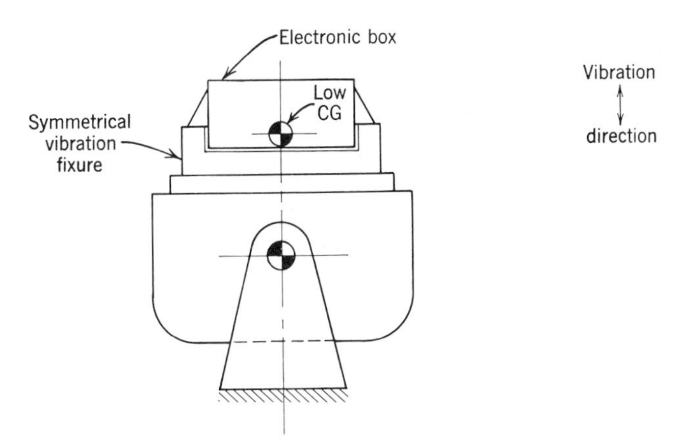

FIGURE 9.9. A symmetrical vibration fixture.

fixture and the electronic box is exactly on the center line of the shaker, it will not stay there at high frequencies. The effective mass of the electronic box is smaller at high frequencies, after passing through major resonance points, while the stiff unsymmetrical fixture maintains its effective mass. This results in a dynamic shift of the CG, which results in rocking modes. There are fewer problems if symmetrical fixtures are designed with a low center of gravity (Fig. 9.9).

Overturning moments in the shaker head on a vibration machine can also be due to nonuniform flexures or springs that support the head. A great deal of time can be saved by knowing the response characteristics of the bare shaker head, which may be out of balance. The vibration characteristics of the bare shaker head should be recorded over the normal frequency bandwidth, to provide a permanent record that can be checked from time to time to determine if any deterioration has taken place.

9.8. Oil-Film-Slider Tables

Most vibration test laboratories make use of oil-film-slider tables when vibration tests must be run in the horizontal plane. A complete electronic system mounted on an oil-film slider plate is shown in Fig. 9.10. An oil-film slider, as the name implies, consists of a large flat plate that slides on a film of oil. The plate usually rests on a very rigid foundation of concrete, steel, or granite while one edge of the plate is bolted to the shaker head (Fig. 9.11).

The test specimen, which consists of the vibration fixture and a test package, is normally bolted to the slider plate. The test specimen can

FIGURE 9.10. A complete electronic system, including a radar antenna with several electronic boxes, mounted on an oil-film-slider plate for vibration testing (courtesy Norden division United Aircraft).

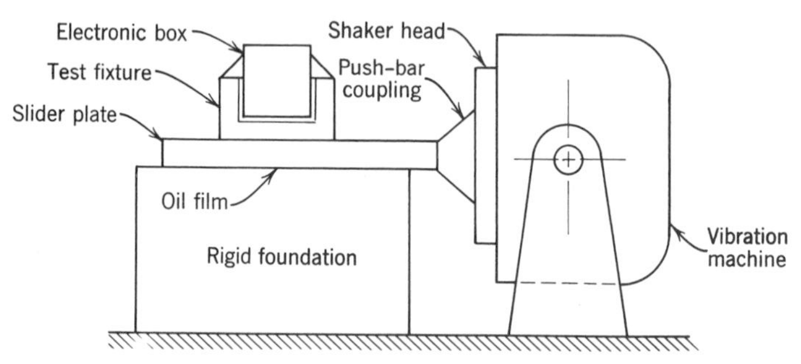

FIGURE 9.11. Vibration machine with an oil-film-slider plate.

usually be rotated 90° without changing the slider plate to permit vibration along both horizontal axes.

Oil-film-slider tables are capable of very pure translational motion with very little cross talk (rotational modes in either the horizontal or vertical planes). This is due to the viscous nature of the oil film acting on a very large surface area. Overturning moments in the vertical plane and rotational moments in the horizontal plane can be effectively damped out in most cases.

If an attempt is made to vibrate a very tall test specimen with a high CG, the surface area of the plate may have to be increased and a thicker slider plate may have to be used in order to prevent lifting and slapping of the slider plate. Counterweights may also be required.

Before slider plates were used, horizontal vibration tests were run using flexure tables and suspension systems. These systems are very difficult to control during resonant conditions and their use is not recommended if an oil film slider can be used instead.

9.9. Vibration-Fixture Counterweights

Oil-film-slider plates can eliminate many difficulties related to overturning moments developed during vibration in the horizontal plane. However, when tall masses with a high CG must be vibrated, counterweights may have to be used to lower the CG. Counterweights would normally be made of a dense material, such as steel or lead, to keep the overall size down.

Counterweights may be good for the static balance on a vibration system but they may not be good for the dynamic balance, due to the lack of dynamic similarity to the test specimen. Consider a tall nose cone that must be vibrated in a direction perpendicular to the axis of the cone. A severe resonance in the nose cone could shift the CG so that a counterweight would not be able to compensate for the overturning moment.

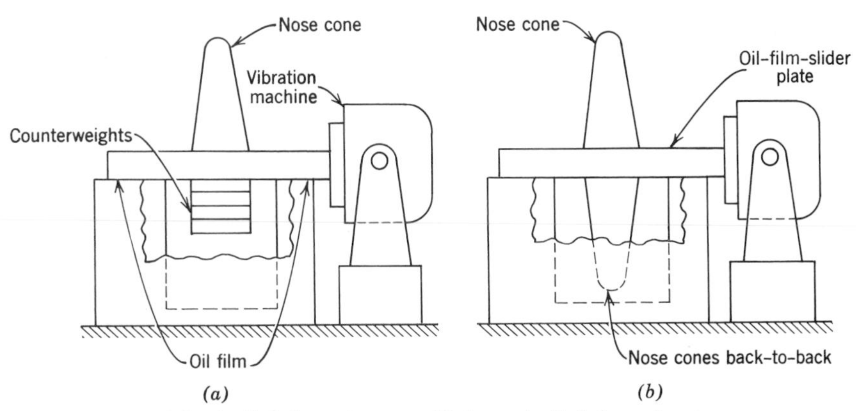

FIGURE 9.12. (*a*) Statically balanced system; (*b*) dynamically balanced system.

Under these conditions, dynamic similarity could be obtained if two nose cones were vibrated back to back simultaneously. Any dynamic change in one nose cone would be duplicated in the opposite nose cone so that severe overturning moments can be reduced (Fig. 9.12).

9.10. A Summary for Good Fixture Design

Vibration test fixtures will provide adequate performance if a few simple rules can be followed:

1. Understand the test specimen and the test specification.
2. Analyze preliminary designs and try to keep the lowest natural frequency about 50% higher than the highest forcing frequency.
3. Avoid sharp changes in the cross-section.
4. Consider the stiffness-to-mass ratio for optimum design.
5. Keep fixtures as small as possible.
6. Avoid bolted fixture assemblies, except where ribs may be required for stiffness and damping.
7. Keep fixture designs simple.
8. Design symmetrical fixtures.
9. Design for dynamic similarity.
10. Consider the effective length of the bolt thread engagement when calculating the effective bolt spring rate.
11. Torque all bolts.
12. Use fine threads, instead of coarse threads, on bolts wherever possible.
13. Balance fixtures with the test specimen.

9.11. Suspension Systems

Vibration tests are often required on large electronic control consoles which are used on ships and submarines. These units may weigh several thousand pounds; this makes them very difficult to handle without special equipment. Very often a special facility must be designed and fabricated just to perform vibration tests.

Oil-film-slider tables may not be practical in this case, so a suspension system might be considered. A special A-frame may have to be designed to provide a rigid structure to support the electronic control console during vibration.

Special vibration holding fixtures may also have to be provided to shake a large electronic console, which is generally fabricated of thin (12- to 18-gage) sheet metal. The vibration holding fixture would normally

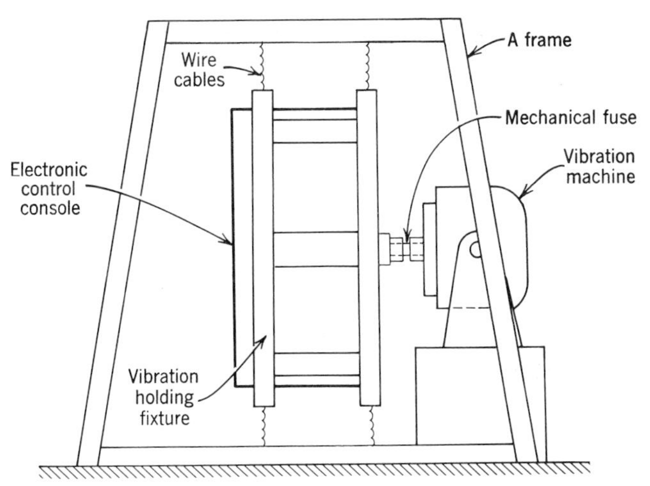

FIGURE 9.13. Vibration machine with a large suspended electronic console.

fasten to the console at the same points the console is fastened to the ship's structure.

The console-and-holding-fixture assembly would be suspended in the A-frame by wire cables (Fig. 9.13).

A mechanical fuse would be advisable for this type of installation. Since there is very little damping in a wire suspension system, overturning moments may become quite severe. Large angular displacements in the electronic console can force the shaker head to rotate enough to permit the electrical coil windings in the shaker head to rub against the stationary field windings; this would eventually cause an electrical short circuit in the system.

9.12 MECHANICAL FUSES

A mechanical fuse is a safety device that permits vibrational motion to be transferred from the shaker head to the test specimen unless severe overturning moments are developed. These moments then break the fuse before the shaker head is damaged. A typical fuse is shown in Fig. 9.14.

The spring rate of the fuse can be determined by considering it to be equivalent to a rod in tension and compression. Equation 9.3 can then be used directly.

When an overturning moment is applied, the bending stress in the fuse is

$$S_b = \frac{Mc}{I} \tag{9.5}$$

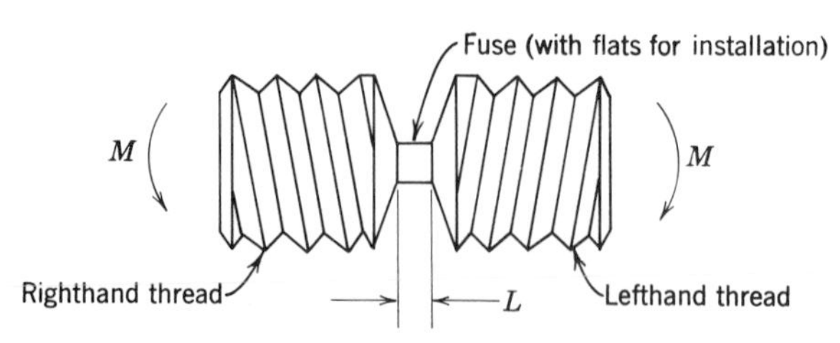

FIGURE 9.14. Enlarged view of a typical mechanical fuse.

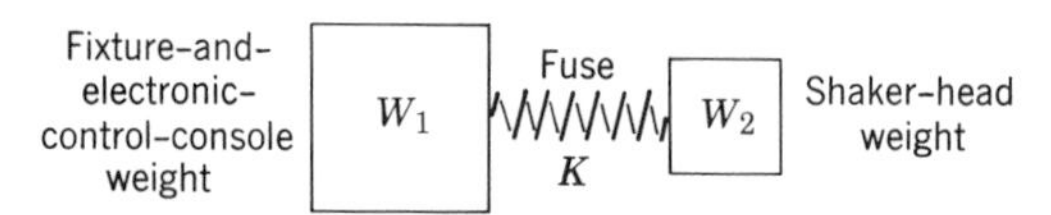

FIGURE 9.15. Simulating the suspension system as a two mass system.

where M = bending moment
I = moment of inertia
c = radius of the fuse section

The natural frequency of the suspension system can be approximated by considering it as a two-mass system with a spring between the masses (Fig. 9.15).

Using the methods shown in Chapter 3, the natural frequency equation is

$$f_n = \frac{1}{2\pi}\left[\frac{Kg}{W_1}\left(1+\frac{W_1}{W_2}\right)\right]^{1/2} \tag{9.6}$$

If the weight of the console, W_1, is very large with respect to the weight of the shaker head, W_2, then the equation becomes

$$f_n = \frac{1}{2\pi}\left(\frac{Kg}{W_2}\right)^{1/2} \tag{9.7}$$

This is similar to the single-degree-of-freedom equations shown in Chapter 2.

9.13. Distinguishing Bending Modes from Rocking Modes

There are occasions when it is rather difficult to determine what is taking place during a vibration test unless proper instrumentation is provided. The case of a large rigid vibration fixture during vibration in the vertical

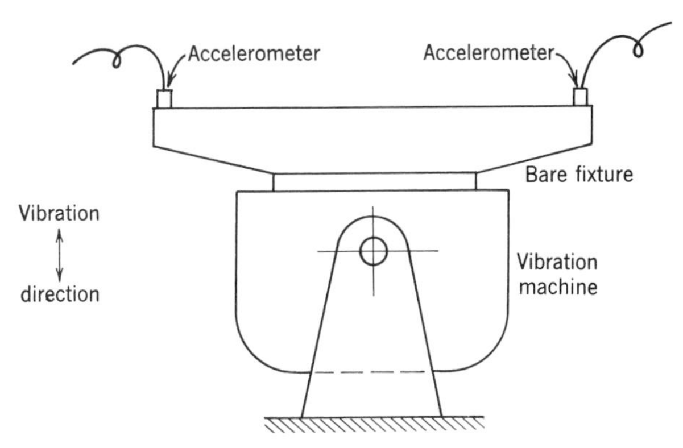

FIGURE 9.16. Vibration machine and test fixture with only two accelerometers.

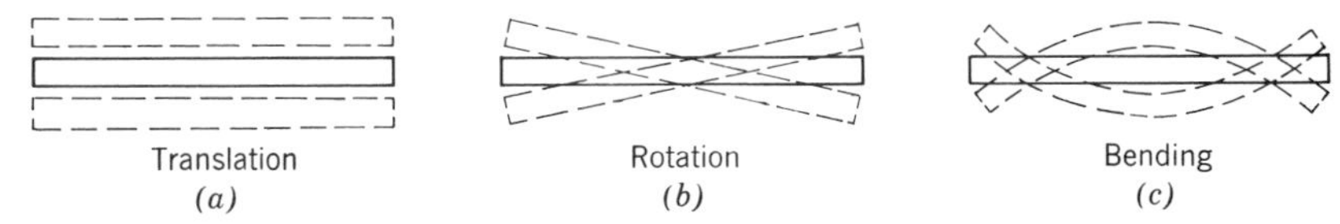

FIGURE 9.17. Three different vibration modes on a test fixture.

direction, with one accelerometer placed at each end, is shown in Fig. 9.16.

With an input of $2G$, both end accelerometers read $10G$. What is the vibration mode of the fixture?

There are three different vibration modes that can have these characteristics: pure translation, rotation, and bending (Fig. 9.17).

Unless more accelerometers are used or more points are monitored, it is very difficult to tell what is happening without the use of a strobe light.

9.14. PUSH-BAR COUPLINGS

Oil-film-slider plates are usually bolted directly to the shaker head. This is generally adequate if the test frequencies are below about 500 Hz and if the test specimen is relatively light. When the test frequencies go up to 2000 Hz and the test specimen becomes very heavy, the few bolts holding the slider plate to the shaker head may not have a spring rate high enough to produce the required force at the high frequencies. What often happens is that the shaker head has an output of $30G$ but the slider plate receives only $3G$ because of a soft bolt–spring system.

In order to increase the stiffness of the bolt system, a push-bar coupling is often used. The coupling permits the use of many more bolts at the junction. Although another bolted interface is added, the increase in the number of bolts will increase the overall stiffness of the system.

Consider the vibration system shown in Fig. 9.11. If the slider plate is bolted directly to the shaker head, the interface would probably appear as shown in Fig. 9.18.

Considering the bolt-hole pattern as shown in Fig. 9.19, only five bolts could be used in the slider plate.

The bolted interface system shown in Fig. 9.18 is nonlinear, since a pulling force on the slip plate places the bolts, which have a small cross-sectional area, in tension. A pushing force, however, acts directly on the slider plate and places that member, which has a large cross-section, in compression. A load deflection diagram for this type of system is similar to that shown in Fig. 9.20.

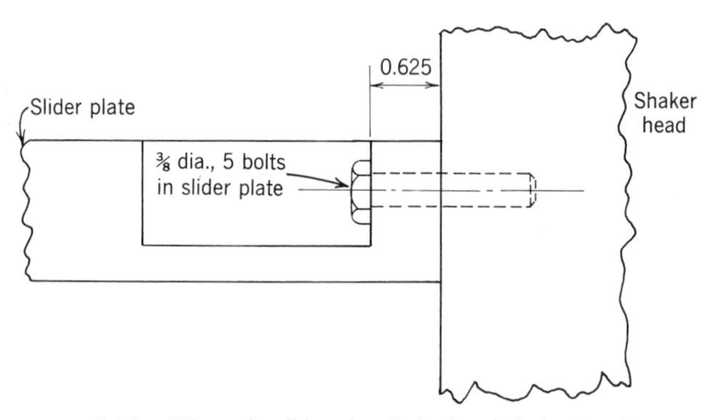

FIGURE 9.18. View of a slider plate bolted to a shaker head.

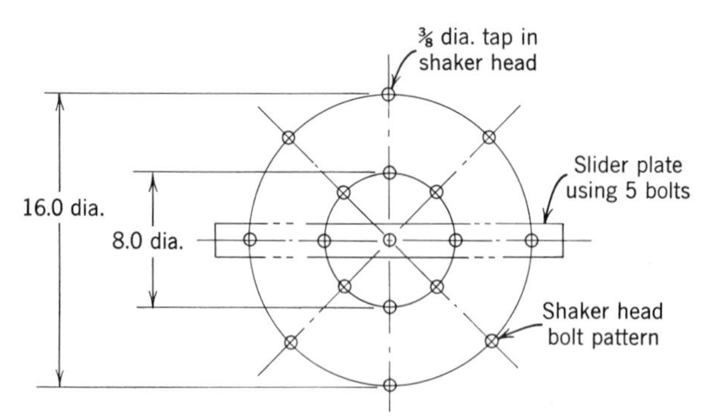

FIGURE 9.19. A typical bolt pattern in a shaker head.

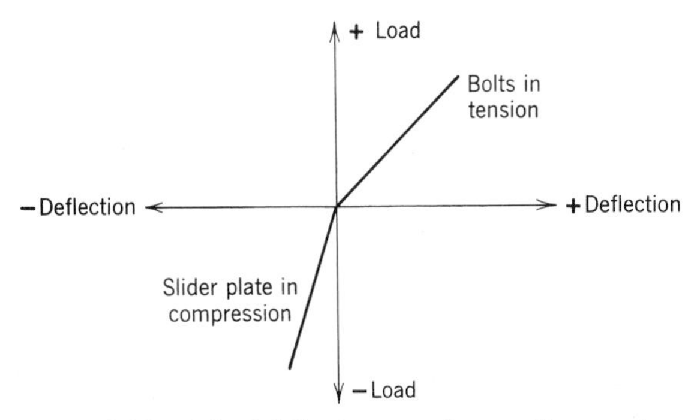

FIGURE 9.20. A load-deflection curve for a nonlinear spring.

Nonlinear systems do not have the same response characteristics as linear systems and the natural frequency equations are not the same. However, if the nonlinear system is approximated as a linear system, then linear-system equations can be used to estimate the probable performance characteristics, as shown in the following section.

The natural frequency of the vibration system shown in Fig. 9.11 can be approximated by lumping the masses into a two-mass system with one spring where the bolts act as the spring (Fig. 9.21).

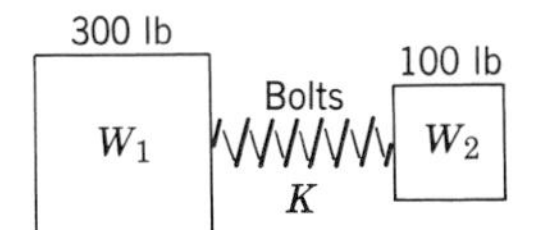

FIGURE 9.21. Simulating the slider plate as a two mass system.

Weight W_1 is as follows:

Slider plate	150 lb
Vibration fixture	90
Electronic box	60

$$W_1 = 300 \text{ lb} \tag{9.8}$$

Weight W_2 is the shaker head. Assuming an MB vibration machine Model C-126, which has a rating of 9000 force pounds, the shaker head weighs.

$$W_2 = 100 \text{ lb} \tag{9.9}$$

The spring rate of the bolts can be determined from Eq. 9.3.

$$K = \frac{A_s E}{L}$$

where $E = 30 \times 10^6$ lb/in.^{2} steel bolts
$L = 0.625 + 1$ bolt diameter (see Fig. 9.18)
$L = 0.625 + 0.375 = 1.00$ in. length
$A_s = 0.0876$ in.^{2} stress area for a $\frac{3}{8}$ dia. fine thread on one bolt

For five bolts, the spring rate is

$$K = \frac{5(0.0876)(30 \times 10^6)}{1.00} = 13.10 \times 10^6 \text{ lb/in.} \tag{9.10}$$

Substituting Eqs. 9.8–9.10 into Eq. 9.6

$$f_n = \frac{1}{2\pi}\left[\frac{Kg}{W_1}\left(1 + \frac{W_1}{W_2}\right)\right]^{1/2} = \frac{1}{2\pi}\left[\frac{(13.1 \times 10^6)(3.86 \times 10^2)}{3.00 \times 10^2}\left(1 + \frac{300}{100}\right)\right]^{1/2}$$

$$f_n = 1310 \text{ Hz} \tag{9.11}$$

Since the system is really nonlinear, the natural frequency would be somewhat higher than this. However this system would not be adequate to perform vibration tests as high as 2000 Hz if high G forces are required. Since the spring system is relatively soft, a high G force in the shaker head produces a relatively small G force, at frequencies well above the resonance of the bolted system, in the slider plate.

Even if the spring rate of the bolts indicates that a resonance might occur in the vibration oil-film-slider system, a high preload in the bolts may not permit a resonance to occur. In order for a resonance to occur, the dynamic loading on the oil-film-slider plate must exceed the preload in the bolts.

If steel bolts with a yield strength S_y of 35,000 psi are used to fasten the slider plate directly to the shaker head, then the maximum allowable preload in each bolt is

$$P = A_s S_y = (0.0876)(35{,}000) = 3060 \text{ lbs} \tag{9.12}$$

The installation torque required on each bolt can be determined from Eq. 9.4.

$$T = 0.2\,DP = 0.2(0.375)(3060) = 230 \text{ in. lb} \tag{9.13}$$

The total preload on five bolts is:

$$P_t = 5(3060) = 15{,}300 \text{ lb} \tag{9.14}$$

The dynamic load acting on the slider plate can be approximated from the input G force, the weight, and the transmissibility at resonance.

$$P_d = G_{\text{in}} W_1 Q \tag{9.15}$$

Assuming a 10-G-peak sinusoidal-vibration input with a transmissibility

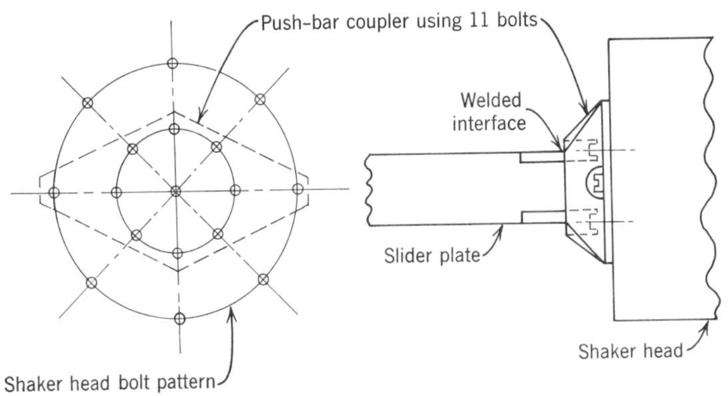

FIGURE 9.22. A push bar coupler with eleven bolts to increase the stiffness.

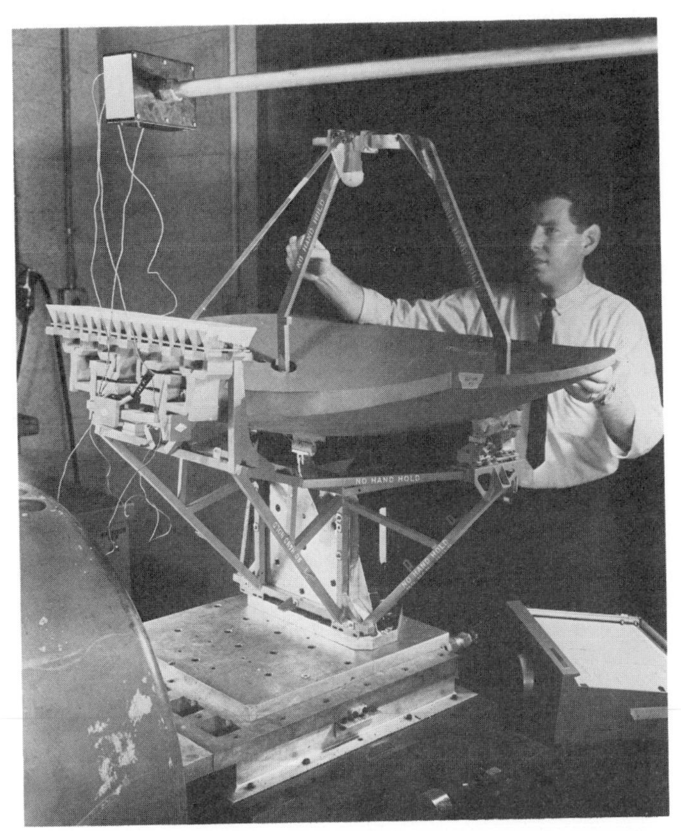

FIGURE 9.23. An airborne radar antenna mounted on an oil-film-slider plate for vibration testing (courtesy Norden division of United Air Craft).

of about 6, the dynamic load is

$$P_d = 10(300)(6) = 18{,}000 \text{ lb} \tag{9.16}$$

The dynamic load shown by Eq. 9.16 is greater than the total preload on the bolts shown by Eq. 9.14; this means the bolt preload is not high enough to prevent a resonant condition from occurring at about 1310 Hz as shown by Eq. 9.11.

A push-bar coupling could be designed to pick up 11 bolts on the shaker head as shown in Fig. 9.22. (For maximum rigidity, the slider plate should be welded to the push-bar coupler.)

All bolted interfaces should load the bolts in tension and not in shear to obtain the maximum effective stiffness.

Bolted interfaces loaded in shear can slip and slide with respect to each other at high frequencies. Even if several large press-fit dowels are used in the interface to prevent relative motion, it can still occur at high G loads. Therefore the bolts should be loaded in tension for the best results.

9.15. Slider-Plate Longitudinal Resonance

A slider plate can develop resonances during vibration in the longitudinal direction if the plate is long enough. The natural frequency for a structure under these conditions can be determined from Eq. 9.1. Since the width of the plate is not important during longitudinal vibration, consider a unit width of 1.0 in. and a length of 30.0 in. (Fig. 9.24). The following information is then required to determine the resonant frequency:

$$\begin{aligned} A &= 1.0 \text{ in.}^2 \text{ cross-sectional area} \\ E &= 10.5 \times 10^6 \text{ lb/in.}^2 \text{ modulus for aluminum} \\ g &= 386 \text{ in/sec}^2 \text{ gravity} \\ W &= (1)(1)(30)(0.1) = 3.0 \text{ lb} \\ L &= 30 \text{ in.} \end{aligned}$$

Substituting into Eq. 9.1, the longitudinal natural frequency becomes

$$f_n = \frac{1}{4}\left(\frac{AEg}{WL}\right)^{1/2} = \frac{1}{4}\left[\frac{(1.0)(10.5\times 10^6)(3.86\times 10^2)}{(3.0)(30)}\right]^{1/2}$$

$$f_n = 1680 \text{ Hz}$$

This represents the natural frequency of the bare slider plate. If the masses of the fixture and electronic box are included, the natural frequency will be lower. A more comprehensive analysis of the slider plate, vibration fixture, electronic box, push-bar coupling, and shaker head must treat this combination as a multiple spring–mass system similar to those shown in Chapter 8.

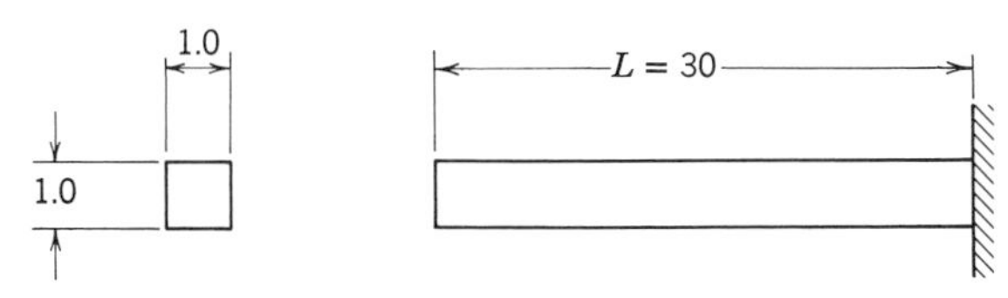

FIGURE 9.24. A unit width on an oil-film-slider plate.

FIGURE 9.25. A "head-up display" electronic box mounted on an oil-film-slider plate for vibration testing (courtesy Norden division of United Aircraft).

9.16. Acceleration-Force Capability of Shaker

Vibration machines are rates in peak force pounds based on sinusoidal vibration, which is determined from the maximum allowable power that can be dissipated in the machine. In order to calculate the maximum G force acceleration possible in a system, the force rating of the machine must be known along with the entire weight of the moving mass. The equation then becomes

$$G_{\text{peak}} = \frac{\text{Rated force}}{\text{Shaker head} + \text{Slider} + \text{Specimen}} \tag{9.17}$$

The peak G force capability for the system shown in Section 9.14 can be determined for the MB C-126 vibration machine, which has a peak rating of 9000 force pounds. Substituting Eqs. 9.8 and 9.9 into Eq. 9.17, the peak G force capability becomes

$$G_{\text{peak}} = \frac{9000}{100 + 300} = 22.5G \tag{9.18}$$

9.17. Positioning the Servo-Control Accelerometer

The acceleration G-force transmitted from the shaker head to the vibration test specimen can be controlled directly from the shaker head or from an accelerometer placed near the test specimen. When the acceleration G force is controlled from the shaker head, a constant G force can be maintained on the shaker head over the entire frequency band. If a resonance should develop in the slider plate or the vibration fixture, the G forces will build up at resonance so that the input to the test specimen will be far greater than the output of the shaker head. At frequencies above the resonance the opposite situation occurs, and the input to the test specimen will be far less than the output of the shaker head.

When the acceleration G force is controlled from an accelerometer placed near the test specimen, the vibration machine can vary the output of the shaker head to hold a constant G force on the control accelerometer over the entire frequency band. If the control accelerometer is placed on a resonant structure, the output of the shaker head will be reduced when the structure passes through its resonance. At frequencies above the resonance, the output of the shaker head will be increased to hold a constant G force on the control accelerometer.

Standard piezoelectric accelerometers, the same types that are generally used to monitor and record acceleration data, are normally used as servo-control accelerometers. These small devices are usually equipped with screw studs to fasten them to the test equipment. Many testing laboratories use adhesives such as dental cement, Eastman 910 quick-drying cement, and even double-backed tape to mount accelerometers. These methods all work quite well on accelerometers that are used to monitor acceleration data, but they should *never* be used to fasten a control accelerometer to a vibrating system.

The servo-control accelerometer actually controls the acceleration G force output of the shaker head. If the servo-control accelerometer should fall off during the vibration test, the shaker head will lose the feedback provided by this accelerometer. This results in a rapid build-up of the acceleration force until the limit displacement switch of the shaker head cuts off the power. By this time the damage has already been done:

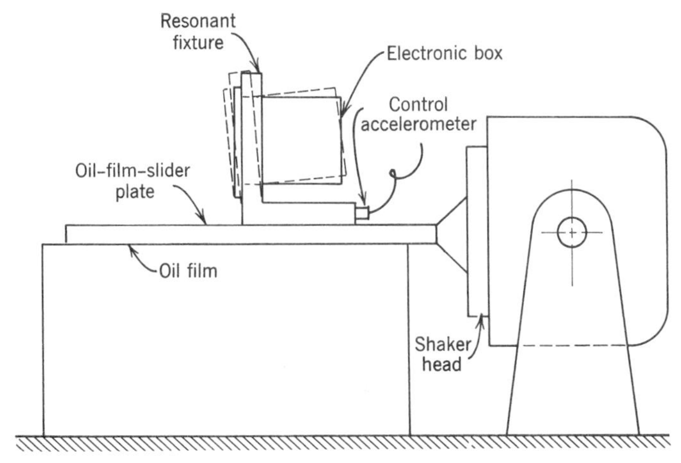

FIGURE 9.26. A resonant fixture on an oil-film-slider plate.

very high G forces can be developed very rapidly. Once an expensive electronic system has been subjected to acceleration loads that far exceed its specifications, there is no good way to determine the actual and potential damage. In most cases where this has happened, and it has happened at many places, the electronic equipment has to be rebuilt because the customer will not accept an item that may fail shortly after it is put into service.

Since the position of the servo-control accelerometer can determine the actual G force received by the test specimen, the location selected for this control accelerometer can be quite important to the success of a vibration test. A resonant fixture on a slider plate is shown in Fig. 9.26.

If the top of the resonant fixture has a transmissibility of 10, then it will see $20G$ at its resonant peak when the control accelerometer is held at $2G$. This means the electronic box will probably see an average of about $11G$ at the fixture resonance point instead of the required input of $2G$.

If the control accelerometer is mounted at the top of the resonant fixture, the bottom of the fixture will see only $0.2G$ at the resonant peak. This means that the electronic box will probably see an average of about $1.1G$ at the fixture-resonance point instead of the required input of $2G$.

The best solution to this problem, of course, is to design resonance free fixtures. Since this is not always possible, a compromise is about the only thing that may keep everyone happy, so the control accelerometer will probably be placed halfway up the fixture.

CHAPTER 10

STRUCTURAL FATIGUE

10.1. The Fatigue Mechanism

Materials can fracture when they are subjected to repeated stresses that are considerably less than their ultimate static strength. The failure appears to be due to submicroscopic cracks that grow into visible cracks and lead to a complete rupture without warning under repeated loadings.

The growth of fatigue cracks in metals depends upon the crystalline structure and the environment, as well as the stress distribution around the local area of the crack. These cracks usually grow slowly in the early stages, then proceed more rapidly in the advanced stages.

The sign of a small crack does not always mean that a failure will occur. Sometimes cracks just seem to stop growing, or grow so slowly that a failure does not occur. Since there is no good way of determining whether a crack will continue to grow, it is generally best to be suspicious and to assume that any crack will eventually result in a fatigue failure. If the cracked element is a major structural member, it should be replaced or repaired to prevent a failure.

The general appearance of a fatigue crack is similar to the fracture of a brittle material. It would appear that a crack starts at some discontinuity, not necessarily a fault in the material but perhaps only the unfavorable orientation of a crystal. The crack would be submicroscopic at this stage but it would grow a little with each stress cycle without causing extensive plastic deformations. Once the crack starts, it appears to take the path of least resistance, which often carries it through various faults, inclusions, and other areas with discontinuities. The small radius at the tip of the crack seems to create a high stress concentration that drives the crack deeper with each stress reversal. A typical example of this characteristic is the crack that may appear in a thick sheet of plastic. This crack continues to grow with each stress cycle. However, if a small hole is drilled at the tip of the crack, the crack growth can often be stopped. Although the drilled hole may not be very large, its radius is so much greater than the crack radius that the stress concentration is effectively reduced.

Eventually the crack, or cracks, will reduce the effective cross-sectional area of the structure to the point where the stress is high enough to produce a sudden and complete rupture of the part.

10.2. Considerations in Design and Analysis

Most electronic boxes are fabricated by making use of several different construction methods. For example, it is quite common to find welded, riveted, and bolted subassemblies in one box. Also, cast and dip-brazed heat exchangers are being used more often to support electronic equipment while removing the heat the same time. These are all complex structures which generally have nonuniform cross-sections and odd shapes to fit into odd corners. These odd shapes usually make it quite difficult to determine the dynamic loads in the various structural elements. Even if the dynamic loads can be determined by tests, the dynamic stresses in odd-shaped structural members are also difficult to determine without the use of strain gages. In addition to all this, there is still the problem of determining stress-concentration factors to be used at corners, notches, and holes. After all of this has been accomplished, the type of loading must be evaluated because the fatigue characteristics for a completely reversed load are not the same as the fatigue characteristics for a load that is not completely reversed.

If there is no actual hardware to test because the design exists only on paper, it would appear to be an impossible task to determine accurately the fatigue strength of major structural elements in a proposed system. However, if a weight penalty can be tolerated, it is possible to design a system for a given environment by being generous with the use of extra material in areas that are suspected of having high stresses. This technique reduces stresses rapidly but also adds weight rapidly.

If the reliability of the structure is a more important consideration than the weight, then a conservative approach should be taken. A worst-case analysis can usually be made by determining the most probable maximum and minimum coefficients associated with a given set of parameters and then using the values which will result in the highest stresses. For example, in the analysis of printed-circuit boards, the transmissibility at resonance for many types of circuit boards can be approximated as being equal to the square root of the natural frequency. However the general range of the transmissibility will normally vary from about 0.50 to about 2.0 times the square root of the natural frequency. Then, for the dynamic analysis, the highest value of 2.0 should be used in order to develop the highest acceleration loads, deflections, and stresses in the circuit board.

The same approach should be taken in the area of stress concentrations. Assume the worst case that is consistent with the material being used and the geometry of the stress raiser. Then use the maximum theoretical K_t values associated with these parameters to determine the maximum dynamic stresses that can occur for a given dynamic environment of vibration and shock.

Fatigue S–N curves for the various structural materials should always be consulted to make sure the dynamic stresses are below the endurance limit for ferrous metals or below the fatigue strength at 500 million cycles for nonferrous metals.

If the weight is just as important as the reliability, the method of analysis outlined above should not be used because it will result in a heavier electronic box. Instead, much more emphasis must be placed on actual test data to determine more accurately the transmissibilities, stresses, and fatigue lives of major structural members. Without the test data, it is virtually impossible to optimize the structural design of an electronic box that will be exposed to a severe vibration environment. The test data can be obtained from mock-ups that are fabricated just to simulate the general structural characteristics of the electronic box, or from tests that were run on boxes that are very similar to the proposed new design.

Stress concentrations are probably the major cause of structural failures in electronic boxes exposed to a vibration environment. This area is usually overlooked because so much more emphasis and manpower are usually expended in two other important areas: electrical and thermal. There is no question about it, if an electronic box does not function electrically, the box is no good no matter how well it is designed thermally and structurally. Also, if the box does not work well after it has warmed up, because it is too warm and many hot spots cause drifting of electrical signals or failures in integrated circuits, transistors, resistors, diodes, and capacitors, then again the box is unsatisfactory no matter how good its structural design.

Maintenance is another important consideration in the design of electronic equipment. Airlines, for example, cannot afford to have giant jets with a full load of passengers and cargo waiting around while some mechanic takes a couple of hours to replace an electronic box. Many compromises have to be made to satisfy all the requirements associated with an electronic system.

While stress concentrations cannot be eliminated in a typical electronic box, they can be minimized by the proper selection of materials, fabrication techniques, and geometry. It is impossible to analyze every notch, every hole, every rivet, every bend, every weld, and every screw in every section of an electronic box. Time and money will permit the examination and the analysis of only major structural members in the system. Therefore it is necessary to recognize the most probable primary and secondary load paths during the preliminary design phase of the program. These load paths should be kept as free as possible from local stress raisers. When this cannot be avoided because of such items as cables, connectors, screws, rivets, and lightening holes, then proper reinforcements should be added to reduce the localized high stresses.

10.3. Endurance Limit and Fatigue Life

The endurance limit is generally defined to be the highest alternating stress value that a structural element can withstand, without failure, regardless of the number of applied loading cycles. The endurance limit is usually based on a complete stress reversal, where the maximum positive stress equals the maximum negative stress and the neutral position has a zero stress. This type of test can easily be set up in a laboratory by rotating a small beam, with a circular cross-section, under the action of a constant bending moment. This develops an alternating tensile and compressive stress condition, of equal value, in the outer fibers of the beam. The stress changes sign every half revolution of the beam, so that the number of stress cycles is equal to the number of revolutions of the rotating beam.

McClintock[15] shows that steels and titaniums have well-defined endurance-limit stresses below which no fracture will occur, regardless of the number of stress-reversal cycles. Nonferrous metals such as aluminum, magnesium, and copper usually show no clearly defined endurance limit.

The fatigue life is defined as the number of fatigue cycles required to produce a failure at a given stress level or under a given set of conditions. This life can be sharply affected by the gaseous atmosphere of the environment. Achter[14], for example, shows that the fatigue life of copper can be 20 times greater in a vacuum of 7×10^{-6} mm.Hg than in air. However there seems to be no effect on the fatigue life of gold when placed in a vacuum.

It would appear that an oxygen atmosphere accelerates crack propagation by attacking the atoms at the tip of the crack. In a vacuum, the cracked surfaces have a chance to weld themselves together during the compression cycle, which retards the growth rate of the crack.

In general, there are two basic types of fatigue. The first type is the alternating stress, with a zero mean stress, as shown in Fig. 10.1. Here the maximum tensile stress is equal to the maximum compressive stress.

The second type of fatigue is the alternating stress, where the mean stress is not zero. This may be considered as a combination type of loading, where a steady stress is superimposed on an alternating fatigue stress, as shown in Fig. 10.2. Here the maximum and minimum stress conditions may both be in tension or in compression.

There appears to be a considerable amount of disagreement over which type of stress is the most severe; the complete alternating stress with a zero mean or the alternating stress with a constant mean. A survey of the literature (15, 16, 18, 20–22, 24–26) seems to indicate that more authors

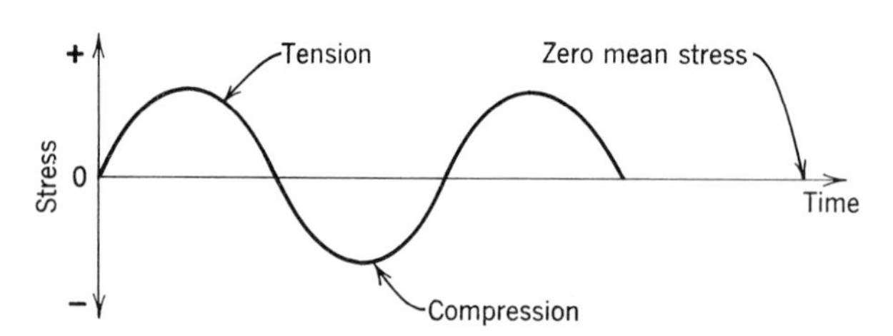

FIGURE 10.1. An alternating stress with a zero mean stress.

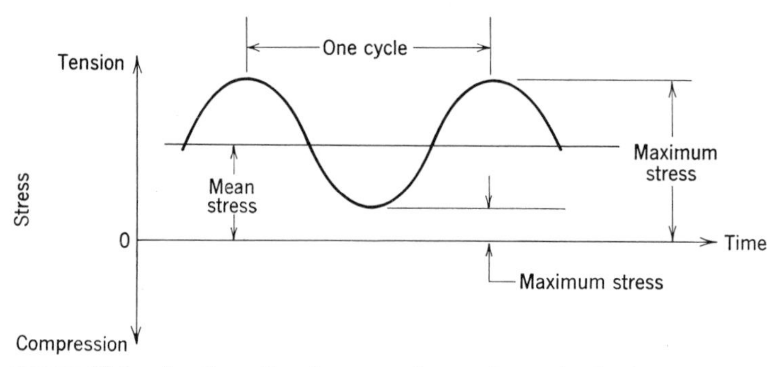

FIGURE 10.2. An alternating stress superimposed on a steady stress.

appear to believe the presence of a constant mean stress, with an alternating stress, will have the effect of reducing the fatigue life below that for a zero mean stress with an alternating stress.

This area requires more investigation because test data can be cited from the above sources which make it appear that either condition is more severe than the other.

10.4. S–N Curves

The fatigue characteristics of various metals can be determined by subjecting a large group of specimens to a series of tests with an alternate loading condition, to generate a complete stress reversal, using a different maximum stress in each test specimen. If the maximum alternating stress S is plotted against the number of cycles N required to produce a failure, the result will be an S–N diagram.

When the mean stress is zero, as shown in Fig. 10.1, the S–N curves can often be approximated by straight lines on log-log paper, as shown in Fig. 10.3.

The fatigue life in the high-stress or low-cycle area, up to point N_1 in Fig. 10.3, will generally range from about 100 cycles for ferrous alloys to about 1000 cycles for nonferrous alloys. In this area, the fatigue strength is approximately equal to the ultimate tensile strength of the material.

Polished ferrous metals reach their endurance limit at a value of about 10 million cycles, at which point the S–N curve becomes flat as shown at point N_2 in Fig. 10.3. It has been found that in most cases, if a steel part can survive this number of cycles without failing, it will never fail. The endurance limit at point N_2 can often be related to the ultimate strength of the steel. For example, experimental test data show that the endurance limit of steel, in reversed bending, is equal to about 50% of the ultimate tensile strength. The endurance limit for an alternating axial load is equal

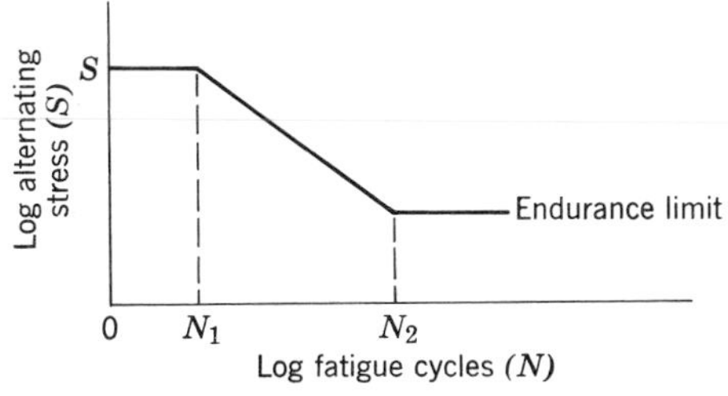

FIGURE 10.3. A typical S-N fatigue curve for metals.

to about 43% of the ultimate tensile strength. The endurance limit for an alternating shear load is equal to about 50% of the ultimate shear strength. These endurance limits are only approximate and should not be used, unless there is no other fatigue test data available on the steel being considered.

The general fatigue S–N curves for polished steel specimens can be approximated as shown in Fig. 10.4.

Nonferrous metals generally do not have a true endurance limit, so it is

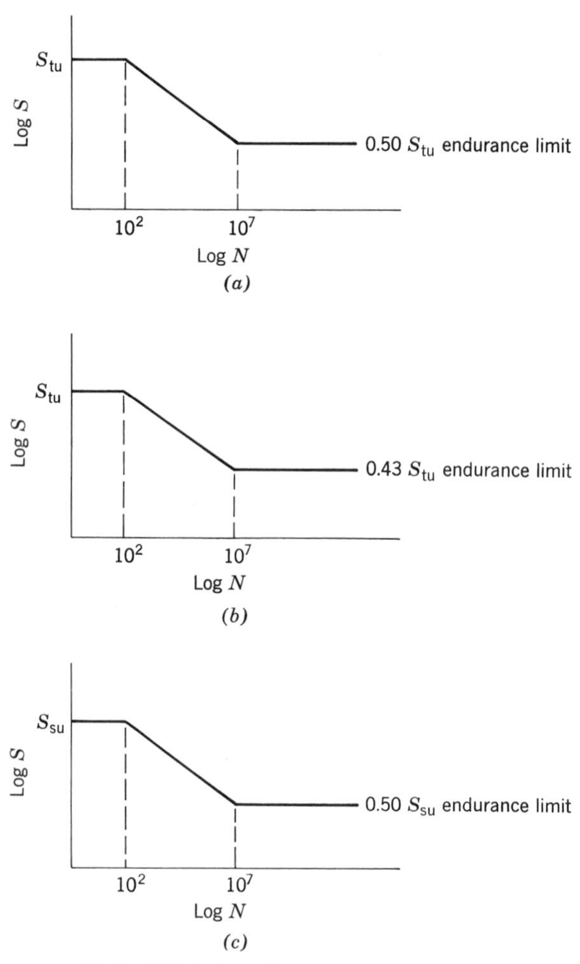

FIGURE 10.4. Polished steel specimens: (*a*) bending; (*b*) axial; (*c*) shear.

customary to show their fatigue strength at some arbitrary long lifetime such as 500 million cycles. Although the S–N curve does not really become flat at this point, the slope of the curve does change quite rapidly, so a flat curve is often shown at point N_2 in Fig. 10.3 for nonferrous metals. For polished aluminum specimens, the fatigue strength at 500 million reversed bending cycles is equal to about 33% of the ultimate tensile strength. The fatigue strength for an alternating axial load is also equal to about 33% of the ultimate tensile strength. The fatigue strength

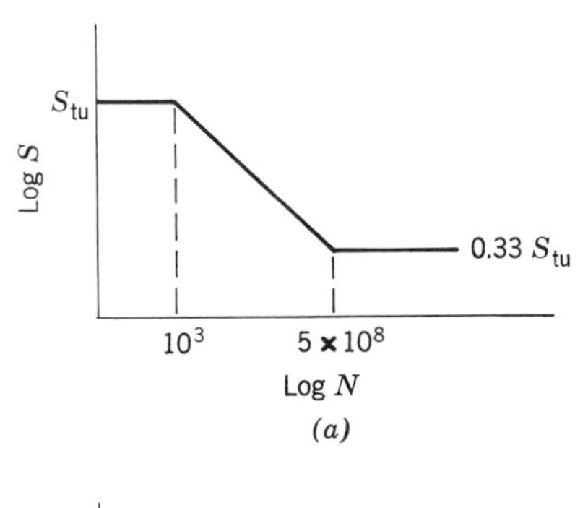

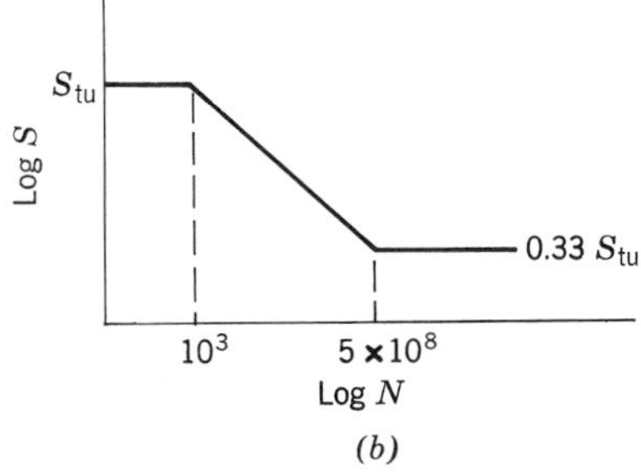

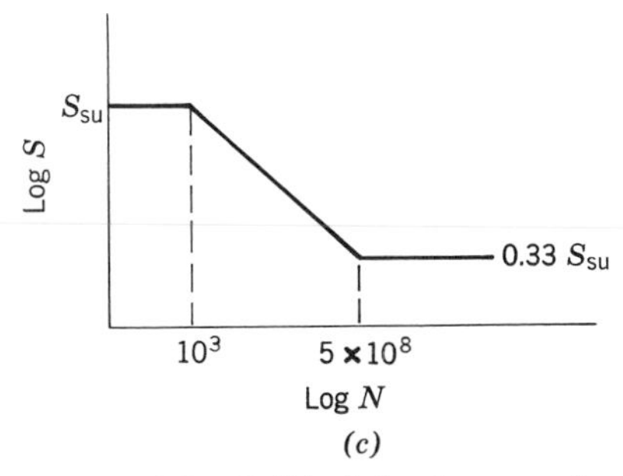

FIGURE 10.5. Polished aluminum specimens: (*a*) bending; (*b*) axial; (*c*) shear.

for an alternating shear load is equal to about 33% of the ultimate shear strength. Again, these fatigue-strength values are only approximate and should not be used unless there is no other fatigue test data available on the aluminum being considered.

The general S–N curves for polished aluminum specimens can be approximated as shown in Fig. 10.5.

10.5. Goodman Diagrams

When the mean stress is not zero, as in Fig. 10.2, then the Goodman diagram is often very convenient for determining the allowable stress range, using the ultimate tensile stress and the endurance limit as shown in Fig. 10.6. Grover[22] and Lipson[21] show a number of typical curves.

The mean-stress scale on the horizontal axis is the same as the operating-stress scale on the vertical axis, which forms a square pattern. A diagonal line across the corners of the square pattern then becomes a mean-stress line, which is normally centered between the maximum and minimum stresses as shown in Fig. 10.6.

For any given mean stress at point A, the permissible operating range can be determined by drawing a vertical line which intersects the maximum and minimum stress lines. The maximum stress line should be limit-

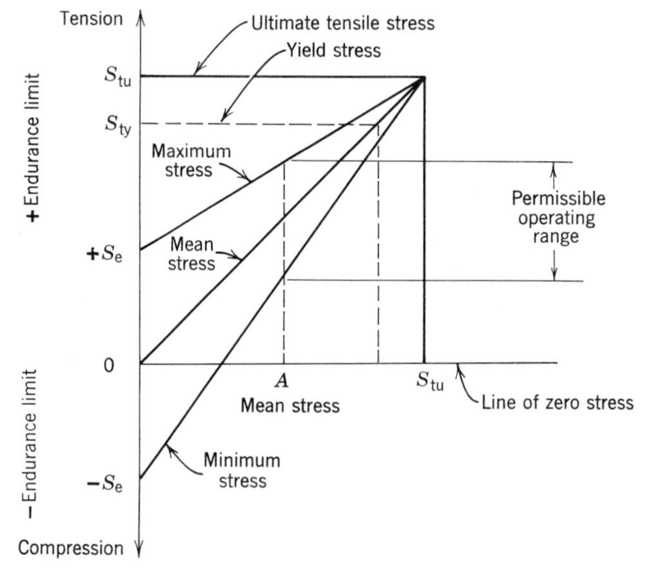

FIGURE 10.6. A typical Goodman diagram for an alternating stress superimposed on a steady stress.

ed by the yield stress of the material, since it is not desirable to exceed the yield point at any time, even if it is for only one cycle.

10.6. Stress Concentrations

Electronic boxes will usually have many areas that contain stress concentrations because of holes, notches, roughly machined surfaces, and sharp changes in cross-sections. These discontinuities, or stress raisers, are most severe when they are subjected to repeated or completely alternating loads. Cracks will often develop and grow with each stress cycle until a sudden and complete rupture develops.

These same members, if they are ductile, can be subjected to a steady load of a greater magnitude than the repeated loads without developing a failure. It appears that stress concentrations are of little or no importance under a static loading condition for ductile metals. This seems to be due to plastic yielding that occurs in the overstressed areas; this greatly reduces the magnitude of the maximum stress.

The magnitude of the stress concentration is usually related to the type of load, the geometry of the discontinuity, and the notch sensitivity of the material. Under ordinary conditions, stress-concentration factors are not applied to ductile materials subjected to steady loads. Temperature cycling, for example, represents an alternating load condition, so stress concentrations should be considered at stress raisers.

The most common stress-concentration factor is the geometric or theoretical stress-concentration factor K_t, which is related to the shape, size, location, and the type of stress such as bending, tension, and torsion. The magnitudes of these stress raisers are usually determined from the theory of perfect materials considering discontinuities such as notches, holes, and shoulders. Polarized light is often used, with transparent plastic models, to study the fringe patterns that are developed in odd-shaped structures subjected to different types of loads.

Another type of stress-concentration measure sometimes used is the fatigue factor K_f, which is slightly less than the geometric or theoretical stress-concentration factor. The fatigue stress-concentration factor is normally based on test data for discontinuities, in order to compensate for normal imperfections in the basic material plus the addition of notches and holes. Even closely controlled metals are not homogeneous. Therefore, if a notch is added to a metal that already contains many small discontinuities, the notch is really only another discontinuity among many. Its effect is not really as great as it would be in a perfectly homogeneous metal, which is shown by the geometric or theoretical stress concentration factor K_t.

The typical electronic box will usually have a very complex structure because of odd shapes required by a combination of electronic components and installation areas. These odd shapes make it very difficult to analyze accurately the dynamic loads and stresses that will be developed in major structural members. A conservative approach is often desirable because of this, so that calculations will tend to show stresses that are slightly high rather than slightly low. Under these circumstances, it is better to use the geometric or theoretical stress-concentration factor K_t, since it is greater than the fatigue stress-concentration factor K_f. Another consideration is the notch sensitivity, which is often related to such factors as ductility, percentage elongation, area reduction, hardness, and strength. Notch sensitivity is sometimes defined in terms related to the fatigue stress-concentration factor with respect to the geometric or theoretical stress-concentration factor. In general, information available on notch sensitivity is quite limited and difficult to use, except perhaps for a few heat-treated steels. The maximum notch-sensitivity factor is normally considered to be one. At this point the fatigue stress-concentration factor is equal to the geometric or theoretical stress-concentration factor.

Stress concentrations are often likened to the flow of fluids, where the force path is similar to the fluid flow path. This analogy can be used to visualize the lines of force in the area of the stress concentration. For example, consider a flat bar with a grooved section subjected to an axial load (Fig. 10.7).

In order for the force-path lines to remain within the part, they are forced to crowd together as they pass the area of the grooves. This crowding, or concentrating, near the grooves is what leads to the stress concentration in the area of the grooves. In areas a relatively short distance away from the grooves, the force-path lines are again uniformly distributed, indicating a uniformly distributed stress across the cross-section.

Stress concentrations can occur in any area, and will often occur in every area, of an electronic box. If the structure does not carry much of a dynamic load, then the stress concentration is not important. If the structure carries a substantial load, then the dynamic stress multiplied by the

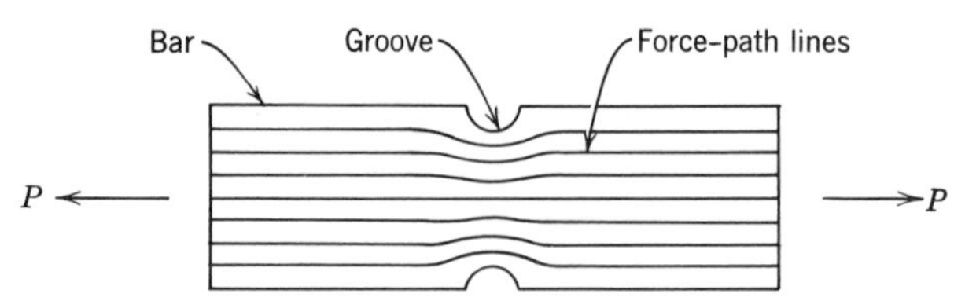

FIGURE 10.7. Stress concentrations at the notches in a bar subjected to an axial load.

appropriate geometric or theoretical stress-concentration factor will determine the peak stress. If this value is too high, then the geometry of the stress raiser should be changed.

In general, it is a poor practice to use sharp notches, square holes, or sharp bends in structures, even if they do not appear to be carrying much of a load. When an electronic box bends and twists during a vibration or shock test, the load distribution can often change enough to transfer loads to members that are not designed to carry these loads. Stress concentrations in these members may result in unsuspected fatigue failures.

The surface finish can also have a substantial effect on the endurance limit. For example, steel test specimens turned in a lathe show endurance limits about 11% lower than the same specimens that are polished after being turned.

Cast iron specimens do not seem to be affected to any great extent by local stress concentrations under an alternating-load condition. For example, a V-notch with a bottom radius of 0.06 in. appears to reduce the endurance limit about 8%. This small reduction seems to be accounted for by the large number of stress concentrations which are caused by the graphite flakes in the cast iron. The V-notch represents just one more stress concentration in a metal that already has many internal discontinuities.

Special surface preparations such as shot-peening, surface rolling, grinding, and polishing can improve the fatigue life of the specimen. These processes develop a compressive residual stress in the surface of the metal, which appears to retard the crack development and propagation.

Since most cracks start at the surface, it is important to devote a considerable amount of time to the surface condition of parts that will be subjected to an alternating-load condition. If the outer shell of a specimen has a greater ultimate strength than the core, the fatigue life will be greater than a homogeneous specimen of the basic core material. This can be accomplished by surface-hardening processes such as carburizing and nitriding. The opposite is also true. If the outer shell of a specimen has a lower ultimate strength than the core, the fatigue life will be lower than that of a homogeneous specimen of the basic core material. This condition occurs with clad aluminum, where the cladding is softer than the core, so that the fatigue life is reduced. A similar effect is observed in decarburized steel.

10.7. Sample Problem

Consider an electronic box that contains relays in the power-supply section. Relays are not generally used in severe-vibration environments because the electrical contacts tend to chatter if internal resonances are excited. However relays are inexpensive and can be quite useful in an electrical circuit if they are properly applied.

One of the first rules for mounting electronic components is to keep away from cantilevered mounts. These mounts usually have low resonant frequencies and high displacements, which often lead to high stresses and rapid fatigue failures. Thus it is very poor practice to mount relays on a cantilevered mount. However this is still done by some manufacturers of electronic equipment (Fig. 10.8).

Determine the probable dynamic bending stress and fatigue life of the 6061-T6 aluminum bracket when the system is subjected to a 5-G-peak sinusoidal-vibration input.

The natural frequency of the relay bracket can be approximated by considering a concentrated mass at the end of a cantilevered beam. Then from Eq. 2.10

$$f_n = \frac{1}{2\pi}\left(\frac{g}{\delta_{st}}\right)^{1/2}$$

where $g = 386$ in/sec^2 gravity

$W = 0.25$ lb relay weight

$L = 1.50$ in. effective length

$E = 10.5 \times 10^6$ lb/in.^{2} aluminum

$$I = \frac{bt^3}{12} = \frac{(1.5)(0.125)^3}{12} = 0.244 \times 10^{-3} \text{ in.}^4$$

$$\delta_{st} = \frac{WL^3}{3EI} = \frac{(0.25)(1.5)^3}{3(10.5 \times 10^6)(0.244 \times 10^{-3})} = 0.110 \times 10^{-3} \text{ in.}$$

Substituting into the above equation

$$f_n = \frac{1}{2\pi}\left(\frac{3.86 \times 10^2}{1.10 \times 10^{-4}}\right)^{1/2} = 298 \text{ Hz} \tag{10.1}$$

The transmissibility at resonance must be known in order to determine the dynamic load P_d at resonance. If there is no test data available, the transmissibility must be approximated from the geometry of the structure, using past experience with similar structures. An examination of the construction indicates there are many rivets used to fasten the bracket to the wall of the chassis. The slight relative motion at the interface during the resonant condition will therefore dissipate energy by changing it into heat because of friction. This energy loss will reduce the transmissibility at resonance.

Past experience with this type of structure shows that the transmissibility can often be related to the natural frequency. Considering the number

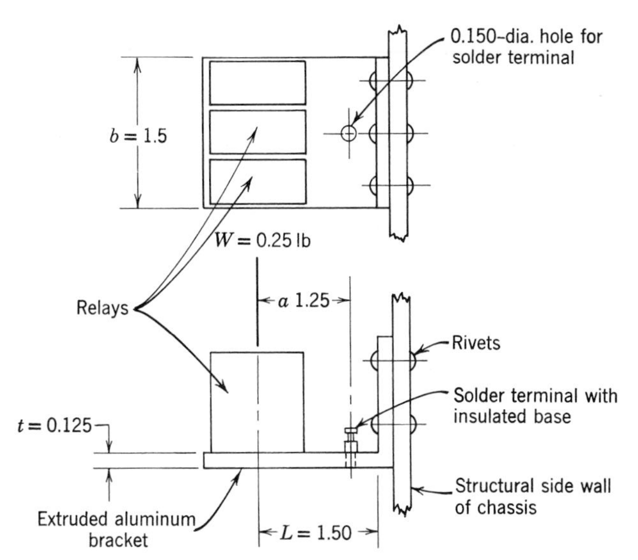

FIGURE 10.8. Relays mounted on a cantilevered bracket.

of fasteners and the number of interfaces, the transmissibility would probably range from about 1.5 to 2.0 times the square root of the natural frequency. Taking an average value of 1.75 for this system, the transmissibility can then be approximated as follows:

$$Q = 1.75(f_n)^{1/2} = 1.75(298)^{1/2} = 30 \tag{10.2}$$

The dynamic load acting on the cantilevered beam at the resonant condition is

$$P_d = WG_{\text{in}}Q \tag{10.3}$$

where $W = 0.25$ lb. relay weight
$G_{\text{in}} = 5.0\ G$ peak input
$Q = 30$ (see Eq. 10.2)

Substituting into the above equation:

$$P_d = 0.25(5.0)(30) = 37.5 \text{ lb} \tag{10.4}$$

The dynamic bending stress can be determined from the standard beam equation, with the addition of the geometric or theoretical stress concentration factor K_t, due to the hole for the Teflon-base solder terminal pressed into the aluminum bracket, as shown in Fig. 10.8.

$$S_b = K_t \frac{Mc}{I} \tag{10.5}$$

where $M = P_d a = 37.5(1.25) = 46.9$ lb in.

$$c = \frac{t}{2} = \frac{0.125}{2} = 0.0625 \text{ in.}$$

$$I = 0.244 \times 10^{-3} \text{ in.}^4$$

$$\left.\begin{aligned} d/b &= 0.150/1.50 = 0.10 \\ d/t &= 0.150/0.125 = 1.20 \end{aligned}\right\} \text{ required for } K_t$$

$$K_t = 2.01 \text{ from Peterson [18]}$$

Substituting into Eq. 10.5 for the bending stress at the hole

$$S_b = 2.01 \left[\frac{46.9(0.0625)}{0.244 \times 10^{-3}} \right] = 24{,}100 \text{ lb/in.}^2 \tag{10.6}$$

The fatigue life of the bracket can be approximated from the S–N curve for the 6061-T6 aluminum alloy with an alternating bending stress of 24,100 psi. Figure 10.14*b* shows a life of about

$$N = 7.0 \times 10^5 \text{ cycles} \tag{10.7}$$

If a qualification test is to be run on the electronic box and if the test requires a resonant dwell at 298 Hz, the estimated time to fail can be approximated from Eq. 4.14 as follows:

$$t = \frac{N}{f_n} \quad \text{(see Eq. 4.14)}$$

where $N = 7.0 \times 10^5$ cycles to fail (see Eq. 10.7)

$f_n = 298$ Hz (see Eq. 10.1)

$$t = \frac{7.0 \times 10^5 \text{ cycles to fail}}{298 \text{ cycles/sec} \times 60 \text{ sec/min}}$$

$$t = 39.2 \text{ min to fail} \tag{10.8}$$

10.8. Corrosion Fatigue

Many structural and machine elements are subjected to an alternating-load condition while operating in a corrosive atmosphere. This condition often occurs with equipment used aboard ships and equipment used along the seashore. Marine propellers, propeller shafts, and water pumps are some examples of parts that are subjected to corrosion fatigue. Lipson [21] and Hoyt [20] show that the endurance limit of many carbon steels can be reduced more than 50% when the steels are tested in tap water. These reductions are even greater in salt water. The fatigue strength of some aluminum alloys can be reduced more than 66% when they are tested in tap water. Again, these reductions are even greater in salt water.

MIL-E-5400 and MIL-T-5422 describe qualification tests on electronic equipment involving vibration, shock, humidity, and salt spray. It is possible to have a test sequence that requires the vibration and shock tests to be run after the humidity and salt-spray tests. If the structural elements in the electronic box are not adequately protected and corrosion has taken place during the humidity and salt-spray tests, the fatigue life of the structure can be reduced enough to produce unexpected fatigue failures during the vibration and shock tests.

Corrosion can reduce the fatigue life of a structure in many ways. The most obvious is the roughening of the surface and pitting, which can reduce the cross-sectional area and become stress raisers at the same time. The area reduction tends to increase the nominal stress at the cross-section while the stress raisers increase the localized stresses that can lead to cracks. Grover[22] points out that the damaging effects of corrosion are more severe when an alternating stress is involved. The corrosive medium not only attacks the walls at the base of the crack, but small corroded particles can separate and fall into the crack as it opens and act as a wedge the crack closes. This action can sharply decrease the fatigue life of any structure.

10.9. Cumulative Fatigue Damage

Structural members are often subjected to alternating loads that will change with time; this results in alternating stresses that will also change with time. Some of these members will fail rapidly while other members never seem to fail. Experience has shown that the maximum stress level and the number of fatigue cycles both have a strong influence on the fatigue life of a member. One method that is often used to explain fatigue failures is to assume that every structural member has a useful fatigue life and that every stress cycle uses up a part of this life. When enough stress cycles have been accumulated, the effective life is used up, and the member will fail. This is a very simple way of explaining a complex phenomenon.

Miner[23] suggested the use of a ratio to determine the fraction of the life that is used up. This ratio compares the actual number of stress cycles, n, at a specific stress level, to the number of cycles, N, required to produce a failure at the same stress level, using an alternating stress. For example, if a structural member is subjected to several different alternating stresses for different periods, the time for a failure can be estimated using the fatigue-cycle ratio R_n as follows:

$$R_n = \frac{n_1}{N_1} + \frac{n_2}{N_2} + \frac{n_3}{N_3} + \ldots = \sum \frac{n_i}{N_i} = 1.0 \qquad (10.9)$$

In the above equation n_1, n_2, and n_3 represent the actual number of completely reversed stress cycles accumulated at stress levels S_1, S_2, and S_3. The numbers of fatigue cycles required to produce failures are shown by N_1, N_2, and N_3. These correspond to the same stress levels S_1, S_2, and S_3, and they can be obtained from an S–N curve for the particular material. If the sum of all the fractions in the summation for the fatigue-cycle ratio R_n is greater than 1.0, failure is supposed to occur.

Miner's cumulative fatigue-damage rule is only approximate in predicting the life of a structural member. The order in which the stresses are applied appears to be important. For example, Crandall [1] points out that when a low stress is applied first followed by a high stress, the fatigue-cycle ratio for SAE 1030 steel is about 1.1–1.3. However if the high stress is applied first followed by the low stress, the fatigue-cycle ratio falls to about 0.8. Also, if an occasional high load is superimposed during an endurance test, Miner's rule suggests the fatigue life will be decreased. Actual tests, however, show the fatigue life is really increased. There are also some limited test data which show that Miner's fatigue-cycle ratio can be made to vary from 0.3 to 3.0 depending upon the manner and order in which the high- and low-stress loads are applied.

Miner's rule appears to work best when the fatigue-cycle ratio is computed at many different stresses. Although the rule may show some wide fluctuations, it is still considered to be very useful in determining the cumulative fatigue-damage effects on a structure, from which the approximate fatigue life can be estimated.

Although Miner's rule is often shown in equations to have a value of 1.0 when failure occurs, a smaller number is usually used to determine the effective life of a structure. Fatigue-cycle ratios as low as 0.3 have been proposed to determine the useful life of a structure. However, for analyzing an electronic system, a ratio as low as this usually connotes too much extra weight. Therefore, when weight is important, a fatigue-cycle ratio of about 0.7 is recommended. A higher ratio means higher working stresses or more fatigue cycles so that extra care must be taken to reduce stress concentrations.

Recommended:

$$R_n = 0.70 \tag{10.10}$$

10.10. Sample Problem

An electronic box will be subjected to a series of vibration tests at different acceleration G levels, in order to demonstrate its reliability before similar boxes can be used in the field. The series of vibration tests is broken up into phases, and each phase must be passed before the next phase can begin. The actual number of fatigue cycles n for each test phase is determined from the 160-Hz fundamental resonant frequency of the electronic box and the time required for each test phase. These values are shown in Table 10.1, along with the input acceleration G values and the expected dynamic stresses in one critical area of the chassis structure.

TABLE 10.1

Test Condition	Input Force (G)	Actual No. Stress Cycles (n)	Stress in Structure (psi)
1 Production acceptance	3	1.92×10^4	22,000
2 Prequalification	3	4.8×10^4	22,000
3 Safety of flight	4	1.44×10^5	29,400
4 Reliability demonstration	4	2.88×10^5	29,400
5 Full qualification	5	3.45×10^5	36,700

The approximate number of fatigue cycles N required to produce a failure at each of the five test conditions is determined from the S–N fatigue curve for the 2014-T6 aluminum alloy shown in Fig. 10.12a as follows:

When $S_1 = 22{,}000$ psi, $N_1 = 7.0 \times 10^7$ cycles to fail
When $S_2 = 22{,}000$ psi, $N_2 = 7.0 \times 10^7$ cycles to fail
When $S_3 = 29{,}400$ psi, $N_3 = 4.0 \times 10^6$ cycles to fail
When $S_4 = 29{,}400$ psi, $N_4 = 4.0 \times 10^6$ cycles to fail
When $S_5 = 36{,}700$ psi, $N_5 = 5.0 \times 10^5$ cycles to fail

Miner's cumulative fatigue cycle ratio R_n can then be determined from Eq. 10.9 as follows:

$$R_n = \frac{n_1}{N_1} + \frac{n_2}{N_2} + \frac{n_3}{N_3} + \frac{n_4}{N_4} + \frac{n_5}{N_5} \tag{10.11}$$

$$R_n = \frac{1.92 \times 10^4}{7.0 \times 10^7} + \frac{4.8 \times 10^4}{7.0 \times 10^7} + \frac{1.44 \times 10^5}{4.0 \times 10^6} + \frac{2.88 \times 10^5}{4.0 \times 10^6} + \frac{3.45 \times 10^5}{5.0 \times 10^5}$$

$$R_n = 0.0003 \times 0.0007 + 0.036 + 0.072 + 0.690$$

$$R_n = 0.799 \tag{10.12}$$

Since this R_n value is greater than the maximum recommended value of 0.70 shown by Eq. 10.10, the structure in this critical area is marginal and a failure can be expected during these tests. A design change should be made to improve the fatigue life of the structure.

10.11 FATIGUE CURVES

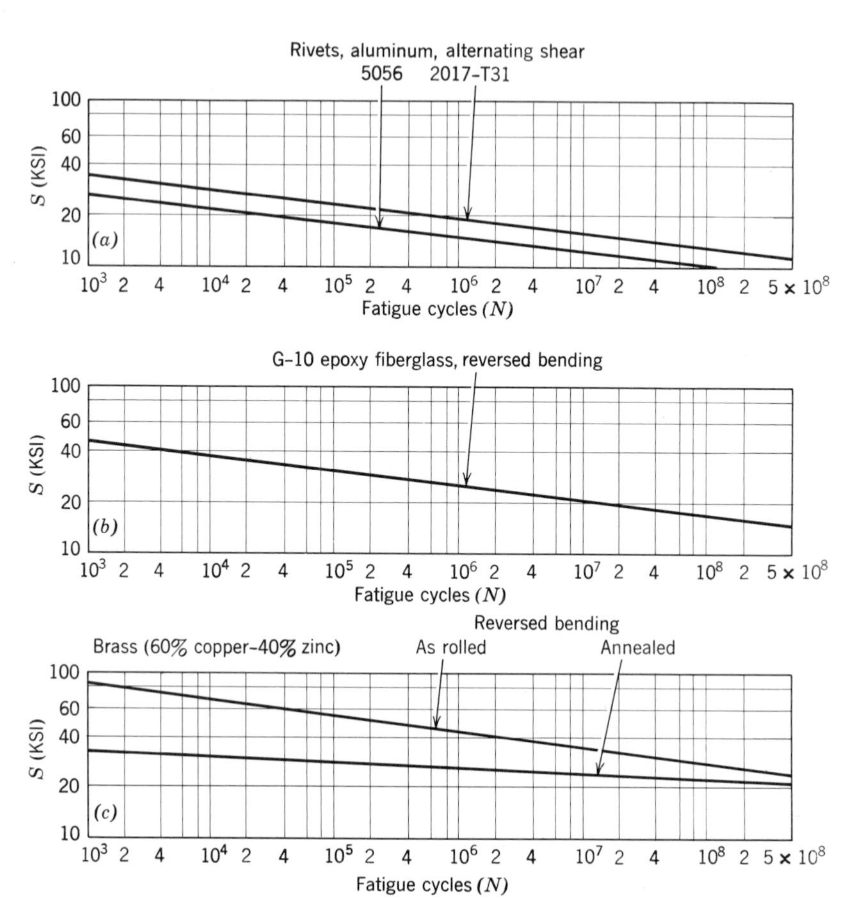

FIGURE 10.9. Fatigue curves.

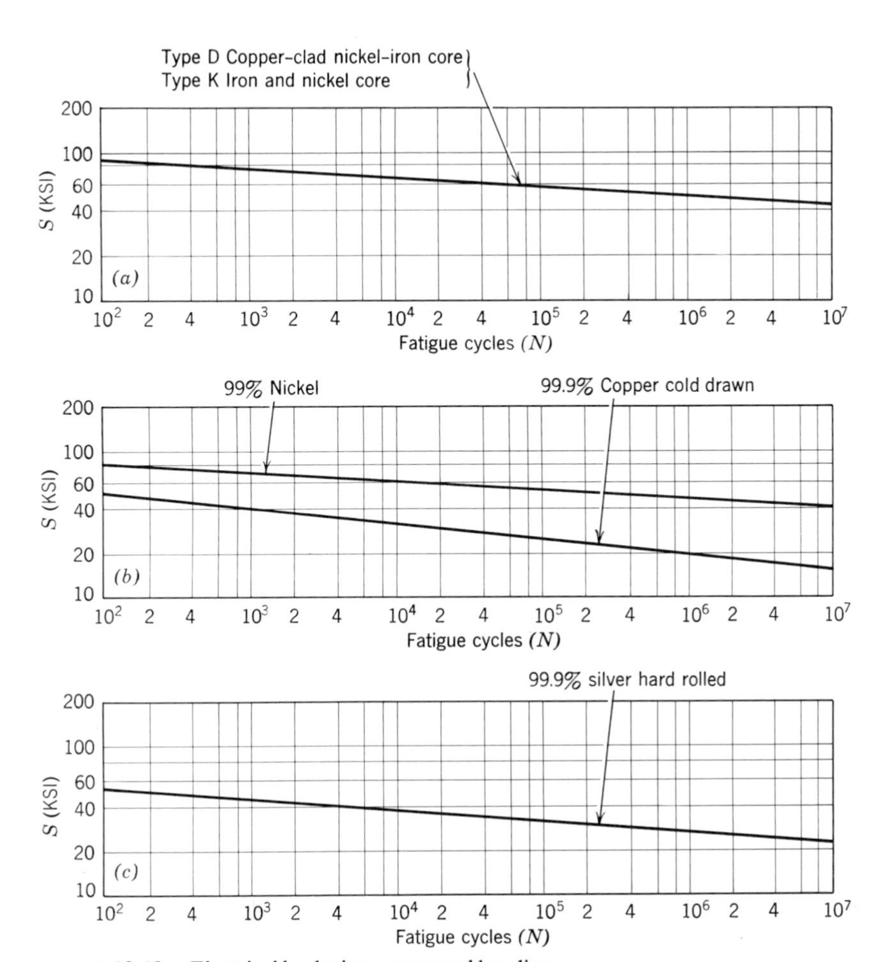

FIGURE 10.10. Electrical lead wires – reversed bending.

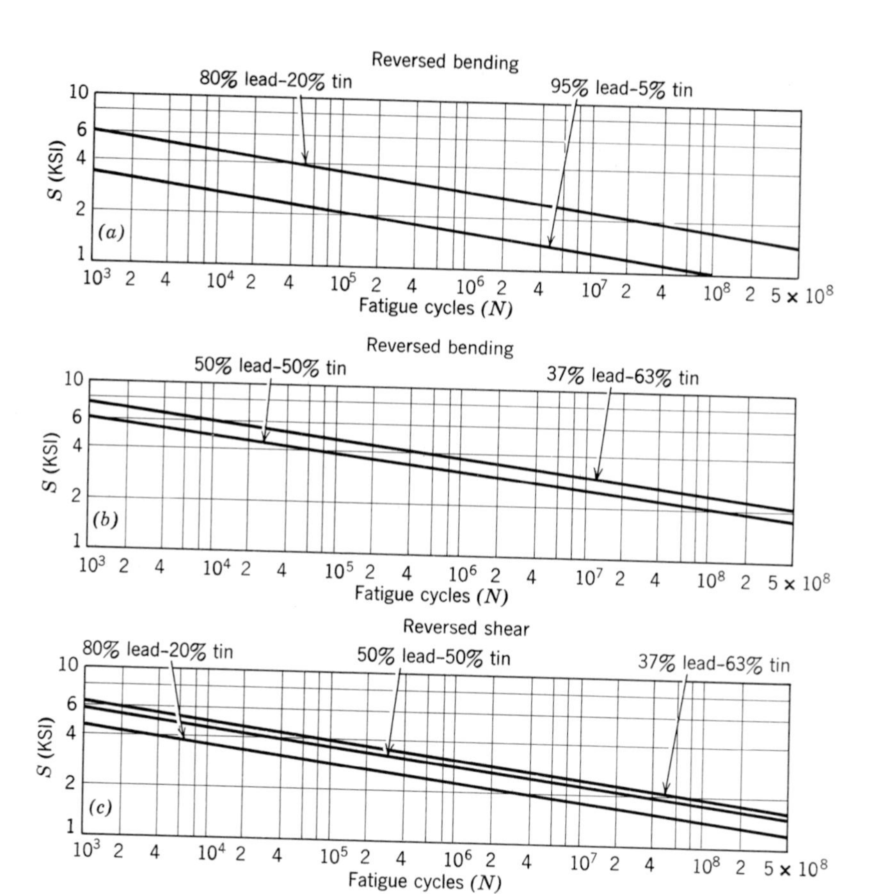

FIGURE 10.11. Soft solder.

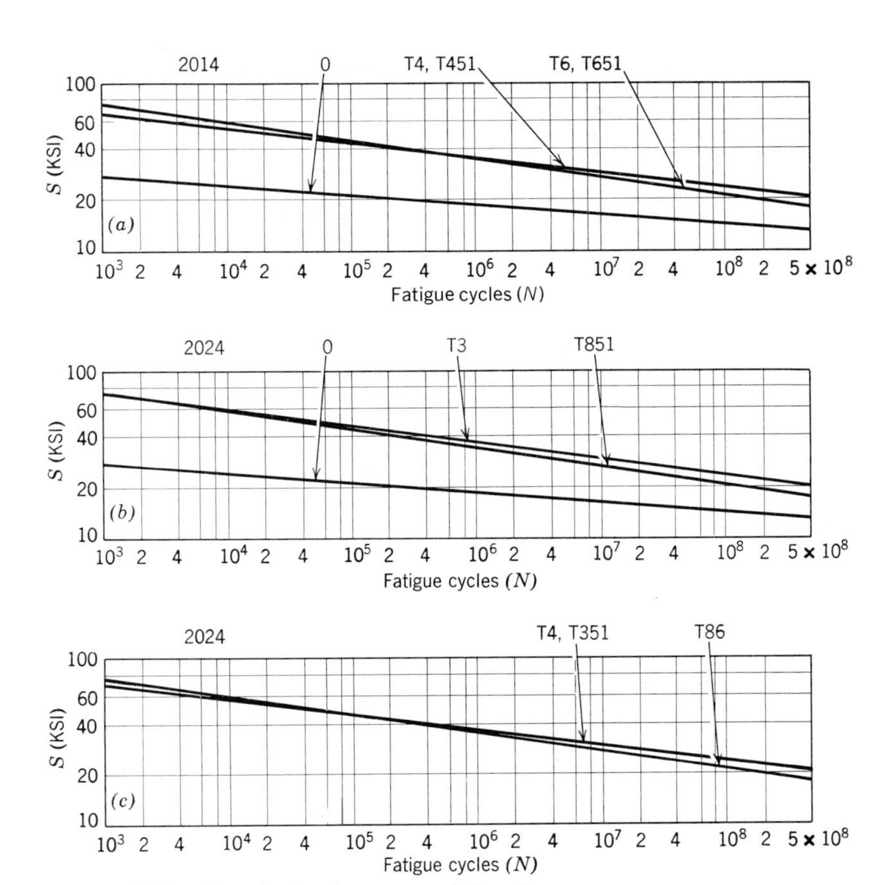

FIGURE 10.12. Wrought aluminum – reversed bending.

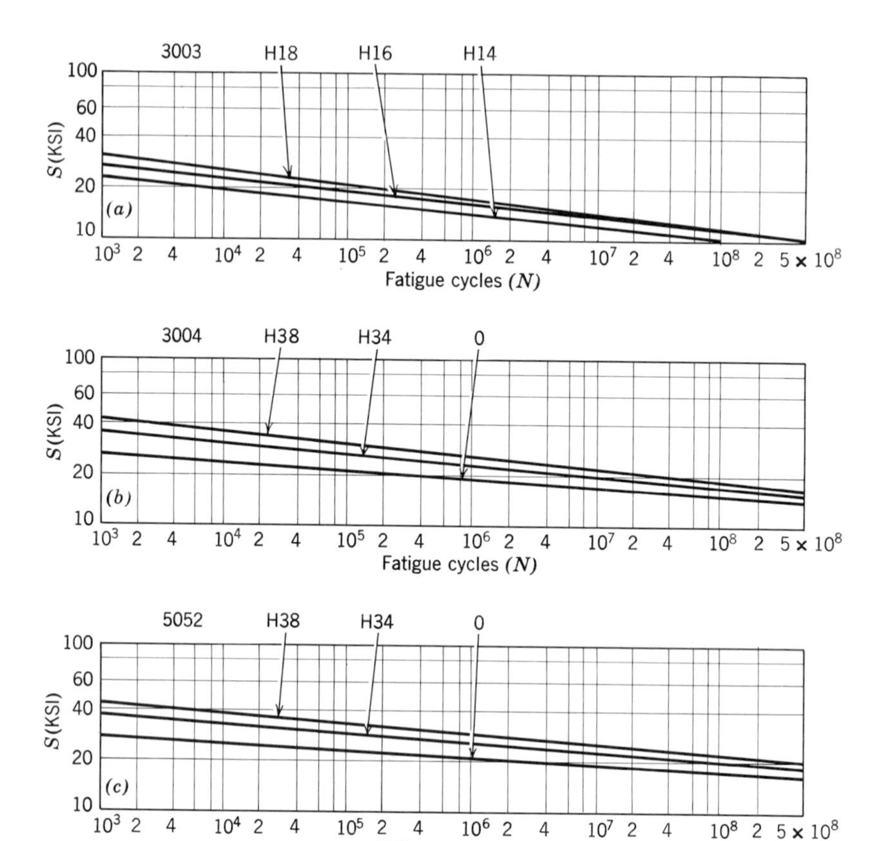

FIGURE 10.13. Wrought aluminum—reversed bending.

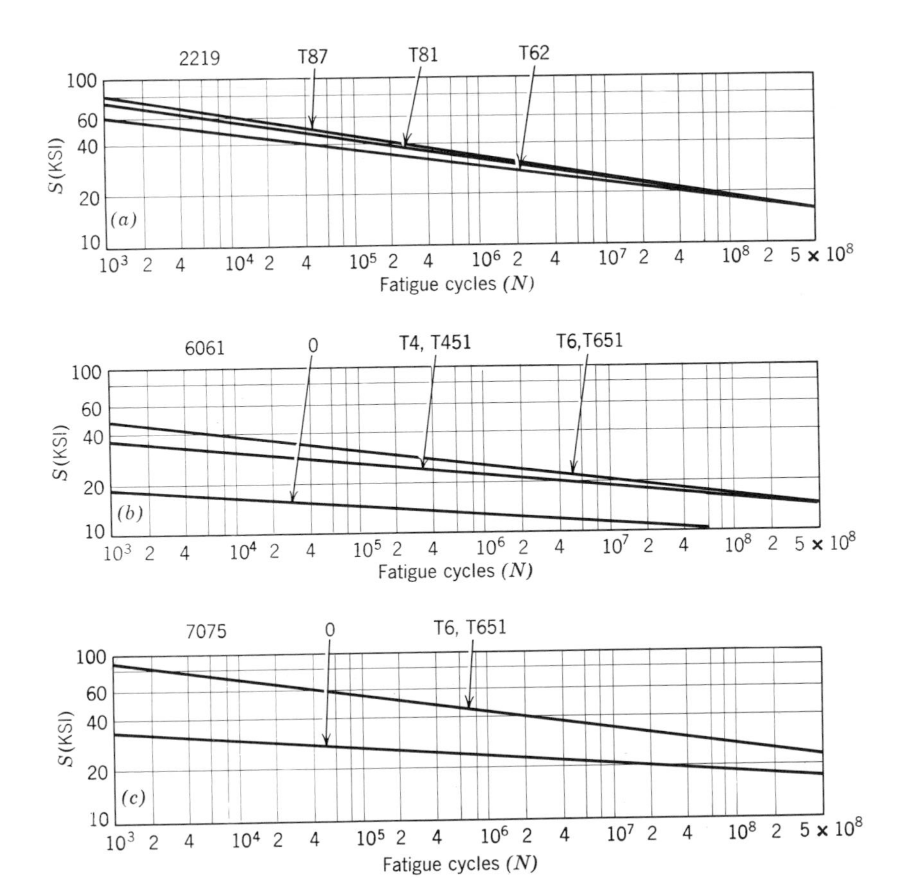

FIGURE 10.14. Wrought aluminum – reversed bending.

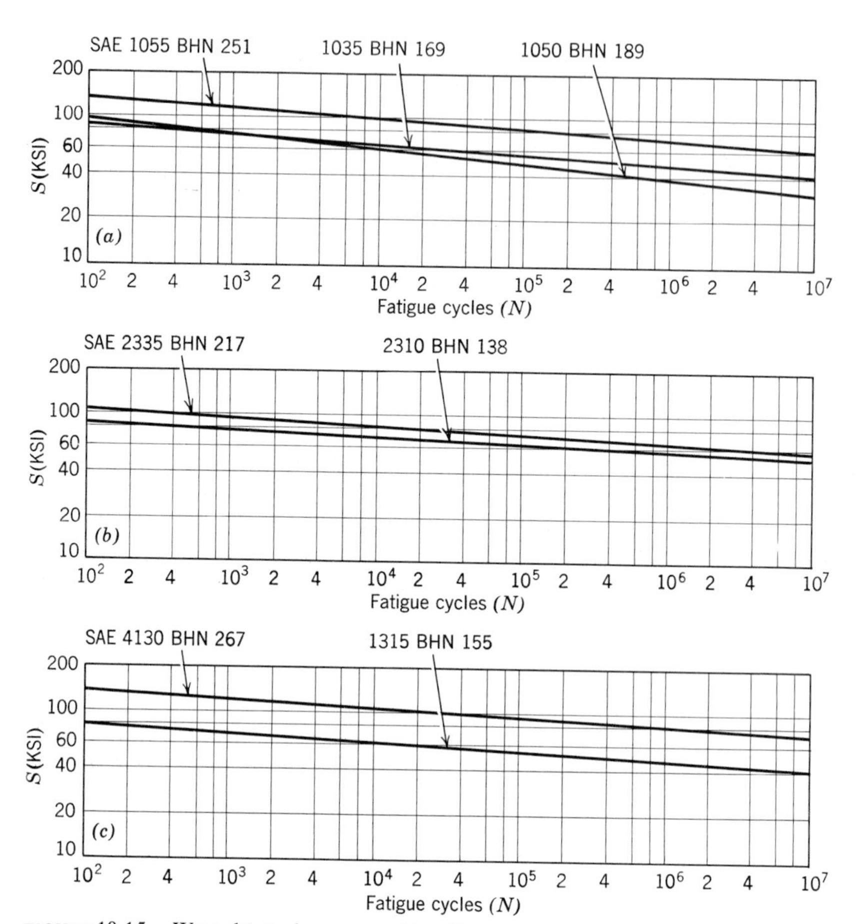

FIGURE 10.15. Wrought steels – reversed bending.

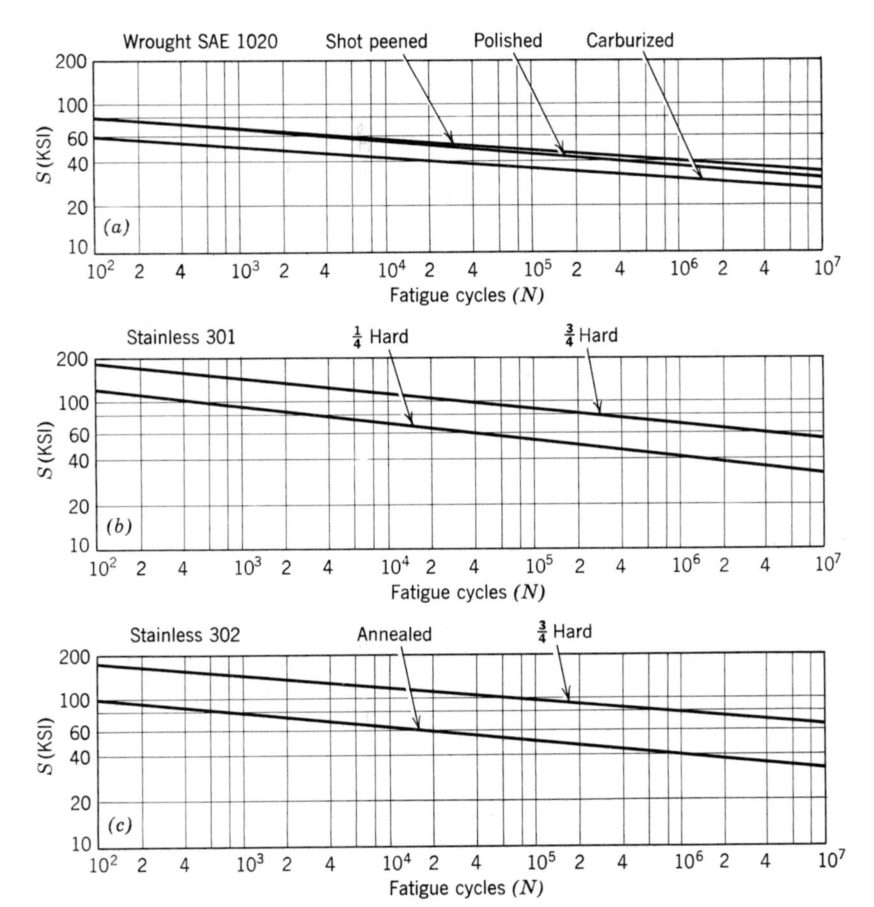

FIGURE 10.16. Steels-reversed bending.

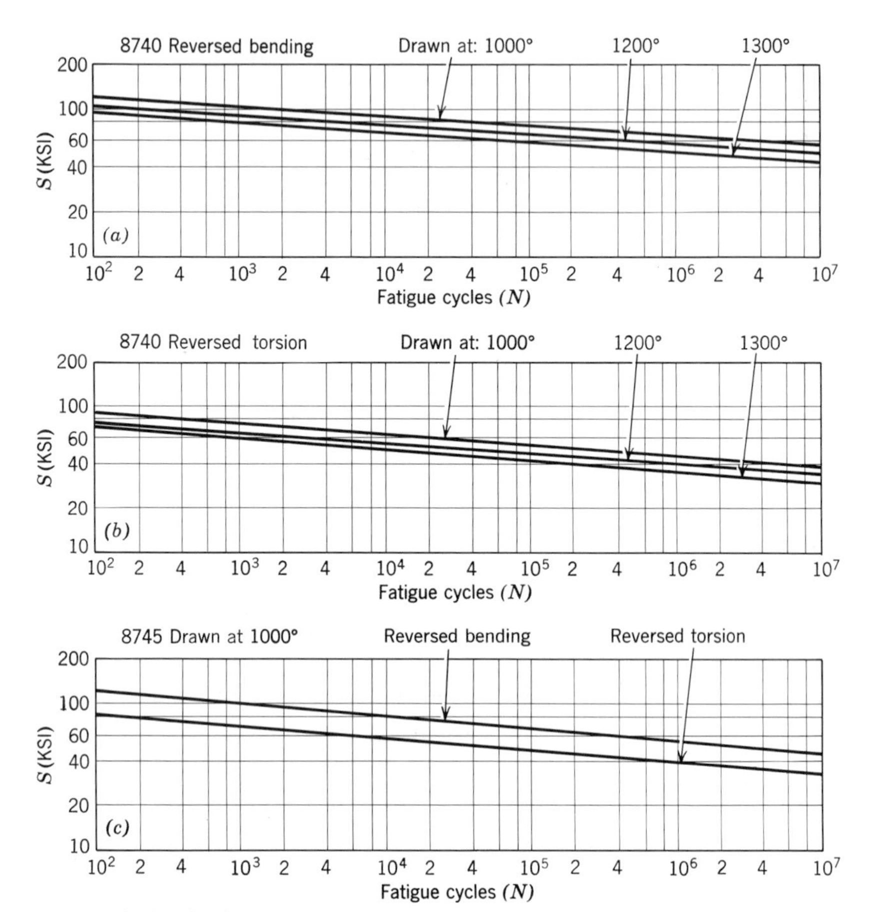

FIGURE 10.17. Steels.

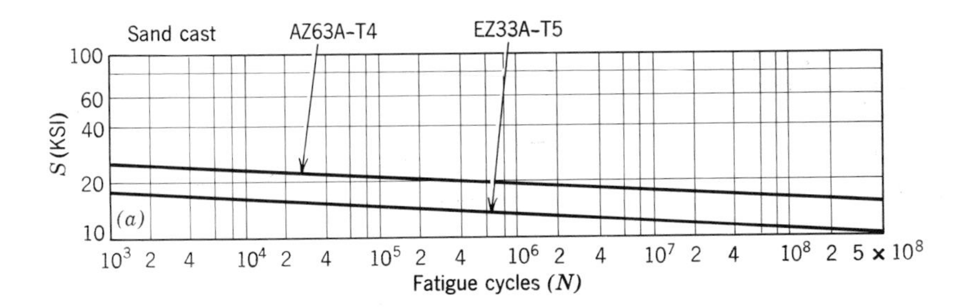

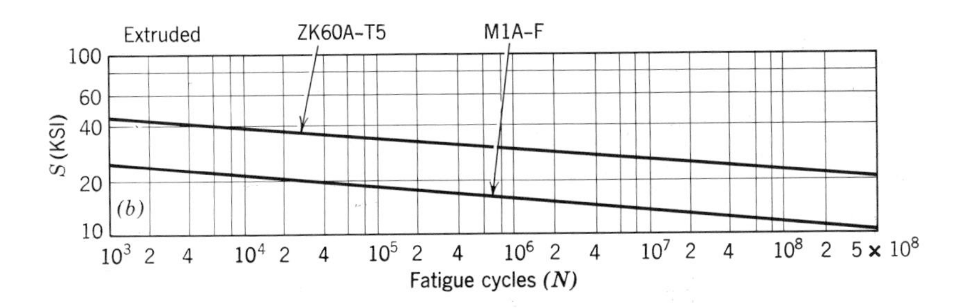

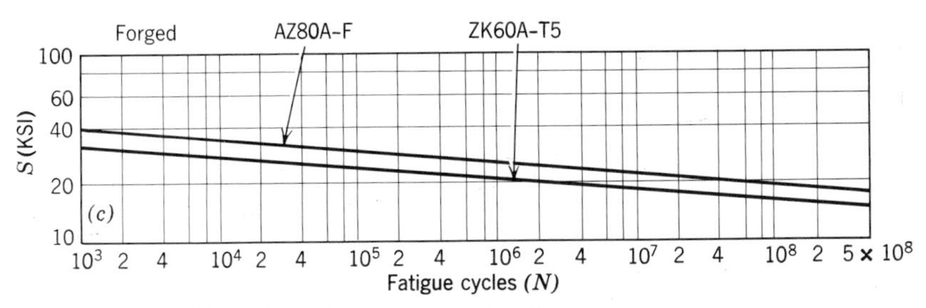

FIGURE 10.18. Magnesium alloys—reversed bending.

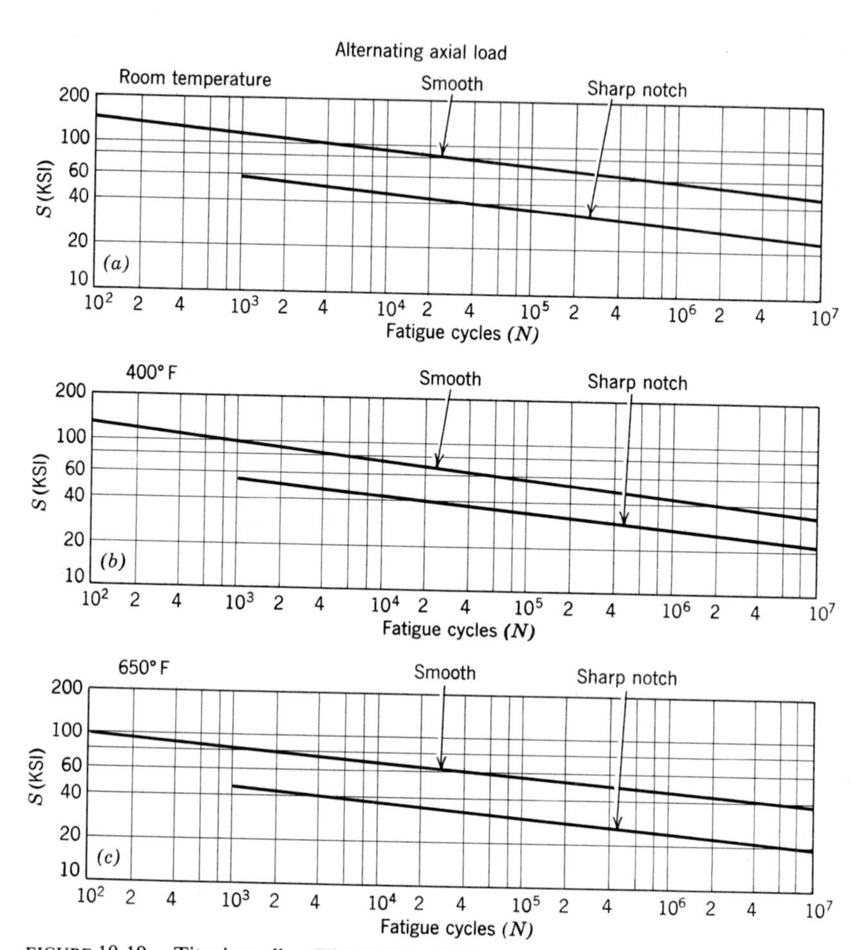

FIGURE 10.19. Titanium alloy (Ti-8Al-1Mo-1V sheet).

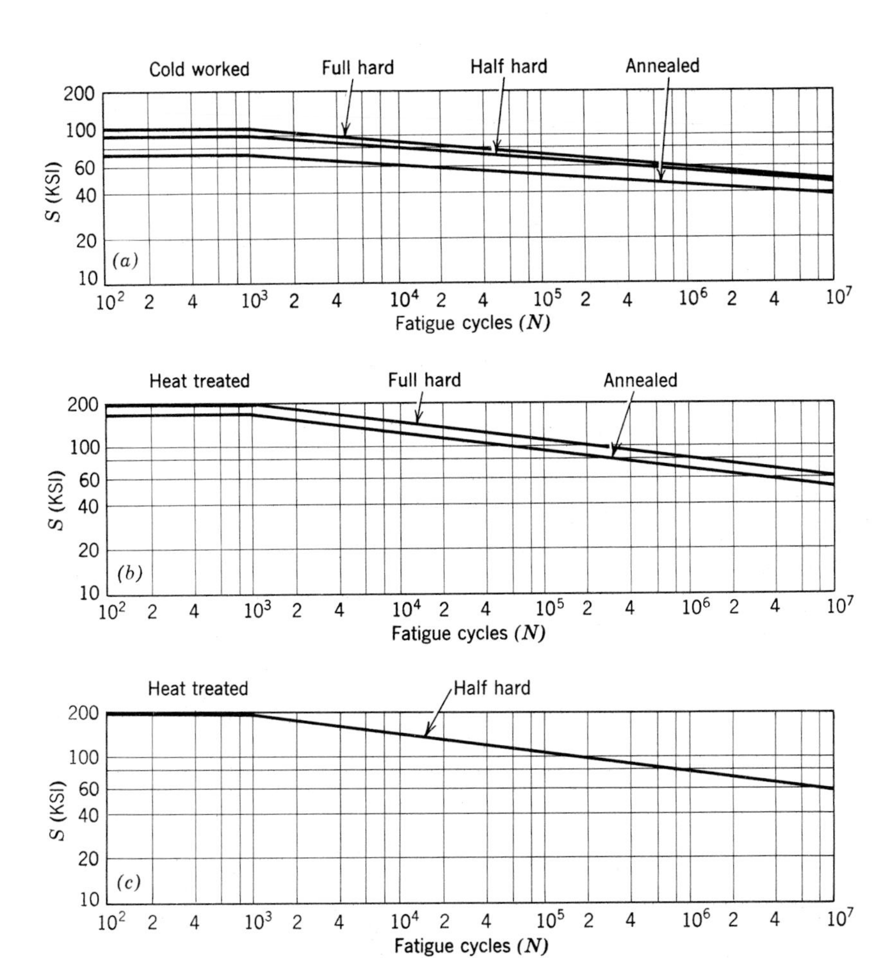

FIGURE 10.20. Beryllium copper (2.25% beryllium) – reversed bending.

REFERENCES

1. Stephen H. Crandall, *Random Vibration*, Technology Press, John Wiley & Sons, 1958.
2. Charles E. Crede, *Vibration and Shock Isolation*, John Wiley & Sons, 1957.
3. Edward J. Lunney and Charles E. Crede, *The Establishment of Vibration and Shock Tests for Airborne Electronics*, Wright Air Development Center, Technical Report 57–75, January 1958, ASTIA Document AD142349.
4. Arthur W. Leissa, *Vibration of Plates, Nat'l. Aeronautics and Space Admin.*, 1969.
5. Norman E. Lee, *Mechanical Engineering as Applied to Military Electronic Equipment*, Coles Signal Laboratory, Red Bank, N.J., June 22, 1949.
6. Geoffrey W. A. Dummer and Norman B. Griffin, *Environmental Testing Techniques for Electronics and Materials*.
7. Charles E. Crede and Edward J. Lunney, *Establishment of Vibration and Shock Tests for Missile Electronics as Derived from the Measured Environment*, Wright Air Development Center Technical Report 56-503, December 1, 1956, ASTIA Document AD118133.
8. Rough road tests, performed at A.C. Spark Plug, Division of General Motors, Milwaukee, Wis., July 15, 1959.
9. W. C. Stewart, "Determining Bolt Tension," *Mach. Des. Mag.*, November 1955.
10. R. W. Dicely and H. J. Long, "Torque Tension Charts for Selection and Application of Socket Head Cap Screws," *Mach. Des. Mag.*, September 5, 1957.
11. Kent's Mechanical Engineer's Handbook, Design and Production Volume, John Wiley & Sons Inc. 1950.
12. B. Saelman, "Calculating Tearout Strength for Cantilevered Beams," *Mach. Des. Mag.*, January 1954.
13. E. F. Bruhn, *Analysis and Design of Aircraft Structures*, Tristate Offset Co., Cincinnati, Ohio, 1952.

14. M. R. Achter, *Effects of High Vacuum on Mechanical Properties*, U.S. Naval Research Laboratory, Washington, D.C., 1961.
15. McClintock and Argon, *Mechanical Behavior of Materials*, Addison-Wesley, 1966.
16. A. M. Freudenthal, *Fatigue in Aircraft Structures*, Academic Press, 1956.
17. *Marks Handbook*, McGraw-Hill Book Co., 1951.
18. R. E. Peterson, *Stress Concentration Design Factors*, John Wiley & Sons, 1959.
19. R. J. Roark, *Formulas for Stress and Strain*, McGraw-Hill, 1943.
20. S. L. Hoyt, *Metals and Alloys Data Book*, Reinhold Publishing, 1943.
21. C. Lipson and R. Juvinal, *Handbook of Stress and Strength*, Macmillan, 1963.
22. H. J. Grover, S. A. Gordon, and L. R. Jackson, *Fatigue of Metals and Structures*, Bureau of Aeronautics, Department of the Navy, 1954.
23. M. A. Miner, "Cumulative Damage in Fatigue," *J. Appl. Mech.*, **12** (Sept. 1945).
24. Reynolds Metals Co., *Structural Aluminum Design Handbook*, 1968.
25. F. B. Stulen, H. N. Cummings, and W. C. Schulte, "A Design Guide Preventing Fatigue Failures," *Mach. Des. Mag.*, Part 5, June 22, 1961.
26. F. R. Shanley, *Strength of Materials*, McGraw-Hill, 1957.
27. MIL-Handbook-5A, *Metallic Materials and Elements for Aerospace Vehicle Structures*, Dept. of Defense, Washington, D.C.
28. J. W. S. Rayleigh, *The Theory of Sound*, Dover Publications, 1945.
29. S. Timoshenko and S. W. Krieger, *Theory of Plates and Shells*, McGraw-Hill, 1959.
30. R. W. Little, Master's Thesis, Univ. Wisconsin, 1959.
31. G. B. Warburton, "The Vibrations of Rectangular Plates," *Proc. Inst. Mech. Eng.*, **168**, (12) (1954).
32. F. J. Stanek, *Uniformly Loaded Square Plate with No Lateral or Tangential Edge Displacements*, Ph.D. Thesis, Univ. Illinois, 1956.
33. P. A. Laura and B. F. Saffel, Jr., "Study of Small Amplitude Vibrations of Clamped Rectangular Plates Using Polynomial Approximations," *J. Acoust. Soc. Amer.*, **41** (4) (1967).
34. M. Vet, "Vibration Analysis of Thin Rectangular Plates," *Mach. Des. Mag.*, April 13, 1967.
35. *Road Shock and Vibration Environment for a Series of Wheeled and Track-Laying Vehicles*, Report DPS-999, March–June, 1963.
36. Navships 900, 185, *A Design of Shock and Resistant Vibration Electronic Equipment for Shipboard Use.*
37. D. S. Steinberg, "Circuit Components vs. *G* Forces, *Mach. Des. Mag.*, October 14, 1971.

INDEX